THE PHYSICS OF EVERYDAY
Phenomena

THE PHYSICS OF EVERYDAY

Phenomena

A Conceptual Introduction to Physics

W. THOMAS GRIFFITH

Pacific University

Book Team

Editor *Jeffrey L. Hahn*
Developmental Editor *Robert B. Fenchel*
Production Editor *Diane E. Beausoleil*
Photo Editor *Robin Storm*
Permissions Editor *Gail Wheatley*
Visuals/Design Consultant *Marilyn Phelps*
Visuals/Design Freelance Specialist *Barbara J. Hodgson*

 Wm. C. Brown Publishers

President *G. Franklin Lewis*
Vice President, Publisher *George Wm. Bergquist*
Vice President, Operations and Production *Beverly Kolz*
National Sales Manager *Virginia S. Moffat*
Group Sales Manager *Vincent R. Di Blasi*
Vice President, Editor in Chief *Edward G. Jaffe*
Marketing Manager *John W. Calhoun*
Advertising Manager *Amy Schmitz*
Managing Editor, Production *Colleen A. Yonda*
Manager of Visuals and Design *Faye M. Schilling*
Production Editorial Manager *Julie A. Kennedy*
Production Editorial Manager *Ann Fuerste*
Publishing Services Manager *Karen J. Slaght*

WCB Group

President and Chief Executive Officer *Mark C. Falb*
Chairman of the Board *Wm. C. Brown*

Cover photo © Photore Inc./Tom Sanders/The Stock Market

Copyeditor Nicholas Murray

Interior and cover design Burmar Technical Corporation

Illustrations rendered by Burmar Technical Corporation

The credits section for this book is on page 478 and is considered an extension of the copyright page.

CONTENTS

Electricity and Magnetism

Wave Motion and Optics

PREFACE

The excitement that students can feel in understanding the physical explanation for some phenomenon that they see or experience almost daily is one of the best motivators for building scientific literacy. This book provides a framework that allows non-science majors to experience that excitement. It is intended for a one-semester or two-quarter course in conceptual physics involving minimal use of mathematics. It is written in a narrative style that should also allow its use by any adult who is interested in exploring the nature of physics and the explanations of everyday physical phenomena.

This book had an unusual genesis; it grew out of the author's work in writing clear conceptual questions designed to probe the understanding of basic physical principles addressed in introductory physics courses at all levels. Many of our conceptual objectives for these courses are the same for any introductory physics course, although we expect more mathematical and problem-solving sophistication in courses taught for science majors. Even students in the calculus-based sequence often have major misunderstandings of physical concepts both before and after taking physics. Instruction would be more effective in all introductory courses if a good battery of conceptual questions were available for regular use on both tests and home assignments.

An unusual feature of this book, therefore, is the carefully worded conceptual questions at the end of each chapter, which are closely tied to the conceptual objectives stated in the outline near the beginning of each chapter. Many of these have been classroom-tested on quizzes and assignments at different levels of introductory physics. Their format is similar: they all call for a short objective response regarding the direction, relative size, or existence of some effect, followed by a brief written explanation. If these questions are used as an integral part of the course, they will help students to clarify their understanding in a manner that is not often possible with the usual numerical exercises or the more vague and open-ended questions that many instructors are reluctant to assign.

Another unusual feature of this book is that each chapter begins with an illustration from everyday experience that motivates the introduction of the relevant physical concepts. The abstractness of physics is one of the barriers to learning for many students. The use of everyday phenomena dissipates that abstractness with concrete examples that are already familiar. Each chapter also includes an "Everyday Phenomenon" box that analyzes some common phenomena in more detail. Many of these examples are drawn from sports, but lightning, automobile collisions, the operation of

a flat-plate solar collector or a television set, and other phenomena are also featured.

Mathematics is a formidable barrier for many students, particularly non-science majors, as they approach a course in physics. There have been some attempts to teach conceptual physics without the use of *any* mathematics, but students thus lose an opportunity to build their confidence in using and manipulating simple quantitative relationships. The use of mathematics should be carefully limited, however, and clearly subordinate to the physical concepts being addressed. This book therefore uses minimal mathematics and carefully explains and illustrates the benefit and application of the mathematical expressions or operations that are introduced. No mathematics prerequisite is assumed. A discussion of the basic ideas of simple algebra is provided in appendix A, along with some practice exercises.

In addition to the questions at the end of each chapter, there are also a limited number of simple numerical exercises as well as a few "Challenge Problems" that are somewhat more involved. It is recommended that the exercises be assigned sparingly to help students get a feeling for the quantities involved and for performing simple computations involving physical concepts. The "Challenge Problems" are designed to give students who are more comfortable with quantitative ideas an opportunity to explore these ideas more thoroughly; they are probably best assigned on an extra-credit basis.

Since many courses taught for non-science majors do not have a laboratory component, the text includes a few "Home Experiments and Observations" at the end of each chapter. Their aim is to get students to explore the behavior of physical objects using things that are readily available, such as rulers, string, paper clips, balls, toy cars, flashlight batteries, and so on. They are no substitutes for a well-conceived laboratory course, but they can serve the purpose of putting students into the exploratory and observant frame of mind that is important to scientific thinking. This should also be an important objective in developing scientific literacy.

The organization of topics in this book is traditional, with some minor variations. Energy, for example, is introduced in chapter 8, following mechanics and just prior to the chapters on heat and thermodynamics in order to provide greater coherence to the ideas of energy and thermodynamics. Wave motion is presented in chapter 14, following electricity and magnetism and prior to the chapter on optics (15), rather than in the mechanics section. The first fifteen chapters are designed to introduce students to the major ideas of classical physics and can be covered in a one-semester course without being badly rushed. The complete twenty chapters could easily support a two-quarter course and even, with minimal augmentation, a two-semester course in which the ideas are treated thoroughly and carefully. Chapters 16 and 17, on atomic and nuclear phenomena, will be considered essential by many instructors, even in a one-semester course, but if they are to be included in such a course, it would be best to curtail coverage in some other area to avoid overload.

The supplementary chapters are separated from the others because they are not essential to the continuity of the material in the first seventeen chapters and could be inserted at several different points in a course. Chapter 19 on fluids would probably best be inserted between chapters 8 and 9. Chapter 18 on relativity might best come in at the end of the mechanics section (after chapter 8) or just prior to the modern physics material. It has very little to do with everyday phenomena, of course, but it is included because of the high interest that it generally holds for students. The final chapter (20) introduces a variety of modern topics, including particle physics, cosmology, semiconductors, computers, and superconductivity that can be used to stimulate interest at various points in a course.

My only request to instructors and to students using this book is that you not try to cram too much material into too short a time. These ideas are most enjoyable when enough time is spent in discussion and in consideration of the conceptual questions to develop a real understanding. Trying to cover too much too quickly defeats the conceptual learning and leaves students in a dense haze of words and definitions. I have worked hard to hold this book to a reasonable length while still covering the core of what usually constitutes an introduction to physics. Less can be more if instructors and students agree to do it right.

ACKNOWLEDGMENTS

A great many individuals have contributed to this project, either directly or indirectly. I extend a special thanks to those who participated in the reviews of the manuscript for their insight and thoughtful suggestions. The reviewers include:

John W. Snyder
Southern Connecticut State University

John E. Crew
Illinois State University

Paul Varlashkin
East Carolina University

Ervin Poduska
Kirkwood Community College

Joseph A. Schaefer
Loras College

Ruth Howes
Ball State University

Stephen J. Shulik
Clarion University of Pennsylvania

Van E. Neie
Purdue University

Leo Takahashi
Pennsylvania State University

In addition, I wish to acknowledge the contributions of the WCB editorial and book-team members whose commitment to quality and teamwork were most gratifying throughout the development of the text. I also owe a debt of gratitude to my family, particularly to my wife, Adelia, for their assistance and forbearance, and to my colleagues in the Department of Physics at Pacific University, Mary Fehrs, Juliet Brosing, and Jurgen Meyer-Arendt for their helpful suggestions and tolerance of times when this project interfered with other activities.

THE PHYSICS OF EVERYDAY
Phenomena

1 The Nature of Physics

Imagine that you are riding your bike on a country road on an Indian-summer afternoon. The sun has come out after a brief shower, and as the rain clouds move on, a rainbow appears in the east (fig. 1.1). A leaf flutters to the ground and an acorn, shaken loose by a squirrel, misses your head by only a few inches. The sun is at your back, and you are at peace with the world around you.

No knowledge of physics is needed to savor the moment. However, if you have an active curiosity, some questions may come to mind, the answers to which might provide an even greater appreciation of your experience. Why, for example, does the rainbow appear in the east during the afternoon rather than in the west, where another rain squall is visible? What causes the colors to appear? Why does the acorn fall more rapidly than the leaf? Why, indeed, does your bicycle remain vertical while you are moving, but fall over when it is not moving?

You may already know the answers to some of these questions from your previous exposure to science. Curiosity about questions such as these has motivated scientists for hundreds of years. Being able to devise and apply theories or models that can be used to understand, explain, and predict such phenomena can be a rewarding intellectual game; crafting an explanation and testing it with simple experiments or observations can be fun. That enjoyment is too often missed in science courses that focus on the accumulation of facts that is one of the products of scientific investigation, but not the main point.

The purpose of this book is to enhance your ability to enjoy the phenomena that are part of everyday experience by being able to produce your own explanations and simple experimental tests. The questions posed above lie in the realm of physics, which is the focus of this book, but the spirit of inquiry and explanation is found throughout science as well as in other areas of human activity. Science, and physics in particular, is an activity whose highest reward is the fun and excitement that come from understanding something that has not been understood before. This is true whether we are talking about your own understanding of how rainbows are formed or that of a physicist making a breakthrough on the ever-changing frontiers of science.

Figure 1.1   A rainbow on a summer afternoon in the Columbia River Gorge. How can this phenomenon be explained?

Chapter Objectives

The main purpose of this chapter is to provide you with an understanding of what physics is and where it fits in the broader scheme of the sciences. A secondary purpose is to acquaint you with some of the features of this book and provide some tips on how to use it most effectively.

Chapter Outline

❶ *The nature of the scientific enterprise.* Is there such a thing as scientific method? If so, what is it? How do scientific explanations differ from other types of explanation?

❷ *The scope of physics.* What is physics, and how is it related to the other sciences and to technology?

❸ *The role of measurement and mathematics in physics.* Why are measurements so important, and why is mathematics so extensively used in physics? Can physics be done without mathematics? What is the metric system?

❹ *Physics and everyday phenomena.* How is physics related to everyday experience and common sense? What are the advantages of understanding common experience in terms of physics?

❺ *Using the features of this book.* What are the organizational features of this book? How can they help you to acquire a good understanding of physical principles?

1.1 THE NATURE OF THE SCIENTIFIC ENTERPRISE

How do scientists go about explaining something like a rainbow? How does a scientific explanation differ from other types of explanations? Can we count on the scientific method to provide us with a prescription for explaining almost anything? Your own sophistication in answering questions like these is a key measure of your scientific literacy. Since science and technology play such an enormous role in the modern world, it is important to know what science can and cannot do.

Philosophers have devoted millions of hours and pages to questions related to the nature of knowledge, and of scientific knowledge in particular. Our ideas on this subject have changed radically from those of the early Greek philosophers, and the growth of science has been an important stimulant for these changes. Science as we now know it is a very recent phenomenon in the history of human development. Its rapid growth has occurred largely during the last two hundred years. Isaac Newton, who is often considered to be the father of modern physical thought, made his contributions just three hundred years ago.

Developing Scientific Explanations

What is it about the nature and methods of science that have permitted this rapid growth in knowledge, which has brought such revolutionary changes to our lives? To make this question more concrete, we can consider a specific example of how a scientific explanation is generated. Where, for example, would you turn for an explanation of how rainbows are formed? If this were assigned as a class project, you might turn to an encyclopedia or a textbook on physics, look up *rainbow* in the index, and read and report on what you find.

Are you behaving like a scientist in so doing? The answer is both yes and no. Certainly many scientists would do the same thing if they were unfamiliar with the explanation. You are appealing to the authority of the textbook author and those who preceded the author in inventing that explanation. Appeal to authority is one means of gaining knowledge, but you are at the mercy of your source for the accuracy and validity of your explanation. You are also

depending on the hope that someone has previously raised the same questions and done the necessary work to create and test an explanation.

But suppose you place yourself back three hundred years or more in time and try the same approach. One book might tell you that a rainbow is the painting of the angels; another might refer to myths involving pots of gold and leprechauns (fig. 1.2). Yet another might speculate on the nature of light and its interactions with raindrops but be quite tentative in its conclusions. All of these books might have seemed quite authoritative in their day. Where, then, do you turn? Which explanation will you accept?

If you are behaving like a scientist, you might start by studying the ideas of other scientists on the nature of light and then try to test these ideas against your own observations of rainbows. You would carefully note the conditions under which rainbows appear, the position of the sun relative to you and to the rainbow, and the position of the rain shower associated with the rainbow. What is the order of the colors in the rainbow, and have you observed that same order in other phenomena? Then you might invent an explanation that utilized current ideas on the nature of light and your own guess about how the sunlight was interacting with the raindrops. You might even devise experiments in an attempt to produce artificial rainbows with water drops or glass beads in order to test your explanation. (See chapter 15 for a modern view of how rainbows are formed.)

If your explanation is consistent with your observations on rainbows and with your experiments, then you might report it in the form of a paper or talk addressed to scientific colleagues. They, in turn, may criticize your explanation, suggest modifications, and perform their own experiments to

Figure 1.2   Leprechauns and their rainbow. Do such stories provide satisfactory explanations of rainbows?

confirm or refute your claims. If different people obtain similar results that confirm your explanation, then it will gain support and eventually become a part of a broader theory on phenomena involving light. The experiments done by you and others will probably also lead to the discovery of new aspects of the behavior of light, which, in turn, will beg for refined theories and explanations.

What are the critical aspects of this process we are describing? One is certainly the value of careful observation. Another is the idea of testability; an acceptable scientific theory or hypothesis must always suggest some means by which its predictions can be checked or tested by observations or experiments. The statement that rainbows are the paintings of angels may, in some poetic sense, be regarded as true, but it is certainly not testable by mere humans. It would therefore be rejected as a scientific explanation.

Another important feature of the process is a social one: the communication of your theory and the results of your experiments to colleagues (fig. 1.3). The submission of your ideas to the criticism (at times unnecessarily blunt) and experimental checks of your peers is crucial to spreading the word of your accomplishment. The communication process is also important, however, in assuring your own care in performing the experiments and interpreting the results. The scathing attack of someone who has found an important error or omission in your work can be a strong incentive for being more careful the next time. One person, working alone, could not hope to think of all of the possible ramifications, alternative explanations, or potential mistakes in an argument or theory. The explosive growth of science has depended heavily upon the communication process.

Scientific Method

Is there something we could call *scientific method* within this description, and if so, what is it? The process we have just described certainly has features that would fit the conventional textbook descriptions of scientific method. Although there are variations on the theme, this method is often laid out as a stepwise process of the form shown in box 1.1.

These steps are all involved in our description of developing an explanation for rainbows. Careful observations may lead to some empirical rules for when and where rainbows appear. An empirical rule or law is just a generalization based upon experiments or observations. An example in this case might be the statement that rainbows are always observed with the sun shining from behind us as we look at the rainbow. This should be an important clue in developing our theory, which should, of course, be consistent with this rule. The theory, in turn, might suggest ways of producing rainbows artificially that could lead to experimental tests of the theory.

This description of scientific method is not bad, although we should recognize that very few scientists engage in the full cycle of these steps. Theoretical physicists, for exam-

Figure 1.3 A scientific meeting. Communication and debate are important to the development of scientific explanations. The speaker is Albert Einstein.

ple, may spend all of their time with step 3; although they are interested in experimental tests and points of contact with their theories, they may never be directly involved with experimental work. Also, very little science is done in the pure exploratory sense implied by step 1; most observations and experiments take place within the context of existing theory. The majority of experimental scientists are therefore involved in some manner with step 4. We should also recognize that, although the method is presented as a stepwise process, in reality these steps usually occur simultaneously, with much cycling back and forth between steps.

It is also important to recognize that the method is only applicable in situations where experimental tests or other forms of consistent observation of nature are feasible. Such tests are crucial for weeding out bad theories; without them, rival theories may compete endlessly for acceptance. To some extent, this is the situation in some of the social sciences. We cannot, for example, repeat history under controlled conditions; no two situations are ever the same. The revealed truths of religion, which are based essentially on faith and appeal to authority, are also not subject to scientific tests or methods of analysis.

Traditional science courses focus on describing the results of the scientific process, rather than on the story of how these results were obtained. This explains why science is often seen by the general public as being merely a collection of facts and established theories. To some extent, that charge could be made against this book, which describes theories (and their application) that have resulted from the work of others, without providing the full picture of their development. Unfortunately, the full historical picture would take more pages and time than any of us could handle. Besides, being able to build upon previous work without needing to repeat the mistakes and unproductive approaches of the past is a necessary condition of human and scientific progress.

At the same time, however, this book attempts to engage you in the process of making observations and developing your own explanations based upon the ideas we discuss. By doing some home experiments, constructing explanations of the results, and defending your interpretation in discussions with your friends, you will learn to appreciate the give and take that is a crucial part of the scientific process.

Whether or not we are aware of it, we all use the scientific process to some degree in our everyday activities. The case of the malfunctioning coffee pot described in box 1.2 provides an example of scientific reasoning applied to an ordinary trouble-shooting problem.

1.2 THE SCOPE OF PHYSICS

Where does physics fit within the broader context of the sciences? Since this book is concerned with physics, rather than biology, chemistry, geology, or some other science, it is reasonable to ask where we draw the lines between the disciplines. It is not easy to make clear distinctions among the disciplines or to provide a definition of physics that will satisfy everyone. It is probably easier to give a sense of what physics is and does by way of example—that is, by listing some of its subfields and exploring their content. But first let us at least consider a definition, however incomplete.

Some Defining Characteristics of Physics

Physics is often defined as the study of the basic nature of matter and the forces that govern its behavior. It is the most fundamental of the sciences; its principles and theories can be used to explain the fundamental interactions that are involved in chemistry and biology, as well as in other sciences at the atomic or molecular level. Modern chemistry, for example, uses the physical theory called *quantum*

mechanics to explain how atoms combine to form molecules. Quantum mechanics was developed primarily by physicists during the early part of this century, but chemists have also contributed greatly to its development. Energy concepts, which arose initially in physics, are now used extensively in chemistry, biology, and other sciences.

The general realm of science is often divided into the life sciences and the physical sciences. The life sciences include the various subfields of biology and health-related disciplines that deal with living organisms. The physical sciences deal with the behavior of matter in both living and nonliving objects. In addition to physics, the physical sciences include chemistry, geology, astronomy, oceanography, and meteorology (the study of weather). Physics underlies all of them to some extent; the basic theories and models of these fields are built upon physical principles.

Physics is also generally regarded as being the most quantitative of the sciences, because it makes heavy use of mathematics and numerical measurements to develop and test its theories. This aspect of physics has often made it seem less accessible to students, even though in many ways the models and ideas of physics can be described more simply than those of other sciences. You will learn more about this aspect in the next section.

The Subfields of Physics

The primary subfields of physics are listed and identified in box 1.3. The field of mechanics, which deals with the study of the motion (or lack of motion) of objects under the influence of forces, was the first for which a comprehensive theory was developed. This was Newton's theory of mechanics, which emerged in the last half of the seventeenth century. It was the first full-fledged physical theory that made extensive use of mathematics, and it became a prototype for subsequent theories in physics.

The first four subfields listed in box 1.3 were well developed by the beginning of this century, although all have continued to advance since then. These subfields are sometimes grouped under the heading of *classical physics.* The latter four subfields are often grouped under the heading of *modern physics,* even though this term is somewhat misleading since all of the subfields are part of the modern practice of physics. The distinction is made because the last four subfields have all emerged during the twentieth century and did not really exist, except in very rudimentary forms, before the turn of the century.

The various photographs in this section (figs. 1.5–1.8) are intended to illustrate characteristic activities or equipment associated with the different physics subfields. The invention of the laser has been an extremely important factor in the rapid advances now taking place in optics (fig. 1.5). The development of the infrared camera, which involves optics as well as thermodynamics, has provided an important tool for the study of heat flow from buildings (fig. 1.6). The

Box 1.2

Everyday Phenomenon:
The Case of the Malfunctioning Coffee Pot

The Situation. It is Monday morning and you are, as usual, only half-awake and feeling at odds with the world. You are looking forward to reviving yourself with a freshly brewed cup of coffee when you discover that your coffee pot refuses to function. What do you do?

Action Alternatives

1. Pound on it with the heel of your hand (fig. 1.4).
2. Search desperately for the instruction manual, which you most likely threw away two years ago.
3. Call a friend who knows about these things.
4. Apply the scientific method.

Figure 1.4   Fixing a malfunctioning coffee pot—alternative 1.

The Analysis. All of the above alternatives have some chance of success. The sometimes positive response of electrical or mechanical appliances to physical abuse is well documented. The second two actions are both forms of appeal to authority that could produce results. The fourth, however, may be the most productive in the least time, barring success with alternative 1.

How would we apply the scientific approach to this problem? We might try applying the sequence of steps outlined in box 1.1. Step 1 involves calmly observing the symptoms of the malfunction. Let us suppose that the coffee maker simply refuses to heat up; when the switch is turned on, no sounds of warming water are noted. As a general rule, no matter how many times you turn the switch on and off, no heat results. This is the kind of simple generalization called for in step 2.

We are now in a position to develop some hypotheses about the cause of the malfunction, as required in step 3. Some possible candidates follow:

1. The coffee pot is not plugged in.
2. The external fuse or circuit breaker has tripped.
3. The power is off in the entire house.
4. An internal fuse in the coffee maker has blown.
5. A wire has come loose or burned through within the coffee maker.
6. The internal thermostat within the coffee maker is stuck in an open position.

No detailed knowledge of electrical circuits is required to check these possibilities, although the last three involve a little more sophistication (and trouble to check) than the first three. The first three are the easiest to test (step 4 of our method), and should be looked into first. A simple remedy like plugging in the pot or flipping on a circuit breaker may put you back in business and start the coffee gurgling. If the power is off in the entire building, then other appliances (lights, clocks, etc.) will not work either, which provides an easy test. There is little that you can do in this case, but at least you have identified the problem; abusing the coffee pot would not be productive.

The pot may or may not have an internal fuse, and if it is blown, a trip to the hardware store or small-appliance shop may be necessary to get things working again. A problem like a loose wire or burnt-out connection is usually obvious from a visual inspection after removing the bottom of the pot (or wherever the cord comes in and the electrical connections are made). Of course, you should unplug the pot before making such an inspection. If one of these alternatives is the case, you have identified the problem, but the repair job is likely to take more time or expertise. The same is true of the final alternative.

Regardless of what you find, this systematic (and calm) approach to the problem is likely to be more productive and satisfying than the other approaches. Trouble-shooting, if done in this manner, is an example of applying the scientific method on a small scale to an ordinary problem. We are all scientists, in a sense, if we approach problems in this manner.

rapid growth in consumer electronics, as seen in the availability of pocket calculators, home computers, and other gadgets, for example, has been made possible by developments in solid-state physics (fig. 1.7). Particle physicists use expensive particle accelerators (fig. 1.8) to study the interactions of subatomic particles in high-energy collisions.

Science and technology depend upon each other for progress. Physics plays an extremely important role in the education and work of engineers, whether they specialize in electrical, mechanical, nuclear, or other engineering specialties. In fact, people with physics degrees often end up

Box 1.3

The Major Subfields of Physics

Mechanics. The study of forces and motion.

Thermodynamics. The study of temperature, heat, and energy.

Electricity and magnetism. The study of electric and magnetic forces and electric current.

Optics. The study of light.

Atomic Physics. The study of the structure and behavior of atoms.

Nuclear Physics. The study of the nucleus of the atom.

Particle Physics. The study of subatomic particles (quarks, etc.).

Solid-State Physics. The study of the properties of matter in the solid state.

Figure 1.6   An infrared photograph showing patterns of heat loss from a house: an application of thermodynamics.

Figure 1.7   An integrated circuit employing semiconductor devices developed from knowledge of solid state physics. Magnification: ×50

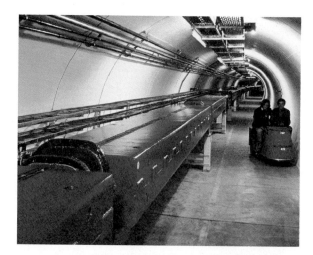

Figure 1.8   A Super-Proton-Synchrotron (SPS) particle accelerator used to study interactions of subatomic particles at high energies at CERN, the European particle physics laboratory outside Geneva.

Figure 1.5   An optics experiment using a laser.

working as engineers when they are employed in industry. The lines between physics and engineering or research and development are often hard to draw. Physicists are generally more involved with developing a fundamental understanding of nature, and engineers with applying that understanding to practical tasks or products, but these functions often overlap.

One final point might be made: physics, at least to its practitioners, is fun. Being able to discuss how a bicycle works or how a rainbow is formed has a fundamental appeal that we hope will draw you into the group that understands why physics is fun. There is no tiresome detail in this, only the thrill of gaining increasing insight into the basic workings of the universe. We can all be physicists in this sense.

1.3 THE ROLE OF MEASUREMENT AND MATHEMATICS IN PHYSICS

If you were to go into your college library, find a volume of *Physical Review* or some other major physics journal, and open it at random, you would most likely find a page covered with almost as many mathematical symbols and formulas as words. It would probably be totally incomprehensible to you. In fact, even most physicists who were not specialists in the particular subfield covered by the article would probably have difficulty making sense of that page because they would not be familiar with the particular symbols and definitions being used.

The Role of Mathematics

Why do physicists make such extensive use of mathematics in so much of their work? Is knowledge of mathematics essential to an understanding of the physical ideas being dis-

cussed? Could the same ideas be developed without the use of mathematics? As mentioned in the previous section, the use of mathematics in physics is one reason that many people avoid physics courses or have painful memories of the experience of studying physics if they could not avoid it.

This book uses mathematics in a very limited and gentle manner and makes it clear that physical ideas can be discussed without extensive use of mathematics. At the same time, it attempts to show why and how mathematics is used to make it easier to discuss the ideas of physics, as well as to organize and manipulate the quantitative measurements that are an important part of most work in physics. Mathematical symbols are a shorthand for representing physical quantities and for stating relationships among those quantities. Once you are familiar with the shorthand, its mystery disappears, and its utility becomes more obvious.

The Role of Measurement

The rapid growth and successes of physics began when the idea of making precise measurements as a test of our theories was accepted. Without careful measurements, vague predictions and explanations may seem reasonable, and making clear choices between different explanations of the same phenomenon may not be possible. A quantitative prediction, on the other hand, can be tested against reality, and an explanation or theory can be accepted or rejected on the basis of the results of a measurement. If, for example, one theory predicts that a cannonball will land 100 meters from us and another predicts a distance of 200 meters, firing the cannon and carefully measuring the actual distance the cannonball traveled can provide persuasive evidence for one theory or the other (fig. 1.9).

Everyday living is full of situations in which measurements can be important, as well as the ability to state and

Figure 1.9 Cannonballs and a measuring tape: the proof lies in the measurement.

use relationships between measurements. Suppose, for example, that you normally prepare pancakes on Sunday morning for three people, but on a particular Sunday there is an extra mouth to feed. What will you do—double the recipe and feed the rest to the dog? Or will you attempt to figure out by just how much the recipe should be increased in order to come out about right?

Suppose that the normal recipe calls for 1 cup of milk. How much milk will you use if you are increasing the recipe to feed four instead of three people? Perhaps you can solve this problem in your head, but others might find that process difficult. (Let's see, 1 cup is enough for three people, so ⅓ cup is needed for each person, and 4 times ⅓ equals ⁴⁄₃, or 1 ⅓ cups; see figure 1.10.)

If you had to describe this operation to someone else, for the milk as well as for the other ingredients, you might find yourself using a lot of words. If you looked closely at the person to whom you were talking, you might also notice his eyes glazing over and a look of confusion settling in. You could avoid this by devising a statement that worked for all the ingredients, thus avoiding the need to repeat yourself. Setting up a ratio or proportion involving the quantities needed would be useful. For example, you might say, "The quantity of each ingredient needed for four people is related to the quantity needed for three people as 4 is to 3." That still takes quite a few words and might not be very clear unless the person to whom you were describing the process was familiar with this way of stating a proportion. A piece of paper and a pencil could solve the communication crisis more neatly. The same statement could be written as follows:

Quantity for four:Quantity for three = 4:3.

To make the statement even briefer, you could use the symbols Q_4 to represent the quantity of any given ingredient needed for four people and Q_3 to represent the quantity of that same ingredient needed for three people. The statement then becomes a mathematical statement of the following form:

$$\frac{Q_4}{Q_3} = \frac{4}{3}.$$

Using symbols is simply shorthand for saying the same thing in words. This shorthand also has the advantage of making manipulations of the stated relationship easier. If, for example, you multiplied both sides of this proportion by Q_3, you would have the following relationship:

$$Q_4 = \left(\frac{4}{3}\right) Q_3 = \left(1 \frac{1}{3}\right) Q_3.$$

This says that the quantity of each ingredient needed for four people is 1 ⅓ (or ⁴⁄₃) times the quantity required for three people, and you could use it to quickly find the proper amount for each ingredient: simply multiply the quantity given in the recipe by ⁴⁄₃.

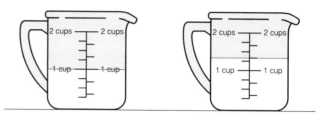

Figure 1.10 Two measuring cups containing enough milk for recipes to feed three people and four people.

There are two points to this example. The first is the idea that making measurements is both a routine and important part of everyday experience. The second is the idea that using symbols in a mathematical statement to represent quantities is a shorter way of expressing an idea involving numbers or measurements than the equivalent statement in words. Using symbols also makes it easier to manipulate the relationship and put it into a more convenient form for your purpose. These are essentially the reasons that physicists (as well as other scientists) find mathematical statements convenient.

Despite the brevity and apparent clarity of the mathematical statement, however, many people are still more comfortable with words. This is a matter of personal choice and is often related to a "math block," or fear of mathematics, that dates to earlier experiences. For this reason, word statements are given along with most of the simple mathematical statements that we will use. These word statements together with the mathematical shorthand (and, often, associated drawings and pictures) should all help to give you a feeling for the concepts we will be discussing. Students who take courses in engineering physics, who are usually quite comfortable with mathematics, sometimes depend so heavily on mathematical descriptions that they lose sight of the underlying physical ideas. That will not be a problem in this book.

The Use of Metric Units. We should return, for a moment, to the idea of units of measurement. The units are really an essential part of the description of any measurement; we do not convey adequate information if we just state the number. In the recipe problem, for example, suppose you talked about adding 1 ⅓ of milk. This statement would be incomplete; you need to indicate whether you are talking about cups, pints, liters, or milliliters. The liter and milliliter are metric units of volume, whereas cups, pints, quarts, and gallons are holdovers from the older British system of units.

Most countries have now adopted the metric system, which has several advantages over the British system that is still extensively used in this country. The primary advantage of the metric system is the use of common prefixes to represent different multiples of 10, thus making unit conversions within the system very simple. The basic unit of length in the metric system, for example, is the meter (m),

Table 1.1

Commonly Used Metric Prefixes

Prefix	Meaning
kilo	$1000 = 10^3 = 1$ thousand.
mega	$1\,000\,000 = 10^6 = 1$ million.
giga	$1\,000\,000\,000 = 10^9 = 1$ billion.
centi	$0.01 = 10^{-2} = \dfrac{1}{100} = 1$ hundredth.
milli	$0.001 = 10^{-3} = \dfrac{1}{1000} = 1$ thousandth.
micro	$0.000\,001 = 10^{-6} = \dfrac{1}{10^6} = 1$ millionth.
nano	$0.000\,000\,001 = 10^{-9} = \dfrac{1}{10^9} = 1$ billionth.

which is slightly longer than 1 yard (39.4 inches as opposed to 36 inches). A kilometer (km) is 1000 meters, and a centimeter (cm) is $^1\!/_{100}$ of a meter. 30 cm is therefore 0.30 m; all we have to do is move the decimal point two places to divide by 100. Table 1.1 provides a table of the prefixes commonly used in the metric system. (See appendix B for a discussion of the power-of-10 (scientific) notation for describing very large and very small numbers.)

The basic unit of volume in the metric system is the liter, which is slightly larger than one quart (1 liter = 1.057 quarts). A milliliter is $^1\!/_{1000}$ of a liter and is a more convenient size for discussing quantities in recipes. One milliliter is also equal to 1 cm^3, or one cubic centimeter, so there is also a simple relationship between the length and volume units. Such simple relationships are hard to find within the British system. A cup, for example, is $^1\!/_4$ quart, and a quart is 67.2 cubic inches.

The metric system predominates in this book. British units will be used occasionally because they are familiar, and this familiarity can be helpful in introducing new concepts. Most of us still relate to distances in miles more readily than to distances in kilometers. The fact that there are 5280 feet in one mile is a nuisance, however, compared to the tidy 1000 meters in one kilometer. Becoming more familiar with the metric system is a worthy objective; it might even help to reduce our balance of trade problems if more of us were comfortable with the system used in most of the world!

1.4 PHYSICS AND EVERYDAY PHENOMENA

The study of physics can and will lead us to ideas as earth-shaking as the fundamental nature of matter and energy, on one hand, and the structure of the universe on the other. With ideas like these available, why spend time on more mundane matters like explaining why a bicycle stays upright or how a flashlight works? Why not just plunge directly into the truly far-reaching issues regarding the fundamental nature of reality?

Our approach, emphasizing the understanding of everyday, or common, phenomena, is partly a matter of choice, but there are excellent reasons for selecting that path. Our understanding of the fundamental nature of matter, or of the universe, is based upon concepts like force, mass, energy, and electric charge that are, by their very nature, abstract and not directly accessible to our senses. It is possible to learn some of the words associated with these concepts, and to read and discuss ideas involving them, without ever acquiring a good understanding of their meaning. This is one risk of playing with the grand ideas without laying a proper foundation.

Using everyday experience to raise questions, introduce concepts, and practice devising explanations using physical principles has the advantage of dealing with examples that are familiar and not abstract. These examples should also appeal to your natural curiosity about how things work, which, in turn, should motivate you to understand the underlying concepts. If we can clearly describe common happenings, we gain confidence in dealing with more abstract ideas in other explanations. With familiar examples, the concepts are based upon firmer ground and their meaning becomes more real.

For example, the question of why a bicycle stays upright while moving but falls over when at rest involves the concept of angular momentum, which is discussed in chapter 7. This same concept plays a role in our understanding of atoms and the atomic nucleus, in the realm of the very small, and in our understanding of the motion of galaxies at the opposite end of the scale (fig. 1.11). You are more likely to get a good handle on the concept by discussing it first in the context of bicycles.

An understanding of the principles governing falling leaves and acorns involves the concepts of acceleration, force, and mass, which are discussed in chapters 2, 3, and 4. These concepts are also central to an understanding of atoms and the universe. Likewise, an understanding of how rainbows are formed involves ideas on the fundamental behavior of light, discussed in chapter 15. The behavior of light also plays an important role in the way we have obtained our knowledge of atoms and the nature of the universe.

We will find that our "common sense" can sometimes mislead us in our understanding of ordinary phenomena. Adjusting that common sense to be more consistent with a view that incorporates well-established physical laws and principles is one of the challenges involved in addressing everyday experience. Performing simple experiments, either at home (as is often suggested in this book) or in the laboratory or via lecture demonstrations, can be an important part of the process of building new ideas into your own world view.

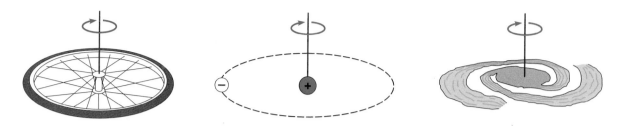

Figure 1.11   A bicycle wheel, an atom, and a galaxy all involve the concept of angular momentum.

Although it may sound like an oxymoron (self-contradictory phrase), everyday experience can be extraordinary. A bright rainbow can be an incredible sight. Understanding how it originates does not detract from the experience; it adds excitement to be able to explain such a beautiful display with just a few simple concepts. (In fact, physicists and others who understand these ideas see more rainbows because they know where to look.) This excitement, and the added appreciation of nature that is part of it, can be accessible to all of us.

1.5 USING THE FEATURES OF THIS BOOK

This book has a number of features that are intended to make it easier for students to organize and acquire the concepts that we will be exploring. These features include the chapter objectives and outline near the beginning of each chapter, the chapter summaries, study hints, special boxed features, and the questions and exercises. How can these features be used to best advantage?

Objectives, Outlines, and Summaries

Research in science education has indicated that students get a better grasp of concepts if they have some structure or framework to aid them in organizing the ideas. Both the chapter objectives and outline at the beginning of each chapter and the summaries at the end are designed to help provide such a framework. As the study hint near the beginning of this chapter indicates, obtaining a clear idea of what you are trying to accomplish before you invest time in reading the chapter will make your reading more effective.

The list of topics and questions contained in the chapter outline can be used as a checklist for measuring your progress as you read. Each topic and the related questions pertain to a specific section within the chapter. These are designed to stimulate your curiosity by providing some blanks (unanswered questions), which your reading of the chapter should fill with answers. Without preparing the blanks, however, your mind has no organizational structure that it can use to store this information. Without this structure, recall is more difficult.

The chapter summaries provide short descriptions of the critical ideas discussed in the chapter and are often cast in the form of answers to questions raised in the chapter outlines. They provide a quick review of key ideas, but they are no substitute for a careful reading of the text. By following the same organizational structure as the outline, however, the summaries do serve to remind you where to find a more complete discussion of these ideas. The function of both the outlines and the summaries is to make your reading more purposeful and effective (fig. 1.12).

Study Hints, Home Experiments, and Examples

Study hints are located at various points in the chapters. These suggest things to do that may help in understanding certain ideas. A passive reading of material like this is often not enough to make the ideas real to you. The study hints generally suggest some process that will get you more actively involved.

If a laboratory experience is not a part of the course you are taking, the home experiments can be crucial to a good understanding. Reading or talking about physical ideas is useful, but there is no substitute for hands-on experience with real cases that illustrate the ideas. You already have a wealth of experience with many of the phenomena that we will discuss, but you have probably not thought about this experience or observed it closely in the context of the physical concepts we will introduce. Trying some simple experiments will make you see things in a way that you have not seen them before. Much of the excitement in studying science arises from this kind of experience. We hope to make you a "born again" observer of nature, in the sense that you will see simple phenomena in a new light.

The examples found within each chapter are simple questions or exercises similar to those at the end of each chapter. They are designed to illustrate how to think about such questions or exercises. Boxes with the heading "Everyday Phenomenon" present more complex examples that lead you to apply the ideas just introduced in order to understand a particular phenomenon. These examples are often drawn from sports, but weather phenomena, automobile accidents, and other illustrations are also used.

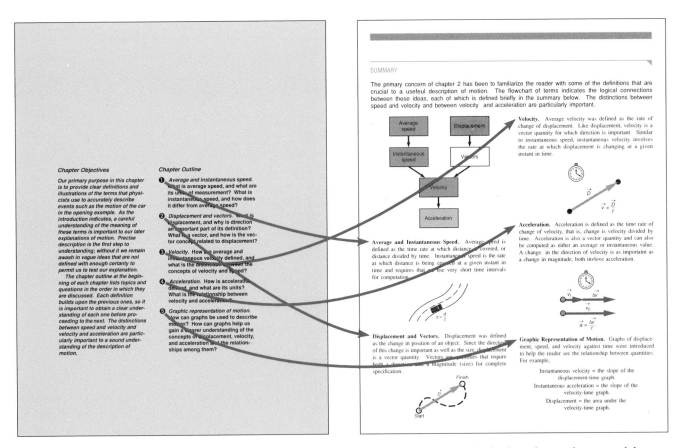

Figure 1.12 The chapter outline and the chapter summary both provide frameworks for learning and summarizing concepts.

Questions, Exercises, and Challenge Problems

At the end of each chapter, you will find a group of questions followed by a group of exercises. It is important that you try some or all of these questions and exercises and carefully write out the answers, either as assigned by your instructor or as independent study. The ideas contained in this and the following chapters cannot be adequately mastered without this kind of practice.

The questions are crucial for helping to consolidate in your mind the concepts and distinctions introduced in the chapter. Each question calls for a short answer as well as an explanation. It is a good idea to write out the explanations in clear sentences when you answer these questions, because it is only through such expression that the ideas become part of you. A sample appears in box 1.4.

The exercises are designed to give you practice in using the ideas and formulas to do simple computations. This also helps to solidify your understanding of the concepts; in addition, it makes you familiar with the units and the sizes of different quantities that are discussed. Even though many of the exercises are simple enough to work in your head without writing much down, we recommend writing out the information given, the information sought, and the solution in the manner shown in box 1.5. This develops habits of

---

Box 1.4

Sample Question

Question: Astrologers claim that various events in our lives are determined by the positions of the planets relative to the stars. Is this a testable hypothesis?

Answer: Yes, it could be tested if the astrologers were willing to make explicit predictions about future events that could be verified by independent observers. In fact, astrologers generally carefully avoid doing this, preferring to cast their predictions in vague statements that are subject only to broad interpretation.

---

careful work that will help you to avoid careless mistakes, which are a frequent pitfall in such exercises.

The challenge problems are more extensive exercises, often involving several parts, which may mix qualitative questions with numerical computations. They are not necessarily more

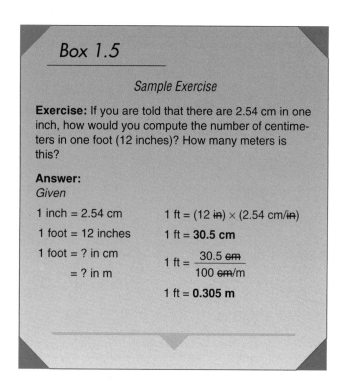

Box 1.5

Sample Exercise

Exercise: If you are told that there are 2.54 cm in one inch, how would you compute the number of centimeters in one foot (12 inches)? How many meters is this?

Answer:
Given

1 inch = 2.54 cm

1 foot = 12 inches

1 foot = ? in cm

     = ? in m

$1\ ft = (12\ \text{in}) \times (2.54\ \text{cm/in})$

$1\ ft = \textbf{30.5 cm}$

$1\ ft = \dfrac{30.5\ \text{cm}}{100\ \text{cm/m}}$

$1\ ft = \textbf{0.305 m}$

difficult than the questions or exercises, but they do take more time and are sometimes used to introduce new applications of ideas that were discussed in the chapter. Doing at least one or two of these in each chapter should build confidence in your ability to apply the concepts learned.

Answers to the odd-numbered exercises are found in the back of the book. Looking up the answer before attempting the exercise, however, is self-defeating; it deprives you of practice in thinking things through independently. Use the answers only to confirm or improve your own thinking.

Students who are not science majors (and even some who are) too often perceive science courses as collections of loosely-connected facts about nature that must be laboriously memorized. In a physics course, however, the connections between ideas are the main point; the facts and details are secondary. To get the maximum benefit from reading this book, you will need to accept the challenge of building the framework that involves these connections. The process works best if, in addition to careful reading, you are actively engaged in doing experiments, working examples, and testing ideas in discussions with your friends. Memorization, if it is involved at all, should play a very minor role.

SUMMARY

This first chapter provides an introduction to the nature of scientific explanation, the nature of physics, and a discussion of the use of mathematics and measurement in physics. It also gives some pointers on how to use this book most effectively in exploring the remaining chapters. The key points are the following:

The Nature of the Scientific Enterprise. Scientific explanations are developed by generalizing from observations of nature and testing by experiment or further observations. This general process of observation, generalization, theory formation, and testing is often called the *scientific method*.

netism, optics, atomic physics, nuclear physics, solid-state physics, and particle physics.

The Role of Measurement and Mathematics in Physics. Much of the progress of physics can be explained by its use of quantitative models, which yield precise predictions that can be tested by making physical measurements. Mathematics is a shorthand for describing and manipulating these results. The concepts of physics can often be described with a minimum of mathematics, however.

Observation
or
experiments

Generalization

Hypothesis
or
theory

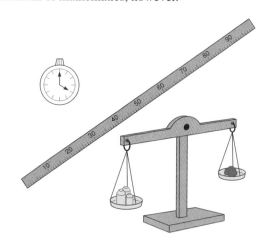

The Scope of Physics. Physics is the most fundamental of the natural sciences in that its theories often form the basis of explanations in the other sciences. Its major subfields include mechanics, thermodynamics, electricity and mag-

Physics and Everyday Phenomena. An appreciation for many of the basic concepts of physics can be gained by considering everyday phenomena, with which we are already familiar. This aspect of physics adds to the enjoyment of understanding physical concepts.

Using the Features of This Book. The features associated with each chapter in this book are designed to help the reader build a structure for learning the basic concepts of physics. These features include chapter outlines and summaries, study hints, special boxes, and the questions, exercises, and challenge problems.

QUESTIONS

Q1.1 Which of the following qualities best distinguishes between explanations provided by science and those provided by religion: truth, testability, or simplicity? Explain.

Q1.2 A person who claims to have paranormal powers claims that he can bend spoons without touching them, simply by exercising his mental powers. Is this a testable hypothesis? Explain.

Q1.3 Suppose that, thinking like a scientist, you have developed a hypothesis or theory about why your car will not start. The hypothesis is that the battery is dead. What would your next step be? Explain.

Q1.4 Which of the three science fields; biology, chemistry, or physics, would you say is the most fundamental? Explain by stating in what sense one of these fields is more fundamental than the others.

Q1.5 Based upon the brief descriptions given in box 1.3, which subfields of physics would you say are involved in the explanation of rainbows? In describing how an acorn falls? Explain.

Q1.6 Suppose that you are told that speed is defined by the relationship $s = d / t$, where s stands for speed, d for distance, and t for time. State this relationship in words, using no mathematical symbols.

Q1.7 Suppose that you are told that momentum is defined as the mass of an object multiplied by its velocity. If we let p be momentum, m be mass, and v be velocity, is $p = m / v$ a correct way of expressing this statement? Explain.

Q1.8 Suppose you are told that the distance an object travels when it is undergoing constant acceleration, starting from rest, is one-half the acceleration multiplied by the square of the time. If we let d be the distance, a the acceleration, and t the time, is $d = \frac{1}{2}at^2$ a correct expression of this statement? Explain.

Q1.9 Is the following statement true or false? "The primary advantage of the metric system of units over the older British system is our familiarity with the units." Explain.

Q1.10 Is the following statement true or false? "The primary advantage of the British system of units over the metric system is that the conversions between units within the system are all achieved by multiplication or division by factors of 10." Explain.

Q1.11 Which of the two systems of units, the metric system or the British system, is used throughout most of the world? Explain.

EXERCISES

E1.1 Suppose that a waffle recipe designed to feed three people calls for 600 ml of flour. How many ml of flour would you use if you wished to extend the recipe to feed five people?

E1.2 Suppose that a cupcake recipe designed to produce twelve cupcakes calls for 800 ml of flour. How many ml of flour would you use if you only wanted to make eight cupcakes?

E1.3 A woman uses her hand to measure the width of a table top. If her hand has a width of 9 cm and she finds the table top to be 14.5 hands wide, what is the width of the table top in cm? In meters?

E1.4 A book is measured to be 185 mm wide. What is this width in cm? In meters?

E1.5 A crate has a mass of 1.25 Mg (megagrams). What is this mass in kilograms? In grams?

E1.6 A large tank holds 3.5 kl (kiloliters) of water. How many liters is this? How many ml?

E1.7 A mile is 5280 ft long. The example exercise in this chapter showed that one foot is approximately 0.305 m. How many meters are there in a mile? How many km?

Part 1

THE NEWTONIAN

Revolution

2 *Description of Motion*

Imagine that you are stopped at an intersection in your automobile. After waiting for cross traffic, you pull away from the stop sign, accelerating eventually to a speed of 56 kilometers per hour (35 miles per hour). Since this is the speed limit, you maintain that speed for a while until a dog runs in front of your car and you hit the brakes, reducing your speed rapidly to 10 kilometers per hour (6 miles per hour) (fig. 2.1). Having missed the dog, you speed up again to 56 kilometers per hour. After another block, you come to another stop sign and reduce your speed gradually to zero.

We can all relate to the preceding description. The idea of measuring speed in miles per hour (MPH) is more familiar than the use of kilometers per hour (km/hr), but the speedometers in modern cars now show both. The use of the term accelerate to describe an increase in speed is also common. To a physicist, however, these concepts take on more precise and specialized meanings, which makes them even more useful in describing exactly what is happening. These meanings are sometimes different than the ones we assume in our everyday usage of such terms. The term accelerate, for example, is used by physicists to describe any situation in which velocity is changing; this includes cases where speed may be decreasing or the direction of the motion may be changing. Physicists also define these quantities in a manner that allows them to be measured and compared.

How would you define the term speed if you were explaining the idea to a younger brother or sister? Does velocity mean the same thing? What do we mean by the term accelerate? Is the notion vague, or does it have a precise meaning? Is it the same thing as velocity? These are areas in which the language used by physicists differs from our everyday language, even though the same words are employed. Clear definitions are essential foundations for the explanations that we will be developing.

Figure 2.1   As the car brakes for the dog, there is a sudden change in speed.

Chapter Objectives

Our primary purpose in this chapter is to provide clear definitions and illustrations of the terms that physicists use to accurately describe events such as the motion of the car in the opening example. As the introduction indicates, a careful understanding of the meaning of these terms is important to our later explanations of motion. Precise description is the first step to understanding; without it we remain awash in vague ideas that are not defined with enough certainty to permit us to test our explanations.

The chapter outline at the beginning of each chapter lists topics and questions in the order in which they are discussed. Each definition builds upon the previous ones, so it is important to obtain a clear understanding of each one before proceeding to the next. The distinctions between speed and velocity and velocity and acceleration are particularly important to a sound understanding of the description of motion.

Chapter Outline

❶ *Average and instantaneous speed.* What is average speed, and what are its units of measurement? What is instantaneous speed, and how does it differ from average speed?

❷ *Displacement and vectors.* What is displacement, and why is direction an important part of its definition? What is a vector, and how is the vector concept related to displacement?

❸ *Velocity.* How are average and instantaneous velocity defined, and what is the distinction between the concepts of velocity and speed?

❹ *Acceleration.* How is acceleration defined, and what are its units? What is the relationship between velocity and acceleration?

❺ *Graphic representation of motion.* How can graphs be used to describe motion? How can graphs help us gain a clearer understanding of the concepts of displacement, velocity, and acceleration and the relationships among them?

2.1 AVERAGE AND INSTANTANEOUS SPEED

Ideas relating to motion are probably most easily grasped in the context of automobiles. The concept of speed, for example, is extremely familiar, since we all have experience in reading a speedometer (or perhaps failing to read it carefully enough in the presence of the law).

Average Speed

What does it mean to say that we are traveling at 55 MPH (89 km/hr)? It means that we would cover a distance of 55 miles in a time of 1 hour if we traveled steadily at that speed. Note carefully the nature of this description: there is a number (55) and some units or dimensions (miles per hour). The miles-per-hour terminology implies that miles are divided by hours in arriving at the speed, which is exactly how we would compute the average speed for a trip. Suppose, for example, that we travel a distance of 260 miles in a time of 5 hours, as shown on the road map of figure 2.2. The average speed is then 260 miles / 5 hours = 52 MPH.

This type of computation is familiar to most of us. If we were to express the definition of average speed used here in a word equation, it would take the following form:

**Average speed = the distance traveled
divided by the time of travel.**

We can represent this same definition with symbols by writing

$$\bar{s} = \frac{d}{t}$$

where the letter s represents speed, d represents distance, and t represents time. The bar over the s is used to indicate that it is an average value, a commonly used notation. As noted in chapter 1, the letters or symbols are a shorthand for saying what can be said with a little more effort in words. You may judge for yourself which is the more efficient way of expressing this definition; most people find the symbolic expression easier to remember and use. Physicists, in particular, prefer the symbolic shorthand.

Another way of looking at the concept of average speed that we have just defined is to regard it as a rate. *Rates* always represent one quantity divided by another; gallons per minute, pesos per dollar, and points per game are all examples of rates. If we are considering a time rate, the quantity that we divide by is time, which is the case, of course, with average speed. We could therefore also define average speed in words as follows:

**Average speed is the average rate at which distance
is covered with time.**

This is just a different way of saying what we have said above, but it may help to make the concept meaningful. The symbolic definition $\bar{s} = d / t$ remains the same, regardless of the words used to express the idea.

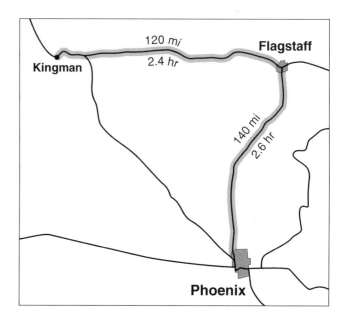

Figure 2.2   A road map showing a trip of 260 miles, with driving times for the two legs of the trip.

We should return for a moment to the idea of units. The units are an essential part of the description; we do not convey adequate information if we just state the number. Suppose you say you were doing 70, without stating the units. That would probably be understood in the United States as 70 MPH, since that is the unit most frequently used. In Europe, on the other hand, people would probably assume that you were talking about the considerably slower speed of 70 km/hr. Failure to state the units is therefore a failure to communicate clearly—a common problem.

It is relatively easy to convert from one unit to another if the conversion factors are known. For example, if we want to convert kilometers per hour, the unit commonly used in Europe and most other parts of the world, to miles per hour, the unit used in the United States, we need to know the relationship between miles and kilometers. A kilometer (km) is roughly 6/10 of a mile (.6214 to be precise). 70 km/hr is therefore equal to 43.5 MPH, as shown in box 2.1. The process is straightforward, involving multiplication or division by an appropriate conversion factor. Box 2.1 also shows the conversion of kilometers per hour to meters per second (m/s), done as a two-step process. As you can see, 70 km/hr can also be expressed as 19.4 m/s. Since this is a convenient size for discussing the motion of real objects, such as cars and people, and is the basic metric unit for speed, most of our examples from here on will use the m/s unit. Table 2.1 shows some familiar speeds expressed in miles per hour, kilometers per hour, and meters per second.

Instantaneous Speed

Returning to our earlier example, if we travel a distance of 260 miles in a time of 5 hours, is it likely that the entire trip

Box 2.1

Unit Conversions

Kilometers per hour to miles per hour:

70 km/hr × 0.6214 miles/km = 43.5 MPH.

Kilometers per hour to meters per second (m/s):

70 km/hr × 1000 m/km = 70 000 m/hr.

$$\frac{70\ 000\ \text{m/hr}}{3600\ \text{s/hr}} = 19.4\ \text{m/s}.$$

   As you can readily see, the numbers change when the units change for the same speed. Notice how the units are canceled (lines drawn through the unit indicate cancellation) in performing these conversions. Notice also that fewer meters are covered in one second than in one hour; this makes gratifying sense and serves as a check on our conversion.

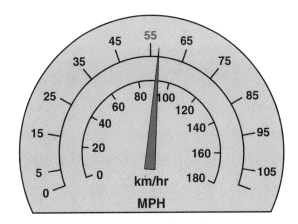

Figure 2.3 **A speedometer with two scales for measuring instantaneous speed: MPH and km/hr.**

Table 2.1

Familiar Speeds in Different Units

20 MPH = 32.2 km/hr =	8.9 m/s.
40 MPH = 64.4 km/hr =	17.9 m/s.
60 MPH = 96.6 km/hr =	26.8 m/s.
80 MPH = 129 km/hr =	35.8 m/s.
100 MPH = 161 km/hr =	44.7 m/s.

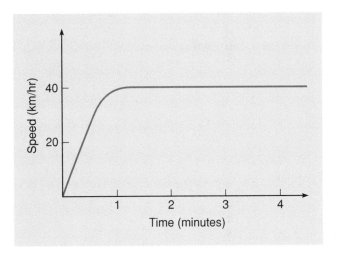

Figure 2.4 **A graph of instantaneous speed varying with time for a car starting from rest and accelerating to a speed of 40 km/hr.**

takes place at a speed of 52 MPH? Of course not. Instead, the speed goes up and down as the road goes up and down, as slower vehicles are overtaken, as rest breaks occur, or as the highway patrol looms on the horizon. If we want to know how fast we are traveling at a given instant in time, we read the speedometer, which displays the instantaneous speed (fig. 2.3).

   How does the instantaneous speed differ from the average speed? The instantaneous speed describes how fast we are going at a given instant but tells us little about how long it will take to travel several miles, unless the speed is held constant. The average speed, on the other hand, permits us to compute how long a trip might take but says little about the variation in speed during the trip. A more complete description of how the speed of a car varies could be provided by a graph such as that shown in figure 2.4. In this case the car starts from a rest position, where the speed is zero, and accelerates to a speed of 40 km/hr, which is then maintained.

   Each point on the curve in figure 2.4 represents the instantaneous speed at the time indicated on the horizontal axis. Figure 2.5 depicts a longer portion of a trip in which the instantaneous speed of the car varies for several reasons. This graph provides a much more complete picture of how the trip took place than would a simple statement of the average speed for the trip. (It would not be easy to compute the average speed from the information provided in the graph of figure 2.5; we would first have to find the total distance traveled by the car.)

   Even though we all have some intuitive sense of what instantaneous speed means from our experience in driving or riding in cars and reading speedometers, defining instantaneous speed in a way that tells us how to compute this quantity presents some problems we did not encounter in defining average speed. We could say simply that instantaneous speed is the rate at which distance is being covered for a given instant in time, but this does not tell us how to compute

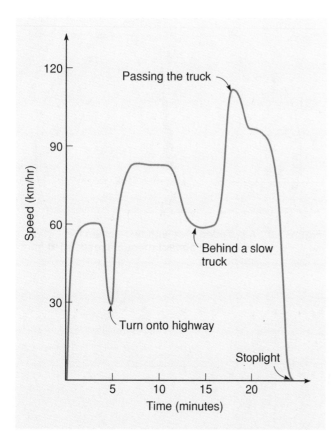

**Figure 2.5   Variations in instantaneous speed for a
portion of a trip on a local highway.**

instantaneous speed. It still represents a rate, for which we
divide a distance by a time, but what time interval should we
use? How long is an instant in time?

Our normal solution to this problem is simply to choose
a very short interval in time during which a short distance is
covered and the speed does not change drastically. If we
knew, for example, that in one second a distance of 20 m was
covered, dividing 20 m by 1 s to give us a speed of 20 m/s
would provide a good estimate of the instantaneous speed if
the speed did not change appreciably during that one second of time. If the speed was changing, however, we would
have to choose an even shorter interval of time. In principle, we can choose these distance and time intervals to be
as small as we wish, but in practice it can be difficult to measure such small quantities. (In fact, the design of a good
speedometer is a nontrivial problem; you might try to imagine how it is done.)

If we were to put these ideas into a simple word definition,
we could state it as follows:

**Instantaneous speed is the time rate at which dis-
tance is being covered at a given instant in time. It
can be computed by finding the average speed for a
very short time interval, during which the speed
does not change appreciably.**

Instantaneous speed is therefore closely related to the concept of average speed, but involves such short time intervals that variations in the rate of travel are not important.

The study of calculus provides us with an elegant and precise mathematical means of dealing with these very small
time intervals, but a knowledge of calculus is not required
to understand the concept of instantaneous speed, and we
will not resort to calculus here. Our familiarity with
speedometers and the definition just given provide a good
beginning basis for understanding the concept. The graphic
descriptions of motion discussed in the last section of this
chapter will reinforce this understanding.

2.2 DISPLACEMENT AND VECTORS

In developing the concepts of average and instantaneous
speed, we have described the rate at which distance is being
covered without concerning ourselves with the direction of
travel. Quite obviously, the direction of travel is important to
a full description of motion; it can dictate, for example,
whether we end up in New York or Hawaii. The concept of
displacement brings direction into the picture and is critically important to the distinction we make between velocity
and speed in the next section. When questions like Which
way? are pertinent, as well as How far? or How fast?, then
vectors are involved.

Displacement

When an object moves, its position changes, and we refer
to such a change in position as a *displacement*. If you were
asked to describe precisely a displacement or change in position, you would need to specify both how far the object had
moved *and the direction* in which it had moved. When
physicists speak of a displacement, then, they specify the
direction as well as the distance.

Displacement is a *vector* quantity, by which we mean that
its direction is important as well as its size, or magnitude.
Both direction and magnitude must be specified for the
description to be complete. To illustrate this idea, figure 2.6

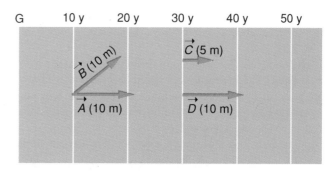

**Figure 2.6   Different displacement vectors shown on a
football field. Displacement B yields a
different result than displacement A even
though their magnitudes are the same.**

shows several different displacements on a football field. The vector quantities $\vec{A}$ and $\vec{B}$ have the same size (10 meters), but they are different displacements because they are in different directions, and therefore result in different final positions, even though the starting point was the same. The arrows over the symbols used here indicate that these quantities are vectors. Arrows are convenient symbols to use in representing vectors; the length of the arrow can represent the size of the vector, and the point of the arrow clearly indicates a direction.

Displacement $\vec{C}$ has the same direction as $\vec{A}$, but it has a different size or length and therefore represents a different vector. Displacement $\vec{D}$ has both the same length and direction as $\vec{A}$, and therefore it is an equal displacement to $\vec{A}$. If $\vec{A}$ and $\vec{D}$ started at the same initial position, their resulting final positions would be identical. In studying figure 2.6, notice also that 10 m is a larger distance than 10 yards, as you can see by looking at the yard markers. One meter is approximately 1.094 yards, so 10 m equals 10.94 yards, or almost 11 yards.

The important point in considering these different displacements is that, even though the distances covered may be the same, the resulting final position is critically dependent upon the direction of the displacement. If a wide receiver runs 10 m straight downfield, expecting to receive the football there from the quarterback, he will be disappointed if the quarterback thought he was going to run 10 m across the field. Clear communication of intent requires mention of direction!

Adding Vectors

Another important point can be made about vectors using displacements on a football field as an example. Displacement $\vec{C}$ in figure 2.7 can be thought of as being the sum of displacements $\vec{A}$ and $\vec{B}$, which we might write as follows:

$$\vec{C} = \vec{A} + \vec{B}.$$

Notice, however, that the magnitude of $\vec{C}$ (its length, or size) is not the same as the sum of the magnitudes of $\vec{A}$ and $\vec{B}$. A greater distance is traveled in following the path represented by displacements $\vec{A}$ and $\vec{B}$, than in following the more direct route represented by the total displacement $\vec{C}$. If we consider only the magnitudes of the vectors, the whole is less than the sum of its parts!

Figure 2.7 also shows that the order in which we add the two displacements $\vec{A}$ and $\vec{B}$ does not matter. If displacement $\vec{B}$ occurs first and then is followed by $\vec{A}$, the result is the same as if $\vec{A}$ occurred first followed by $\vec{B}$. The order in which we choose to add them on our drawing is not important if we only want to know the total displacement.

To show more clearly how we might obtain the sum of two vectors, we have drawn these displacements to a larger scale in figure 2.8: 1 centimeter (cm) is used to represent a length of 1 m in the diagram. If we use a ruler to measure the lengths of $\vec{A}$ and $\vec{B}$, we find that they are approximately 7.1

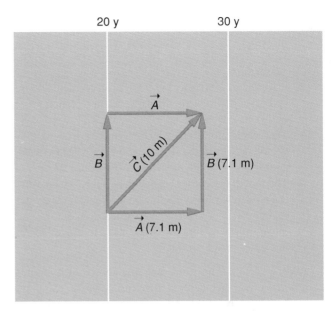

Figure 2.7 **The two displacements $\vec{A}$ and $\vec{B}$ add together to produce displacement $\vec{C}$. The order of addition is not important.**

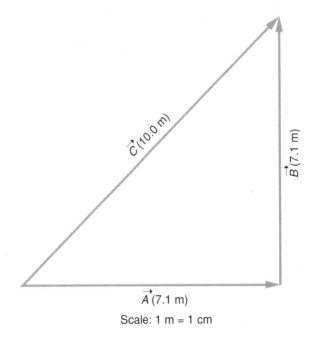

Figure 2.8 **A scale diagram showing the addition of vectors $\vec{A}$ and $\vec{B}$. Vector $\vec{C}$ is drawn from the toe of vector $\vec{A}$ to the head of vector $\vec{B}$.**

cm (corresponding to 7.1 m on the football field), while displacement $\vec{C}$ is 10 cm in length (corresponding to 10 m on the football field). The combined length of displacements $\vec{A}$ and $\vec{B}$ is therefore 7.1 m + 7.1 m = 14.2 m, which is obviously greater than 10 m. As we indicated earlier, vector addition does not generally result in a vector that is equal in magnitude to the numerical sum of the two vectors being added.

The process of adding vector $\vec{B}$ to vector $\vec{A}$ on the graph consists of the following steps:

1. Draw vector $\vec{A}$ to scale (7.1 cm) and in the appropriate direction.

2. Start vector $\vec{B}$ with its toe at the head of vector $\vec{A}$, and draw it to scale in the appropriate direction for $\vec{B}$.

3. The vector sum $\vec{C}$ can then be found by drawing a vector starting at the starting point for $\vec{A}$ (its toe) and ending at the ending point for $\vec{B}$ (its head).

4. The length of vector $\vec{C}$ can then be found by measuring its length on the graph with a ruler and using the scale factor to convert it to meters. Likewise, its direction can be found by measuring its angle (45° in this case) with a protractor.

Adding vectors is not quite the same, therefore, as adding simple numbers. We can estimate the magnitudes, as we have done here, by drawing the vectors to scale and in the appropriate directions, and then measuring the lengths with a ruler. This is a simple procedure, which also has the advantage of providing a picture of the process. (Appendix C contains a more complete description of this process and more examples.) The same results can also be achieved mathematically by using trigonometric functions, but the mathematical operation often obscures the basic simplicity of the process.

As a final comment on vectors and displacement, note that in describing a displacement, we are not really describing the actual path taken, but merely the change in position. Displacement $\vec{C}$ is the same regardless of whether it is achieved by traveling in a straight line, or by the net effect of the two displacements $\vec{A}$ and $\vec{B}$, or by some more circuitous route. In fact, if we traveled all the way around the earth and ended up at the final position indicated by displacement $\vec{C}$, the change in position (displacement) is still equal to $\vec{C}$!

Many of the concepts that we will encounter as we proceed will involve vector quantities. Velocity, acceleration, force, and momentum are all vector quantities that are introduced in this and the next few chapters. Direction is a critical aspect of all of these concepts.

2.3 VELOCITY

Do the words *speed* and *velocity* mean the same thing? They are often used interchangeably in everyday language, but physicists make a very important distinction between the two terms. This distinction is essential to understanding Newton's theory of motion, which is introduced in chapter 4, so it is not just a matter of whim or jargon. It does make a difference!

Average Velocity

The difference between speed and velocity can be simply stated, but is easily missed. *Velocity* is the rate of change of position with time, whereas *speed* is the rate of change of distance covered with time, without regard to direction. Since displacement was just defined in the previous section as a vector representing change in position, average velocity is the *displacement* divided by the time. Velocity is therefore a *vector quantity,* because its direction is important as well as its size.

We can summarize this distinction in symbolic form as follows:

$$\textbf{Average speed: } \bar{s} = \frac{d}{t} = \frac{\textbf{distance covered}}{\textbf{time}}.$$

$$\textbf{Average velocity: } \vec{v} = \frac{\vec{D}}{t} = \frac{\textbf{displacement}}{\textbf{time}}.$$

The symbol $\vec{D}$ is used to represent displacement to avoid confusion with the symbol d, which is used for distance covered. The arrow indicates its vector nature. To help understand how these two quantities are different in practice, we might consider the trip between two cities shown in figure 2.9. The

Box 2.2

Sample Question

Question: When does the magnitude of the vector sum of two displacements equal the numerical sum of the magnitudes of the individual displacements?

Answer: You should be able to convince yourself that the vector sum of two vectors, added as in figure 2.8, will equal the sum of the lengths of the individual vectors only when the two vectors being added are in the same direction. Try it!

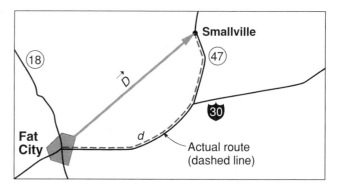

Figure 2.9   A highway map showing both the distance *d* and the displacement $\vec{D}$ for a trip between two towns.

driver, being prudent, stayed on the available highways rather than taking off cross-country. The distance covered, *d,* is considerably longer than the displacement, $\vec{D}$, and the average speed is therefore larger than the average velocity. Think about that carefully and you will begin to appreciate the difference between velocity and speed.

Velocity can change even when the speed of an object remains constant. To see this, consider the car traveling in a circular path pictured in figure 2.10. If the car travels at a constant speed, say 20 km/hr as read on its speedometer, the speed does not change. The velocity, however, is continually changing for this motion because the *direction of motion* is continually changing. For the velocity to remain constant, the car would have to travel in a straight line with constant speed. Both its direction and magnitude are then constant.

Even for straight-line, one-dimensional trips there is a difference between speed and velocity if the traveler reverses direction. Consider the case shown in figure 2.11 for a moment. Here a car goes forward 25 m, then backward 10 m. The total distance covered (without regard to direc-

tion) is 35 m, but the displacement (change in position) is only 15 m. The magnitude of the average velocity is, therefore, again smaller than the average speed.

Instantaneous Velocity

In considering automobile trips, average speed is a much more useful quantity than average velocity. Instantaneous speed is the quantity of interest to the traffic patrol. In considering physical theories of motion, however, instantaneous *velocity* is the quantity that is generally most useful. In a manner similar to that used in defining instantaneous speed, the instantaneous velocity can be defined as follows:

Instantaneous velocity is the rate at which displacement is changing at a given instant in time. It can be computed by finding the average velocity for a very short time interval during which the velocity does not change appreciably.

As with instantaneous speed, in order to compute an instantaneous velocity, we must choose a very short time interval and divide the displacement occurring in that time by the time. The time interval chosen must be short enough so that the velocity does not change appreciably during that time.

Because of its relationship to displacement, instantaneous velocity represents the rate *and direction* in which position is changing with time at a given instant in time. Like displacement and average velocity, instantaneous velocity is a vector quantity; its direction is important. If we know the instantaneous speed and the direction in which an object is traveling at that instant, then we know the instantaneous velocity. The magnitude (size, or numerical value) of the instantaneous velocity *is* the instantaneous speed. This is not necessarily the case for average velocity, as our earlier examples indicated.

The concept of average velocity is useful primarily to help us see the distinction between the concepts of speed and velocity. Instantaneous velocity is the concept that is necessary to understand Newton's theory of motion. As we will see in chapter 4, where Newton's theory is discussed, forces must be exerted in order to change either the magnitude or direction of the instantaneous velocity of an object. The directional aspect is extremely important.

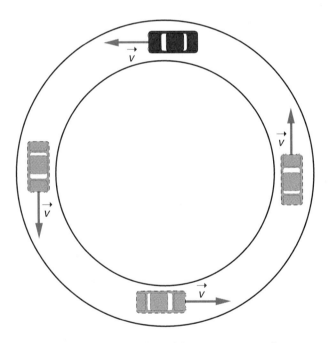

Figure 2.10 A car traveling with constant speed on a circular track. The velocity is continually changing because its direction is changing.

Figure 2.11 A one-dimensional trip showing the difference between total distance and displacement. The distance traveled is larger than the displacement.

2.4 ACCELERATION

The final quantity that we must define in order to provide a full description of motion is probably the most crucial to our understanding of motion. This quantity is *acceleration,* and it deals with the rate at which velocity changes. (Note that we said velocity, not speed.) A general definition of acceleration can be stated as follows:

Acceleration is the rate at which *velocity changes* with time.

Acceleration is a familiar idea, but our everyday notions of its meaning may be vague. We use the term in speaking of the acceleration of a car away from a stop sign or the acceleration of a running back in football. We feel the effects of acceleration in our bodies when a car rapidly increases or decreases its velocity and even more strikingly when an elevator lurches upward, leaving our stomachs slightly behind (fig. 2.12). These are all accelerations, and you can think of your stomach as an acceleration detector. A roller-coaster ride gives it a real workout!

How would we go about providing a quantitative description of an acceleration? Suppose, for a moment, that your car starts from a stop sign and, moving in a straight line, increases its velocity (and yours) from 0 to 20 m/s in a time of 8 seconds. The time required to produce this change in velocity is critical, because here again we are interested in the *rate* of change. If it took 20 seconds instead of 8 seconds, the rate of change, and therefore the acceleration, would be smaller. In either case, the acceleration would be computed by dividing the change in velocity (20 m/s − 0 m/s = 20 m/s) by the time required to produce that change. Thus,

$$a = \frac{20 \text{ m/s}}{8 \text{ s}} = 2.5 \text{ m/s/s}.$$

The unit *m/s/s* is usually written *m/s²* and is read as *meters per second squared*. Its nature is better understood, however, as *meters per second per second*. The car's velocity (which itself is a rate involving division by time) is changing at a rate of 2.5 m/s each second. Other units could also be used for acceleration, but they would all have the same form: distance per unit time per unit time. In discussing the acceleration of a car on a drag strip, for example, the unit *miles per hour per second* is sometimes used.

Average Acceleration

The quantity that we have just computed is the average acceleration of the car. The *average* acceleration involves a time interval that might represent the entire time during which the acceleration process is occurring (8 seconds in the example above). We can express the definition of average acceleration in a word equation as follows:

$$\text{Acceleration} = \frac{\text{change in velocity}}{\text{time required}}.$$

In symbols, this takes the following form:

$$\vec{a} = \frac{(\vec{v_2} - \vec{v_1})}{t},$$

where t is the time required for the velocity to change from $\vec{v_1}$ to $\vec{v_2}$.

The word *change* is all-important in this definition. Acceleration is not velocity over time; it is the *change* in velocity divided by time. We often use the Greek letter Δ (capital delta) as a symbol meaning "change." Using this notation, our definition of average acceleration becomes

$$\vec{a} = \frac{\Delta \vec{v}}{t}.$$

If the velocity of an object does not change, there is no acceleration, no matter how fast the object may be going. Intuitively, we have a tendency to confuse the concepts of velocity and acceleration and to think that a large velocity must mean a large acceleration and vice versa. This is *not* the case; we can have a large acceleration even when the instantaneous velocity is momentarily zero. It is the rate at which velocity is changing that is critical.

Instantaneous Acceleration

Instantaneous acceleration is defined similarly except that, once again, we are concerned with the rate of change at a given instant in time. It is the instantaneous acceleration to which our stomachs respond. Instantaneous acceleration can be defined in words as follows:

Instantaneous acceleration is the rate at which velocity is changing at a given instant in time. It can be computed by finding the average acceleration for a very short time interval, during which the acceleration does not change appreciably.

If the acceleration is changing with time, then again we need to choose a very short time interval in order to compute the instantaneous acceleration.

Acceleration is also a vector quantity; its direction is important. This follows because velocity is a vector quantity,

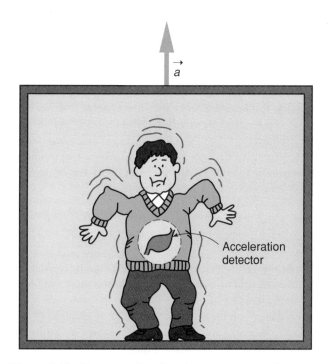

$\vec{a}$

Acceleration detector

Figure 2.12   Your acceleration detector senses the upper acceleration of the elevator.

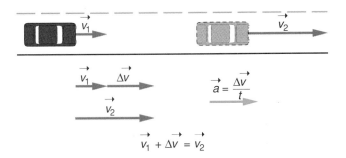

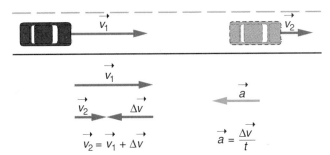

Figure 2.13 The acceleration vector is in the same direction as the velocity vectors when the velocity is increasing.

Figure 2.14 The velocity and acceleration vectors for decreasing velocity. $\Delta \vec{v}$ and $\vec{a}$ are now opposite in direction to the velocity.

and the difference of two vectors is itself a vector quantity (just like the sum of two vectors). We have therefore used the arrow notation in the definition of $\vec{a}$. The direction of the acceleration vector is determined by the direction of the change in velocity, $\Delta \vec{v}$. In our earlier example, where the velocity of the car was increasing, the change in velocity was in the same direction as the velocity vectors themselves, and this was also the direction of the acceleration, as illustrated in figure 2.13. A different situation is represented by the examples in box 2.3 and figure 2.14.

We use the term *acceleration* to describe the rate of any change in an object's velocity. This change could be an increase (as in our initial example), a decrease, or a change

in the direction of the velocity. The use of language is sometimes confusing when we consider decelerations, which, to the physicist, are simply negative accelerations. For example, if a car is braking while traveling in a straight line, its velocity is decreasing, and its acceleration is negative, given our definition. This situation is illustrated in the sample exercise in box 2.3.

The minus sign is an important part of the result of the exercise in box 2.3 because it indicates that the change in velocity is negative; in other words, the velocity is getting smaller. We can call it a *deceleration* if we like, but it is the same thing as a negative acceleration. One word, *acceleration,* can cover all situations in which changes in velocity are occurring.

The acceleration vector in this case is in the opposite direction to the velocity vectors, as indicated in figure 2.14. The change in velocity, $\Delta \vec{v} = \vec{v}_2 - \vec{v}_1$ is also in the opposite direction to the velocity vectors themselves. If we add $\Delta \vec{v}$ to $\vec{v}_1$, we get $\vec{v}_2$, as shown in the vector-addition diagram of figure 2.14. By our definition of acceleration, the acceleration vector has the same direction as that of $\Delta \vec{v}$, the vector representing the change in velocity.

A familiar situation in which the velocity changes because its direction is changing is that of a car going around a curve, as represented in figure 2.15. The speed may be constant, but the car is accelerated because the *direction* of the velocity is changing, and therefore the velocity is changing. The idea that a car can be accelerated when it is traveling with constant speed around a curve is sometimes difficult to grasp and accept, because it goes counter to our everyday usage of the term *acceleration*. This idea is essential, however, to Newton's theory; the speed may be constant, but the velocity is not.

Further discussion of such cases appears in chapter 5, where rotational motion is considered. The point to remember is that acceleration represents the rate of change of velocity with time, regardless of the nature of that change.

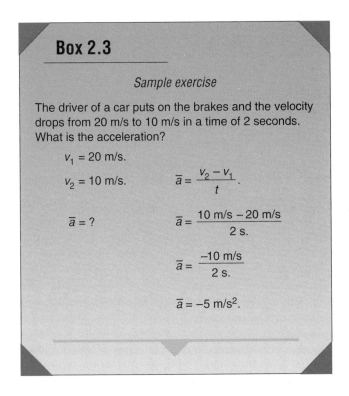

Box 2.3

Sample exercise

The driver of a car puts on the brakes and the velocity drops from 20 m/s to 10 m/s in a time of 2 seconds. What is the acceleration?

$v_1 = 20$ m/s.

$v_2 = 10$ m/s. $\bar{a} = \dfrac{v_2 - v_1}{t}.$

$\bar{a} = ?$ $\bar{a} = \dfrac{10 \text{ m/s} - 20 \text{ m/s}}{2 \text{ s.}}$

$\bar{a} = \dfrac{-10 \text{ m/s}}{2 \text{ s.}}$

$\bar{a} = -5 \text{ m/s}^2.$

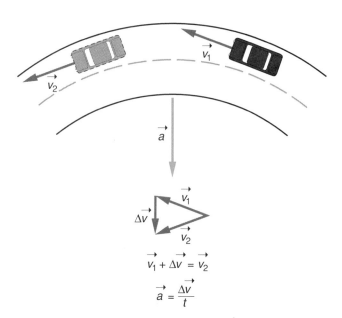

Figure 2.15   A change in the direction of the velocity vector also involves an acceleration.

Figure 2.16   A toy car moving along a meter stick. Its position can be recorded at different times.

2.5 GRAPHIC REPRESENTATION OF MOTION

It is often said that a picture is worth a thousand words, and the same can be said of graphs. Imagine, for example, attempting to describe the motion depicted in the graph of figure 2.5 precisely in words and numbers. We rest our case.

Displacement Graph

How can we produce and use graphs to aid our description of motion? A simple example, in which a battery-powered toy car travels in a straight line along a meter stick (fig. 2.16), can be used to introduce the basic ideas. As an observer, you follow the car's change in position while also keeping track of time with your digital watch. At regular time intervals (say every 5 seconds), you read the value of the car's displacement on the meter stick and record these values on a

Table 2.2
Displacement Data at Different Times for the Toy Car on the Meter Stick

Time	Displacement
0 s	0 cm
5 s	4.1 cm
10 s	7.9 cm
15 s	12.1 cm
20 s	16.0 cm
25 s	16.0 cm
30 s	16.0 cm
35 s	18.0 cm
40 s	20.1 cm
45 s	21.9 cm
50 s	24.0 cm
55 s	22.1 cm
60 s	20.0 cm

sheet of paper or in a laboratory notebook. (Since displacement is the change in position, you can read displacement values directly on the meter stick if you start at 0 cm.) The results might appear something like those shown in table 2.2.

To make a graph of this data, we create evenly spaced intervals on each of two perpendicular axes, one for displacement and the other for time. If we wish to show how displacement varies with time, we usually put time on the horizontal axis and displacement on the vertical axis. Such a graph is shown in figure 2.17, where each data point is

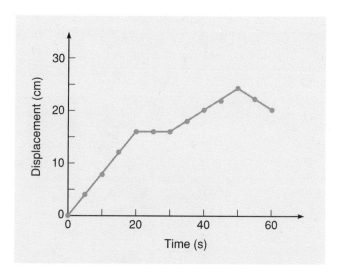

Figure 2.17   Displacement plotted against time for the motion of the toy car. The data points are those listed in table 2.2.

plotted and a smooth line drawn through the points. To make sure that you understand this process, you should choose different points from table 2.2 and find where they are located on the graph.

The graph summarizes the information presented in the table in a visual format that makes it easier to grasp at a glance. It also contains information on the velocity and the acceleration, although that is less obvious. What can we say quickly, for example about the average velocity of the car between 20 seconds and 30 seconds? Is the car moving during this time? A glance at the graph shows us that the displacement is not changing during that time interval, which is to say that the car is *not* moving. The velocity is therefore zero, which is represented by a horizontal line on our graph of displacement versus time.

What about the velocity at other points in the motion? Obviously, the car is moving more rapidly between 0 and 20 seconds than it is between 30 and 50 seconds. A steeper slope to the curve therefore represents a higher velocity; the displacement is increasing more rapidly. If the car traveled backwards, the displacement would decrease, and the curve would go down, as it does between 50 and 60 seconds. We refer to this downward-sloping portion of the curve as having a *negative slope* and often also say that the velocity is negative during this part of the motion.

In fact, the slope of the displacement-versus-time curve at any point on the graph is equal to the instantaneous velocity of the car.* A large slope represents a large instantaneous velocity, a zero slope (horizontal line) represents a zero velocity, and a negative slope represents a negative velocity. The slope indicates how rapidly displacement is changing with time at any instant in time, and the rate of change of displacement with time is the instantaneous velocity, according to the definition given in section 2.3.

Velocity Graph

If we plot instantaneous velocity against time for the same data, we get a graph like the one in figure 2.18. The velocity is constant wherever the slope of the displacement-versus-time graph in figure 2.17 is constant. Any straight-line segment of a graph has a constant slope; when the slope changes in figure 2.17, the velocity changes. If you carefully compare the graph of figure 2.18 to that of figure 2.17, these ideas should become clear.

Since acceleration is the rate of change of velocity, the slope of the velocity graph provides information about acceleration. In fact, by an argument similar to the one just used for displacement and velocity, the slope of the velocity-versus-time graph of figure 2.18 at any point is equal to the instantaneous

*Since the slope is defined mathematically as the change in the vertical coordinate ΔD divided by the change in the horizontal coordinate Δt, the instantaneous velocity, $\Delta D / \Delta t$, where Δt is small, is equal to the slope of the curve at any given point on the graph. It is easy to grasp the concept of slope visually, however, without knowing the mathematical definition.

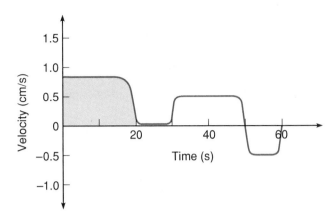

Figure 2.18 Instantaneous velocity plotted against time for the motion of the toy car. The velocity is greatest when the displacement is changing most rapidly.

acceleration of the car at that time. The acceleration turns out to be zero for most of the motion described by our data. Here again, horizontal lines represent no change, and if the velocity is not changing, the acceleration is zero. The velocity changes at only a few points; the acceleration would be large at these points and zero everywhere else.

Since our data do not indicate how rapidly the changes in velocity occur, we do not really have adequate information to determine just how large our acceleration values are. We would need measurements of displacement or velocity every tenth of a second or so in order to determine this. As we will see later (chapter 4), we do know that the changes in velocity cannot occur instantly; some time is required for the changes to take place.

What other information could be gleaned from the graph of figure 2.18? Can we determine the displacement from a velocity-versus-time graph? Think for a moment of how we would go about finding the displacement if we knew the velocity. If the velocity is constant, the average velocity is equal to the instantaneous velocity at any point, and we can get the displacement by multiplying the velocity by the time. In the first 20 seconds of the motion in figure 2.18, for example, the velocity is 0.8 cm/s and the displacement is found as follows:

$$D = vt;$$
$$D = (0.8 \text{ cm/s})(20 \text{ s}) = 16 \text{ cm}.$$

This is just the reverse process of that used in determining the velocity in the first place.

How is this quantity (the displacement) represented on the velocity graph? If you recall simple formulas for computing areas, it is easy to see that D is the area of the shaded rectangle on figure 2.18. We find the area of a rectangle by multiplying the height times the width, and that is exactly what we have done here; the velocity, 0.8 cm/s, is the height, and 20 seconds is the width.

Box 2.4

Everyday Phenomenon: The 100-Meter Dash

The Situation. A world-class sprinter can run 100 meters in a time of a little under 10 seconds. The race begins with the runners in a crouched position in the starting blocks, waiting for the sound of the starter's pistol. The race ends with the runners lunging across the finish line, where their time is recorded by stopwatches or automatic timers.

An analysis of what happens between the start and the finish of such a race might raise several questions. How do velocity and acceleration vary during the course of the race? Can we make some reasonable assumptions about what the velocity-versus-time graph looks like for a typical runner? Can we estimate the maximum velocity of a good sprinter? And (perhaps the most important question for improving athlete's performance) what factors affect the success of a runner in the dash?

The Analysis. If we assume that the runner covers the 100 m in a time of exactly 10.0 s, it is a simple matter to compute the average velocity:

$$\overline{v} = \frac{D}{t} = \frac{100 \text{ m}}{10.0 \text{ s}} = 10.0 \text{ m/s}.$$

Obviously, however, this is not the runner's instantaneous velocity throughout the entire course of the race, since his velocity at the beginning of the race is zero, and it takes some time to accelerate to the maximum velocity.

At the beginning of the race, the runner must accelerate away from the starting blocks. The objective is to reach a maximum velocity as quickly as possible and then to sustain that velocity for the remainder of the race. Success is determined by two things: how quickly the runner can accelerate to this maximum velocity and how large this velocity is. A smaller runner often has better acceleration but a smaller maximum velocity, whereas a larger runner sometimes takes longer to reach full speed but has a larger maximum velocity.

The typical runner does not reach top velocity until he has traveled at least 15 to 20 meters. If his average velocity for the race is 10 m/s, his maximum velocity must be somewhat larger than this value, since we know that his instantaneous velocity will be less than 10 m/s during most of the time that he is accelerating. These ideas are most easily visualized by sketching a graph of the form shown here. We have assumed that the top velocity is reached at approximately 2 seconds into the race.

The average velocity during the time that the runner is accelerating is approximately half of the maximum value if the runner's acceleration is more or less constant. If we assume that the runner's average velocity during the first 2 seconds of the race was about 5.5 m/s (half of 11 m/s), then his velocity through the remaining 8 seconds of the race would have to be about 11.1 m/s in order to give an average velocity of 10 m/s for the entire race. This can be seen by computing the displacement from these values:

Runners in the starting blocks, waiting for the starter's pistol to fire.

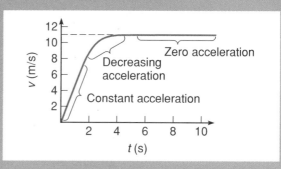

A graph of velocity versus time for a hypothetical runner in the 100m dash.

$$D = (5.5 \text{ m/s})(2 \text{ s}) + (11.1 \text{ m/s})(8 \text{ s}).$$
$$D = 11 \text{ m} + 89 \text{ m} = 100 \text{ m}.$$

All we have done here is to guess at reasonable values that will make the average velocity come out to roughly 10 m/s and then check these guesses by computing the total displacement. This suggests that the maximum velocity of a good sprinter must be about 11 m/s (25 MPH). (For the sake of comparison, a distance runner who can run a 4-minute mile has an average speed of about 15 MPH, or 6.7 m/s.)

The runner's strategy should be to get a good jump out of the blocks at the beginning of the race, keep the body low and leaning forward during the acceleration phase in order to minimize air resistance and maximize the effects of the leg drive, and then to maintain the top velocity for the remainder of the race. A runner who fades near the end needs more conditioning drills to improve endurance. For a given runner with a given maximum velocity, the average velocity for the race depends critically upon how quickly that runner can reach top velocity. This ability to accelerate rapidly depends upon leg strength (which can be improved by working with weights and other exercises) and natural quickness.

It turns out that we can find the total displacement from any velocity-time graph by finding the total area under the velocity curve. This can sometimes be done by breaking the area up into simple shapes such as rectangles and triangles and computing their areas separately. Negative velocities produce decreases in displacement, as is the case in the last 10 seconds of figure 2.18. Areas below the time axis can be regarded as negative, which we then subtract from areas above the axis.

In more complex situations, more exotic techniques must be used for estimating the area. The general mathematical technique for doing this is referred to as *integration* and is a major part of the subject matter of calculus. Computers are often used now to handle this type of problem.

Even without computing the area accurately, however, it is possible to get a quick visual impression of the relative sizes of different areas. If you can mentally identify the area under the velocity-time graph with the displacement, then looking at such a graph can give you information about the displacement for different portions of the motion. Quick visual comparisons can therefore give a rough picture of what is happening without the need for lengthy calculations.

The next chapter returns to these ideas; they are used in treating the motion of falling objects for which the acceleration is constant. As you study the graphs that are presented there, you may wish to return to figures 2.17 and 2.18 in this chapter for comparisons. A good grasp of these ideas is not acquired overnight, but once you are accustomed to interpreting graphs of this nature, you will find that they are a powerful tool for obtaining an overall picture of what is happening.

SUMMARY

The primary concern of chapter 2 has been to familiarize the reader with some of the definitions that are crucial to a useful description of motion. The flowchart of terms indicates the logical connections between these ideas, each of which is defined briefly in the summary below. The distinctions between speed and velocity and between velocity and acceleration are particularly important.

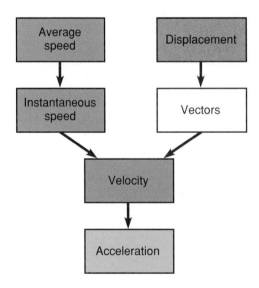

Average and Instantaneous Speed. Average speed is defined as the time rate at which distance is covered, or distance divided by time. Instantaneous speed is the rate at which distance is being covered at a given instant in time and requires that we use very short time intervals for computation.

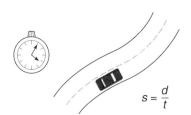

$$s = \frac{d}{t}$$

Displacement and Vectors. Displacement was defined as the change in position of an object. Since the direction of this change is important as well as the size, displacement is a vector quantity. Vectors are quantities that require both a direction and a magnitude (size) for complete specification.

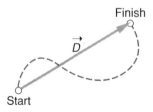

Velocity. Average velocity was defined as the time rate of change of displacement. Like displacement, velocity is a vector quantity for which direction is important. Similar to instantaneous speed, instantaneous velocity involves the rate at which displacement is changing at a given instant in time.

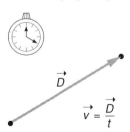

$$\vec{v} = \frac{\vec{D}}{t}$$

Acceleration. Acceleration is defined as the time rate of change of velocity, that is, change in velocity divided by time. Acceleration is also a vector quantity and can also be computed as either an average or instantaneous value. A change in thedirection of velocity is as important as a change in magnitude; both involve acceleration.

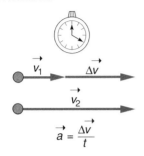

Graphic Representation of Motion. Graphs of displacement, speed, and velocity against time were introduced to help the reader see the relationships between quantities. For example,

Instantaneous velocity = the slope of the displacement-time graph.

Instantaneous acceleration = the slope of the velocity-time graph.

Displacement = the area under the velocity-time graph.

QUESTIONS

It is best to write out answers to these conceptual questions in much the same manner that you might for the numerical exercises. Use full sentences in your explanations.

Q2.1 Suppose that critters are discovered on the planet Mars who measure distance in *doozies* and time in *dings*.

   a. What would the units of speed be in this system? Explain.

   b. What would the units of velocity be? Explain.

   c. What would the units of acceleration be? Explain.

Q2.2 The tortoise and the hare cover the same distance in a race. The hare goes quite fast for brief intervals, but stops frequently. The tortoise plods along steadily and finishes the race ahead of the hare.

   a. Which of the two racers has the greater average speed over the duration of the race? Explain.

   b. Which of the two racers is likely to reach the greatest instantaneous speed at any time during the race? Explain.

Q2.3 A driver in England states that she was doing 90 when stopped by the police. Is there anything lacking in clarity in that statement? Explain.

Q2.4 A football player runs a pass route that takes a right-angle bend, as pictured in the diagram.

   a. Indicate on the diagram the direction of the displacement of the player.

   b. Which has the greater magnitude, his displacement or the distance traveled? Explain.

Q2.5 A car travels from Washington D.C. to Chicago on major highways.

   a. Which quantity, distance traveled or displacement, is larger for this trip? Explain.

   b. Which quantity, average speed or average velocity, is the larger for such a trip? Explain.

Q2.6 The driver of a car steps on the brakes, causing the velocity of the car to decrease. According to the definition provided in this chapter, is the car accelerated in this process? Explain.

Q2.7 At a given instant in time, two cars are traveling at different velocities, one twice as large as the other. Is it possible to say with confidence which of these two cars has the larger acceleration at this same instant in time? Explain.

Q2.8 In the graph shown below, velocity is plotted as a function of time for an object traveling in a straight line.

   a. Is the velocity constant during any time interval? Explain.

   b. Is the displacement of the object constant during any time interval? Explain.

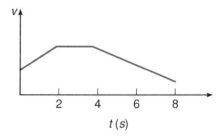

Q2.9 A car moves along a straight section of road so that its position as a function of time is described by the graph at the top of the next page.

   a. Does the car ever go backwards? Explain.

   b. Is the instantaneous velocity at point *A* greater or less than the average velocity between *A* and *B*? Explain.

   c. Is the velocity of the car constant during any time covered by the graph? Explain.

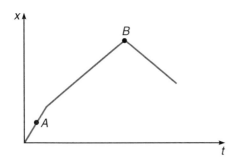

Q2.10 A car moves along a straight section of road so that its velocity as a function of time is described by the graph below.

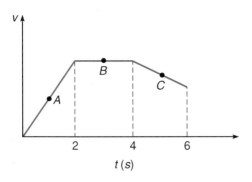

a. At which of the labeled points on the graph is the magnitude of the acceleration the greatest? Explain.

b. Does the car ever go backwards during the time interval pictured? Explain.

c. In which of the equal time segments, 0-2 sec, 2-4 sec, or 4-6 sec, is the distance traveled by the car the greatest? Explain.

Q2.11 A car travels with constant speed along the path sketched in the diagram.

a. Is the velocity of the car constant over this path? Explain.

b. Is the car accelerated at any point along this path? Explain.

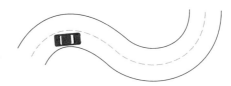

Q2.12 Look again at the displacement-versus-time graph for the toy car in figure 2.17 discussed in the section on graphic representation.

a. Is the instantaneous velocity greater at any time than the average velocity for the entire time shown? Explain.

b. Is the car accelerated when the direction of the car is reversed at $t = 50$ s? Explain.

EXERCISES

The following exercises are designed to build confidence in using numbers with the quantities introduced in this chapter. A calculator could be useful, but is not essential. However, the college student who does not own at least a simple four-function calculator (cost: $10 or less) is not minimally equipped for life in the real world.

E2.1 A traveler covers a distance of 450 miles in a travel time of 9 hr. What is the average speed for this trip?

E2.2 A walker covers a distance of 45 m in a time of 30 s. What is the average speed of the walker for this distance?

E2.3 A person in a hurry averages 60 MPH on a trip covering a distance of 150 miles. How long (what time?) does it take to cover that distance?

E2.4 A hiker walks with an average speed of 1.3 m/s.

a. What distance (in meters) does the hiker cover in a time of 30 s?

b. What distance does the hiker cover in a time of 3 min?

E2.5 A bug covers a distance of 36 cm in a time of 3 min.

a. What is its average speed in cm/min?

b. What is its average speed in m/min?

c. What is its average speed in inches/min? (2.54 cm = 1 inch)

E2.6 A car travels with an average speed of 65 MPH. What is this speed in km/hr? (See box 2.1)

E2.7 A car travels with an average speed of 15 m/s.

a. What is this speed in km/s?

b. What is this speed in km/hr?

E2.8 A walker achieves a displacement of 250 m in a direction due north in a time of 5 min.

a. What is the average velocity of the walker?

b. Is this likely to have the same magnitude as the average speed of the walker? Explain.

E2.9 Starting from rest and moving in straight line, a runner achieves a velocity of 6 m/sec in a time of 4 sec. What is the average acceleration of the runner?

E2.10 A car accelerates at a rate of 4 m/s² for a time of 3 s. Its initial velocity was 10 m/s. What is its velocity at the end of the 3-second acceleration?

E2.11 The velocity of a car decreases from 30 m/s to 10 m/s in a time of 5 s. What is the average acceleration of the car during this process?

CHALLENGE PROBLEMS

CP2.1 The velocity of a car increases with time as shown in the graph below.

  a. What is the average acceleration between 0 s and 4 s?

  b. What is the average acceleration between 4 s and 10 s?

  c. What is the average acceleration between 0 s and 10 s?

  d. Is the result in (*c*) equal to the average of the two values computed in (*a*) and (*b*)? Explain.

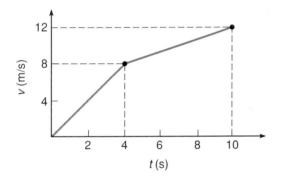

CP2.2 A switching engine moves forward along a straight section of track for a distance of 60 m due west. It then reverses itself and travels 15 m due east. The entire process takes 40 s.

  a. What is the resulting displacement of the engine?

  b. What is the average velocity of the engine during this process?

  c. What is the average speed of the engine during this process?

  d. Which is greater, the average speed or the average velocity? What conditions would be necessary in order for the average speed and the average velocity to have the same value?

CP2.3 Walking on city streets, a person walks a distance of 600 m due north and then turns on a cross street and walks 400 m due east.

  a. Carefully draw the two displacement vectors for this trip to the following scale: 1 cm = 100 m.

  b. Draw the total displacement vector by drawing a straight line from the beginning point of the trip to the end point.

  c. Using your ruler to measure the length of this line, and the scale factor used in drawing the diagram, determine the magnitude of the total displacement.

  d. If the total trip took 15 min, what is the average velocity for this trip in m/min?

  e. What is the average speed for the trip? How does it compare to your result for the average velocity?

HOME EXPERIMENTS AND OBSERVATIONS

HE2.1 How fast do you normally walk? Using a meter stick or a piece of string of known length, lay out a straight course of 40 or 50 meters. Then use a watch with a second hand or a stopwatch to determine the following values:

  a. Your normal walking speed in m/s

  b. Your speed for a brisk walk

  c. Your speed in jogging the same distance

  d. A sprinting speed for the same distance.

How do these speeds compare to one another?

HE2.2 If you have access to a car, take a short trip on city streets or other places where there are traffic lights and stop signs, so that you will be doing a lot of stop-and-go driving. Before starting, make a guess as to what your average speed will be. Then, using the mileage reading from your odometer

and the time of your trip measured with your watch, compute your average speed for the trip. How does this compare to your original guess? (Do not exceed the speed limit and do not look at your watch while you are driving.)

HE2.3 Using a car, estimate your normal average acceleration for the process of starting from a stop sign and reaching a speed of 50 km/hr (about 31 MPH). Use a watch with a second hand or a stopwatch to measure the time that you normally require to reach that speed. (You should enlist a friend either to drive or make the time measurements so that you do not need to look at your watch while you are driving. Repeating the process a few times and using an average value of the time is a good idea.) What is your average acceleration in km/hr/s? What is this value in m/s²?

3 Falling Objects and Projectile Motion

Who among us has not watched a leaf or a rock or some other object fall to the ground? Although our memories of our first five years of life are generally vague, sometime during that period you probably amused yourself by repeatedly dropping some object and watching it fall. As we grow older, that experience becomes so common that we usually do not stop to think about it or ask ourselves why objects fall as they do. Yet this is a question that has intrigued scientists and philosophers for centuries.

To understand nature, we must first carefully observe it. If we attempt to control the conditions under which we make our observations, we are doing an experiment; the observations of falling objects that you performed as a young child were a simple form of experiment. We would like to rekindle that interest in experimentation here, but with more conscious control of conditions than we exercised as children. Progress in scientific knowledge depends upon experimental exploration and testing. Your own progress in understanding nature likewise depends upon your active participation in some simple home experiments. You may be amazed at what you discover!

Look around for some small, reasonably compact objects. A short pencil, a rubber eraser, a paper clip, or a small ball will all do nicely. Holding two such objects at arm's length in either hand, release them simultaneously and watch them fall to the floor (fig. 3.1). Be careful to release them from the same height above the floor without giving either one a push.

How would you describe the motion of these falling objects? Is their motion accelerated? Do they reach the floor at the same time? Does the motion depend on the shape and nature of the object? To explore this last question you might take a small piece of paper and drop it at the same time as the eraser or ball. First drop the paper unfolded and then try folding it or crumpling it into a ball. What difference does this make?

From controlled observations such as these simple experiments, we can draw some conclusions regarding the motion of falling objects as well as the motion of objects that are thrown or fired in some manner. We will find that a downward acceleration is involved in both of these cases, due to the gravitational attraction of the objects to the earth.

Figure 3.1   An experimenter dropping objects of different mass. Do they reach the ground at the same time?

Chapter Objectives

Our main objective in this chapter is to develop an understanding of how objects move under the influence of the acceleration of gravity near the surface of the earth. The concepts and questions listed in the chapter outline indicate the order in which these ideas will be discussed. The nature of constant acceleration and its effects upon the motion of an object are the primary theme. These ideas are very important in the chapters that follow, and they are also useful in describing a host of familiar phenomena.

Chapter Outline

1 *The acceleration of gravity.* What are the nature and magnitude of the acceleration of an object due to the earth's gravitational pull? How is this acceleration measured, and in what sense can it be considered to be constant?

2 *Uniformly accelerated motion.* How do the velocity and displacement of an object change under the influence of a constant acceleration?

3 *Beyond free-fall: constant acceleration applications.* How can these ideas on constant acceleration be used to predict the position and velocity of an object thrown upwards or of an accelerating car?

4 *Projectile motion.* How does the downward acceleration of an object combine with its horizontal motion to determine the behavior of a horizontally fired projectile? How do the velocity and position change in this case?

5 *Hitting a target.* What factors determine the trajectory of a rifle bullet or a football that is launched at some angle to the horizontal in an attempt to hit a target?

3.1 THE ACCELERATION OF GRAVITY

You should already know the answer to one of the questions posed in the introduction: Are the falling objects accelerated? Think for a moment about their velocity. Before you release an object, its velocity is zero, and an instant later it has some value different from zero; thus, there has been a change in velocity. If the velocity is changing, there is, by definition, an acceleration.

But what is the nature of this acceleration? Things happen so rapidly that it is difficult, just by watching the object fall, to say much about the acceleration itself, except that it appears to be large, because the velocity increases rapidly. Does the object reach a large velocity instantly and then continue to fall with constant velocity, or is it accelerated in a more uniform manner, increasing in velocity as it falls? To answer these questions, it is necessary to slow the motion down somehow so that our eyes and brains can keep up with what is happening.

Measuring the Acceleration of Gravity

There are at least two ways to slow down the action. One was pioneered by the Italian scientist Galileo Galilei (1564–1642), who was the first to accurately describe the nature of the acceleration of gravity. Galileo's method was to roll or slide objects down a slightly inclined plane. This has the effect of allowing only a small fraction of the total acceleration of gravity (just the portion that is in the direction of motion) to come into play, thus yielding a much smaller acceleration. A second method involves time-lapse photography using a stroboscope to illuminate the object at regularly spaced time intervals. (This method was not an option, of course, for Galileo.)

If you happen to have a grooved ruler handy and a small ball or a marble, you might try the inclined-plane method yourself. Lift one end of the ruler slightly by placing a pencil or eraser underneath the end, and let the ball or marble roll down the ruler under the influence of gravity (fig. 3.2). Can you see it gradually pick up speed as it proceeds? Is it clearly moving faster at the bottom of the ruler than it was halfway down? Does the distance it covers in equal time intervals

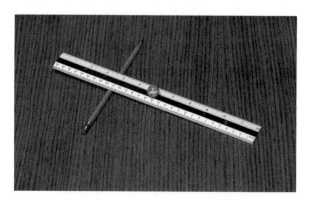

Figure 3.2 **A marble rolling down a ruler serving as an inclined plane. The velocity of the marble increases as it rolls down the incline.**

seem to increase uniformly as it moves down the ruler? These observations should provide answers to some of the questions posed above.

Galileo was handicapped by a lack of accurate timing devices; his own pulse often served as a timer. Despite this limitation, however, he was able to establish that the acceleration is uniform, or constant, with time, and to estimate its value using inclined planes. We are more fortunate; we have devices that allow us to study the motion of a falling object more directly. One such device is the stroboscope mentioned above, which is merely a rapidly blinking light whose flashes occur at regular intervals in time. Figure 3.3 shows a photograph taken using a stroboscope to illuminate an object as it falls; the position of the object is pinpointed every time the light flashes.

From figure 3.3, you can see that the velocity is increasing in the downward direction as the ball falls. The trick in interpreting a strobe photograph such as this is to realize that the flashes occur at regularly spaced intervals in time, so that you are seeing the position of the ball every $1/20$ of a second (or whatever the time interval is between flashes for the particular photograph). In each successive $1/20$ of a second, the ball has traveled a larger distance than in the previous $1/20$ of a second, indicating a larger velocity.

This point can be seen more clearly if we actually compute values of the velocity from numbers taken from a strobe photograph. If the stroboscope light was flashing at a rate of 20 flashes per second, there is a time interval of $1/20$ of a second, or 0.05 s, between flashes. Knowledge of the distance between the grid marks in the photograph can be used to estimate the displacement of the ball from its starting position, but to make these measurements accurately, an enlarged photograph is necessary. Table 3.1 shows displacement values measured from the top downward using an enlarged photograph similar to that in figure 3.3.

To see that the velocity is indeed increasing, we can use these displacement values to compute the average velocity at

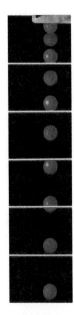

Figure 3.3 **A falling ball is illuminated by a rapidly blinking stroboscope. The stroboscope blinks at regular time intervals.**

Table 3.1		
Displacement values of falling ball		
Flashes	**Time**	**Displacement**
1	0 s	0 cm
2	0.05 s	1.2 cm
3	0.10 s	4.8 cm
4	0.15 s	11.0 cm
5	0.20 s	19.7 cm
6	0.25 s	30.6 cm
7	0.30 s	44.0 cm
8	0.35 s	60.0 cm
9	0.40 s	78.4 cm
10	0.45 s	99.2 cm
11	0.50 s	122.4 cm

different points. For example, between the third and fourth flashes, the downward displacement is given by

$$D = 11.0 \text{ cm} - 4.8 \text{ cm} = 6.2 \text{ cm} \quad \text{(or } 0.062 \text{ m).*}$$

Since the time interval between flashes is 0.05 s, the average velocity for this time interval is then

$$\bar{v} = \frac{D}{t} = \frac{0.062 \text{ m}}{0.05 \text{ s}} = 1.24 \text{ m/s}.$$

* When we are discussing just the magnitude of a vector quantity, we do not use the arrow over the symbol.

If we choose a later time interval, say between the ninth and tenth flashes, a similar computation yields

$$D = 99.2 \text{ cm} - 78.4 \text{ cm} = 20.8 \text{ cm} = 0.208 \text{ m};$$

$$\bar{v} = \frac{D}{t} = \frac{0.208 \text{ m}}{0.05 \text{ s}} = 4.16 \text{ m/s}.$$

The downward velocity has obviously increased over the value computed between the third and fourth flashes. These are average velocities for the time intervals indicated, but since the velocity is increasing in a uniform manner, they also represent the instantaneous velocities at the midpoints of the time intervals selected.

How can we get the acceleration from this information? Acceleration is just the rate of change in velocity, so if we know the increase in velocity and the time required for that increase to occur, we can compute the acceleration. In this case the time interval between the two velocity values is 0.30 s, because the third flash occurred at 0.10 s and the ninth flash at 0.40 s. The acceleration is then given by

$$a = \frac{\Delta v}{t} = \frac{(4.16 \text{ m/s} - 1.24 \text{ m/s})}{0.30 \text{ s}};$$

$$a = \frac{2.92 \text{ m/s}}{0.30 \text{ s}} = 9.73 \text{ m/s}^2.$$

This is the average downward acceleration of the ball between these two points in time.

Although this value of $\vec{a}$ has been computed between two particular points, we would get approximately the same value for any other two points that we might choose. To verify this, you may wish to choose two other points and repeat these calculations. (Exercise 3.1 asks you to do this; doing it will give you a better understanding of the process that we have just gone through.) Performing this computation for several different points and averaging the results yields a value for the acceleration due to gravity of approximately 9.8 m/s². The fluctuations in the individual values are due to inaccuracies in measurement or in the timing of the strobe flashes.

Study Hint

Take a break from your reading and do exercise 3.1 immediately. Going through the operations yourself is a much more effective way to convince yourself that you understand the process than passively reading what someone else has done. It is a good idea, anyway, never to trust a textbook or instructor completely in such matters; do it yourself, and then you will have confidence in the result.

The fact that we get the same value for this acceleration regardless of which data points we choose indicates that the acceleration is constant, or *uniform*, in time. We gain a visual indication of this fact by noting the regularity with which the displacements in figure 3.3 increase. Another way of representing this idea is shown in figure 3.4, where we have plotted the acceleration of a falling object against time. The resulting curve is a horizontal line, indicating that the acceleration is not changing with time.

This constant acceleration of objects near the earth's surface due to gravity is often referred to as the *acceleration of gravity* and is usually given its own symbol, *g*.

$$g = 9.8 \text{ m/s}^2.$$

We could measure its value more accurately than to just the two digits given here, but its value actually varies slightly from point to point on the earth's surface. Variations in altitude and in the density of the earth's crust, as well as other factors, are responsible for these fluctuations. The value of 9.8 m/s^2 is valid anywhere reasonably close (within a few kilometers or miles) to the surface of the earth and at any point on the earth.

Galileo and Aristotle

There is another sense in which *g* is constant, which takes us back to the experiment at the beginning of this chapter. When you dropped objects of different sizes and weights, did they reach the floor at the same time? What differences did you observe for the piece of paper in the unfolded and folded conditions? Except for the unfolded piece of paper, it is

likely that all the other objects you tested, regardless of their weight, reached the floor at the same time when released simultaneously. This fact suggests that the acceleration of gravity, *g*, does not depend on the weight of the object.

Galileo used experiments of this sort to prove this point. These experiments contradicted the view of Aristotle, which had been accepted for several centuries, that heavier objects fall to the earth more rapidly. How could Aristotle's view have been accepted for so long when such simple experiments can disprove it? Part of the answer lies in the fact that experimentation was not part of the intellectual tradition practiced by Aristotle and his followers. Galileo was a pioneer in using experimentation as an aid to his thinking about motion.

The other part of the answer, however, lies in the recognition that Aristotle's view corresponds with our intuition that heavy objects do fall more rapidly than some lighter objects. If, for example, we drop a brick together with a feather or an unfolded piece of paper, as in figure 3.5, the brick will reach the floor first. The paper or feather will not fall in a straight line, but instead will flutter to the floor much as a leaf falls from a tree. What is happening here?

It is not hard to recognize that the effects of air resistance impede the fall of the feather or unfolded paper much more than they do a rock, a paper clip, or a steel ball. When we fold or crumple the same piece of paper and drop it simultaneously with the rock or other heavier object, the two objects reach the floor at approximately the same time. We live at the bottom of a sea of air, and the effects of air resistance and air currents can be substantial for objects like leaves, feathers, or pieces of paper. These effects produce a slower and less regular flight for objects that have a large surface area and a small weight.

If we dropped a feather and a brick simultaneously in a vacuum or in the very thin atmosphere of the moon, they would reach the ground at the same time. Such conditions

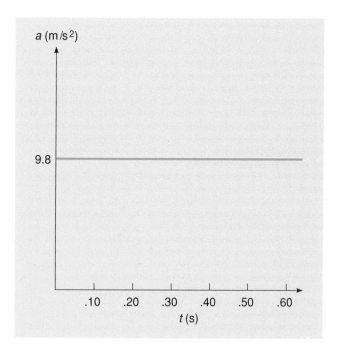

Figure 3.4 **A plot of acceleration against time for a falling object. The horizontal line indicates that the acceleration does not change with time.**

Figure 3.5 **The brick reaches the floor first when a brick and a feather are dropped at the same time.**

are obviously not part of our everyday experience, however, so we are used to seeing feathers fall more slowly than rocks or bricks. Galileo's insight was that the acceleration of gravity was the same for all objects, regardless of their weight, provided that the effects of air resistance were not significant.

Aristotle did not separate the effect of air resistance from that of gravity in his observations. Neither did he do experiments such as those done by Galileo using objects of similar form but different weights. Our preconceptions and the intellectual tradition within which we work inevitably influence what we choose to observe. Careful experimentation was not a part of the tradition of the Greek philosophers.

3.2 UNIFORMLY ACCELERATED MOTION

We have just learned that the acceleration of a freely falling object is constant in time, and that it has the same value for different objects, provided that the effects of air resistance are not important. As we will see in the next chapter, constant accelerations also arise in other circumstances. Suppose that we want to describe how velocity or displacement change with time for an object undergoing constant acceleration. How do we go about this? A closer look at this process, using the descriptive tools developed in chapter 2, will give us better insight into the motion of falling objects as well as other situations involving constant acceleration.

We have already seen from figure 3.3 and from the computations done in the last section that velocity and displacement both increase with time in the downward direction for our falling ball. Suppose, however, that we want to predict the value of the velocity or displacement at some particular point in the motion. For this we need more precise descriptions than the simple statement that these quantities are increasing with time. Graphs or mathematical formulas are far more efficient ways of providing these descriptions than words alone.

Variation of Velocity with Time

You have already seen a graph of acceleration against time in figure 3.4, but it is more interesting to ask what the graph of velocity-versus-time looks like in this case. How does the velocity increase? Since acceleration is the rate of increase of velocity, a constant acceleration implies that the velocity increases at a uniform rate. The *slope* of the velocity-versus-time graph will therefore be constant, since the slope of a curve on a graph represents the rate at which the quantity is increasing or decreasing. A constant slope yields a straight line. If the object starts with an initial velocity of v_0, the graph of our falling object is as shown in figure 3.6.

If the initial velocity, v_0, were zero, then the upward-sloping line would start at the origin where both t and v are zero, rather than at some point on the velocity axis above the origin, as pictured here.

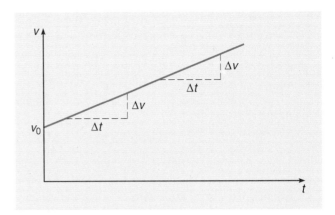

Figure 3.6   Velocity plotted against time for an object moving with constant acceleration. The initial velocity is v_0.

The velocity graph in figure 3.6 provides a clear way of seeing what is meant by constant acceleration, perhaps even clearer than the acceleration graph (figure 3.4) itself. The change in velocity for equal intervals in time is the same regardless of where we are in the motion. This is indicated by the triangles on the graph showing Δv for equal values of Δt at two different points on the graph. If the acceleration were increasing with time, Δv would be larger for the second triangle than for the first, and the curve would bend upward. Likewise, a decreasing acceleration would lead to a downward bend in the curve. Constant acceleration, however, produces a curve that bends neither upward or downward; its slope is constant.

How might we describe these ideas mathematically? Starting with the definition of acceleration, we can find a simple equation for the increase of velocity with time. By our definition of acceleration, $\vec{a} = \Delta \vec{v} / t$, the change in the magnitude of the velocity is given by $\Delta v = at$. All we have to do is add this change in velocity to the initial velocity, v_0, and we have an expression for how velocity changes with time. The result is the simple formula

$$v = v_0 + at,$$

which says that the velocity at time t is equal to the initial velocity plus the change in velocity that results from the acceleration. This expression allows us to compute the velocity v at any time t if we know the initial velocity v_0 and the constant acceleration $\vec{a}$.

This result represents merely a rearrangement of the ideas contained in the definition of $\vec{a}$. The condition that $\vec{a}$ be a constant allows us to use the expression for average acceleration, $\overline{a} = \Delta v / t$, because for constant acceleration the average acceleration is equal to the instantaneous acceleration. Our result, however, can then be used only for situations in which the acceleration is constant. This limitation is worth remembering.

For a simple example of the use of this result, consider the case of a falling object. In this case, the downward acceleration $\vec{a} = \vec{g}$, and if we do not give the object an initial push, its initial velocity, v_0, is zero. The formula $v = v_0 + at$ then becomes $v = gt$, provided that we let the downward direction be positive. Half a second after the ball is released, for example, its velocity is

$$v = gt = (9.8 \text{ m/s}^2)(0.5 \text{ s}) = 4.9 \text{ m/s}$$

in the downward direction. Larger times will obviously lead to larger values of the velocity; after 1 second the ball will have a velocity of 9.8 m/s, and after 2 seconds the velocity will be 19.6 m/s.

Variation of Displacement with Time

What about the displacement? How much has the position of the object changed in this process? In section 2.5 we showed that the displacement is related to the area under the curve of velocity versus time. Figure 3.7 is another version of the graph of figure 3.6 that shows the area involved more clearly. The bottom portion of this area is a rectangle of height v_0 and width t; its area is $v_0 t$. This is just the displacement that would result if the object were traveling with constant velocity v_0 for time t.

The top portion of the area under the curve is a triangle. The area of a triangle is equal to one-half the base times the height, or just half the area of a rectangle of the same height and width. As figure 3.6 shows, the height of this portion is Δv, which we know is equal to at. Since the width is just the time t, the area of this portion is then $\frac{1}{2}(at)t = \frac{1}{2}at^2$. This quantity represents the additional displacement that results because the object is accelerated. Adding this to the first term yields the total area under the curve and thus the total displacement:

$$D = v_0 t + \frac{1}{2}at^2.$$

This result describes the displacement, or change in position, for any time t, given the initial velocity and the constant acceleration.

As a quick illustration of the use of this result, we can find the displacement of our falling object after it has been falling for 0.5 s. Here again, $v_0 = 0$, and $a = g$.

$$D = (0)\,t + \frac{1}{2}\,(9.8 \text{ m/s}^2)(0.5 \text{ s})^2;$$

$$D = \frac{1}{2}(9.8 \text{ m/s}^2)(.25 \text{ s}^2) = 1.23 \text{ m}.$$

This corresponds approximately to the value given in the data table of box 3.1: the ball falls more than a meter in just a half a second.

If we plot displacement versus time, we get a graph like that shown in figure 3.8. The data plotted here is from table 3.1. Because the velocity is increasing, and velocity is the rate of change of the displacement, the slope of the displacement curve increases as time gets larger, causing the graph to curve upward. Mathematically, this upward curvature is due to the t^2 term in our formula for displacement; a doubling of the time produces a quadrupling of this term rather than a simple proportional increase.

A third equation is often used in treating constant-acceleration situations. It can be obtained from the two equations developed here for v and D by eliminating the time t

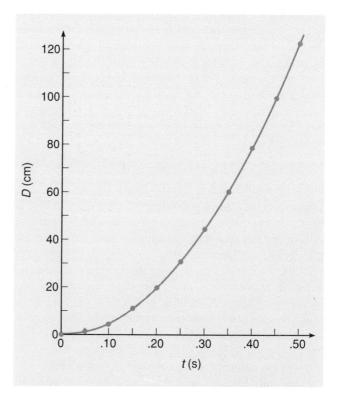

Figure 3.8 Displacement versus time for an object moving with constant acceleration, starting from rest. The data is taken from table 3.1.

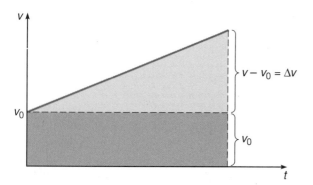

Figure 3.7 The velocity-versus-time graph redrawn to show the area under the curve as having two portions, a rectangle and a triangle.

between these two equations. (This is done by solving one equation for t and substituting that expression in the second equation. Give it a try if you are comfortable with algebra.) The equation that results:

$$v^2 = v_0^2 + 2aD$$

relates the final velocity to the initial velocity, and the acceleration, and the displacement. It is useful in cases where the displacement may be known, but not the time.

A look back at the graphs presented in this section and in section 3.1 provides the overall picture of how acceleration, velocity, and displacement behave for the case of constant acceleration. The acceleration itself does not change, yielding a horizontal line (figure 3.4); the velocity increases steadily with time, producing an upward-sloping straight line when the acceleration is positive (figure 3.6); and the displacement increases ever more rapidly with time, producing an upward-curving graph (figure 3.8). The graphs provide pictures that are often easier to grasp than the mathematical formulas containing the same information.

3.3 BEYOND FREE-FALL: CONSTANT ACCELERATION APPLICATIONS

In the preceding section, our examples dealt with the simple case in which an object was dropped without giving it any initial velocity. This is the *free-fall* case. What happens, however, if the object is thrown, either straight up or down, with an initial velocity different from zero? What goes up must come down, but when and how fast are interesting questions with practical implications.

The constant acceleration equations developed in section 3.2 provide tools for answering these questions. They are also useful in treating other situations in which the constant acceleration has nothing to do with gravity. In this section we analyze a few simple cases involving the constant acceleration equations, paying particular attention to the signs of the acceleration and velocity terms, to gain a better understanding of the meaning of acceleration.

Throwing a Ball Upward

Suppose that you throw a ball straight upward with an initial velocity of $v_0 = 20$ m/s. In section 3.2 we chose down as the positive direction because that was the direction in which the object was moving. In this example, choosing up as positive makes sense because that is the direction in which the ball is moving initially. The ball will travel upward with decreasing velocity, because the acceleration of gravity is downward and thus negative (since we chose up as positive).

At some point as the ball is moving upward, the velocity reaches zero, and the ball then begins to gain downward velocity (fig. 3.9). The formulas developed in the previous section can be used to find the velocity of the ball or the

Figure 3.9 **A ball thrown upward returns to the ground. We can calculate the magnitude and direction of the velocity of the ball at different points in its path.**

position of the ball at any time after it is thrown. The process is straightforward and easily extended to other cases.

Suppose we want to know the velocity 1 second after the ball leaves the hand. The velocity expression developed in section 3.2 is $v = v_0 + at$, but the acceleration in this case is g, the acceleration of gravity, which is downward. Since we have chosen up as positive, this acceleration is negative, and we can write our equation for velocity as follows:

$$v = v_0 + at = v_0 - gt.$$

Putting the numerical values into this expression yields

$$v = 20.0 \text{ m/s} - (9.8 \text{ m/s}^2)(1 \text{ s});$$

$$v = 20.0 \text{ m/s} - 9.8 \text{ m/s} = 10.2 \text{ m/s}.$$

The second term in this expression is the change in velocity, Δv, and we subtract it from the initial velocity of 20 m/s to obtain the resulting velocity of +10.2 m/s. The positive sign of this velocity indicates that the ball is still moving upward, although at a slower velocity than v_0.

If we want to know the position of the ball after 1 second, we use the formula developed for displacement:

$$D = v_0 t + \frac{1}{2} at^2 = v_0 t - \frac{1}{2} gt^2.$$

Here again we have substituted $-g$ for a to indicate that the acceleration is downward and thus negative. Putting in the numerical values yields

$$D = (20.0 \text{ m/s})(1 \text{ s}) - \frac{1}{2}(9.8 \text{ m/s}^2)(1 \text{ s})^2;$$

$$D = 20.0 \text{ m} - 4.9 \text{ m} = 15.1 \text{ m}.$$

This tells us that the ball is 15.1 m above its starting point, since the displacement is positive. The 20 m in the first term is the distance the ball would have traveled if there had been no acceleration; the actual displacement is less than this because the acceleration is negative.

We can readily repeat these computations to see what the ball is doing at some other time, say after 3 seconds. In the case of the velocity;

$$v = v_0 - gt = 20 \text{ m/s} - (9.8 \text{ m/s}^2)(3 \text{ s});$$

$$v = 20 \text{ m/s} - 29.4 \text{ m/s} = -9.4 \text{ m/s}.$$

In this case the change in velocity is larger than the initial velocity, so our result is a negative velocity, indicating that the ball is moving downward. A similar computation for the displacement yields

$$D = v_0 t - \frac{1}{2} gt^2;$$

$$D = (20 \text{ m/s})(3 \text{ s}) - \frac{1}{2} (9.8 \text{ m/s}^2)(3 \text{ s})^2;$$

$$D = 60 \text{ m} - 44.1 \text{ m} = 15.9 \text{ m}.$$

This displacement is positive, so the ball is still above the starting point, but on its way down. Had it been negative, as it would be at 5 seconds, the ball would be below the starting point.

Figure 3.10 shows the flight of the ball with velocity and acceleration indicated at a few points in its motion. Notice that the acceleration is always directed downward and is always equal to g, *regardless of the size or direction of the velocity.* At the instant that the ball's velocity is momentarily zero (at the turnaround point at the top of the path), the acceleration is still −9.8 m/s². The acceleration is constant; it does not change.

How high does the ball go before it turns around and begins its fall? What do we know about the highest point in the flight? We know that the velocity at this point is zero, since it has been decreasing from its initial positive value. The high point is that point at which it has slowed to zero and begins to go negative. Using these ideas and the constant acceleration formula for velocity, we can find the time required to reach the high point as shown in box 3.1.

The ball travels upward for just a little over 2 seconds before starting to fall. This is consistent with our earlier computation indicating that the ball is already in its downward

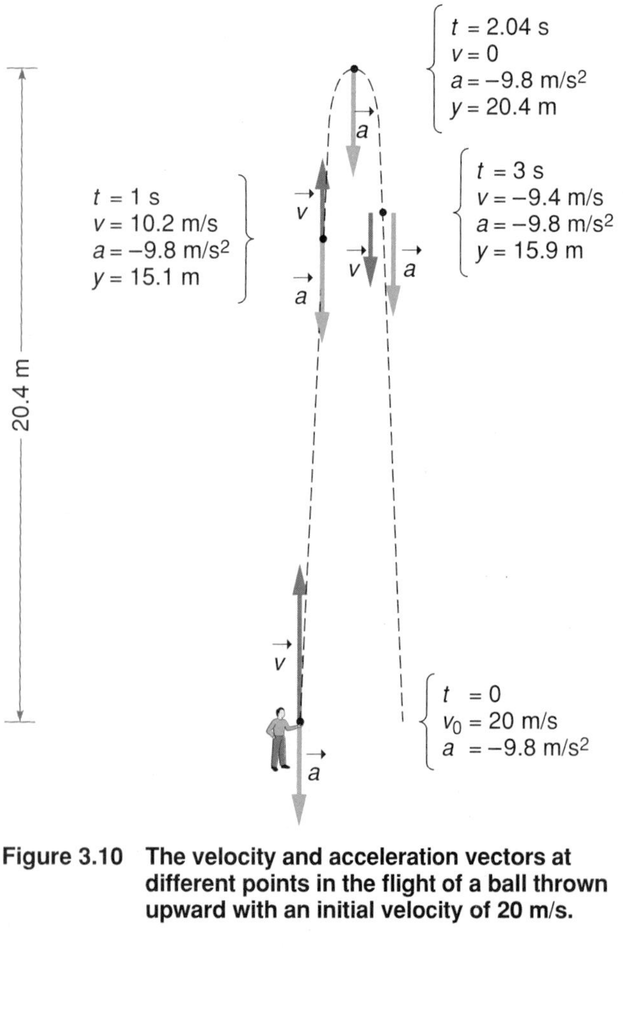

Figure 3.10 **The velocity and acceleration vectors at different points in the flight of a ball thrown upward with an initial velocity of 20 m/s.**

Box 3.1

Sample Exercise

A ball is thrown upward with an initial velocity of 20 m/s. How long does it take to reach the high point of its flight, and how high is this point?

$v_0 = 20$ m/s. $v = v_0 + at.$

$g = -9.8$ m/s². $v = 0 = v_0 - gt.$

$v = 0.$ If $v_0 - gt = 0$, then

$t = ?$ $gt = v_0 = 20$ m/s.

$D = ?$ Dividing both sides by g, we get

$$t = \frac{v_0}{g} = \frac{20 \text{ m/s}}{9.8 \text{ m/s}^2}.$$

$$t = \textbf{2.04 s}.$$

We find the high point as follows:

$$D = v_0 t - \frac{1}{2} gt^2$$

$$D = (20 \text{ m/s})(2.04 \text{ s}) - \frac{1}{2} (9.8 \text{ m/s}^2)(2.04 \text{ s})^2$$

$$D = 40.8 \text{ m} - 20.4 \text{ m} = \textbf{20.4 m}.$$

flight 3 seconds after being thrown. Here again we have substituted –*g* for *a* in box 3.1; the acceleration due to gravity is downward regardless of the direction of the ball's motion (or even if it is *not* moving). Both the time the ball takes to reach the high point and the height that it reaches depend upon the initial velocity; the harder the ball is thrown, the higher it goes.

Notice also that the ball spends a full 2 seconds above the height of 15 m as it passes through the high point at 20.4 m, whereas it took only 1 second to reach the height of 15 m. Because the ball is moving more slowly near the top of its flight, it seems to hang there briefly. The velocity is much larger for the first 15 m of the flight and for the final 15 m on the way down.

You might try throwing a ball upwards with different initial velocities to get an intuitive feeling for this process. If you are indoors, of course, the ceiling places an upper limit on the turnaround point; going outside allows a wider range of initial velocities. Try estimating how high you can throw a ball by estimating the time it takes to reach its turnaround point. Fresh air and a little exercise are fringe benefits of this process.

Stopping a Car

The same ideas that we have treated in detail to analyze the case of the upward-thrown ball can be applied to any other object that experiences a constant acceleration. Applying the brakes in a moving car, for example, can produce a negative acceleration that is approximately constant, as illustrated in box 3.2.

In the velocity computation in box 3.2, the change in velocity ($\Delta v = at$) is –10 m/s, which is subtracted from the initial velocity of 25 m/s to yield the smaller final velocity. The displacement (40 m) is also less than it would have been (50 m) if the brakes had not been applied. If the acceleration had been positive, involving the gas pedal rather than the brakes, the second term in the displacement computation would have been added to the 50 m distance that the car would travel with no acceleration. Acceleration, whether positive or negative, always changes the results that we would expect for uniform motion (constant velocity).

3.4 PROJECTILE MOTION

Suppose that instead of throwing a ball straight upwards, you throw it horizontally from some distance above the ground. What happens then? Does the ball go straight out until it loses all of its horizontal velocity and then start to fall, like the coyote in the Roadrunner cartoons (fig. 3.11)? What does the real path, or *trajectory,* look like?

The answer, of course, is that the cartoons give us a rather misleading impression of what happens. Two things are hap-

---

Box 3.2

Sample Exercise

A car is moving with a velocity of 25 m/s due east along a straight stretch of highway. The driver applies the brakes, producing an acceleration of –5 m/s². What is the velocity after 2 seconds of applying the brakes, and how far does the car travel in this time?

$v_0 = 25$ m/s. $v = v_0 + at.$

$a = -5$ m/s². $v = 25$ m/s $+ (-5$ m/s²$)(2$ s$).$

$t = 2$ s. $v = 25$ m/s $- 10$ m/s.

$v = ?$ $v = \textbf{15 m/s}$ (due east).

$D = ?$

$$D = v_0 t + \frac{1}{2} at^2$$

$$= (25 \text{ m/s})(2 \text{ s}) + \frac{1}{2}(-5 \text{ m/s}^2)(2 \text{ s})^2$$

$$= 50 \text{ m} - 10 \text{ m}$$

$$= \textbf{40 m.}$$

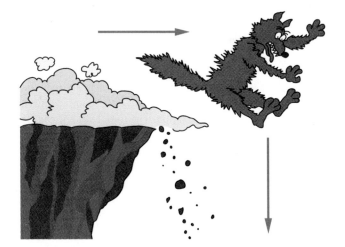

Figure 3.11   A cartoon coyote falling off a cliff. Is this a realistic picture of what happens?

pening at the same time, and this provides a mental challenge in trying to visualize the flight:

1. The ball falls, accelerating downward due to the constant acceleration of gravity.
2. The ball continues to move horizontally with a constant horizontal velocity (if we ignore air resistance).

The second statement follows from the fact that there is no acceleration in the horizontal direction. We need to think about the vertical and horizontal motions separately, which is what creates the mental challenge.

You can perform a simple experiment to help you visualize the path that such an object takes. Take a marble or small ball and roll it along the top of your desk or table, letting it fall off the edge. What does the path of the ball look like as it travels through the air to the floor? Is it like the coyote of figure 3.11? Try rolling the ball at different velocities and observe how the path changes. You might try to sketch the shape of the path after making these observations.

Analyzing the Motion

How do we go about analyzing and describing this motion using the principles stated above? The key is to think about the motions in the horizontal and vertical directions separately and then combine them to get the actual path. If we do this for equal intervals in time, we can use our earlier insights regarding free-fall to good advantage.

We might start with the horizontal motion, since it is the simpler of the two. If we assume that the effects of air resistance are small enough to be ignored, the ball is not accelerated in the horizontal direction once it has left the edge of the desk or table and is no longer in contact with your hand. It therefore travels with a constant horizontal velocity; in equal time intervals it travels equal distances. This is pictured across the top of the diagram in figure 3.12. We can express this fact mathematically as follows for the horizontal displacement:

$$D_x = \Delta x = v_0 t.$$

Here we have used Δx to represent a change in position in the horizontal direction, which is the horizontal displacement. This is a commonly used notation. The initial velocity, v_0, is completely horizontal in this case.

At the same time that the ball is traveling with constant horizontal velocity, it is being accelerated downward by the constant acceleration of gravity, g. Its vertical velocity is therefore increasing in exactly the same manner as that of the falling ball that was photographed for figure 3.3. This is pictured along the left-hand side of the diagram in figure 3.12. As shown earlier in this chapter, the vertical displacement

$$D_y = \Delta y = \frac{1}{2}at^2 = -\frac{1}{2}gt^2,$$

if we choose up as the positive direction again. The symbol Δy is frequently used to indicate vertical displacements or changes in vertical position.

Combining the horizontal and vertical motions, we get the path shown curving downward in figure 3.12. This is found simply by drawing grid lines for equal time intervals (0.05 s each) and finding the position of the ball every 0.05 s at the

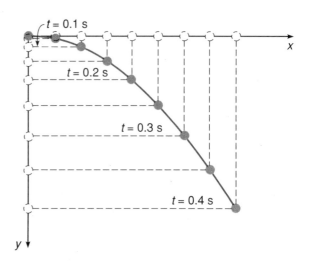

Figure 3.12 The horizontal and vertical motions combine to produce the path of the projected ball.

points where these horizontal and vertical lines intersect. We have used a horizontal velocity of 2.0 m/s in figure 3.12, so that in 0.5 s the ball covers a horizontal distance of 1 m.

If you study figure 3.12 carefully and understand exactly how we have obtained the path of the ball, you are well on your way to understanding projectile motion. The velocity of the ball at each time pictured is in the direction of the path at that point, which is the actual direction of the ball's motion. This can also be thought of as a combination of horizontal and vertical motions. The horizontal velocity remains constant, while the downward vertical velocity gets progressively larger.

When we add the vectors representing these two velocities, we get the total velocity. This is pictured for a few points in the motion in figure 3.13. The process used here for adding the velocity vectors is the same as that used in chapter 2 for adding displacement vectors (see appendix C for other examples). Notice that the total velocity vector points more and more downward as the vertical component of the velocity gets larger.

Time of Flight and Horizontal Distance Covered

The time that the ball is in the air is determined strictly by how high the starting point is above the ground or floor. If, for example, the table top from which the ball is rolled is 1 m above the floor, the time of flight can be found by computing how long it takes an object to fall through a height of 1 m. Since the initial velocity in the vertical direction is zero, the vertical displacement can be expressed as follows:

$$D_y = \Delta y = -\frac{1}{2}gt^2.$$

Setting $\Delta y = -1$ m, since it is also downward, yields

$$-1 \text{ m} = -\frac{1}{2}(9.8 \text{ m/s}^2)(t)^2.$$

This can be readily solved for the time of flight, *t;*

$$t^2 = \frac{1 \text{ m}}{\frac{1}{2}(9.8 \text{ m/s}^2)} = 0.204 \text{ s}^2.$$

Taking the square root, we find that $t = 0.452$ s. In 1 second, the ball would fall a distance of 4.9 m, but it takes almost half that time to go just the first meter. This is consistent with the picture and values provided in figure 3.3.

Once we know the time of flight, we can compute the horizontal displacement Δx:

$$\Delta x = v_0 t;$$

$$\Delta x = (2.0 \text{ m/s})(0.452 \text{ s}) = 0.904 \text{ m},$$

using the value for the initial horizontal velocity of 2 m/s that was given earlier. Obviously, the greater the initial velocity, the greater the horizontal displacement, but the time of flight is the same as long as the starting height is not changed.

Figure 3.14 shows the trajectories and horizontal displacements for balls rolled with different initial velocities. You might compare them to your own observations. Try hitting various points on the floor by varying the initial velocity. A game of this nature will give you a better feeling for the effects of initial velocity than words or even the picture in figure 3.14.

The time of flight is the same for all of the paths pictured in figure 3.14; it is determined strictly by the height of the table top and is completely independent of the horizontal motion. This fact goes against our intuition, and people often have trouble accepting this idea. An understanding of it is crucial, though, to an understanding of projectile motion. You might try rolling two marbles or balls off a table at different initial speeds so that they leave the table top simultaneously. They should also hit the floor simultaneously. You can also try dropping a ball at the same time you throw another one from the same height with a completely horizontal initial velocity (fig. 3.15). Do they hit the ground at the same time?

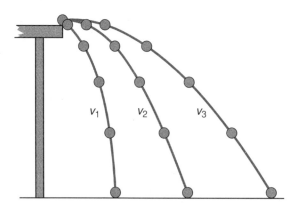

Figure 3.14   Some trajectories for different initial velocities of a ball rolling off a table. v_3 is larger than v_2, which in turn is larger than v_1.

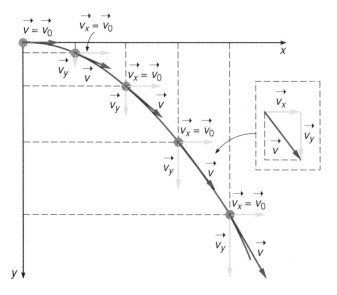

Figure 3.13   The horizontal and vertical velocity components add to give the total velocity at each point. The inset illustrates the vector-addition process.

Figure 3.15   A ball is dropped at the same time another ball is thrown horizontally from the same height. Which ball reaches the ground first?

If the two balls of figure 3.15 do not hit the ground at the same time, it is only because it is difficult to throw the second ball completely horizontally and to release it at the same time that the first one is dropped. Both balls are accelerated downward at the same rate and, as discussed earlier, their vertical velocities at a given time should be identical, regardless of the horizontal motion.

3.5 HITTING A TARGET

Predicting where a projectile such as a cannonball will land after it is fired has been a problem of interest as long as humans have been hunters or warriors. Being able to hit a target such as a bird in a tree or ship at sea has obvious implications for survival. Being able to hit a catcher's mitt with a baseball from center field is also a highly valued skill.

Firing a Rifle
Imagine that you are firing a rifle at a small target some distance away. Assume that the rifle and the target are at the same height above the ground, as pictured in figure 3.16. If the rifle is fired directly at the target in the horizontal direction, will the bullet hit the center of the target?

If you think of the pictures you have seen of the ball rolling off the table, you should be convinced that the bullet will strike the target below its center. Why? As the bullet travels horizontally, it is also falling. Since the time of flight is small, the bullet does not fall very far, but it falls far

enough to miss the center of the target. The farther away the target, the farther the bullet falls, since the time of flight increases with distance.

How do you compensate for the fall of the bullet? The answer is obvious; you aim a little high. You aim properly either through trial and error or, if the target distance is constant, by adjusting the rifle sight so that your aim is automatically a little above center. This actually gives the bullet a small upward velocity initially, causing it first to travel upward and then down to meet the target.

Throwing a Football or Baseball
The same problem occurs in throwing a football or baseball. When making a long throw, the thrower must aim above the target so that the ball does not come to earth too soon. A good athlete does this automatically as a result of trial and error during practice. The harder the throw, the less he needs to aim upwards, because a larger initial velocity causes the ball to reach the target more quickly, giving it less time to fall.

Figure 3.17 shows the flight of a football that has been launched at an angle of 30° above the horizontal. The left side of the diagram plots the vertical position of the ball, just as in figure 3.12. Across the bottom of the diagram, the horizontal position of the ball is shown, assuming that the horizontal velocity is constant as before. Combining these two motions yields the overall path.

As the football climbs, its upward vertical velocity decreases due to the constant downward acceleration of

Figure 3.16   A target shooter fires at a distant target. The bullet falls as it travels to the target.

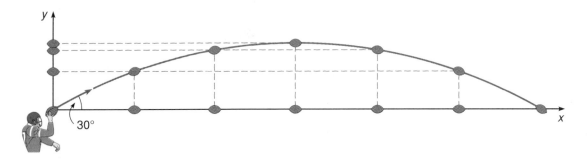

Figure 3.17   The flight of a football launched at an angle of 30° above the horizontal. The vertical and horizontal positions of the ball are shown at regular time intervals.

gravity. At the high point in its flight, the vertical component of the velocity is zero, just as it is for a ball thrown straight upwards. It then begins to fall with increasing downward velocity.

In throwing a ball, there are two quantities that you can vary to help you hit your target. One is the initial velocity, as mentioned above, which is determined by how hard you throw the ball. The other is the launch angle, which is determined by your aim. There is no time like the present to test these ideas. Take a full page of paper and crumple it into a compact ball. Then take your wastebasket and put it on a chair or your desk. Throwing underhand, experiment with different throwing speeds and launch angles to see which is most effective in making a basket. Try to get a feeling for how the launch angle and throwing speed interact in determining a successful shot. A low, flat-trajectory shot should require a greater throwing speed than a higher, arching shot.

Achieving Maximum Distance

In firing a rifle or cannon, the initial velocity of the projectile is usually fixed by the amount of gunpowder in the shell, which leaves only the launch angle to play with in attempting to hit a target. Figure 3.18 shows the paths, or trajectories, for a cannonball at different launch angles and a fixed initial velocity. Notice that the greatest distance is achieved for an intermediate angle; this angle is 45° if the effects of air resistance are negligible. The same considerations are involved in the shot-put event in track and field; the launch angle is very important and will be near 45° for the greatest distance. (Air resistance effects are significant, so the optimal angle is not exactly 45°. The fact that the shot hits the ground below the launch point is also a factor.)

The reason that the optimal angle for maximum distance is about 45° is apparent if we consider what happens to the horizontal and vertical velocities for different launch angles. At 45°, the horizontal velocity is equal to the vertical veloc-ity, as shown in figure 3.19. At a larger launch angle, say 70°, the horizontal component of the velocity is quite small, so that even though the ball may remain in the air for a longer time, it does not travel very far horizontally during that time. If you threw it straight up, of course, it would have no horizontal velocity, and its horizontal displacement would be zero.

At a smaller launch angle, say 20°, the vertical component of the velocity is small (see figure 3.19), and the ball does not remain in the air as long. Even though the ball now has a large horizontal velocity, its short time of flight dictates that it will not travel as far as in the 45° case. The 45° case represents a compromise, in which you are putting relatively large portions of the initial velocity into both the vertical motion and the horizontal motion. The vertical component of the velocity is responsible for keeping the ball in the air long enough for the horizontal component of the velocity to be effective.

Using the principles and techniques developed in the previous two sections, we could compute the time of flight and horizontal distance traveled for different launch angles and velocities. These computations can become rather involved, however, and are not essential to understanding the ideas presented here. The key to understanding the ideas and computations is, again, to treat the horizontal and vertical motions separately and then combine them to yield the overall motion. The downward acceleration of gravity is a constant influence, regardless of whether the projectile is rising or falling.

These ideas are important in football, baseball, basketball, tennis, and almost any game that involves balls flying through the air. A ball thrown or hit with a flat trajectory will get to its destination sooner, but will not go as far as one thrown or launched near a 45° angle. A tennis lob hit at a steep angle also does not go as far, but stays in the air for a longer time. Deciding which of these options to use is part of the strategy of the game.

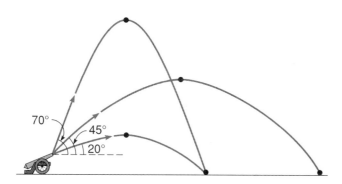

Figure 3.18   Cannonball paths for different launch angles and a constant initial launch speed, v_0.

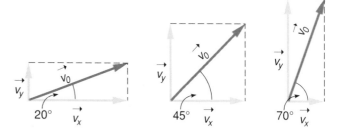

Figure 3.19   Vector diagrams showing the horizontal and vertical components of the initial velocity for the cases shown in figure 3.18.

Box 3.3

The Situation. Whenever you shoot a basketball at the basket, you are unconsciously selecting a trajectory for the ball that you feel will have the greatest likelihood of causing the ball to fall through the basket. The target in this case is (with the exception of dunk shots and sky hooks) above the launch point. (The ball must be on its way down, however, for the basket to count.)

What factors determine the optimal trajectory? When is a high, arching shot desirable, and when might a flatter trajectory be better? Will these factors be different for a free throw than for a shot taken

when a player is being guarded? These are some of the questions that coaches or players might wish to pose to each other —or to the resident sports physicist.

The Analysis. The diameter of the basketball and the diameter of the basket establish limits for the angle at which the basketball can pass cleanly through the basket. This is illustrated in the accompanying drawing, where the range of possible paths is shown for a ball that is coming straight down on the basket as well as for a ball coming in at an angle of 45° to the basket. The shaded area in each case shows how much the center of the ball can vary from the center line if the ball is to pass through the basket. As you can see in the drawing, a wider range of paths is available when the ball is coming straight down.

The drawing illustrates the advantage of the arched shot; there is a larger margin of error in the path that the ball can take and still make it through cleanly. Given the dimensions of a regulation basketball and basket, the angle shown in the drawing must be at least 32° for a clean shot. As the angle gets larger, the range of possible paths increases.

The disadvantage of the arched shot is a little more difficult to see. It lies in the fact that as you get further from the basket, the launching conditions for an arched shot must be very precise if the ball is to travel the desired distance horizontally. If an arched shot is launched from 30 feet, it must travel over a much higher path than an arched shot launched much closer to the basket at the same angle, as shown in the third drawing. Since the ball stays in the air for a much longer time, small variations in either the release speed or the angle can cause relatively large errors in its position at the other end of the path.

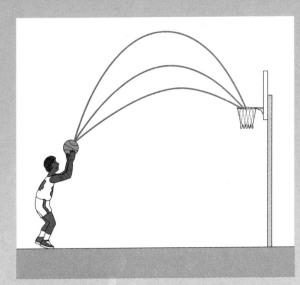

Different possible trajectories for a basketball free throw. Which has the greatest chance of success?

We might conclude, then, that a highly arched shot is more desirable nearer to the basket. Near the basket you can take advantage of the greater range of paths available to the arched shot. As you move away from the basket, the desirable trajectories become gradually flatter, thus permitting more accurate control of the shot. An arched shot is sometimes necessary, of course, from any point on the court in order to avoid blocking attempts.

The spin of the basketball, the height of release, and other factors also play a role in successful shots. A much fuller analysis than that given here can be found in an article by Peter J. Brancazio in the *American Journal of Physics* (April 1981) entitled "Physics of Basketball." A good understanding of projectile motion might improve the game of even an experienced player!

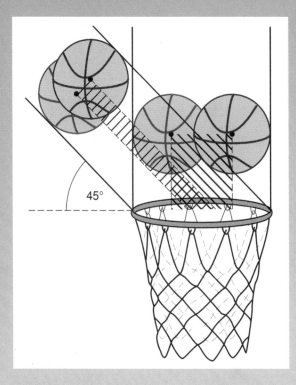

Possible paths for a basketball coming straight down and for one coming in at a 45° angle. The ball coming straight down has a wider range of possible paths.

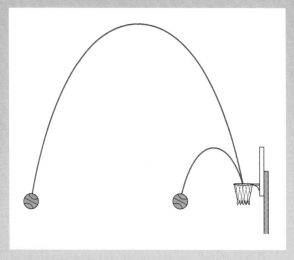

An arched shot launched from a large distance compared to one launched at the same angle from much closer to the basket.

SUMMARY

The aim of this chapter has been to make you familiar with the nature of the acceleration of gravity, *g,* as well as some of the effects of constant acceleration on common types of motion. The key ideas were the following.

The Acceleration of Gravity. Measurements of the position of a dropped object at different times can be used to find the acceleration of gravity. Its value is $g = 9.8$ m/s^2, and it does not vary with time as the object falls. It also has the same value for different objects, regardless of their weight.

Uniformly Accelerated Motion. The definition of acceleration and graphs were used to develop simple expressions that describe how velocity and displacement vary with time for an object that is undergoing constant acceleration.

$$v = v_0 + at$$

$$D = v_0 t + \frac{1}{2} at^2$$

$$v^2 = v_0{}^2 + 2aD$$

Beyond Free-Fall: Constant Acceleration Applications. The motion of a ball thrown straight upward was carefully considered using the constant acceleration expressions. The velocity is zero at the high point, but the acceleration is still downward and equal to 9.8 m/s^2. The constant acceleration equations can also be used to describe the motion of a car.

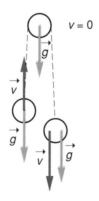

Projectile Motion. If an object is projected or thrown horizontally, it moves with constant horizontal velocity at the same time that it accelerates downward due to gravity. These two motions combine to yield a curved path for the projectile.

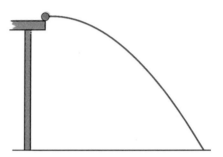

Hitting a Target. The motion of a football or other object launched at an angle in an attempt to hit a target was described. Again, the horizontal and vertical motions combine to form a curved path whose shape depends upon the launching speed and angle.

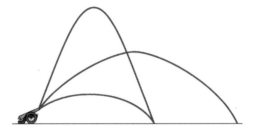

Since gravity is always present in our ordinary experience of motion, it is important that we understand its role in influencing motion. You will gain more insight into the nature of *g* in the next chapter, when we consider forces and Newton's laws of motion, as well as in chapter 5, where Newton's law of universal gravitation is discussed.

QUESTIONS

Q3.1 The diagram shows the positions of a ball moving from left to right at 10-second intervals (as in a photograph taken with a stroboscope). Is the ball accelerated? Explain.

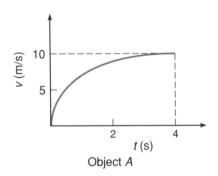

Q3.2 The diagram shows the positions of two balls moving from left to right at 0.05-second intervals. Are either or both of these balls accelerated? Explain.

A
B

Q3.3 A lead ball and an aluminum ball, each one inch in diameter, are released simultaneously and allowed to fall to the earth. Which of these balls, if either, has the greater acceleration due to gravity? Explain.

Q3.4 Two identical pieces of paper, one crumpled into a small ball and the other left uncrumpled, are released simultaneously and allowed to fall to the floor. Which one, if either, reaches the floor first? Explain.

Q3.5 In these two graphs, velocity has been plotted as a function of time for two different moving objects. Which of the two objects, *A* or *B*, covers the greater distance in the equal times shown? Explain.

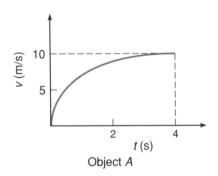

Object *A*

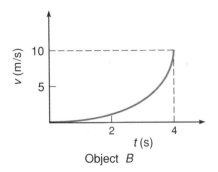

Object *B*

Q3.6 At which point in the motion of object *A* (in the first graph shown for question 3.5) is the acceleration the greatest? Explain.

Q3.7 Is object *B* (shown in the second graph of question 3.5) undergoing constant acceleration? Explain.

Q3.8 A ball rolling rapidly along a table top is allowed to roll off the edge and fall to the floor. A second ball is released from the same height above the floor at the exact instant that the first ball rolls off the edge. Which ball, if either, reaches the floor first? Explain.

Q3.9 Two balls are rolling along a table top, ball *A* with twice the horizontal velocity of ball *B*. Ball *A* starts from further back than *B*, however, so that the two balls roll off the edge of the table at the same instant. Which of the two balls reaches the floor first? Explain.

Q3.10 Do the two balls of question 3.9 have the same acceleration? Explain.

Q3.11 Which of the two balls of question 3.9, if either, hits the floor at the greater horizontal distance from the edge of the table? Explain.

Q3.12 An expert marksman aims a high-speed rifle directly at a distant target. Assuming that the rifle sight is accurately adjusted for near targets, will the bullet hit the distant target above or below the target center? Explain.

Q3.13 An expert marksman aims a high-speed rifle directly at a near target. Assuming that the rifle sight is adjusted to compensate for the effect of gravity for distant targets, will the bullet hit the near target above or below the target center? Explain.

Q3.14 In the diagram shown below, two different possible trajectories are shown for a ball thrown by a center fielder to home plate in a baseball game. Which of the two trajectories, the higher one or the lower one, will result in a longer time for the ball to reach home plate? Explain.

Q3.15 Assuming that the two trajectories in the diagram for question 3.14 represent throws by two different center fielders, which of the two is likely to be thrown by the player with the stronger arm? Explain.

Q3.16 The diagram below shows a wastebasket placed behind a chair. At the left of the diagram, three different directions are shown for the initial velocity of a ball thrown from near floor level. Which of the three directions shown, *A, B,* or *C,* is likely to be the most effective in making a basket? Explain.

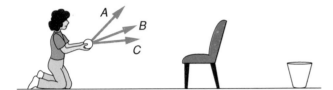

Q3.17 In shooting a free throw, what is the primary advantage that a high, arching shot has over one with a flatter trajectory? Explain.

Q3.18 In shooting a shot from greater than free-throw range, what is the primary disadvantage of a high, arching shot? Explain.

EXERCISES

E3.1 Using the data in table 3.1 and following the examples done in section 3.1, find the average velocity for the following:

  a. The flight of the ball between the first and second flashes.

  b. The flight of the ball between the seventh and eighth flashes.

  c. Compute the acceleration of the ball using the velocities found in (*a*) and (*b*). (The time interval is 0.30 s.)

E3.2 A steel ball is dropped from a diving platform with an initial velocity of zero.

  a. What is the velocity of the ball 1.5 seconds after its release?

  b. What is its velocity 3 seconds after its release?

E3.3 For the ball in exercise 3.2, answer the following questions.

  a. Through what distance does it fall in the first 1.5 seconds of its fall?

  b. Through what distance does it fall in the first 3 seconds of its fall?

E3.4 A ball is thrown downward with an initial velocity of 15 m/s.

  a. What is its velocity 2 seconds after it is thrown?

  b. Through what distance has it traveled in this time?

E3.5 A ball is thrown upwards with an initial velocity of 15 m/s.

  a. What are the magnitude and direction of its velocity 1 second after it is thrown?

  b. What are the magnitude and direction of its velocity 3 seconds after it is thrown?

E3.6 How far above the ground is the ball of exercise 3.5

  a. after 1 second?

  b. after 3 seconds?

E3.7 For the ball of exercise 3.5, answer the following questions.

  a. At what time after the ball is thrown does it turn around and begin to fall? (Remember that its velocity is zero at this point.)

  b. How high does the ball go before beginning to fall?

E3.8 A car moving with an initial velocity of 10 m/s accelerates at a constant rate of 3 m/s^2.

  a. What is the velocity of the car after accelerating for 3 seconds at this rate?

  b. Through what distance does the car travel in this time?

E3.9 A car moving with an initial velocity of 30 m/s slows down at the rate of 4.0 m/s^2.

  a. What is its velocity after 3 seconds of deceleration?

  b. Through what distance does it travel in this time?

E3.10 A rifle is fired horizontally at a target 200 m distant. The initial velocity of the bullet is 800 m/s.

  a. How much time does it take for the bullet to reach the target?

  b. How far does the bullet fall in this time?

E3.11 A ball rolls off of a table with a horizontal velocity of 5 m/s. At what horizontal distance from the edge of the table does the ball land if it takes 0.6 seconds for it to hit the floor?

E3.12 A ball rolls off of a platform that is 4.9 m above the ground with a horizontal velocity of 4 m/s.

  a. How long does it take for the ball to hit the ground?

  b. Through what horizontal distance does it travel in this time?

CHALLENGE PROBLEMS

CP3.1 A ball is thrown straight upward with an initial velocity of 12 m/s.

    a. What is its velocity at the high point in its motion?

    b. How much time is required to reach this high point?

    c. How high above its starting point is the ball at the high point?

    d. How high above its starting point is the ball 2 seconds after it is released?

    e. Is it moving up or down after 2 seconds? Explain.

CP3.2 Two balls are released simultaneously from the top of a tall building. Ball *A* is simply dropped with no initial velocity, and ball *B* is thrown downwards with an initial velocity of 10 m/s.

    a. What are the velocities of the two balls 2 seconds after they are released?

    b. How far has each ball dropped in 2 seconds?

    c. What conclusion can you draw about the difference in the velocities of the two balls at any time after their release? Explain.

CP3.3 Just as car *A* is starting up, it is passed by car *B*. Car *B* travels with a constant velocity of 9 m/s, whereas car *A* accelerates with a constant acceleration of 5 m/s², starting from rest.

    a. Compute the distance traveled for each car for times of 1 s, 2 s, 3 s, and 4 s.

    b. At what time, approximately, does car *A* overtake car *B*?

    c. How might you go about finding this time exactly? Explain.

CP3.4 Two balls are rolled off of a horizontal table top that is 1.2 m above the floor. Ball *A* has an initial horizontal velocity of 2 m/s and that of ball *B* is 4 m/s.

    a. How long does it take each ball to reach the floor?

    b. How far does each ball travel horizontally before hitting the floor?

    c. What conclusion can you draw about the ratio of the horizontal distances traveled by each ball for any height of the table top above the floor? Explain.

CP3.5 A cannon is fired over level ground at an angle of 30° to the horizontal. The initial velocity of the cannonball is 100 m/s, but because of the angle, the vertical component of this velocity is 50 m/s and the horizontal component is 86.6 m/s.

    a. How long is the cannonball in the air? (Hint: the total time is just twice that required to reach the high point.)

    b. How far does the cannonball travel horizontally?

    c. Repeat these computations assuming that the cannon was fired at 60° to the horizontal, resulting in a vertical component of 86.6 m/s and a horizontal component of 50 m/s. How does the distance traveled compare to the earlier result?

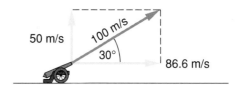

HOME EXPERIMENTS AND OBSERVATIONS

HE3.1 Try dropping a ball from one hand at the same time that you throw a second ball with your other hand. At first, try to throw the second ball horizontally, with no upward or downward component to its velocity. (It will take some practice to make the release as horizontal as possible and simultaneous with the dropping of the first ball.)

    a. Do the balls reach the floor or ground at the same time? (It may be useful to enlist a friend to help make this judgment.)

    b. If the second ball is projected slightly upwards from the horizontal, which ball reaches the ground first?

    c. If the second ball is projected slightly downwards, which ball reaches the ground first?

HE3.2 Take a ball outside and throw it straight up in the air as hard as you can. By counting seconds yourself, or by enlisting a friend with a watch, estimate the length of time that the ball stays in the air. From this information, can you find the initial velocity that you imparted to the ball? (The time required to reach the high point should be just half the total time of flight.)

HE3.3 Take a stopwatch to a football game and estimate the hang time of several punts. Note also for each punt how far it travels horizontally. Do the highest punts have the longest hang times? Do they travel the greatest distances horizontally?

HE3.4 Using a rubber band and a plastic ruler or other suitable support, build yourself a marble launcher. By pulling the rubber band back the same distance each time, you should be able to launch the marble with approximately the same initial speed each time.

    a. Placing your launcher at a number of different angles to the horizontal, launch a marble and note how far it travels over a level surface for each angle. Which angle yields the greatest distance?

    b. Firing the launcher at different angles to the horizontal from the edge of a table or desk, see how far the marble travels before hitting the floor. Which angle yields the greatest distance?

4 Newton's Laws: Explaining Motion

A large, cantankerous person gives you a shove, and you move in a direction related to that push. A child pulls a toy wagon with a string, and the wagon lurches along in an irregular way. An athlete kicks a football or soccer ball, and the ball is launched towards the goal. These are familiar examples that involve forces in the form of pushes or pulls that cause changes in motion. We all have an intuitive sense of how motion is influenced by such forces because it is part of our everyday experience, but that intuitive sense is probably not sufficient to permit a clear explanation of what is happening.

To pick a less complicated example, imagine yourself pushing a chair across a wood or tile floor (fig. 4.1). Why does the chair move? Will it continue its motion if you stop pushing? What determines the velocity of the chair? If you push harder, will the chair's velocity increase? Up to this point, we have used some tools for describing motion, but we have not talked about what causes motion or changes in motion. Explaining motion is a more challenging project than simply describing it.

You already have some intuitive notions about the causes of the chair's motion. Certainly the push that you exert on the chair has something to do with it. But is the strength of that push more directly related to the velocity of the chair or to its acceleration? It is at this point that intuition often serves us poorly.

Over two thousand years ago, the Greek philosopher Aristotle attempted to provide answers to some of these questions. Many of us would find that his explanations match our intuition for the case of the moving chair but are less satisfactory for the case of a thrown object. Aristotle's ideas were widely accepted until they were replaced by the theories that Isaac Newton introduced in the seventeenth century. Newton's theory of motion has proved to be a much more complete and satisfactory explanation of motion and allows us to make quantitative predictions that were largely lacking in Aristotle's ideas. This chapter introduces Newton's theory of motion and applies it to some familiar situations.

Figure 4.1 Moving a chair. Will the chair continue to move after the person stops pushing?

Chapter Objectives

The primary objective of this chapter is to provide an understanding of the meaning of each of Newton's three laws of motion and to illustrate how they are applied in familiar situations. A good understanding of these ideas will permit you to analyze and explain what is happening in almost any simple motion. Such insight can be useful in driving a car, moving heavy objects, and a host of other everyday activities.

The chapter outline indicates the approach to this task. The chapter begins with a brief historical sketch, proceeds to state Newton's laws of motion, and then illustrates their application. The concepts of force, mass, and weight play important roles in this discussion. The chapter will also focus your attention on the contrast between Newton's ideas and those of Aristotle. Since Aristotle's thinking comes closer in many respects to your initial common-sense ideas about motion, recognizing where they differ from Newton's formulation should help to counter some of your own misconceptions.

Chapter Outline

1 *Historical background: Aristotle, Galileo, and Newton.* Where do our ideas and theories about motion come from? What role was played by people such as Aristotle, Galileo, and Newton?

2 *Newton's first and second laws.* What is the meaning of Newton's first two laws of motion, and how are they related to one another? What are the roles of the concepts of force and mass in these laws?

3 *Mass and weight.* What is the distinction between the concepts of mass and weight? How can we define mass independently of the concept of weight?

4 *Newton's third law.* What is Newton's third and final law of motion, and how is it applied? How is it involved in completing our definition of force?

5 *Applications of Newton's laws.* How can Newton's laws be applied to some familiar situations such as pushing a chair, skydiving, throwing a ball, and pulling connected carts across the floor?

4.1 HISTORICAL BACKGROUND: ARISTOTLE, GALILEO, AND NEWTON

How has our understanding of motion developed? Did some genius, sitting under an apple tree, concoct a full-blown theory of motion in a sudden, blinding flash of inspiration? Not quite. The story of how theories are developed and gain acceptance involves many players over long periods of time, and it is impossible to do justice to this history in a textbook devoted primarily to other objectives. A few key people, however, whose insights produced major advances in our understanding, can be briefly highlighted. A glimpse of this history should help you to accept and appreciate the concepts we will discuss. It is important, for example, to know whether a theory was just proposed yesterday or has been tried and tested over a long period of time.

Aristotle

Questions regarding the causes of motion and of changes in motion had perplexed philosophers and other observers of nature for centuries prior to the work of Isaac Newton. For many centuries the prevailing view was that of the Greek philosopher Aristotle (384–322 B.C.), who was an astute and careful observer of nature. Aristotle conceived of the idea of force much as we have talked about it to this point, as a push or a pull on an object. He believed that a force had to be present in order for an object to move, and that the velocity of the object was related to the strength of the force. These ideas seem quite reasonable in the context of our discussion of the chair.

Aristotle was an observer of nature rather than an experimenter, and he did not attempt to make quantitative predictions that could be checked by experiment, as is the practice now. Even without such checks, however, some problems with his basic notions were troubling to Aristotle himself as well as to later thinkers. For example, in the case of a ball or other thrown object, the force that causes it to move initially is no longer in effect after the ball leaves the hand that provides the push. The ball keeps moving, however, and to Aristotle that meant that a force must be acting. How else could the ball continue to move? Aristotle theorized that the force that keeps the ball moving once it leaves the hand is provided by the air rushing around to fill the vacuum in the spot where the ball has just been (fig. 4.2). This idea, however, does not fit easily into our common-sense view of motion.

Galileo

Aristotle's ideas were challenged many centuries later by the Italian scientist and philosopher Galileo Galilei (1564–1642). By this time, much of Aristotle's natural philosophy had been incorporated into the teachings of the Roman Catholic church, which gave it an official standing not enjoyed by modern scientific theories. Galileo chal-

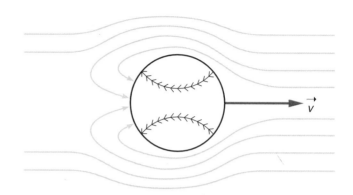

Figure 4.2 Aristotle pictured air rushing around a thrown object to push the object forward. Does this picture seem reasonable?

lenged Aristotle's thinking in several areas, including astronomy, and these challenges were at times a matter of serious concern to church authorities. The problems that this conflict caused for Galileo were substantial; at one point he was tried by the church, placed under house arrest, and forced to retract some of his views on the nature of the solar system.

Galileo studied the motion of falling and rolling objects and directly disputed some of Aristotle's thinking on these subjects. In particular, as noted in chapter 3, he showed that objects of different weights fall at the same rate, contradicting Aristotle's prediction that heavier objects should fall faster due to a stronger force. Galileo also argued that the natural tendency of a moving object is to continue moving in a straight line and that no force is required to sustain this motion. (Think about the chair again; does this statement seem to make sense in that situation?) Galileo did a number of simple experiments with moving objects and, although his measurements were crude by today's standards, developed quantitative predictions that agreed with his experimental results. He was a pioneer in using experiments to bolster his scientific arguments.

Newton

Isaac Newton (1642–1727; fig. 4.3), building upon some of Galileo's thinking, proposed a theory of the causes of motion that could explain the motion of any object—the motion of ordinary objects such as the chair as well as the motion of heavenly bodies such as the moon and the planets. The central ideas in his theory are his three laws of motion (to be introduced shortly) and his law of universal gravitation.

Not only did Newton's theory successfully explain such motions, but it also provided a framework for many new studies in physics and astronomy, some of which led to predictions of phenomena that had not previously been observed. For example, calculations applying Newton's theory to irregularities in the orbits of the planets led to predictions of the existence of additional planets, which were later

Figure 4.3   A portrait of Isaac Newton (1642–1727).

confirmed by observation. Confirmed predictions such as this are one of the marks of a successful theory. Newton's theory served as the basic theory of mechanics for over two hundred years and is still used extensively in physics and engineering.

Newton developed the basic ideas of his theory around 1665, when he was a young man. In order to avoid the plague, he had returned to his family's estate in England, where he found time to engage in serious thought with little interruption. He may, in fact, have spent some time sitting under apple trees, and some flashes of sudden insight or inspiration surely were a part of the process. The story has it that seeing an apple fall led to his insight that the moon was also falling towards the earth, and that the force of gravity was involved in both cases.

Although Newton developed the framework of the theory and many of its details in 1665, he did not formally publish his ideas until 1687. One of the reasons for his delay involved the need to develop some of the mathematical techniques required to calculate the effects of his proposed gravitational force on large objects such as planets. (He is generally credited with being the coinventor of what we now call *calculus*.) The English title of his 1687 treatise was *The Mathematical Principles of Natural Philosophy* (*Philosophiae Naturalis Principia Mathematica* in Latin), which is often referred to as Newton's *Principia*.

Physical theories, like Newton's mechanics, do not just emerge in an intellectual vacuum. They are products of their time and the state of knowledge current at that time. They usually replace earlier and often cruder theories. In Newton's day, the accepted theory of motion was that of Aristotle,

already partially discredited by Galileo's work. Although Aristotle's ideas are now considered unsatisfactory and are essentially worthless for making quantitative predictions regarding motion, they do have an intuitive appeal that corresponds to some of our own untrained notions about motion. For this reason, we often speak of the need to replace Aristotelian ideas about motion with Newtonian concepts as we learn mechanics, despite the fact that our own naive ideas about motion are not usually as fully developed as those of Aristotle. In any case, you may find that some of your common-sense notions will require modification.

Newton's theory, in turn, has been superseded by more sophisticated theories that provide more accurate descriptions of reality. These include Einstein's theory of relativity as well as the theory of quantum mechanics, both of which were developed in the early part of this century. Although the predictions of these theories differ substantially from those of Newton's theory in the realm of the very fast and the very small respectively, they differ insignificantly in the realm most interesting to us here, namely the motion of ordinary objects at speeds much less than the speed of light. Thus we continue to use Newton's theory, recognizing its limitations, because of its ease of application to ordinary motion.

The following sections introduce and carefully explain Newton's three laws of motion, indicating also how they differ from Aristotle's ideas. We then proceed to apply Newton's laws to some common situations, including that of the chair mentioned in the introduction. Discussion of the other basic postulate of Newton's theory, his law of universal gravitation, appears in chapter 5, which describes and explains planetary motion.

4.2 NEWTON'S FIRST AND SECOND LAWS

What causes objects to move or stop moving? Newton's first two laws address these questions and, in the process, provide part of our definition of force. The first law tells us what happens in the absence of any force, and the second deals with the effects of an applied force.

We discuss the first two laws together because the first law is actually a special case of the more general second law. Newton felt the need to state the first law separately, however, in order to refute some strongly held Aristotelian ideas. In doing so, Newton was following the lead of Galileo who had actually stated the first law in much the same terms several years earlier.

Newton's First Law

In language not too different from that used by Newton, the first law is expressed as follows:

An object remains at rest, or in uniform motion in a straight line, unless it is compelled to change that state by an externally imposed force.

In other words, unless there is a force acting on an object, its velocity will not change; if it is initially at rest, it will remain so; if it is moving, it will continue to do so with constant velocity (fig. 4.4). Note that our use of the term *constant velocity* here implies that neither the direction nor the magnitude of the velocity are changing. When the object is at rest, its velocity, of course, is zero, and that value also remains constant in the absence of a force.

Although this law seems simple enough, it is important to recognize that it directly contradicts Aristotle's ideas and perhaps your own intuition as well. As mentioned earlier, Aristotle believed that a force is required in order to keep an object moving. His views make intuitive sense if we are talking about pushing a heavy object (like the chair) across the floor. They run into trouble, however, if we consider the motion of a ball or other thrown object that continues to move even after the initial push. Newton (and Galileo before him) made the strong statement that no force is needed to keep an object moving.

How can Aristotle's ideas be so different from those of Newton and Galileo and yet seem so reasonable in some situations? The key to answering that question is to recognize the existence of frictional forces. The chair does not move very far after you stop pushing because frictional forces between the chair and the floor cause the velocity to decrease. Even a ball would not keep moving indefinitely if it did not fall to the ground because the frictional force of air resistance is pushing against it. It is really quite difficult to find a situation in which there are *no* forces acting on an object. Aristotle did not take frictional forces, such as air resistance, directly into account.

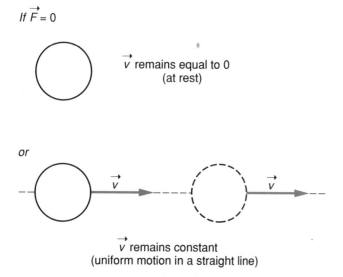

If $\vec{F} = 0$

$\vec{v}$ remains equal to 0
(at rest)

or

$\vec{v}$ remains constant
(uniform motion in a straight line)

Figure 4.4   Newton's First Law: In the absence of force, an object remains at rest or moves with constant velocity in a straight line.

Newton's Second Law

Newton's second law is a more general statement about the effect of an imposed force on the motion of the object. Putting it in terms of the concept of acceleration, the second law makes the following assertion:

The acceleration of an object is in the direction of the externally imposed force and is directly proportional to the magnitude of the imposed force and inversely proportional to the mass of the object.

This statement is most easily grasped when written in symbolic form:

$$\vec{a} \sim \frac{\vec{F}}{m}.$$

Here we have used the symbol ~ to mean "proportional to"; $\vec{F}$ represents the imposed force, and m represents the mass of the object. Since the acceleration is directly proportional to the imposed force, if we double the force acting on the object, we double the acceleration of the object. If we double the mass of the object, on the other hand, we halve the acceleration.

Note also that it is the *acceleration* that is directly related to the imposed force, not the velocity. Aristotle did not make a clear distinction between acceleration and velocity and thus did not think directly in these terms. Unfortunately, most of us also fail to make that distinction when we discuss or think about motion informally. In Newton's theory, that distinction is critical.

Newton's second law is *the* central idea of his theory. According to this law, the acceleration of an object is determined by two quantities: the total force acting on the object and the mass of the object. The total force acting upon the object is the cause of its acceleration. In fact, the concepts of force and mass are, in part, defined by the second law. Figure 4.5 pictures two different masses acted upon by identical forces. The larger mass, m_1 has a smaller acceleration than the smaller mass, m_2; a larger force would be necessary to give it the same acceleration as m_2.

Mass, therefore, is a quantity that tells us how much resistance an object has to a change in its motion. We could define mass as follows:

Mass is a measure of the *inertia* of an object: that property which causes it to resist a change in its motion.

The basic metric unit of mass is the kilogram (kg). More will be said about the definition of mass and its relationship to weight in the next section.

If we express force in appropriate units, we can write our symbolic form of Newton's second law as an equation rather than a proportionality:

$$\vec{a} = \frac{\vec{F}}{m}.$$

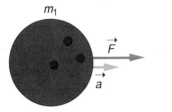

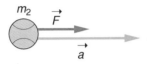

Figure 4.5 Objects of different mass experience different accelerations when acted upon by identical forces.

Multiplying both sides of this equation by the mass, *m*, puts it in the form in which it is most often expressed:

$$\vec{F} = m\vec{a}.$$

From this form we can see that the appropriate unit for force in the metric system would be that of mass times acceleration, in other words, kilograms times m/s². Because it is used so often, this unit is given its own name: a *newton* (N). Thus;

$$1 \text{ newton} = 1 \text{ N} = 1 \text{ kg·m/s}^2.$$

Also implicit in the statement of the second law is the idea that we are talking about the *total* external force acting on an object, not just one of several that may be involved. Thus the second law is sometimes written as follows:

$$\Sigma\vec{F} = m\vec{a}.$$

where the Greek letter Σ represents a summation (addition) process. In words, we often use the term *net force* to describe this summation. Since forces are vectors with an obvious directional character, they must be added as vectors. Examples later in this section will illustrate this process.

Study Hint

In studying physics, students often have difficulty in isolating the fundamental principles from the many examples that are discussed to illustrate their application. Clearly Newton's laws, and the second law in particular, are central ideas and deserve careful reading and rereading. Before you leave this section, you should convince yourself that you have a clear concept of what the second law is saying. Try to explain it to a friend, if only in an imagined conversation. If you can explain it, there is a good chance that you understand it.

A final point should be understood about Newton's first and second laws of motion: The first law is actually contained within the second law as a special case of the second law. This can be seen by asking what happens according to the second law if the total force acting on an object is zero. In

Box 4.1

Sample Exercise

Suppose that a block is being pulled across a table top under the influence of a force of 10 N applied via a string attached to the front end of the block (see fig. 4.6). The block has a mass of 5 kg, and the frictional force exerted by the table on the block is 2 N. What is the acceleration of the block?

$F_s = 10$ N. $\qquad \Sigma\vec{F} = m\vec{a}.$
$f = 2$ N.
$m = 5$ kg. $\qquad$ Since the two forces are in opposite
$a = ?$ $\qquad$ directions,

$$\Sigma F = 10 \text{ N} - 2 \text{ N} = 8 \text{ N}.$$

Therefore,

$$\Sigma F = 8 \text{ N} = ma.$$

Dividing both sides of this equation by *m* yields

$$a = \frac{\Sigma F}{m} = \frac{8 \text{ N}}{5 \text{ kg}} = \textbf{1.60 m/s}^2.$$

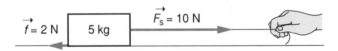

Figure 4.6 A block being pulled across a table. Two horizontal forces are involved.

this case the acceleration of the object must also be zero, and if the acceleration is zero, then the velocity of the object is not changing. The object therefore moves with constant velocity (in a straight line), which is what the first law states. Newton's first law simply addresses the special case in which $\Sigma\vec{F} = 0$.

The sample exercise in box 4.1 illustrates the application of Newton's second law. The solution is straightforward, but it does involve a simple application of vector addition (see appendix C). What is the total force acting on the block in figure 4.6? Is it the numerical sum of the two forces, 10 N + 2 N = 12 N? The diagram should convince you that this cannot be the case; the frictional force retards the motion, whereas the force exerted by the string causes the block to move forward. Since these two forces are obviously in opposite directions, the net force is found by subtracting the frictional force from the force exerted by the string, as is done in box 4.1.

Example 1

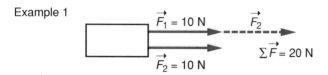

Example 2

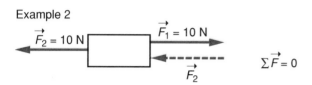

Example 3

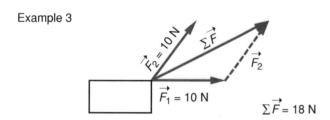

Figure 4.7 Vector addition of two equal-magnitude forces with different directions. The resulting sums are quite different in these different situations.

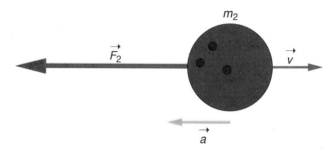

Figure 4.8 Stopping a bowling ball and a ping-pong ball. A much larger force is required to produce the same rate of change of velocity for the larger mass.

The fact that forces are vectors and that we must therefore take their directions into account in adding them is an important aspect of the second law that is easily missed in a first reading. Figure 4.7 provides a few examples of the different results that can be obtained when two forces of the same magnitude but different directions are added. When the two forces are not parallel, we use the head-to-toe method of addition introduced in chapter 2 (and discussed in more detail in appendix C). If the vectors are drawn to scale and in the proper directions, then the magnitude of the resulting sum can be estimated with a ruler. Try it and see if you agree with the numbers shown.

4.3 MASS AND WEIGHT

What exactly is weight? Is it the same thing as mass, or is there a difference in the meanings of these two terms? Clearly mass plays an important role in Newton's second law—a role implied in the definition provided in the preceding section. Weight is also a familiar concept that is often used interchangeably with mass in everyday language. Here again, however, physicists make a distinction between mass and weight that is important to Newton's theory.

From the role that mass plays in Newton's second law, it is not difficult to establish methods of comparing masses. Mass is defined as the property that determines how much an object resists a change in its motion. The greater the mass,

the greater the *inertia,* or resistance to *change* in motion, and the smaller the acceleration will be for a given applied force. Imagine, for example, trying to change the motion of a bowling ball and a ping-pong ball moving with equal velocities. Because of its greater mass, a greater force is required to decelerate the bowling ball at a given rate than is required for the ping-pong ball (fig. 4.8).

In effect, we are using Newton's second law to define mass. We can then devise experiments by which to compare masses, simply by arranging to have the same force act on different masses and measuring the resulting acceleration. By Newton's second law,

$$m_1 a_1 = m_2 a_2 = F.$$

Rearranging this equation, we get

$$\frac{m_2}{m_1} = \frac{a_1}{a_2}.$$

If we choose m_1 to be our standard—equal, say, to 1 kilogram—and then measure accelerations produced by the same force for m_1 and any other mass, m_2, this ratio could be used to compute the value of m_2 in kilograms.

Although in principle this procedure provides a way of comparing masses directly from Newton's second law, in practice it is not a convenient procedure. As you are probably aware, the more common method of comparing masses is to weigh the objects on a balance, or scale (fig. 4.9). What we are doing in this case is comparing the force of gravity acting on the object whose mass we wish to determine to that acting on some standard mass. The force of gravity acting on an object is what physicists commonly refer to as the *weight* of the object. It is a force and has different units (newtons) than those of mass (kilograms).

How is the force of gravity acting on an object related to the object's mass? From our discussion of the acceleration of gravity in the last chapter, we know that objects of different mass experience the same acceleration due to gravity near the earth's surface, namely $g = 9.8$ m/s^2. Assuming that this is the only force acting on the object, and using Newton's second law, we can write

$$\vec{F} = m\vec{a} = m\vec{g}.$$

The force of gravity, or weight of the object, must then be just

$$\vec{W} = m\vec{g},$$

where we have used the symbol W to represent weight.

Suppose, for example, that a woman has a mass of 50 kg. To find her weight, we multiply her mass by the acceleration of gravity:

$$W = mg = (50 \text{ kg})(9.8 \text{ m/s}^2) = 490 \text{ newtons}.$$

In the British system of units, weights are commonly expressed in pounds (lb), where 1 lb = 4.448 N. Dividing 490 N by this conversion factor yields

$$W = \frac{490 \text{ N}}{4.448 \text{ N/lb}} = 110 \text{ lb}.$$

Try finding your own weight in newtons and your mass in kg to make these quantities more meaningful. Will your weight in newtons be a larger or smaller number than your weight in pounds?

Notice that weight is proportional to mass, but it also depends on the acceleration of gravity, g. Since g varies slightly from place to place on the earth, and has a much lower value on the moon or on smaller planets, the weight of an object depends on where the object is located. The mass, on the other hand, is a property of the object related to the quantity of matter making up the object. The mass changes only when material is added to or subtracted from the object; it does not depend on the object's location. If we transported the woman we were just discussing to the moon, her weight would decrease to about 18 lb or 82 N due to the lower value of g, but her mass would still be about 50 kg (provided that the trip did not take too much out of her).

An interesting point arises from the relationship between mass and weight. Let us return to the case of a falling object and consider its motion from the perspective of Newton's second law. Reversing the argument given earlier, we can say that the force acting on the object is its weight, $W = mg$. From the second law, then,

$$\vec{F} = m\vec{g} = m\vec{a},$$

or

$$\vec{a} = \vec{g}.$$

Looking at the situation in this way, we see that the reason the acceleration due to gravity for different objects has the same value is that the mass cancels out of this equation. The force of gravity is proportional to the mass, but by Newton's second law, the resulting acceleration is inversely proportional to mass and these two effects cancel one another.

Force and acceleration are *not* the same thing, although they are closely related through Newton's second law. Objects of different mass experience different forces due to gravity (their weights), but the same acceleration (fig. 4.10). It is because the force of gravity is dependent upon mass, but mass also determines the rate of acceleration, that we find the same acceleration of gravity for different objects.

All of this depends, however, on the assumption that the mass that determines the size of the gravitational force is the same quantity as the *inertial* mass that goes into Newton's second law. Why should this be true? As far as New-

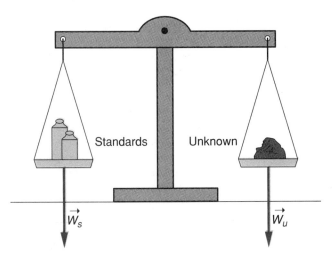

Figure 4.9 Comparing an unknown mass to standard masses on a balance.

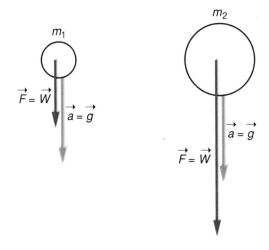

Figure 4.10 Different forces act upon falling objects of different mass, but because acceleration is inversely proportional to mass, the objects have the same acceleration.

ton's theory is concerned, there is no reason, except that things do seem to work this way. Is this a curious coincidence or a fundamental fact of nature? This was a question that bothered Newton as well as many who followed in his footsteps, including Albert Einstein. Einstein provided an answer to this question in his general theory of relativity, an answer that has far-ranging implications but takes us beyond the scope of this discussion.

4.4 NEWTON'S THIRD LAW

Where do forces come from? If you push on a chair to move it across the floor, does the chair also push back on you? If so, how does that push affect your motion? Questions like these are important to a complete understanding of what we mean by *force*. Newton's third law furnishes some of the answers.

Newton's third law of motion is a very important part of his definition of force and is an essential tool in analyzing the motion or lack of motion of real objects. It is frequently misunderstood, however, and its importance is easily missed. For this reason, this section provides a careful explanation of its meaning and use.

Statement of the Third Law

Newton's third law deals with the idea that forces are caused by the interaction of two objects, each exerting a force on the other. It can be stated as follows:

If object *A* exerts a force on object *B*, then there is an equal and oppositely directed force exerted by object *B* on object *A*.

The third law is sometimes referred to as the *action/reaction principle* (for every action, there is an equal but opposite reaction). The important point to recognize, which we have made explicit in the above statement, is that the two equal and opposite forces that we are discussing always act on two *different* objects, never on the same object. Newton's definition of force thus includes the notion of an *interaction* between objects.

To illustrate this principle, let us return to the example of the chair. If you exert a force, $\vec{F_1}$, on the chair with your hand in order to move it across the floor, the chair must push back on your hand with an equal but opposite force, $\vec{F_2}$ (fig. 4.11). You can, in fact, feel the effects of force $\vec{F_2}$ in the pressure on your hand; it partly determines your own motion, but it has nothing to do with the motion of the chair. Of this pair of forces, the only one that affects the motion of the chair is the one acting on the chair, $\vec{F_1}$.

Newton's third law can be stated in symbolic form for the above situation (or any other) as follows:

$$\vec{F_2} = -\vec{F_1}.$$

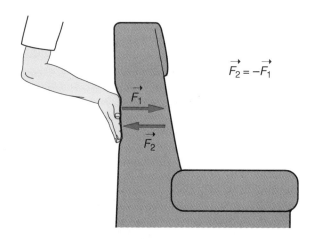

$$\vec{F_2} = -\vec{F_1}$$

Figure 4.11   Newton's third law at work in the interaction of a chair and a hand. The chair pushes back on the hand with a force $\vec{F_2}$ that is equal in magnitude, but opposite in direction to the force $\vec{F_1}$ exerted by the hand on the chair.

The minus sign indicates that the two forces are oppositely directed. Again, remember that these two forces act on different objects and are used to define the interaction between the two objects. That is the essence of the third law.

Our definition of force is now complete. Newton's second law tells us how the motion of a single object is affected by a force, and his third law tells us where forces come from, namely from interactions with other objects. With a suitable definition of mass, which also depends upon the second law, we know how to measure forces by the size of the acceleration that they produce on a given mass ($\vec{F} = m\vec{a}$).

Applications of the Third Law. Analysis of the motion or the lack of motion of an object must start with an identification of the various interactions (and hence forces) that the object is experiencing with other objects. As a simple example, consider the forces acting upon a book that is at rest on a table (fig. 4.12). The book experiences two significant forces. First, there is the downward pull of gravity (the book's weight, $\vec{W}$). The other object involved in that interaction is the earth itself; the book and the earth are attracted toward one another (via gravity) with equal and opposite forces that form a third-law pair. The earth pulls down on the book, and the book pulls up on the earth. The effect of this upward force on the earth is extremely small, of course, because of the enormous mass of the earth.

The second force acting on the book is an upward force exerted on the book by the table. (This force is often referred to as the *normal force*, where the word *normal* means "perpendicular" in this context rather than "usual." The normal force, $\vec{N}$, is always perpendicular to the surfaces of contact.) The book, in turn, exerts an equal but oppositely directed downward force on the table. These forces, that of the table on the book and of the book on the table, form another third-

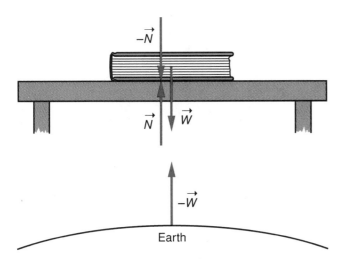

Figure 4.12 Two forces, $\vec{N}$ and $\vec{W}$, act upon a book resting on a table. The third-law reaction forces $-\vec{N}$ and $-\vec{W}$ act upon different objects.

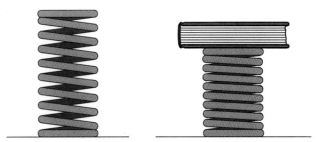

Figure 4.13 An uncompressed spring and the same spring supporting a book. The compressed spring exerts an upward force on the book.

Figure 4.14 A mule and a cart. Does Newton's third law prevent the mule from moving the cart?

Figure 4.15 The car pushes against the road, and the road, in turn, pushes against the car.

law pair. These forces result from the mutual compression of the book and table as they come into contact. You could think of the table as a large and very stiff spring that compresses ever so slightly when the book is placed on it (fig. 4.13).

It happens that the two forces acting on the book, that of gravity and that due to the table, are also equal and opposite to one another. How do we know this? We know that the acceleration of the book is zero, since it is remaining at rest. According to Newton's *second* law, then, the net force $\vec{N} + \vec{W}$ must also be zero, since $\Sigma \vec{F} = m\vec{a}$. The only way in which the total force can be zero is for the two contributing forces to cancel one another; they must be equal in magnitude and opposite in direction for their sum to be zero.

Even though equal and opposite in direction, however, these two forces do not constitute a third-law action/reaction pair, since they both act on the same object. As described above, each one is due to an interaction with other objects (the earth and the table). The fact that, in this case, they are equal and oppositely directed is a consequence of the second law, not the third. If these forces were not equal in magnitude, the book would have to be accelerating according to the second law.

As another example, consider the story of the stubborn mule who, having had brief exposure to physics, argued to his handler that there was no point in his pulling on the cart to which he was connected. According to Newton's third law, the mule argued, the harder he pulls on the cart, the harder the cart pulls back on him and the net result is nothing (fig. 4.14). Is he right, or is there a fallacy in his argument? The fallacy is simple but perhaps not obvious; only one of the two forces that he is talking about acts on the cart, and it alone (of these two forces) is related to the motion of the cart. Try to place yourself in the role of the handler and explain the fallacy to the mule (without resorting to violence).

The reaction force to a push or pull exerted by an object is often extremely important in describing the motion of the object itself. To illustrate this claim, we need only consider the acceleration of a car. The engine does not push the car because it is part of the car. The engine drives the rear (or front) axle of the car, which causes the tires to rotate. They, in turn, push against the road surface via the force of friction between the tires and the road (fig. 4.15). According to Newton's third law, the road must then push against the tires with an equal but oppositely-directed force, and this is the external force that causes the car to accelerate. Obviously, friction is desirable in this case; without friction, the tires would spin, and the car would go nowhere.

Similar arguments could be made for the processes of walking or running (see question 4.15 at the end of the chapter). Think about it the next time you find yourself walking. What external force causes you to accelerate as you start out? What is your role in producing that force?

To figure out what forces are acting on an object, then, we need to identify the other objects that are interacting with the object in question. Some of these forces will be obvious; any object in direct contact with the object of interest will presumably contribute a force. Others, such as air resistance or gravity, may be less obvious but still easy to identify. The third law is the principle underlying the identification of any of these forces.

4.5 APPLICATIONS OF NEWTON'S LAWS

We have stated Newton's laws of motion and discussed the definitions of force and mass that are contained in these laws. As they stand, they form a self-consistent basis for explaining how and why objects move or fail to move. We cannot begin to appreciate their utility, however, without testing them in the context of situations we are familiar with. Do they provide a satisfactory picture of what is going on? The following examples help us answer that question.

Pushing a Chair

What about the chair being pushed across the tile floor that we described at the beginning of this chapter? How would we analyze this example in terms of Newton's laws? The first step is to identify the forces that act upon the chair. As shown in figure 4.16, four forces act on the chair, which are due to four separate interactions:

1. The force of gravity (weight) due to interaction with the earth, $\vec{W}$
2. The upward (normal) force exerted by the floor due to compression, $\vec{N}$
3. The force exerted by the hand of the pusher, $\vec{P}$
4. The frictional force exerted by the floor, $\vec{f}$

Two of these forces ($\vec{N}$ and $\vec{f}$) are actually due to interaction with a single entity, the floor, but since they are due to different effects and are perpendicular to one another, they are usually treated separately.

Just as with the book on the table discussed in section 4.4, two of these forces, $\vec{W}$ and $\vec{N}$, cancel one another. Again, we know this because it is clear from the nature of its motion that the chair is not being accelerated in the vertical direction. The sum of the vertical forces must therefore be zero, by Newton's second law. The other two forces, the push of the hand, $\vec{P}$, and the frictional force, $\vec{f}$, do not necessarily cancel, however. These two forces together determine the

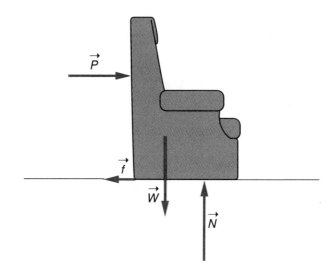

Figure 4.16 Four forces act upon a chair being pushed across the floor.

horizontal acceleration of the chair, according to Newton's second law.

In the most likely scenario for moving the chair, you first exert a push with your hand that is larger than the frictional force, thereby producing a net horizontal force ($P - f$) in the forward direction that accelerates the chair. Once you have accelerated it to a comfortable velocity, you reduce the strength of $\vec{P}$ so that it just equals the frictional force. In this condition, the net horizontal force is zero and (by the second law) the acceleration is also zero. If you sustain the push at this level, the chair moves across the floor at a constant velocity. Finally, you remove your hand (and $\vec{P}$), and the chair quickly decelerates to zero velocity under the influence of $\vec{f}$, the frictional force. If you happen to have a chair and a smooth floor handy, try to produce this motion and see if you can feel differences in the force that you are exerting with your hand at various points in the motion.

The strength of the push that you need to move the chair is largely determined, then, by the strength of the opposing frictional force. This in turn is influenced by the weight of the chair and the condition of the floor surface. If the floor has been recently waxed, the push required may be considerably less than that required with an unwaxed floor. You might also find, however, that the chair goes farther than you expected after you remove the push. (Frictional forces can be useful.)

Notice that you need to apply a continuous force in order to keep the chair moving with constant velocity because of the presence of the frictional force. If you failed to recognize the importance of the frictional force, you might be led, like Aristotle, to think that a force is always needed to keep an object moving. Frictional forces are almost always present, but are not as obvious to us as the forces that we apply directly.

Skydiving

As discussed in the previous chapter as well as in the section on mass and weight in this chapter, objects released near the earth's surface fall with a constant acceleration, *g*, provided that we ignore the effects of air resistance. The force producing this acceleration is the gravitational attraction exerted on the object by the earth, $\vec{W} = m\vec{g}$, which is the weight of the object. Do objects falling over large distances just continue to accelerate at this rate, gaining larger and larger downward velocities? A sky diver should be able to answer this question.

A falling object for which the forces of air resistance are negligible experiences just the single force due to gravity and does indeed continue to accelerate. It turns out, however, that the effects of air resistance get larger as the velocity of the object increases. The sky diver initially has an acceleration of *g*, but as her velocity increases, the force of air resistance becomes significant, and the acceleration decreases.

This situation is illustrated in figure 4.17. For small velocities, the force of air resistance is very small, and we can ignore it. As the velocity increases, the air-resistance force $\vec{R}$ gets larger, causing the total downward force $(W - R)$ to be less than $\vec{W}$. Ultimately, as the velocity of the object continues to increase, the force of air resistance reaches a value equal to the force of gravity. The net force in this case is then zero, and the object is no longer accelerated. We say that it has reached its *terminal velocity.*

Sky divers or any other objects that fall a long distance through the atmosphere do reach a terminal velocity (between 150 and 200 km/hr for a sky diver, depending upon body profile and clothing). Although this terminology has a certain ominous ring to it, there is little danger, provided that the parachute works. Once the cord is pulled and the parachute extended, the air resistance is greatly increased, and the sky diver decelerates to a much smaller terminal velocity. She decelerates again (somewhat more abruptly) on contact with the ground. The upward force of the ground on the person is the cause of the final deceleration.

Here again frictional or resistive forces play an important role in a complete analysis of the motion. Aristotle did not have the opportunity to try skydiving (nor have many of us), so this example was not a part of his experience. He probably did observe the terminal-velocity effect, however, with very light objects such as feathers or leaves. The weight of such objects is small, and the surface area is relatively large; and hence the air-resistance force $\vec{R}$ becomes equal to the weight much more readily. Try tearing a small corner from a piece of paper and watching it fall. Does it appear to reach a constant (terminal) velocity? You can see why Aristotle concluded that heavier objects fall faster than light objects.

Throwing a Ball

Since Aristotle had trouble explaining the motion of a thrown object or projectile such as a ball by using his theory, we should consider this example from a Newtonian perspective. Do we need a force to keep the ball moving? Not according to Newton's first law. Three forces are involved in the flight of the ball, however: the initial push exerted by the thrower on the ball, the force of gravity, and (once again) air resistance (fig. 4.18).

With Newton's framework, it is most convenient to break the motion down into two different periods of time. The first is that of the throwing process itself, during which the hand is in contact with the ball. The force exerted by the hand on the ball, $\vec{P},$ dominates the process during this time interval; the other forces can be ignored because they are much smaller. The force $\vec{P}$ accelerates the object to a velocity that we often refer to as the *initial* velocity. The magnitude and direction of $\vec{P}$ and the length of time that it is in effect determine the magnitude and direction of the initial velocity. Since this force usually varies with time, a full analysis of the throwing process can be quite complex.

Once the ball has left the hand, however, we are in the second time period, in which the forces of gravity and air resistance are the primary causes of changes in the ball's velocity. From this point on, the problem becomes one of projectile motion, which we discussed in chapter 3. The gravitational force accelerates the ball downwards, and the

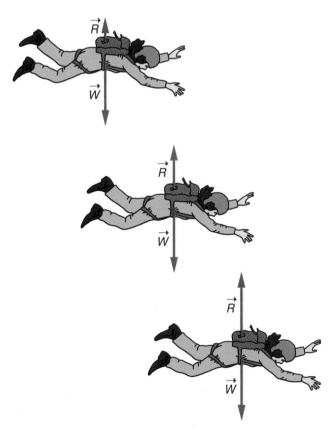

Figure 4.17   The force of air resistance acting upon a sky diver increases as the velocity increases.

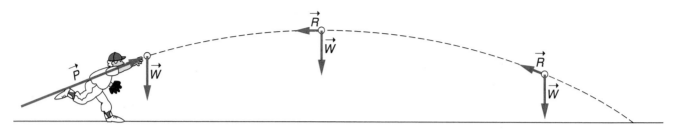

Figure 4.18   The forces acting upon a thrown ball at different points in its path.

Figure 4.19   Two connected carts being accelerated by a force, $\vec{F}$, applied via a string.

force of air resistance acts in a direction opposite to the velocity, gradually decelerating the ball.

Contrary to Aristotle's viewpoint, no forces are needed to keep the object moving once it has been thrown. In fact, if the object were thrown in space, where air resistance and gravitational forces are negligible, it would simply keep moving in a straight line with constant velocity as stated in Newton's first law. (This is why it pays to be careful with your tools when engaged in extravehicular activity in space.)

Connected Carts

Because the force of air resistance or the push exerted by a person throwing an object often vary with time, we have avoided trying to work out numerical examples for these situations. We can identify the forces involved (and their causes due to third-law interaction with other objects) and provide a satisfactory qualitative picture of what is happening. (What is the other object involved for the force of air resistance?) With more time and mathematics, satisfactory numerical predictions can be made for all of these examples.

Initial verification of Newton's laws, however, and of the methods of analysis that we have been discussing, came from simpler examples that can be set up in the laboratory. (The value of keeping things simple in testing ideas was apparent to Newton and Galileo, just as it is to most of us.) Numerical tests and predictions are almost always more convincing than descriptive arguments.

One situation that is easy enough to set up in a physics laboratory (or even on the floor wherever you may be) involves two connected carts being accelerated by the pull of a string (fig. 4.19). Insights regarding both Newton's second and third laws can be gained from this example. In the spirit of keeping things simple, we will assume that the carts have

excellent wheel bearings, so that they roll with very little friction. We will also assume that we have balances available, so that we can readily measure the masses of the two carts.

Suppose that the mass, m_1, of the first cart with its load is 10 kg and that of the second cart m_2, is 8 kg, so that the total mass of the two-cart system is 18 kg. If we apply a horizontal force of 36 N (chosen to keep the numbers simple) via the string, we can easily predict what the acceleration of the two-cart system will be from Newton's second law:

$$\vec{a} = \frac{\Sigma \vec{F}}{m} \; ;$$

$$a = \frac{36\ \text{N}}{18\ \text{kg}} = 2.0\ \text{m/s}^2.$$

If the force $\vec{F}$ is kept constant, the acceleration will also be constant, and we could verify this prediction by measuring the displacement for a given time, t, and using the constant acceleration formulas developed in chapter 3 to compute a. The force could be measured by inserting a small spring balance between the first cart and the string. (These are often found in physics labs, but are unlikely to be available in your kitchen.)

The acceleration that we have just calculated is that of the entire system, but it is also that of each cart separately, since they move together. What is the force required to accelerate just the second cart? By Newton's second law,

$$F = m_2 a = (8\ \text{kg})(2.0\ \text{m/s}^2) = 16\ \text{N}.$$

We could verify this value by placing a second spring balance between the two carts (provided its mass is negligible compared to that of the carts), but we could also check it by considering the acceleration of the first cart. The net force acting on the first cart should be

$$\Sigma F = m_1 a = (10\ \text{kg})(2.0\ \text{m/s}^2) = 20\ \text{N}.$$

By Newton's third law, the force pulling back on the first cart due to its connection to the second cart should be −16 N, since it must be equal but opposite in direction to the force exerted on the second cart by the first cart (fig. 4.20).

The net force acting on the first cart is then

$$\Sigma F = 36\ \text{N} - 16\ \text{N} = 20\ \text{N},$$

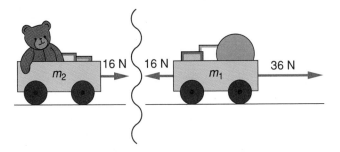

Figure 4.20 **The interaction between the two carts pictured in figure 4.19. How is Newton's third law involved here?**

which is in agreement with that calculated from the acceleration. We can see that our analysis using Newton's laws pro-

vides a consistent picture for both carts. Notice that a larger (20 N) net force is required to accelerate the larger mass, m_1, than that required (16 N) to accelerate the smaller mass, m_2.

We could try many variations on the above theme in the laboratory to see if the results agreed with our predictions from Newton's laws. We should realize, however, that even with careful experimental technique and the use of precise stopwatches and balances, our experimental results are unlikely to agree exactly with calculations because of the subtle effects of friction and our inability to make measurements with infinite precision. The art of the experimentalist is to reduce these inaccuracies to a minimum as well as to predict what their effects will be. Newton's laws have been verified many times over by careful experiments such as these.

Box 4.2

Everyday Phenomenon:
Riding an Elevator

The Situation. We have all probably had the experience of riding an elevator and feeling sensations of heaviness or lightness as the elevator accelerates upward or downward. The feeling of lightness as the elevator accelerates downward is usually the most striking, particularly if the acceleration is not very smooth.

Do we really weigh more or less than normal in these situations? If you took a bathroom scale into the elevator, would it read your true weight while the elevator was accelerating? Our challenge here is to see how we can carefully apply Newton's laws of motion to address these questions.

The Analysis. The first step in analyzing any situation from the perspective of Newton's laws is to isolate the body in question and to carefully identify the forces that act on just that body. Different choices are possible for which body or objects to isolate; some choices will be more productive than others. In this case, it makes sense to isolate the person on the scale, since her weight is the focus of our concern. The drawing on the next page shows a *free-body* diagram of the woman that indicates just those forces that act upon her.

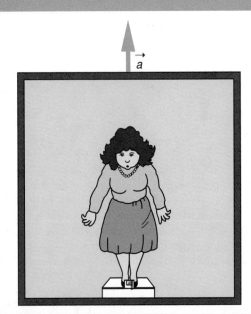

A woman standing on a bathroom scale inside an accelerating elevator. Will she read her true weight on the scale?

Continued

Box 4.2 Continued

In this case, just two other objects interact with the woman; thus there are two forces. The earth pulls downward on the woman via the force of gravity. By our earlier definition of weight (section 4.3), this force is the weight of the woman, $\vec{W}$. The other object interacting with the woman is the scale, which exerts an upward force, $\vec{N}$, upon her feet. The vector sum of these two forces will determine the acceleration of the woman.

If the elevator is accelerating upward with the given acceleration $\vec{a}$, then the woman must also be accelerating upward at the same rate. By Newton's second law, the net force acting upon the woman must be equal to her mass times her acceleration. Since this acceleration is directed upward, the net force must also be directed upward, and this means that $\vec{N}$ must be larger than $\vec{W}$. Using signs to indicate direction and letting the positive direction be upward, we have the following:

$$\Sigma F = N - W = ma.$$

What about the scale reading? By Newton's third law, the woman must exert a downward force on the scale equal to N in magnitude but opposite in direction. Since this is the force pushing downward on the scale, this should be the reading on the scale dial. (Actually, the reading may be slightly larger than N because the force exerted on the bottom of the scale by the elevator must be slightly larger than N in order for the scale to accelerate upward. A detailed analysis would require that we take the scale apart and determine the forces acting upon the spring in the scale. If the masses of the spring and footplate are small, however, we should get a reading close to N in value.)

The woman's true weight has not changed, but her apparent weight, as indicated by the scale reading, has increased by an amount equal to ma ($N = W + ma$). Suppose the elevator is accelerating downward; what happens then? In that case, the net force acting upon the woman must also be downward, and that requires $\vec{W}$ to be larger than $\vec{N}$. The scale reading, being approximately equal to N, will then be less than the woman's true weight, presumably producing a smile rather than a scowl.

If the elevator cable breaks, we have a special case of particular interest. In this situation, the downward acceleration of the elevator (and the woman) will be equal to $\vec{g}$, the acceleration of gravity. The net force required to produce an acceleration of $\vec{g}$ for the woman is just her true weight, $\vec{F} = m\,\vec{g} = \vec{W}$. The force $\vec{N}$ will therefore be zero, and the woman will likewise exert no downward force on the scale, which will read zero. She will be in a condition of apparent weightlessness!

The sensation of weight is produced, in part, by the pressure upon our feet and the forces in our muscles that are required to maintain our posture. The woman will feel weightless in the situation just described even though her true weight (the force of gravity pulling her downward) has not changed. In fact, she will be able to float around in the elevator just as astronauts do in the space shuttle; the shuttle is also falling towards the earth as it orbits. This happy scenario will come to a crashing halt for the woman in the elevator, however, when the elevator reaches the bottom of the shaft. The shuttle, on the other hand, remains in orbit because of its lateral motion, despite the fact that it is being continually pulled toward the earth.

A free-body diagram of the woman in the elevator. Why is the force $\vec{N}$ drawn as being larger than the force $\vec{W}$?

SUMMARY

In 1685, Isaac Newton published his Principia, in which his three laws of motion were introduced and applied. These laws form the backbone of Newton's theory of motion, which has served as an extremely useful model for explaining the causes of motion and predicting how objects will move. The basic features of Newton's theory are summarized below.

Historical Background: Aristotle, Galileo, and Newton. Newton's theory was constructed upon groundwork laid by Galileo and replaced the much earlier and less quantitative theory developed by Aristotle to explain motion. Newton's theory had much greater predictive power than that of Aristotle's, and although we now recognize limits to its validity, it is still used extensively for explaining motion.

Newton's First and Second Laws. Newton's second law states that the acceleration of an object is proportional to the net external force acting upon the object and inversely proportional to the mass of the object. The first law is a special case of the second and describes what happens when the net force (and therefore the acceleration) is zero.

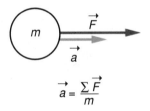

$$\vec{a} = \frac{\Sigma \vec{F}}{m}$$

Mass and Weight. Newton's second law defines inertial mass as the property of an object that causes it to resist a change in its motion. The weight of an object is the force of gravity acting upon the object and is equal to the mass times the acceleration of gravity.

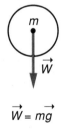

$$\vec{W} = m\vec{g}$$

Newton's Third Law. Newton's third law completes the definition of force by indicating that forces result from the interaction between objects. If object *A* exerts a force on object *B*, then object *B* exerts an equal but oppositely directed force upon object *A*.

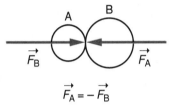

$$\vec{F}_A = -\vec{F}_B$$

Applications of Newton's Laws. In analyzing the motion of an object using Newton's laws, the first step is to identify the forces that act upon the object in question via interactions with other objects. The strength and direction of these forces determine how the object's motion will change. These ideas were illustrated in explaining the motion of a chair being pushed across the floor, a thrown ball, a sky diver, connected carts, and a person in an elevator.

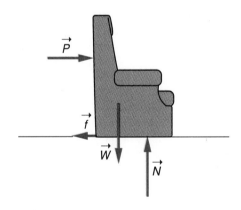

QUESTIONS

The following questions are designed to help you test your understanding of Newton's laws of motion and the concepts of force, mass, and weight.

Q4.1 Two equal forces act on two different objects, one of which has a mass 10 times as large as the other. Will the more massive object have an equal acceleration, a larger acceleration, or a smaller acceleration than the less massive object? Explain.

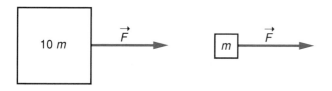

Q4.2 A 2-kg block is observed to accelerate at a rate twice that of a 4-kg block. Is the net force acting on the 2-kg block therefore twice as large as that acting on the 4-kg block? Explain.

Q4.3 Two equal-magnitude horizontal forces act on a box as shown in the diagram below.

   a. Is the object accelerated in the horizontal direction? Explain.

   b. Is it possible that the object is moving, given the fact that the two forces are equal in magnitude? Explain.

Q4.4 Two forces are acting on an object in the directions pictured in the diagram below.

   a. Using a diagram, indicate the direction of the total, or *net,* force acting on the object.

   b. Will the object be accelerated? Explain.

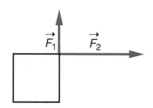

Q4.5 An object moving horizontally across a table is observed to slow down.

   a. Is there a net force acting on the object? Explain.

   b. Is the force of gravity involved in accelerating or decelerating the object in this situation? Explain.

Q4.6 A car changes direction as it rounds a curve with constant speed.

   a. Is the acceleration zero in this process? Explain.

   b. Is there a nonzero net force acting on the car in this case? Explain.

Q4.7 A ball hangs from a string attached to the ceiling, as shown in the diagram.

   a. What forces act on the ball? How many are there?

   b. What is the total (net) force acting on the ball?

   c. For each force listed in your answer to (*a*), what is the reaction force given by Newton's third law?

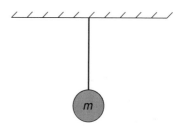

Q4.8 The acceleration of gravity on the moon is approximately one-sixth of the value that it has near the earth's surface (9.8 m/s^2).

   a. If a ball is transported to the moon, will its mass change? Explain.

   b. Will the weight of the ball change in this process? Explain.

Q4.9 Two blocks with the same mass are connected by a string and are being pulled across a frictionless surface by a constant force, $\vec{F}$, exerted by another string (see diagram).

   a. Will the two blocks move with constant velocity? Explain.

   b. Will the tension in the connecting string be greater than, less than, or equal to the force F? Explain.

   c. Which block, if either, will have the greater total force acting upon it? Explain.

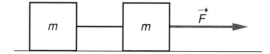

Q4.10 Two equal masses, connected by a string, are placed upon a frictionless pulley as shown.

a. Will the system accelerate? Explain.

b. Will the tension in the connecting string be greater than, less than, or equal to the weight of either mass? Explain.

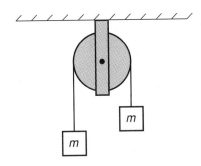

Q4.11 Suppose that a sky diver wears a specially lubricated superfly suit that reduces air resistance to a small constant force that does not increase as the velocity increases. Will this sky diver ever reach a terminal velocity before opening his chute? Explain.

Q4.12 A large crate sits at rest on the floor.

a. Is there a net horizontal force acting on the crate? Explain.

b. What two vertical forces act on the crate? Do these two forces constitute an action/reaction pair as defined in Newton's third law? Explain.

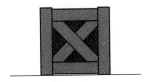

Q4.13 The engine of a car is internal to the car and therefore cannot push directly on the car in order to accelerate it.

a. What external force on the car is actually responsible for its acceleration? Explain.

b. Why is it difficult to accelerate (or decelerate) a car on an icy road?

Q4.14 Two identical cans are filled with different substances: one with lead shot and the other with feathers. The cans are dropped by a student standing on a chair.

a. Which can, if either, experiences the greater force due to the gravitational attraction of the earth? Explain.

b. Which can, if either, experiences the greater acceleration due to the gravitational attraction of the earth? Explain.

c. Which can, if either, would require the greater force to decelerate it after it has fallen a certain distance? Explain.

Q4.15 A sprinter accelerates at the beginning of a 100-meter race and then attempts to maintain maximum speed throughout the rest of the race.

a. What external force is responsible for accelerating the runner at the beginning of the race. Explain carefully how this force is produced.

b. Once the runner reaches her maximum velocity, is it necessary to continue pushing against the track in order to maintain that velocity? Explain.

Q4.16 Suppose that a bullet is fired from a rifle out in space where there are no appreciable forces of gravity or air resistance. Will the bullet slow down as it gets farther away from the rifle? Explain.

EXERCISES

The following exercises are designed to provide numerical experience with Newton's second law in simple situations. Working these exercises should build your confidence in your understanding of the second law.

E4.1 A single force of 15 N acts upon a 10-kg block. What is the magnitude of the acceleration of the block?

E4.2 A ball with a mass of 3 kg is observed to accelerate at a rate of 5 m/s². What is the net force acting on this ball?

E4.3 Two forces, one of 50 N and the other of 40 N, act in opposite directions on a box as shown. What is the mass of the box if it is observed to accelerate at a rate of 2.0 m/s² under the influence of these two forces?

E4.4 A 4-kg block being pulled across the table by a horizontal force of 30 N is retarded by a frictional force of 10 N. What is the acceleration of the block?

E4.5 A 2-kg block is acted upon by three horizontal forces as shown in the diagram (10 N, 5 N, and 25 N respectively).

a. What is the net horizontal force acting on the block?

b. What is the horizontal acceleration of the block?

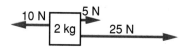

E4.6 What is the weight of a 20-kg mass?

E4.7 Joe has a weight of 150 lb.

a. What is his weight in newtons? (1 lb = 4.45 N)

b. What is his mass in kilograms?

E4.8 The author of this text has a weight of 600 N.

a. What is his mass in kilograms?

b. What is his weight in pounds?

E4.9 At a given instant in time, a 10-kg rock that has been dropped from a high cliff experiences a force of air resistance of 40 N. What are the magnitude and direction of the acceleration of the rock? (Do not forget the force of gravity!)

E4.10 At a given instant, a 5-kg rock is observed to be falling with an acceleration of 7.0 m/s². What is the magnitude of the force of air resistance acting upon the rock at this instant?

E4.11 A 0.5-kg book rests on a table. A downward force of 3 N is exerted on the book by a hand resting on it. What is the upward force exerted on the book by the table? (Do not forget the weight of the book. Is the book being accelerated?)

E4.12 An upward force of 20 N is applied via a string to lift a ball with a mass of 1 kg.

a. What is the net force acting upon the ball?

b. What is the acceleration of the ball?

CHALLENGE PROBLEMS

CP4.1 A constant horizontal force of 20 N is exerted by a string attached to a 5-kg block as shown. The block also experiences a frictional force of 8 N due to its contact with the table.

a. What is the horizontal acceleration of the block?

b. If the block starts from rest, what will its velocity be after 2 seconds?

c. How far will it travel in 2 seconds?

d. Must the forces acting on the block be constant in order for your answers to (b) and (c) to be valid? Explain.

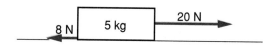

CP4.2 A rope exerts a constant horizontal force of 150 N to pull a 30-kg crate across the floor. The velocity of the crate is observed to increase from 1 m/s to 7 m/s in a time of 2 seconds under the influence of this force.

a. What is the acceleration of the crate?

b. What is the net force acting upon the crate?

c. What is the magnitude of the frictional force acting upon the crate?

d. What force would have to be applied to the crate via the rope in order for it to move with constant velocity? Explain.

CP4.3 Two blocks, tied together by a horizontal string are being pulled across a table by a horizontal force of 50 N. The

2-kg block has a 6-N frictional force exerted on it by the table, and the 4-kg block has an 8-N frictional force exerted upon it.

a. What is the net force acting upon the two-block system?

b. What is the acceleration of the system?

c. What force is exerted on the 2-kg block by the connecting string? (Consider only the forces acting upon this block.)

d. Does your answer to (c) provide the correct acceleration for the 4-kg block also? (This serves as a check on your work.)

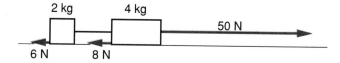

CP4.4 An 80-kg man is in an elevator that is accelerating downward at the rate of 2 m/s².

a. What is the true weight of the man in newtons?

b. What is the net force acting upon the man?

c. What is the force exerted on the man's feet by the floor of the elevator?

d. What is the apparent weight of the man in newtons?

e. How would your answers to (a) through (d) change if the elevator were accelerating upward at the rate of 2 m/s²? Explain.

HOME EXPERIMENTS AND OBSERVATIONS

HE4.1 Collect a variety of small objects such as coins, pencils, erasers, and keys. Ice cubes, if they are available, will also be excellent test objects. Try sliding these objects across a smooth surface such as a desk or table top, being as consistent as possible in the initial velocity that you give to them.

a. Do the objects slide the same distance after they leave your hand? What differences are apparent, and how are they related to the nature of the surfaces of the objects? Which come closest to demonstrating Newton's first law of motion?

b. Try placing a weight, such as a book, on top of a few ice cubes and sliding it across the table. Does this arrangement come closer to showing first-law behavior than the book or the ice cubes by themselves?

c. Can you think of other ways of reducing frictional forces so that you could approximate first-law behavior for sliding objects. Give them a try!

HE4.2 Tie a string to a rubber band, and fix the other end of the string to your textbook or a similar object. Looping the other end of the rubber band over your finger, try pulling the book across a smooth surface such as a desk top, counter top, or table. The amount of stretch of the rubber band will give you a rough idea of the strength of the force you are applying.

a. Try pulling the book with constant velocity across the surface. Does the force required depend upon the velocity?

b. Try accelerating the book at different rates. How does the required force compare to that for constant velocity?

c. Try adding mass to the book and repeating (*a*) and (*b*). How are the results affected?

d. Experiment with different objects. How does the nature of the surfaces affect the required forces?

HE4.3 Falling objects whose surface area is large compared to their weight will reach terminal velocity more readily than a ball or a book.

a. Fill a balloon with air and, standing in an area with minimal air currents, drop the balloon. Does it appear to fall with constant velocity?

b. Try other objects, such as small pieces of paper, plant parts (leaves, flowers, or seeds), or whatever you think might work. Do these objects appear to fall with constant velocity? How does the rate of fall compare for different objects when they are dropped simultaneously?

c. Which of the various objects tested produces the clearest demonstration of terminal velocity?

HE4.4 Use an elevator in your dormitory or other building to observe the effects of acceleration. Most elevators accelerate relatively quickly as they start and again as they are stopping (decelerating).

a. Note the sensation of change in weight as the elevator accelerates. If you have access to a bathroom scale, see how much your apparent weight differs from your true weight when the elevator is accelerating. Could you estimate the rate of acceleration from this information?

b. Try holding your arm out away from your body and maintaining it in this position as the elevator accelerates. It is almost impossible to do so if the elevator accelerates quickly. Why?

5 Circular Motion, the Planets, and Gravity

"The car failed to negotiate the curve." How many times have you seen a phrase like that in an accident report in the newspapers? Either the road surface was slippery, or the driver was simply driving too fast for the degree of curvature. In either case, poor judgment and probably a poor feeling for the physics of the situation were at work (fig. 5.1).

When a car goes around a curve, the direction of its velocity changes. A change in velocity implies acceleration, and by Newton's second law, an acceleration requires a force. The situation has a lot in common with that of a ball being twirled in a circle at the end of a string (and many other examples of circular motion).

What forces are involved in keeping a car moving around a curve? How does the required force depend upon the speed of the car and the degree of curvature of the road? What other factors are involved? Finally, what does a car rounding a curve have in common with a ball on a string or the motion of the planets about the sun?

The motion of the planets about the sun and of the moon about the earth played an important role in the development of Newton's theory of mechanics. Newton's law of universal gravitation is the final piece of that theory; gravity is the force that keeps the planets and the moon moving in curved paths. Circular motion is therefore a very important special case of motion in two dimensions, both in the history of physics and in our everyday experience.

Figure 5.1   This car failed to negotiate a curve: Newton's first law at work.

Chapter Objectives

Using the example of a ball on a string, we will first examine the acceleration involved in changing the direction of the velocity in circular motion (centripetal acceleration). Our purpose will be to see how this acceleration depends upon the velocity and the radius of the curve. We will then consider the forces involved in producing a centripetal acceleration in different cases; understanding the factors involved in a car negotiating a curve will be a key objective.

Planetary motion is also a case of particular interest. Newton's law of universal gravitation will be introduced to explain the motion of the planets, but we will also see how this gravitational force is related to the weight of an object and the acceleration of gravity near the earth's surface.

Chapter Outline

❶ *Centripetal acceleration.* What is the acceleration that is required to change the direction of a velocity? How does the magnitude of this acceleration depend upon the speed of the object and the radius of the curve?

❷ *Centripetal forces.* What types of forces are involved in producing centripetal accelerations in different situations? In particular, what are the forces involved for a car moving around a curve?

❸ *Planetary motion.* What is the nature of the motion of the planets about the sun? How has our understanding of planetary motion developed historically?

❹ *Newton's law of universal gravitation.* What is the fundamental nature of the gravitational force, according to Newton? How does this force help to explain planetary motion, and how is it related to the force and acceleration of gravity near the earth's surface?

❺ *The moon and other satellites.* What is the nature of the moon's orbit? How do the orbits of artificial satellites differ from that of the moon and from each other? What does the moon have to do with the tidal cycles?

5.1 CENTRIPETAL ACCELERATION

Suppose that we attach a ball to a string and twirl the ball in a horizontal circle (fig. 5.2). With a little practice, it is not hard to keep the ball moving with a constant speed, but of course the direction of its velocity is continually changing. A change in velocity implies an acceleration, but just what is the nature of the change in velocity (and the related acceleration) that occurs in this process?

The key to analyzing this situation involves taking a careful look at what is happening to the velocity vector as the ball moves in its circle. In the preceding chapters we have been preparing you for this moment by repeatedly mentioning that the velocity can change by changing its direction, even if its magnitude (the speed) remains constant. The time has now come to take a close look at the actual magnitude and direction of this change in velocity.

The Change in Velocity, $\Delta \vec{V}$

Figure 5.3 presents the situation as seen from above, with the ball moving in a horizontal circle. The velocity vectors are drawn on the circle at two different positions separated by a very short time interval. In other words, the velocity $\vec{v_2}$ is that which occurs a short time after the velocity $\vec{v_1}$ as the ball moves in a counterclockwise direction around the circle. These two vectors are drawn with the same length, indicating that the speed of the ball is unchanged.

Obviously, the direction of the velocity vector has changed in this short time interval. The change in velocity, $\Delta \vec{v}$, that this represents can be found by subtracting the initial velocity, $\vec{v_1}$, from the later velocity, $\vec{v_2}$. This is shown also in figure 5.3; subtraction is accomplished by adding $-\vec{v_1}$ to $\vec{v_2}$ in the usual head-to-toe manner (see appendix C).

Another way of looking at this process is to notice that if $\Delta \vec{v} = \vec{v_2} - \vec{v_1}$, then $\vec{v_2} = \vec{v_1} + \Delta \vec{v}$. This is also illustrated in the vector triangle showing these three vectors in figure 5.3; $\Delta \vec{v}$ adds to the original velocity, $\vec{v_1}$, to yield the new velocity, $\vec{v_2}$.

Notice that the vector $\Delta \vec{v}$ has a direction that is different from either of the velocity vectors themselves. If we choose a small enough time interval between the two positions, then $\Delta \vec{v}$ points toward the center of the circle. This then is the direction of the *change* in velocity, which is also the direction of the instantaneous acceleration of the ball. (Acceleration is defined as $\Delta \vec{v} / \Delta t$ and thus has the same direction as the vector $\Delta \vec{v}$.) The ball is therefore being accelerated towards the center of the circle in this process.

But how large is this acceleration, and how does it depend upon the speed of the ball and the radius of the curve? We can use the vector diagrams to explore these questions. For example, a larger speed means that the velocity vectors will be longer and that the ball will have traveled a greater distance around the circle in a given time interval. Both of these factors increase the size of the difference vector, $\Delta \vec{v}$, as is shown in figure 5.4. In fact, doubling the length of the velocity vector *and* thus doubling the distance that the ball has traveled in time interval Δt have the combined effect of quadrupling the length of the vector $\Delta \vec{v}$. This suggests that the acceleration will be proportional to the *square* of the speed.

The Radius Effect

Changing the radius of the curve also has an effect. Figure 5.5 shows two circles of different radii, but the speed of the ball is the same for the two cases, so that the distance traveled and the length of the velocity vectors are the same. Obviously, the length of the vector $\Delta \vec{v}$ is greater for the circle with the smaller radius. Since a larger radius produces a smaller change in velocity, this suggests that the acceleration

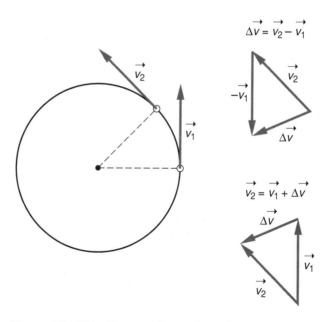

$$\Delta \vec{v} = \vec{v_2} - \vec{v_1}$$

$$\vec{v_2} = \vec{v_1} + \Delta \vec{v}$$

Figure 5.3 This diagram shows the velocity vectors for two positions of a ball moving in a horizontal circle; $\Delta \vec{v}$ is the change in velocity between these two positions.

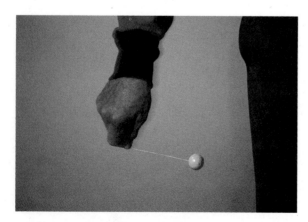

Figure 5.2 A ball being twirled in a horizontal circle. Is the ball accelerated?

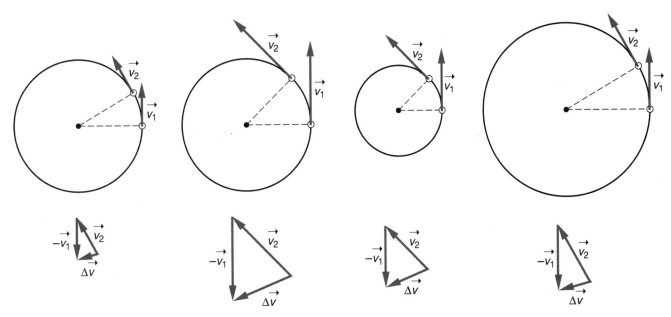

Figure 5.4 **Doubling the speed of the ball quadruples the magnitude of the change in velocity; $\Delta \vec{v}$, as shown in these two diagrams.**

Figure 5.5 **Increasing the radius of the circle decreases the magnitude of the change in velocity $\Delta \vec{v}$, as shown in these vector diagrams.**

should be *inversely proportional* to the radius of the circle. (Other relationships might be possible, but a mathematical analysis yields the suggested inverse proportion.)

Our vector diagrams suggest, then, that the acceleration involved in changing the direction of the velocity vector should be proportional to the square of the speed and inversely proportional to the radius of the curve. In symbols, this can be expressed as follows:

$$a_c = \frac{v^2}{r},$$

where a_c stands for *centripetal acceleration,* v is the speed of the ball (the magnitude of the velocity), and r is the radius of the circle.

> **Centripetal acceleration is the rate of change in velocity of an object that is associated with the change in *direction* of the velocity. It is always perpendicular to the velocity itself and towards the center of the curve.**

The ball moving in a circle is therefore accelerated, even though its speed remains constant. This conclusion follows directly from our definition of acceleration as the rate of change in velocity. To change the direction of the velocity vector is to change the velocity, therefore an acceleration is involved. People often resist this idea, however, because we use the term *acceleration* in everyday language to refer to situations in which speed is increasing without considering the effects of change in direction. The more general concept of acceleration needed for Newton's theory of mechanics and

his second law of motion, however, includes the effects of change in direction.

According to Newton's second law of motion, this centripetal acceleration must be caused by a force acting on the ball in the direction of the acceleration. In the case of the ball on the string, this force is provided by the tension in the string, which is attached to the ball. The string pulls the ball towards the center of the circle, causing the direction of the velocity vector to continually change. Using the second law, the magnitude of the required force can be found simply by multiplying the mass of the object times the centripetal acceleration, $F = ma_c$.

What would happen if this force were not present? According to Newton's first law of motion, an object will continue to move in a straight line with constant speed if there is no net force acting on it. If the string breaks, or if we let go of the string, this is exactly what will happen; the ball will fly off in the direction in which it was traveling at the instant the string broke (fig. 5.6). Without the force due to the string pulling on it, the ball will move in a straight line as seen from above. (It will also be pulled downward, of course, by the force of gravity.)

In the next section we examine several examples of circular motion in which different forces produce a centripetal acceleration. Just as with the ball on the string, if the direction of the velocity changes, a force must be acting to produce this change; otherwise the object would move in a straight line rather than in a curved path, according to Newton's first law of motion.

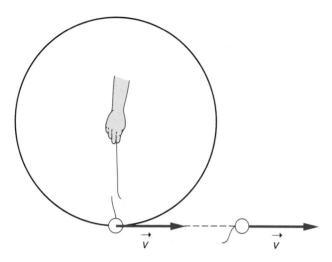

Figure 5.6 If the string breaks, the ball flies off in a straight-line path in the direction in which it was traveling at the instant the string broke.

5.2 CENTRIPETAL FORCES

The discussion in the previous section should have convinced you that an object moving in a circle is accelerated, even though its speed may remain constant. The acceleration involved is called the *centripetal acceleration* and has the magnitude v^2 / r. It is a common feature of motions like that of the ball on the string, a car moving around a curve, or any object moving in a circle. The only differences among these situations come from the nature of the force or forces involved in producing the centripetal acceleration.

The *net* force involved in producing a centripetal acceleration is often referred to as the *centripetal force,* but this terminology is frequently a source of confusion because it implies that somehow a special force is involved. In fact, centripetal force is just any force, or combination of forces, that act on an object in a particular situation to produce centripetal acceleration. It can be any of the types of force that we have already considered: pulls from strings, pushes from contact with other objects, friction, and so forth.

The Ball on a String Revisited

In the case of the ball on the string, the force in question is that which the string exerts on the ball due to the tension in the string. Since the string is not completely within a horizontal plane, this tension has both horizontal and vertical components. The horizontal component of the tension, T_h, is what pulls the ball towards the center of the horizontal circle and produces the centripetal acceleration. Figure 5.7 illustrates this and also shows the vertical component of the tension, T_v.

The vertical component of the tension must equal the weight of the ball, since the net force in the vertical direction should be zero. This follows from the fact that the ball is

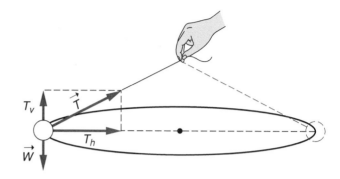

Figure 5.7 The horizontal component of the tension is the force that produces the centripetal acceleration. The vertical component of the tension is equal to the weight of the object.

not accelerated in the vertical direction; it stays in the horizontal plane of the circle around which it is moving. Using these ideas, we can compute the magnitude of the tension that is required in a typical situation for the rotating ball.

Suppose that the ball has a mass of 50 g (0.050 kg) and is rotating at the relatively slow rate of one revolution per second (1.0 rev/s) in a circle with a radius of 40 cm. We can find the constant speed of the ball for this motion by dividing the distance traveled in one revolution by the time required for one revolution. The distance traveled is the circumference of the circle, $2\pi r$; thus,

$$v = \frac{2\pi r}{t} = \frac{2\,(3.14)(40\text{ cm})}{1\text{ s}} = 251 \text{ cm/s, or 2.51 m/s.}$$

The centripetal acceleration is then

$$a_c = \frac{v^2}{r} = \frac{(2.51 \text{ m/s})^2}{0.40 \text{ m}} = 15.8 \text{ m/s}^2,$$

and the horizontal component of the tension, which is the centripetal force in this case, can be found from Newton's second law:

$$T_h = \Sigma F_x = ma_c = (0.050 \text{ kg})(15.8 \text{ m/s}^2) = 0.788 \text{ N}.$$

The vertical component of the tension is equal to the weight of the ball, so

$$T_v = W = mg = (0.050 \text{ kg})(9.8 \text{ m/s}^2) = 0.490 \text{ N}.$$

Even at this relatively slow rate of revolution, then, the horizontal component of the tension is larger than the vertical component.

As the ball is twirled at a faster rate, the centripetal acceleration increases even more rapidly because it is proportional to the square of the speed of the ball. This causes the horizontal component of the tension to become much larger than the vertical component, which is still equal to just the weight of the ball. As the horizontal component of the tension becomes larger, the string flattens out, coming closer to being completely in the horizontal plane, as shown in figure 5.8. A small vertical component of the tension must still be present, however, in order to counter the weight of the ball. You can easily observe these effects with your own ball and string; give it a try!

A Car Rounding a Curve

Our analysis of the ball on the string can be used as a prototype for other situations that may have even greater practical relevance. What about a car going around a curve, for example; what forces are involved in producing the centripetal acceleration in that situation? It turns out to depend on whether or not the curve is banked. The easiest situation to analyze is that in which the curve is not banked, so that we are dealing with a flat road surface (fig. 5.9).

A Flat Road Surface. For a flat road surface, we are totally dependent upon friction to produce the necessary centripetal acceleration. The size of a frictional force depends upon whether or not there is motion along the surfaces of contact producing the friction. If there is no motion, we refer to the force as the *static* force of friction. If there is motion along the surface of contact, it is the *kinetic*, or sliding, force of friction that is involved. Usually the kinetic force of friction will be less than the maximum possible static force of friction, so the existence of motion along the surface of contact becomes an important factor.

Unless the car has already begun to skid, it is the *static* force of friction that is involved. This is because the car (and

its tires) are not moving in the direction of the centripetal acceleration and its required force, which is toward the center of the curve. Thus, if the car travels around the curve without skidding, the tires do not move in the direction of the frictional force.

How large is the required frictional force? It depends upon the speed of the car and the radius of the curve. From Newton's second law, we know that

$$\Sigma F = f_s = ma_c = \frac{mv^2}{r},$$

where f_s is the force of static friction. The maximum possible value of the frictional force, f_s is roughly proportional to the normal force pushing the surfaces together, which in turn is related to the mass and weight of the car. Since the required centripetal force also depends upon the mass of the car, as well as the square of the speed and the radius of the curve, the mass appears on both sides of the equation and ends up not being an important factor. The speed of the car, since it is squared in the expression for the required force, is extremely important, however.

If the mass times the centripetal acceleration, mv^2 / r, is greater than the maximum possible frictional force, then we are in trouble! The speed is obviously a critical factor. Doubling the speed would require a frictional force four times as large as that for the lower speed. A sharper curve with a smaller radius, r, also requires either a larger frictional force or a lower speed. Since the frictional force can be only so large and depends upon factors like the dryness of the road surface and the condition of the tires, disaster lurks if the speed is too great.

What happens if mv^2 / r, the required centripetal force, is larger than the maximum possible frictional force? The frictional force is then insufficient to produce the necessary centripetal acceleration, and the car begins to skid. Once it

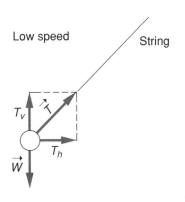

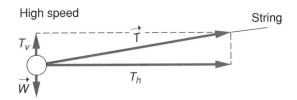

Figure 5.8 **At higher speeds, the string comes closer to lying in the horizontal plane of the ball because a large horizontal component of the tension is necessary to provide the required centripetal force.**

Figure 5.9 **The centripetal acceleration of a car rounding a level curve is produced by frictional forces between the tires and the road surface.**

is skidding, the force of *kinetic* friction is involved. Since the force of kinetic friction is generally less than that of static friction, the frictional force decreases, and the skid gets worse. The car, of course, like the ball on the broken string, is just following its natural tendency to move in a straight line in the absence of a net force, according to Newton's first law of motion.

Any factor that reduces the force of static friction will also cause problems. Wet or icy road surfaces are the usual culprits. In the case of ice, the force of friction may be reduced almost to zero, and an extremely slow speed will be required to negotiate a curve. There is nothing like driving on a real icy road to make you appreciate the value of friction. Newton's first law is constantly illustrated.

A Banked Curve.   If the road surface is properly banked, we are not totally dependent upon friction, as we are with a flat road surface. For the banked curve, the normal force between the car's tires and the road surface can be of some use, as shown in figure 5.10. The normal force is always perpendicular to the surfaces involved, so it points in the direction shown in the diagram. The total normal force is the sum of those for each of the four tires.

Since the car is not accelerated in the vertical direction, the net force in the vertical direction must again be zero, which implies that the vertical component of the normal force must be equal in magnitude to the weight of the car. This fact determines how large the normal force will be. The horizontal component of the normal force, N_h, on the other hand, is in the appropriate direction to produce the centripetal acceleration. Using Newton's second law, as we have before, we find:

$$\Sigma F = N_h = ma_c = \frac{mv^2}{r}.$$

The banking angle and the weight of the car determine the size of the normal force, and thus also define the magni-

tude of the horizontal component of the normal force. For some appropriate speed, then, the horizontal component of the normal force of the road pushing against the wheels is all that is needed to produce the centripetal acceleration. The higher the speed, the steeper the banking angle should be in order to provide a larger horizontal component for the normal force. Fortunately, since both the normal force and the required centripetal force depend on the mass of the car, the same banking angle will work for vehicles of different mass.

A banked curve is designed, then, for a particular speed. Since friction is also usually present, the curve can be negotiated at a range of speeds on either side of the design speed. If the road is icy and there is no friction, the curve can still be negotiated at the appropriate design speed. Speeds higher than the design speed would cause the car to fly off, as in our discussion of the flat road surface. Speeds too low, on the other hand, would cause the car to slide down the banked incline towards the center of the curve.

The Ferris Wheel

As a final example, consider what goes on when you are riding on a Ferris wheel. In this case, the circular motion is in a vertical plane rather than the horizontal circles that we have been considering. The weight of the rider and the normal force of the seat pushing on the rider are the forces that produce the centripetal acceleration.

Figure 5.11 shows the forces acting on the rider at the bottom of the wheel. At the bottom of the cycle, the normal force acts upward and the weight, of course, downward. The normal force in this case must be larger than the weight of the rider in order to produce a net force that is upwards,

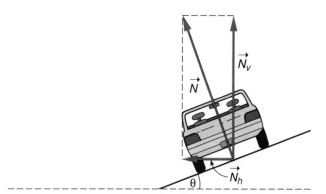

Figure 5.10   The horizontal component N_h of the normal force exerted by the road on the car can help to produce the centripetal acceleration if the road is banked.

Figure 5.11   The weight of the rider and the normal force exerted by the seat combine to produce the centripetal acceleration for a rider on a Ferris wheel.

toward the center of the circle. The centripetal force in this situation is therefore the difference between the normal force and the weight of the rider:

$$\sum F = N - W = ma_c = \frac{mv^2}{r}.$$

Since the normal force is larger than your weight, you feel heavy in this position, you are pushed against the seat with a greater force than that of your weight alone.

At the sides of the cycle, there must be a horizontal component of the normal force in order to keep the rider moving in a circle. This horizontal component may be provided by the seat back on the left-hand side of the cycle, or by a seat belt or hand bar on the right side of the cycle. The latter case is somewhat more exciting! The most interesting position, however, is probably at the top of the cycle.

At this position, the weight of the rider is the only force (other than a possible seat-belt force) that is in the appropriate direction to produce the centripetal acceleration. Again, from Newton's second law, the difference between the weight and the upward normal force must equal the mass of the rider times the centripetal acceleration:

$$\sum F = W - N = ma_c = \frac{mv^2}{r}.$$

As the speed gets larger, the normal force must get smaller in order to produce a larger net force. Usually the top speed of the Ferris wheel is adjusted so that the normal force approaches zero when the rider is right at the top of the cycle. Momentarily the rider feels weightless; no force is exerted by the seat on the rider.

If the opportunity is available nearby, take a break and go ride a Ferris wheel. There is nothing like the direct experience to bring home the ideas we have just described. As you ride, try to sense the direction and magnitude of the normal force. The weightless feeling at the top and the sense of plunging outward on the downward portion of the cycle are what the price of the ride is all about.

5.3 PLANETARY MOTION

From the standpoint of the historical development of ideas, the most important example of centripetal acceleration involves the motion of the heavenly bodies: the sun, the moon, and the planets. Certainly these objects are a part of our everyday experience, and yet many of us are surprisingly unaware of the details of their motions. How many of us have given it much thought since the third grade?

Observing the heavens was probably a more popular pastime when there were fewer roofs over our heads. If you have ever spent a night in a sleeping bag under the stars somewhere in the mountains, you may recall the sense of wonder and amazement at all of those bright objects out there. If you did so night after night, you might notice, as

the ancients did, that some of the brightest objects move relative to the other stars; their relative positions change.

These wanderers are the planets. The so-called fixed stars always maintain the same relative position to one another as they move across the sky. The Big Dipper never changes its shape, for example. But the planets roam about in a regular but curious fashion, and these motions excited the curiosity of ancient observers of the heavens. They were still a point of scientific debate in Galileo's day.

Early Attempts at Explanation

Suppose that you were an early philosopher/scientist trying to make sense of these motions. What kind of model might you develop? Some features seem very simple and regular. The sun, for example, moves across the sky each day, from east to west, as if it were at the end of an enormously long and invisible rope tethered at the center of the earth. The stars follow a similar pattern; their apparent rotation about the earth could be described by picturing them as lying upon a giant sphere that rotated about the earth. This, of course, is an earth-centered, or *geocentric,* view of things, a natural view for us to take (fig. 5.12).

The moon also moves across the sky in an apparently circular orbit about the earth, but unlike the stars, it does not return to the same position each successive night. Instead, it goes through a series of regular changes in position and phase which have a cycle of approximately thirty days. How many of us can provide a clear explanation of the phases of the moon?

Early models of the motions of the heavenly bodies, developed by Greek philosophers, involved a series of con-

Figure 5.12   The heavenly spheres: a geocentric view of the universe showing earth, water, air, and fire at the center. The sun is in the fourth sphere beyond the fire sphere.

centric spheres centered upon the earth. The fixed stars were on the outermost sphere, and the sun, the moon, and the planets had their own spheres. These spheres were viewed as rotating about the earth in a manner that could explain the positions of the objects relative to one another. Plato and others of his time viewed spheres and circles as ideal shapes, which would be appropriate, therefore, to the heavens.

There were a few problems with this simple view involving spheres, however. Some of the planets, for example, did not behave as though they were on a continuously moving sphere. Mars, Jupiter, and Saturn all undergo *retrograde* motion relative to the fixed stars; their paths trace loops in the sky when viewed against the background of the fixed stars. It takes a few months for Mars to traverse one of these loops, as illustrated in figure 5.13.

In order to explain the retrograde motions of the outer planets, Ptolemy, working in the second century A.D., devised a somewhat more sophisticated model than that used by the Greek philosophers. Ptolemy's model used circular orbits rather than spheres, but it was still geocentric. For the outer planets, he invented the idea of *epicycles,* which were circles that rolled upon the larger circle that was the basic orbit about the earth. These epicyles produced the loops of the retrograde motion (fig. 5.14).

Ptolemy's model could accurately predict where the planets would be found on any given evening of a year. Various refinements were needed, however, to improve the accuracy of prediction as more accurate observations became available. In some cases this meant putting epicycles upon epicycles, but the basic scheme of circles was retained. Ptolemy's system became a part of accepted knowledge and was incorporated into the teachings of the early Roman Catholic church.

Ptolemy's model, of course, is not the one that you were introduced to in elementary school. During the Renaissance, in the sixteenth century, a Polish astronomer, Nicolaus

Copernicus (1473–1543) put forth a sun-centered, or *heliocentric,* view, which was later championed by Galileo. Copernicus was not the first to suggest such a model, but earlier versions had not taken hold. Copernicus put the idea forward tentatively, to avoid charges of heresy, but Galileo promoted it somewhat more vigorously. For his pains, Galileo did end up in trouble with the church, a problem that was not to be taken lightly at that time. People had been burned at the stake for lesser offenses!

The heliocentric view of Copernicus places the sun at the center of the circular orbits of the planets, and demotes the earth to the status of just another planet. In addition, it requires that the earth rotate on an axis through its center in

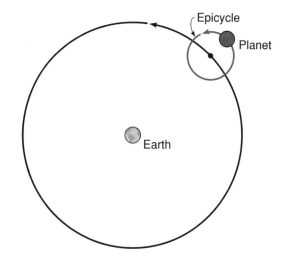

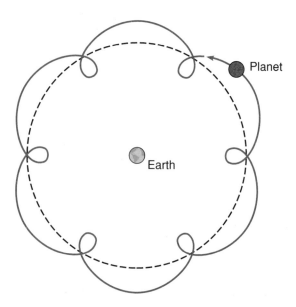

Figure 5.14 Ptolemy's epicycles were circles rolling upon circular orbits for the outer planets. This model produced the observed retrograde motion for these planets.

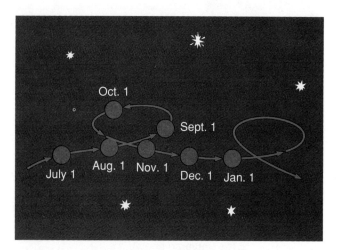

Figure 5.13 The retrograde motion of Mars relative to the background of fixed stars.

order to explain the daily motions of the sun and the other heavenly bodies (including the fixed stars). We are now used to this view, but it was a revolutionary idea at the time. Would we not be blown away by the enormous winds that such a rotation would produce?

The advantage of the Copernican view was that it did not require epicycles to explain retrograde motion. Retrograde motion comes about instead from the fact that the earth is also moving relative to the sun. (To understand this, try to imagine how the position of Mars as seen from the earth appears to change as the earth, in its orbit closer to the sun, overtakes and passes Mars as both move in the same direction about the sun.) It was therefore a simpler model since it could predict, without the need for epicycles, the same facts that Ptolemy's model could. The Copernican model could also explain variations in the brightness of the planets that could not be explained by Ptolemy's model.

The price of accepting the Copernican model was the need to give up our egocentric view of the universe and to accept what seemed to some to be an absurd proposition: that the earth is rotating with the seemingly rapid frequency of one cycle per day. Since the radius of the earth was approximately known (6400 km), this rotation implied that we must be moving along at roughly 1680 km/hr (or just over 1000 MPH) if we are standing on the earth's surface! We certainly do not feel that motion.

Because Copernicus assumed the orbits to be circles, the accuracy of prediction of his model was no greater than that of Ptolemy's model. In fact, it required some adjustments just to make it agree with known astronomical data of that time. The controversy that was generated by the competition between the two models suggested a need for more accurate observations, a project that was undertaken by a Danish astronomer, Tycho Brahe (1546–1601).

Brahe developed a large quadrant that could be used to make very accurate sightings of the positions of the planets and stars at any given time (fig. 5.15). It was capable of measuring these positions to an accuracy of $1/60$ of a degree, which was considerably better than the previously available data. Brahe spent several years painstakingly collecting data on the precise positions of the planets and other bodies.

Kepler's Laws of Planetary Motion

The task of analyzing the data collected by Brahe fell to his assistant, Johannes Kepler (1571–1630). The magnitude of the task was enormous; it required the transformation of the data to coordinates about the sun, and then numerical trial and error in an attempt to find regularities in the orbits. It became apparent, through Kepler's work, that the orbits were not perfect circles. After considerable time and effort, Kepler was able to show that the orbits of the planets about the sun were more accurately described as ellipses, with the sun at one focus.

An ellipse is a curve that can be drawn by attaching a string between two fixed foci and then moving a pencil around the perimeter of the path allowed by the string (fig. 5.16). A circle is a special case of an ellipse in which the two foci coincide. The orbits of most of the planets are quite close to being circles, but the precision of Brahe's data was sufficient to show the difference between a perfect circle, on one hand,

Figure 5.15 Tycho Brahe's large quadrant permitted accurate measurement of the positions of the planets and other heavenly bodies.

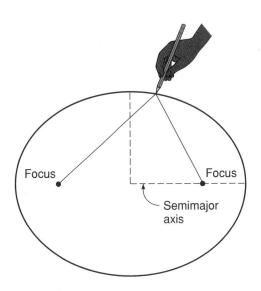

Focus

Focus

Semimajor axis

Figure 5.16 An ellipse can be drawn by fixing a string at two points (foci) and moving a pencil around the path permitted by the string.

Box 5.1

Kepler's Laws of Planetary Motion

1. All of the planets have elliptical orbits about the sun, with the sun at one focus of the ellipse.
2. The radius vector drawn from the sun to the planet sweeps out equal areas in equal times.
3. The cube of the radius about the sun for each planet is proportional to the square of the period of the orbit.

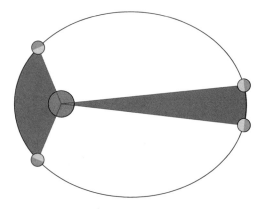

Figure 5.17 **Because planets move faster when nearer to the sun, the radius line for each planet sweeps out equal areas in equal times (Kepler's second law).**

and an ellipse with the two foci close together. The fact that the orbits of the planets are more accurately described as ellipses became Kepler's first law of planetary motion.

Kepler's other two laws of planetary motion came after much more laborious efforts at playing numerical trial-and-error games with Brahe's data (see box 5.1). The second law describes the fact that the planets move faster when they are nearer to the sun, so that the radius vector sweeps out equal areas in equal times regardless of where it is in its orbit (fig. 5.17). The third law states a relationship between the radius of the orbit and the time taken for one complete cycle around the sun (the *period* of the orbit). Throughout this work, Kepler had to deal with the problem of a mentally ill mother, and with the opinions of some of his contemporaries that he was a little weird himself.

Kepler found the third law by showing that the ratio of r^3 to T^2 is the same for all of the planets (T is the period of the orbit, the time taken for one complete cycle about the sun). Since the radius vector changes during the orbit, the appropriate distance to use for r is actually the semimajor axis, the distance from the center of the orbit to the edge measured along the longer axis of the ellipse. This distance is pictured in figure 5.16.

Kepler's laws, based as they were upon Brahe's data, brought a significant advance in the accuracy with which the positions of the planets could be predicted. Since Kepler's laws represented a sun-centered, or heliocentric, view, following the ideas of Copernicus and Galileo, the heliocentric view was strengthened by his work. More importantly, however, Kepler's laws provided a new set of precisely stated regularities that begged explanation. The stage was set for Isaac Newton to incorporate these regularities into a grand theory that could explain both celestial mechanics (the motion of the heavenly bodies) as well as the more mundane motion of everyday objects near the earth's surface.

5.4 NEWTON'S LAW OF GRAVITATION

If you have followed the story to this point, then the next question should be obvious. If the planets are moving in curved paths around the sun, what is the nature of the force that must be present to produce the centripetal acceleration? You probably know that gravity is involved, but this was not at all obvious when Newton began his work.

Whether or not the idea came to Newton while he was sitting under an apple tree and was struck by falling fruit, at some point early in the development of his theory he arrived at an important insight. He realized that there is a similarity between the motion of a projectile launched near the earth's surface and the orbit of the moon. A very famous drawing in Newton's *Principia* makes this case (fig. 5.18).

The idea is simple, but earthshaking. Imagine, as Newton did, a projectile being launched horizontally from a very high mountain. The larger the launch velocity, the farther away from the base of the mountain the projectile will land. At very large launch velocities, the curvature of the earth begins to be a significant factor in the result. In fact, if the launch velocity is large enough, the projectile may never reach the surface of the earth; it keeps falling, but the earth's curvature falls off also, resulting in a circular orbit about the earth for the projectile.

The Law of Gravitation

The insight that came to Newton was the idea that the moon is actually falling towards the earth but never gets there because of its lateral velocity and the curvature of the earth itself. The moon, of course, is at a distance from the earth that is much higher than the height of any mountain, but the concept is the same. Newton recognized that the moon falls under the influence of gravity just as a projectile does. The same force that accounts for the acceleration of objects near the earth's surface, as described by Galileo, can be used to explain the orbit of the moon!

But what is the nature of this force? From Galileo's work, Newton knew that near the earth's surface the force of gravity is proportional to the mass of the object; $F = mg$. It was natural, then, to think that mass might also be involved in a more general expression for the gravitational force. The major question was how the force might vary with distance.

Figure 5.18 **In a diagram similar to this in his *Principia*, Newton pictured projectiles fired from an imaginary (and incredibly high) mountain. If fired with a large enough horizontal velocity, the projectile falls towards the earth but never gets there.**

At this point, Kepler's laws of planetary motion and the concept of centripetal acceleration came into play. Newton was able to prove mathematically that Kepler's first and third laws of planetary motion could be derived from the assumption that the gravitational force between the planets and the sun falls off with the inverse square of the distance. In other words, if the gravitational force were proportional to $1 / r^2$, where r is the distance between the centers of the planet and the sun, then Newton could prove that the orbits must be ellipses with the sun at one focus, and that the ratio of r^3 to T^2 must be the same for all of the planets. The proof involved setting the assumed force equal to the required centripetal force in Newton's own second law of motion.

The idea that a force could act between two masses that are separated by a large distance was difficult to accept in Newton's day (and, in some ways, even now). If such a force exists, it is natural to suspect that this force "acting at a distance" decreases in strength as the distance increases. There may have been other reasons for suspecting that the decrease might be proportional to $1 / r^2$, as shown in figure 5.19, but the most compelling reason for Newton was the fact that he could prove Kepler's laws by assuming that the force of gravity behaved in this manner.

This line of reasoning led to the statement of Newton's law of gravitation. The law of gravitation and Newton's three laws of motion are the fundamental postulates of his theory of mechanics. The law of gravitation can be stated as follows:

The gravitational force between two objects is proportional to the mass of each object and inversely proportional to the square of the distance between the centers of the masses. The direction of the force is attractive and lies along the line joining the centers of the two masses.

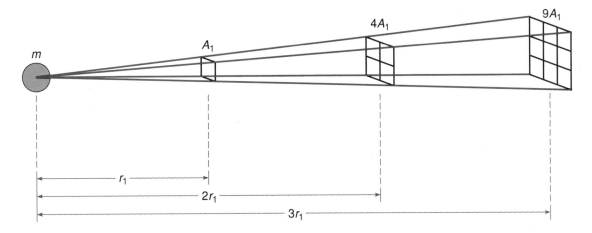

Figure 5.19 **If lines are drawn radiating outward from a point mass, the areas intersected by these lines increase in proportion to r^2. Does this fact suggest that the force exerted by *m* on a second mass might decrease in proportion to $1/r^2$?**

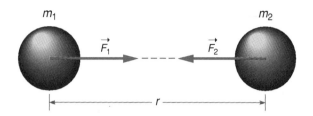

Figure 5.20 **The gravitational force is attractive and acts along the line joining the centers of the two masses. (It obeys Newton's third law of motion.)**

Actually, for this statement to be completely valid, the masses in question must be either point masses or perfect spheres (fig. 5.20). Newton spent several years proving that if the masses were spherically symmetrical (as the planets and the sun are approximately), then the mass can be treated as though it were all located at the center of the object. This proof required the development of new mathematical techniques that we now call *calculus*; Newton is generally credited with being one of the coinventors of calculus.

We can write the law of gravitation in symbols as follows:

$$F = \frac{Gm_1m_2}{r^2},$$

where G is the universal gravitational constant. Newton did not actually know the value of this constant, because he did not know the masses of the earth, the sun, and the other planets. Its value was determined over one hundred years later in an experiment performed by Henry Cavendish (1731–1810) in England. Cavendish measured the very weak gravitational force between two massive lead balls for different distances of separation.

In the metric units that we are using,

$$G = 6.67 \times 10^{-11} \text{ N·m}^2/\text{kg}^2.$$

The power-of-ten notation is useful here because G is a very small number (see appendix B). It means that the decimal point is located eleven places to the left in the number as written, or

$$G = 0.0000000000667 \text{ N·m}^2/\text{kg}^2.$$

Because of the small size of this constant, the gravitational force between two ordinary-sized objects, such as people, is extremely small and not usually noticeable. Cavendish's experiment required real ingenuity to measure such a weak force.

Weight and the Law of Gravitation

Suppose, however, that one of the objects in question is a planet or other very massive object. The force of gravity

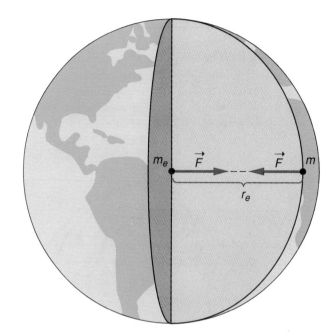

Figure 5.21 **For an object near the earth's surface, the distance between the centers of the two objects is equal to the radius of the earth.**

then can be quite large, in spite of the small value of G, because one of the masses is so large. As an example, we might consider the force on a 50-kg person standing on the surface of the earth. The magnitude of the gravitational force could then be written as follows:

$$F = \frac{Gmm_e}{r_e^2} = mg,$$

where m_e is the mass of the earth and r_e is the radius of the earth. As figure 5.21 illustrates, the distance between the centers of the two objects is essentially the radius of the earth in this case.

The equations just given express the gravitational force in two ways: (1) in terms of Newton's law of gravitation, and (2) in terms of g, the acceleration of gravity ($F = mg$). We could use either of these two expressions for F to compute the magnitude of the force. It is also apparent that the acceleration of gravity near the earth's surface, g, must be related to the universal gravitational constant, G, by the following relationship:

$$g = \frac{Gm_e}{r_e^2}.$$

This is found by simply dividing m, the mass of the smaller object, out of the dual expressions for the force. The acceleration of gravity, g, near the earth's surface is obviously *not* a universal constant; it will be different on different planets and even slightly different at different points on the earth due to variations in the radius of the earth and other factors.

It is easier to use the expression $F = mg$ to compute the magnitude of the force on the 50-kg person standing on the surface of the earth. We find that

$$F = mg = (50 \text{ kg})(9.8 \text{ m/s}^2) = 490 \text{ N},$$

a force that is certainly felt. If we knew the mass of the earth, the more general expression could also be used; the calculation would then take the following form:

$$F = \frac{Gmm_e}{r_e^2};$$

$$F = \frac{(6.67 \times 10^{-11} \text{ N} \cdot \text{m}^2/\text{kg}^2)(50 \text{ kg})(5.98 \times 10^{24}\text{kg})}{(6\,370\,000 \text{ m})^2};$$

$$F = 491 \text{ N}.$$

The difference in these two results arises because we have written g to only two-figure accuracy in the initial computation.

The mass of the earth (5.98×10^{24}kg) is, of course, a very large number; to write it out without using the power-of-ten notation, we must move the decimal point twenty-four places to the right:

$$m_e = 5\,980\,000\,000\,000\,000\,000\,000\,000 \text{ kg}.$$

This mass was initially determined from knowledge of the the radius of the earth, the acceleration of gravity near the earth's surface, and Cavendish's determination of the universal gravitational constant, G. In a sense, Cavendish weighed the earth by making that measurement!

If we wanted to know the force of gravity on a 50-kg person in a space capsule several hundred kilometers above the surface of the earth, we would have to use the more general expression represented by Newton's law of gravitation. The expression $F = mg$ is valid only near the surface of the earth. Likewise, if we wanted to know the weight of a 50-kg person who was standing on the moon, we would need to use the mass and radius of the moon in our calculation. This is done in the example in box 5.2.

This weight is only about $\frac{1}{6}$ the value of 490 N that we computed for the same person standing on the earth. The acceleration of gravity on the surface of the moon is therefore only about $\frac{1}{6}$ that found near the earth's surface. The weight of an object on the moon is thus considerably smaller than it is on earth; the mass, on the other hand, is an intrinsic property of the object which does not change when we move it to the moon.

The weaker force of gravity on the moon (compared to that on earth) is due to the smaller mass of the moon. This is partially compensated for by the smaller radius, but the effect of mass is dominant. Since our muscles are adapted to conditions on earth, we would find that our smaller weight on the moon would make possible some amazing athletic feats. It is indeed possible to leap small buildings in a single bound

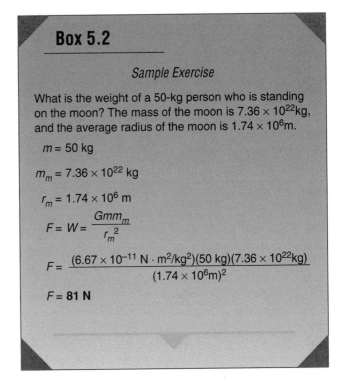

Box 5.2

Sample Exercise

What is the weight of a 50-kg person who is standing on the moon? The mass of the moon is 7.36×10^{22}kg, and the average radius of the moon is 1.74×10^6m.

$m = 50$ kg

$m_m = 7.36 \times 10^{22}$ kg

$r_m = 1.74 \times 10^6$ m

$$F = W = \frac{Gmm_m}{r_m^2}$$

$$F = \frac{(6.67 \times 10^{-11} \text{ N} \cdot \text{m}^2/\text{kg}^2)(50 \text{ kg})(7.36 \times 10^{22}\text{kg})}{(1.74 \times 10^6\text{m})^2}$$

$$F = 81 \text{ N}$$

on the moon! The smaller force of gravity for objects near the surface of the moon also explains why the moon has essentially no atmosphere; the gas molecules can escape the gravitational pull of the moon much more readily than that of the earth.

5.5 THE MOON AND OTHER SATELLITES

The moon has been an object of fascination for as long as people have existed and wondered about things. In this century, we have actually finally visited the moon and brought back samples from its surface. That has not seemed to dull the romance that the moon holds for us, but may have slightly reduced its mystery. How much mystery does the moon hold for you? Do you understand its phases and its changes in location? Do Kepler's laws hold for the moon? A closer look at the motion of the moon and other earth satellites may be enlightening.

The Moon

The moon itself was the only earth satellite available for Newton and his predecessors to contemplate. As we have already indicated, it played a crucial role in Newton's thinking and in the development of his law of gravitation. Observations of the moon and its phases, however, go much farther back than Newton's day. The moon played an important role in many early religions and rituals, so its behavior must have been carefully followed even in prehistorical times.

How do we explain the phases of the moon? Does the time that the moon rises in the evening have anything to do with whether it will be a full moon or not? These features of the moon's behavior are fairly easy to understand given our current picture.

The light that comes to us from the moon is, of course, reflected sunlight, so an understanding of the moon's phases needs to take into account the relative positions of the sun, the moon, and the observer. Figure 5.22 should help you to get this perspective. When the moon is full, it is on the opposite side of the earth from the sun and we see the side of the moon that is fully illuminated by the sun. It rises in the east, therefore, about the same time that the sun sets in the west, these events being determined by the rotation of the earth.

Because the earth and the moon are both small relative to the distances between the earth, the moon, and the sun, they do not usually get in the way of light coming from the sun. When they do, however, we have an eclipse. During a lunar eclipse, the earth casts a full or partial shadow upon the moon. From figure 5.22, we can see that this can only occur during the full-moon phase. The rarer solar eclipse occurs when the moon is exactly in the right position to cast a shadow upon the earth. During what phase of the moon will this occur?

At other times during the 27.3-day period of rotation of the moon about the earth, we do not see all of the illuminated side of the moon; we see a crescent or a half moon or some other phase in between (fig. 5.23). The new moon occurs when the moon is on the same side of the earth as the sun and is therefore invisible. When we are a few days on either side of the new moon, we see the familiar crescent. When will the moon rise and set for this condition?

When the moon is not at either the full-moon or new-moon phase, it can often be seen during daylight hours. In particular, when it is near the half-moon phase, it should rise around noon and set around midnight (or vice versa depending upon which side of the earth it is on). When it is too near to the position of the sun in the sky, it will be hard to see,

but when it is well separated from the sun, it can be clearly visible during the day.

All of this is familiar, but perhaps often not given much thought. The next time you see the moon, try thinking about where it is in the sky, when it will rise and set, and how this is related to its phase. Better yet, try explaining this to a friend. You too can be the wizard who explains the motions of the heavens!

Kepler's Laws and the Moon

The moon's orbit about the earth is actually a little more complicated than that of the planets about the sun. The reason for this is that the moon is strongly influenced by two bodies rather than just one; the earth and the sun (fig. 5.24). The earth is by far the larger influence because it is much closer to the moon than is the sun. On the other hand, the sun has a much larger mass than either the earth or the moon, so its effect is still appreciable. As a first approximation, though, we can ignore the effect of the sun and consider the motion of the moon about the earth as just a two-body problem.

The basic physics of the problem is the same as that for the orbits of the planets about the sun. The gravitational attraction between the moon and the earth provides the centripetal acceleration that keeps the moon moving in a roughly circular orbit. If we treat the earth as stationary, with the moon orbiting around the earth, we can write

$$F = m_m\, a_c = \frac{Gm_m\, m_e}{r^2}.$$

In this case, r is the distance between the center of the moon and the center of the earth. This relationship has the same form, however, as that between the planets and the sun.

Just as with the planets, this relationship predicts that the orbit will be elliptical, but with the earth at one focus of the ellipse rather than the sun. The force exerted by the third body, the sun, on the moon distorts the ellipse, however, causing the path of the moon to oscillate about a true elliptical path. Calculating the exact nature of these distortions was a problem that kept mathematical physicists busy for many years.

Figure 5.22 **The phases of the moon depend upon the relative positions of the sun, the moon, and the earth.**

Figure 5.23 **Photographs of different phases of the moon. When during the day will each of these rise and set?**

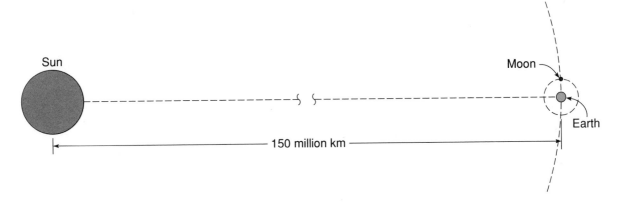

Figure 5.24 The moon is influenced by gravitation attraction to both the earth and the sun. The distances and sizes are not drawn to scale; the moon is but a speck compared to the size of the sun.

Kepler's first law is approximately true for the moon, therefore, providing that we substitute the earth for the sun in its statement. The second law also holds approximately, ignoring the distortions caused by the sun. It turns out that the second law is a consequence only of the fact that the gravitational force, which provides the centripetal acceleration, is directed along the line joining the centers of the two objects, the earth and the moon in this case. The second law would be true regardless of how the force depended upon distance and is a consequence of the principle of conservation of angular momentum, an idea that will be discussed in chapter 7.

Kepler's third law provides some differences, however. Using the expression for centripetal acceleration, $a_c = v^2/r$, and Newton's law of gravity, we can derive the relationship between T^2 and r^3. The expression that results for the ratio of these two quantities is

$$\frac{T^2}{r^3} = \frac{4\pi^2}{Gm_e}.$$

The interesting point is that this ratio depends upon the mass of the earth for the case of the moon orbiting about the earth. For the planets orbiting about the sun, the mass of the sun takes the place of the mass of the earth in this expression. In other words, the ratio is not the same for the moon's orbit about the earth as for the orbits of the planets about the sun.

Other Satellites

Any other satellite orbiting about the earth, however, would have the same value for the ratio T^2/r^3 as that of the moon. Kepler's third law could be said to hold, therefore, as long as we recognize that this ratio will not have the same value as it does for the orbits of the planets. The value of the ratio for satellites orbiting about the earth can be easily calculated from knowledge of the moon's orbit. The average distance between the moon and the earth is 3.82×10^8 m or 382 000 km. The period of the moon's orbit, or the time taken for one complete cycle, is 27.3 days.

If we put these numbers into the ratio and do some unit conversions to make the number a reasonable size, the ratio of r^3 to T^2 can be expressed as follows:

$$\frac{r^3}{T^2} = \frac{(3.82 \times 10^8 \text{ m})^3}{(27.3 \text{ days} \times 24 \text{ hr/day})^2};$$

$$\frac{r^3}{T^2} = 130 \text{ Mm}^3/\text{hr}^2.$$

A megameter (Mm) is 1 million meters, or 1000 kilometers. Since this ratio should hold for any satellite orbiting about the earth, we can use it to find the appropriate period or average distance for various artificial satellites.

For example, suppose that we want to know the appropriate height for a satellite with a synchronous orbit, one that orbits with a period of 24 hours, so that it stays at the same point above the earth at all times. Using the ratio resulting from Kepler's third law, we find that

$$r^3 = (130 \text{ Mm}^3/\text{hr}^2) \times T^2;$$

$$r^3 = (130 \text{ Mm}^3/\text{hr}^2)(24 \text{ hr})^2 = 74\,900 \text{ Mm}^3;$$

$$r = 42.2 \text{ Mm} = 42\,200 \text{ km}.$$

Since the radius of the earth is approximately 6400 km, this would put the satellite about 35 800 km above the surface of the earth. This is quite a ways up, but still a lot closer than the moon.

Most artificial satellites are even closer to the earth than this; the original Russian satellite, Sputnik, for example, had a period of about 90 minutes, or 1.5 hours. If we put this value into our ratio, we find a value for the average distance r of 6640 km. Since the radius of the earth is 6370 km, this is only 270 km above the surface of the earth.

The orbits of different satellites can be designed to meet different objectives; some are close to being circular; others

Box 5.3

Everyday Phenomenon:
Explaining the Tides

The Situation. A college freshman returns to her home on the coast for summer vacation. Her father operates a fishing boat and is familiar with the moods and ways of the ocean but somewhat suspicious of the "book learning" that he is paying so dearly for his daughter to obtain. "If you are so smart," he tells her, "explain to me why the tides behave as they do."

The facts are well known to both our student and her father. Roughly twice a day the tides go in and go out again. (The actual cycle of two high tides and two low tides is closer to 25 hours.) Sometimes high tide is higher and low tide is lower than at other times, and these times correspond to full-moon or new-moon conditions. The times at which high tides and low tides occur during the day shift from day to day because of the 25-hour cycle, but the pattern repeats on a monthly cycle. Tidal charts can be used to predict these times for any given day. The task is similar to that faced by Newton and others of his day in using the new theory to explain a well-known phenomenon.

Can Newton's laws (including the law of gravitation) be used to explain tidal behavior?

The Analysis. The monthly cycle and the correlations of the highest tides with the phase of the moon suggest a lunar influence. Both the moon and the sun exert gravitational forces upon the earth; the sun exerts the stronger force because of its much larger mass. The moon, on the other hand, is much closer than the sun, and variations in its distance from different points on the earth may be important. Since the strength of the gravitational force depends upon $1/r^2$, and r is much smaller for the earth-moon interaction than for the earth-sun interaction, the variation in strength of the gravitational force of the moon from one side of the earth to the other could be a factor.

Since water is a fluid (except where frozen), the mass of water that makes up the oceans can move with respect to the more rigid crust of the earth. The primary force acting upon the water is the gravitational

High tide   Low tide

High tide and low tide produce different water levels at the dock.

are much more elongated ellipses. The plane of the orbit can pass through the poles of the earth (polar orbit) or take any desired orientation between the poles and the equator (fig. 5.25). It all depends upon the mission of the satellite.

Artificial satellites have become a routine feature of today's world that did not exist prior to 1958. Their uses include communications, surveillance, weather observations, and possible defense applications such as are contemplated in the Strategic Defense Initiative ("Star Wars") plans. The basic physics of their behavior is nicely accounted for by Newton's theory. If Newton could return, he might be amazed at the developments, but for him the analysis would be routine!

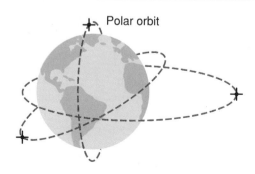

Polar orbit

Figure 5.25 The orbits of different artificial satellites can have different orientations and elliptical shapes.

attraction of the earth, which keeps the water on the surface of the earth. The gravitational force (per unit mass) exerted by the moon on the water, however, is greatest on the side of the earth closest to the moon and weakest on the opposite side of the earth, because of the greater distance.

This variation in strength of the moon's gravitational pull produces a bulge in the water surface on both sides of the earth. The bulge on the side nearest the moon can be thought of as resulting from the fact that the water is pulled towards the moon by a stronger force per unit mass than that exerted upon the rest of the earth. This produces a high tide; the water will be nearer to the top of the dock.

On the opposite side of the earth, we can think of the *earth* being pulled towards the moon with a stronger force per unit mass than the water on the far side. This also produces a high tide since the earth is pulled away (slightly) from the water. Thus we get two high tides during each rotation of the earth (and two low tides at times roughly halfway between the high tides). The forces exerted by the moon on the water are small, of course, compared to the force that the earth exerts on water, but they are still large enough to account for the tidal effect.

When the sun and the moon are both lined up on the same side of the earth during the new-moon phase, the sun also contributes to this force difference and produces its own bulges on either side of the earth, which add to those produced by the moon. This is why the highest tides occur during a full moon or a new moon. This model then explains both the twice-daily cycle and the correlation of the highest tides with the phase of the moon. But why the 25-hour cycle?

A little further thought provides the answer. The high tide bulges occur on either side of the earth along the line joining the moon and the earth. The earth rotates underneath this bulge with a period of 24 hours. But in one day's time the moon also moves, since it orbits the earth with a period of 27.3 days. In one day, therefore, the moon has moved through roughly $1/27$ of its orbital cycle, so that the time when the moon again lines up with a given point on the earth is a little longer than one day. This additional time would be approximately $1/27$ of 24 hours, or a little less than an hour.

This model of tidal behavior was developed initially by Newton and accounts neatly for the major features of the tides. The variation of the gravitational force with distance is the key to the explanation. A complete analysis would also include the tidal effects of the sun.

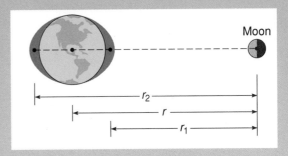

Because of its dependence on distance, the gravitational force per unit mass exerted by the moon on the water or the earth gets weaker as we move from the side of the earth nearest the moon to the opposite side.

SUMMARY

This chapter has shown that objects moving in circular motion are accelerated and that this centripetal acceleration is related to the speed of the object and the radius of the circle. This idea was used to explain the motion of a ball on a string, cars on curves, a Ferris wheel, and finally the planets about the sun and the earth about the moon. The force providing the centripetal acceleration in the case of planetary motion is described by Newton's law of gravity, the final general law in his theory of mechanics.

The following key concepts are involved.

Centripetal acceleration. Centripetal acceleration is the acceleration involved in changing the direction of velocity and is related to the speed of the object and the radius of the curve.

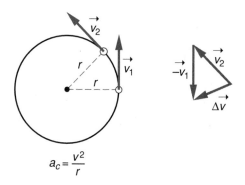

$$a_c = \frac{v^2}{r}$$

Centripetal force. A centripetal force is any force or combination of forces that act on a body to produce a centripetal acceleration. It is related to centripetal acceleration by Newton's second law.

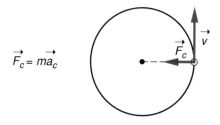

$$\vec{F}_c = m\vec{a}_c$$

Banking a curve. A curve is banked so that the normal force between the road surface and the tires can help to produce the centripetal acceleration. Friction also contributes.

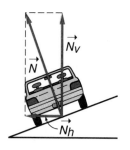

Planetary motion. Kepler's three laws of planetary motion describe the orbits of the planets about the sun. The orbits are ellipses, equal areas are swept out in equal times, and there is a relationship between the period of the orbit and the distance of the planet from the sun.

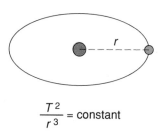

$$\frac{T^2}{r^3} = \text{constant}$$

Newton's law of gravitation. Newton's law of gravitation states that the gravitational force between two masses is proportional to the masses and inversely proportional to the square of the distance between the masses. Using this with his other laws of motion, Newton could derive Kepler's laws of planetary motion.

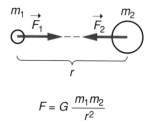

$$F = G \frac{m_1 m_2}{r^2}$$

The moon and other satellites. The moon's motion about the earth can also be described by Kepler's laws provided that we substitute the mass of the earth for that of the sun in the expression for the period, T. Artificial earth satellites have the same ratio of T^2/r^3 as that for the moon.

QUESTIONS

Q5.1 A car travels around a curve with constant speed.

    a. Does the velocity of the car change in this process? Explain.

    b. Is the car accelerated? Explain.

Q5.2 Two cars travel with constant speed around the same curve, one at twice the speed of the other. Which car, if either, experiences the larger change in velocity? Explain.

Q5.3 A car travels at the same constant speed around two curves, one with twice the radius of curvature of the other. For which of these curves is the rate of change in velocity of the car greater? Explain.

Q5.4 A ball on the end of a string is whirled in a horizontal circle with constant speed. At point *A* in the circle, the string breaks. Which of the curves sketched in the diagram most accurately represents the path that the ball will take after the string breaks (as seen from above)? Explain.

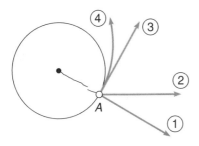

Q5.5 Before the string breaks in the situation of question 5.4, is there a net force acting upon the ball? If so, what is its direction? Explain.

Q5.6 A car travels around a nonbanked (flat) curve with constant speed.

    a. Sketch a diagram showing all of the forces acting upon the car.

    b. What is the direction of the net force acting upon the car? Explain.

Q5.7 Is there a maximum speed at which the car in question 5.6 will be able to negotiate the curve? If so, what factors determine this maximum speed? Explain.

Q5.8 If a curve is banked, is it possible for a car to negotiate the curve even when there is no frictional force at all because of an icy road surface? Explain.

Q5.9 If a ball is whirled in a vertical circle with constant speed, at what point in the circle, if any, is the tension in the string the greatest? Explain. (Hint: Compare this situation to that of the Ferris wheel described in section 5.2.)

Q5.10 In what way was the heliocentric view of the solar system proposed by Copernicus more simple than the geocentric view of Ptolemy? Explain.

Q5.11 Did Ptolemy's view of the solar system involve motion of the earth, rotational or otherwise? Explain.

Q5.12 How did Kepler's view of the solar system differ from that of Copernicus? Explain.

Q5.13 Consider the method of drawing an ellipse pictured in figure 5.16. How would you modify this process to make the ellipse into a circle, which is a special case of an ellipse? Explain.

Q5.14 Is there a net force acting upon the planet earth? If so, what is the nature of this force? Explain.

Q5.15 Are the planets accelerated in their motions about the sun? Explain.

Q5.16 Two masses are separated by distance *r*. If this distance is doubled, is the force of interaction between the two masses doubled, halved, or decreased to a value smaller than half the original value? Explain.

Q5.17 Three *equal* masses are located as shown in the diagram below. What is the direction of the net force acting upon m_2? Explain.

Q5.18 A painter depicts a portion of the night sky as shown in the diagram below, showing the stars and a crescent moon. Is this view possible? Explain.

Q5.19 At what times during the day or night would you expect the crescent moon shown in question 5.18 to rise and set? Explain.

Q5.20 Does Kepler's third law hold for artificial satellites orbiting the earth? Explain.

Q5.21 Since the earth rotates on its axis once every 24 hours, why do high tides not occur exactly twice every 24 hours, given our explanation of the tides? Explain.

Q5.22 If the gravitational force did not depend upon the distance between objects, would there be tides? Explain.

EXERCISES

E5.1 A ball is moving initially with a velocity of 3 m/s due north. Two seconds later, the ball is moving with the same speed (3 m/s), but now it is traveling due east.

    a. Using a vector diagram drawn to scale, estimate the magnitude of the change in velocity of the ball. (Hint: Subtract the two vectors by adding $-\vec{v}_1$ to $\vec{v}_2$ using the head-to-toe method.)

    b. Using this estimate, find the *average* acceleration of the ball.

E5.2 A ball is traveling at a constant speed of 3 m/s in a curve with a radius of 2 m. What is the centripetal acceleration of the ball?

E5.3 A car with a mass of 800 kg is moving around a curve with a radius of 30 m at a constant speed of 20 m/s (about 45 MPH).

    a. What is the centripetal acceleration of the car?

    b. What is the magnitude of the force required to produce this centripetal acceleration?

E5.4 A car with a mass of 1000 kg travels around a banked curve with a constant speed of 27 m/s (about 60 MPH). The radius of curvature of the curve is 50 m.

    a. What is the centripetal acceleration of the car?

    b. What is the magnitude of the horizontal component of the normal force that would be required to produce this centripetal acceleration in the absence of any friction?

E5.5 A Ferris wheel at a carnival has a radius of 20 m and rotates at a rate such that the speed of the riders is 8 m/s.

    a. What is the magnitude of the centripetal acceleration of the riders?

    b. What is the magnitude of the net force required to produce this centripetal acceleration for a rider with a mass of 70 kg ?

    c. When the rider is at the top of the cycle, what force or forces combine to provide this centripetal force?

E5.6 What is the ratio of the earth's orbital period about the sun to the earth's period of rotation about its axis?

E5.7 Oliver Lundy has a weight of 600 N (about 135 lb) when he is standing on the surface of the earth.

    a. What would his weight be if he was twice as far from the center of the earth as when he is standing on the earth's surface?

    b. What would his weight be if he were three times as far from the center of the earth?

E5.8 Two 1000-kg masses are separated by a distance of 1 meter.

    a. Using Newton's law of gravity, find the magnitude of the gravitational force exerted by one mass on the other.

    b. If the distance between the two masses is increased to 3 m, what is the magnitude of the force of interaction?

E5.9 The acceleration of gravity at the surface of the moon is approximately 1/6 that at the surface of the earth (9.8 m/s^2). What is the weight of an astronaut standing on the moon whose weight on earth is 180 lb?

E5.10 The acceleration of gravity on the surface of Jupiter is 26.7 m/s^2. What is the weight on Jupiter of a woman whose weight on earth is 110 lb? (Hint: How much larger is the acceleration of gravity on Jupiter than on the earth?)

E5.11 The mass of Venus is 4.87×10^{24} kg, and its radius is 6050 km (6.05×10^6 m). What is the acceleration of gravity on the surface of Venus? ($g = Gm_p / r_p^2$).

E5.12 The time separating high tides is 12 hr and 25 minutes. If high tide occurs at 3:10 P.M. one afternoon,

    a. at what time will high tide occur the next afternoon?

    b. when would you expect low tides to occur the next day?

CHALLENGE PROBLEMS

CP5.1 A Ferris wheel with a radius of 30 m makes one complete rotation every 15 seconds.

a. Using the fact that the distance traveled by a rider in one rotation is $2\pi r$, the circumference of the wheel, find the speed with which the riders are moving.

b. What is the magnitude of their centripetal acceleration?

c. What is the magnitude of the centripetal force required to keep a rider with a mass of 40 kg moving in a circle? Is the weight of the rider large enough to provide this centripetal force at the top of the cycle?

d. What would happen if the Ferris wheel went so fast that the weight of the rider was not sufficient to provide the centripetal force at the top of the cycle?

e. What is the magnitude of the normal force acting on this rider when the rider is at the bottom of the cycle?

CP5.2 A car with a mass of 900 kg is traveling around a curve with a radius of 80 m with a constant speed of 25 m/s (56 MPH). The curve is banked at an angle of 20°.

a. What is the magnitude of the centripetal acceleration of the car?

b. What is the magnitude of the net force required to produce this acceleration?

c. What is the magnitude of the vertical component of the normal force acting upon the car to counter the weight of the car?

d. Draw a careful diagram of the car (as in figure 5.10) on the banked curve. Draw to scale the vertical component of the normal force. Using this diagram, find the magnitude of the total normal force, which is perpendicular to the surface of the road.

e. Using your diagram, estimate the magnitude of the horizontal component of the normal force. Is this component sufficient to provide the centripetal force?

f. Is it possible for the car to negotiate the curve at this speed? Explain.

CP5.3 The mass of the sun is 1.99×10^{30}kg, of the earth 5.98×10^{24}kg, and of the moon 7.36×10^{22}kg. The average distance between the moon and the earth is 3.82×10^8m, and the average distance between the earth and the sun is 1.50×10^{11}m.

a. Using Newton's law of gravity, find the average force exerted on the earth by the sun.

b. Find the average force exerted on the earth by the moon.

c. What is the ratio of the force exerted on the earth by the sun to that exerted by the moon? Will the moon have much of an impact on the earth's orbit about the sun?

d. Using the earth-sun distance as the average distance between the moon and the sun, find the average force exerted on the moon by the sun.

e. What is the ratio of the force exerted on the moon by the earth to that exerted on the moon by the sun? Will the sun have much impact upon the orbit of the moon about the earth?

CP5.4 The period of the moon's orbit about the earth is 27.3 days, but the average time between full moons is approximately 29.3 days. The difference is due to the motion of the earth about the sun.

a. Through what fraction of its total orbital period does the earth move in one period of the moon's orbit?

b. Sketch the sun, the earth, and the moon with the moon in the full-moon condition. Sketch the moon's position again 27.3 days later for the new position of the earth. If the moon is in the same position relative to the earth as it was 27.3 days earlier, is this a full moon?

c. How much farther would the moon have to go to reach the full-moon condition? Show that this represents approximately two extra days.

Chapter 5

100

HOME EXPERIMENTS AND OBSERVATIONS

HE5.1 Tape a string, half a meter or so in length, securely to a small rubber ball. Practice whirling the ball in both horizontal and vertical circles and make the following observations.

a. For horizontal motion of the ball, how does the angle that the string makes with your hand vary with the speed of the ball?

b. If you let go of the string at a certain point in the circle, what path does the ball follow after release?

c. Can you feel differences in tension in the string for different speeds of the ball? How does the tension vary with speed?

d. For a vertical circle, how does the tension in the string vary for different points in the circle? Is it greater at the bottom than at the top when the ball moves with constant speed?

HE5.2 Attach a small paper cup to a string at two points near the rim, as shown in the diagram. Take a marble or other small object and place it in the cup.

a. Whirl the cup in a horizontal circle. Does the marble stay in the cup?

b. Whirl the cup in a vertical circle. Does the marble stay in the cup? What keeps the ball in the cup at the top of the circle?

c. Try slowing the cup down. At what point does the marble start to fall?

d. If you are feeling brave, put some water in the cup and repeat some of these observations.

HE5.3 Using strips of cardboard and tape, build yourself a banked curve for a small toy car. Make it flexible enough so that you can vary both the radius of curvature and the banking angle.

a. Try propelling the car at different initial speeds around the curve. Is there an optimal speed for the car to make it all the way around the curve?

b. What happens if the car is started around the curve too slowly or too fast?

c. Vary the radius of the curve without changing the banking angle. How does this affect the observations of (*a*) and (*b*)?

d. Try varying the banking angle without changing the radius of the curve. How does this affect your observations?

HE5.4 Using a pencil, a piece of string, and a piece of paper, try drawing ellipses in the manner illustrated in figure 5.16.

a. How does the shape of the ellipse change if you move the two foci, to which the string is attached, closer together or farther apart?

b. What happens if the two points of attachment coincide?

HE5.5 Observe the position and phase of the moon on several days in succession and at regularly chosen times during the day and evening. (Depending on where the moon is in its phase cycle, morning observations may be more appropriate than evening observations. Think about this before you choose your times.)

a. Sketch the shape of the moon on each successive day. Does this shape change for different times in the same day?

b. Can you devise a method for accurately noting changes in the position of the moon at a set time, say 10:00 P.M., on successive days? A fixed sighting point, a meter stick, and a protractor may be useful.

c. By how much does the position of the moon change from one day to the next at your chosen time?

HE5.6 If you are located near enough to an ocean (or one of the Great Lakes) to make this feasible, observe the height of the tides on successive days. A dock support or some other vertical piling makes a good point of reference. Tidal charts are useful for determining the exact times of high and low tides.

a. How much variation do you see in the height of the high tides on different days?

b. Are the low tides the lowest on the same days that high tides are the highest?

6

Momentum and Impulse

The word *momentum* is overused by sports announcers to signify changes in the flow of a game, and is therefore part of our everyday language. That "old mo" that the pundits talk about does bear a metaphorical relationship to the physical concept of momentum. Plenty of real examples of changes in momentum of balls, people, and other objects are available, however, in both the world of sports and the world more generally.

Take, for example, a collision between a hard-charging fullback and a fearless defensive back on the football field (fig. 6.1). If they meet head-on, the velocity of the fullback is sharply reduced, although the two players might continue moving in the original direction of the fullback's velocity briefly before hitting the ground. If the defensive back is also moving before the collision, his velocity also changes abruptly. Strong forces must be at work to produce these accelerations, but these forces act for only an instant. How can we use Newton's laws to analyze this kind of situation?

The concepts of *impulse, momentum,* and *conservation of momentum* play key roles in any discussion of collisions. It turns out that the initial momenta of both the fullback and defensive back are involved in predicting what will happen after the collision. But how is momentum defined and what, if anything, does conservation of momentum have to do with Newton's laws? Finally, how can the idea of conservation of momentum be useful in predicting what happens in collisions?

These questions will be addressed as we examine a variety of collisions and other high-impact events.

Figure 6.1   A collision between a fullback and a defensive back. How will the two players move after the collision?

Chapter Objectives

The primary aim of this chapter is to define impulse and momentum and to examine how these concepts are useful in analyzing events such as the collision between the fullback and defensive back. The law of conservation of momentum is introduced and the limits of its validity explained. A number of examples illustrate the utility of these concepts, particularly that of conservation of momentum.

The order of topics and the questions addressed under each are indicated, as usual, in the chapter outline. The concept of momentum is central to all of these topics; you should find that it is a powerful tool for understanding a lot of life's sudden changes!

Chapter Outline

1 *Impulse and momentum.* How are the concepts of impulse and momentum defined and how are they related to Newton's second law of motion?

2 *Conservation of momentum.* What is the law of conservation of momentum, and when is it valid? How does this law follow from Newton's laws of motion?

3 *Recoil.* How are momentum ideas useful in explaining the recoil of a rifle or shotgun? How are these situations similar to what happens in the firing of a rocket?

4 *Elastic and inelastic collisions.* How can collisions be analyzed using conservation of momentum? What is the difference between an elastic and an inelastic collision?

5 *Collisions involving motion in two dimensions.* How can we extend momentum ideas to two dimensions? How is what happens in the game of pool similar to what happens in automobile collisions?

6.1 IMPULSE AND MOMENTUM

Imagine a baseball heading towards the catcher's mitt when its flight is rudely interrupted by the impact of a bat. In a very short interval of time, the velocity of the ball changes direction, and the ball is accelerated in the direction opposite to that of its original path. Similar changes occur when a tennis racket hits a ball or when a ball bounces off of a wall or the floor. There are many everyday situations in which a brief impact causes a rapid change in an object's velocity.

Consider the seemingly simple example of dropping a ball. The ball is initially accelerated downward by the force of gravity. When it reaches the floor, however, its velocity quickly changes direction, and the ball heads back up toward you (fig. 6.2). Obviously, a force is exerted on the ball by the floor in the brief instant that they are in contact. This force provides the upward acceleration necessary to change the direction of the ball's velocity.

If we could use a high-speed camera to catch the action during the time that the ball is in contact with the floor, we would see that the ball's shape is distorted during the process (fig. 6.3). The ball actually behaves like a spring, first being compressed as it is still moving downward, and then expanding (springing back) as it begins to move upward. A quick test—squeezing the ball with your hands—would convince you that a strong force is required to distort the ball very much.

What we have, then, is a strong force acting for a very brief time to produce an acceleration that changes the ball's velocity from a downward direction to an upward one. The magnitude of the ball's velocity must therefore decrease rapidly to zero and then increase equally rapidly in the oppo-

site direction. This all happens in a time so short that we would miss it if we blinked.

We have described this collision of the ball with the floor in terms of force and acceleration, and we could use Newton's second law to predict how the velocity actually changes. The problem with this approach is that the time of action is very short, and the force itself varies during this short time, so that it is difficult to describe it accurately. It turns out to be more productive to look instead at the total change in motion produced in this brief interaction.

Definitions of Impulse and Momentum

If we write Newton's second law in the usual form, $\vec{F} = m\vec{a}$, and replace $\vec{a}$ with the definition of acceleration as the rate of change in velocity, the second law can be cast in a different form. Since $\vec{a} = \Delta\vec{v} / \Delta t$ (see chapter 2), then

$$\vec{F} = m\vec{a} = m\left(\frac{\Delta\vec{v}}{\Delta t}\right).$$

Multiplying both sides of this equation by the time interval, Δt allows us to express the second law as follows:

$$\vec{F}\Delta t = m\Delta\vec{v}.$$

This is still Newton's second law, but writing it in this manner provides a different way of looking at events.

The product on the left side of this equation, $\vec{F}$ times Δt, is called the *impulse*. The impulse is merely the force acting on the object multiplied by the time interval during which the force acts. If the force is varying during this time interval, as it often is, then we must use the average value of the force, $\overline{\vec{F}}$, over the time that it acts. The more general definition of impulse, then, would take this form:

The impulse is the average force multiplied times its time of action, or impulse = $\overline{\vec{F}}\Delta t$.

The quantity on the right side of our rewritten second-law expression, $m\Delta\vec{v}$, is the mass of the object multiplied by the

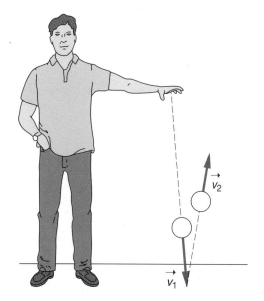

Figure 6.2 A ball bouncing off the floor. There is a rapid change in the direction of the ball's velocity when it hits the floor.

Figure 6.3 A high-speed photograph of a ball hitting the floor. The ball is compressed like a spring.

change in its velocity that is produced by the impulse. This quantity represents the change in the motion of the object. Since $\Delta\vec{v}$ can be expressed as the difference between the final velocity and the initial velocity, $\vec{v}_f - \vec{v}_i$, we can recast $m\Delta\vec{v}$ as follows:

$$m\Delta\vec{v} = m(\vec{v}_f - \vec{v}_i) = m\vec{v}_f - m\vec{v}_i.$$

The product of the mass times the velocity, $m\vec{v}$, is called the *momentum* and is often represented by the symbol $\vec{p}$. Momentum can be regarded as a measure of the *quantity of motion*, which was the term used by Newton himself in discussing his second law of motion.

Momentum is defined as the product of the mass, *m*, of an object times its velocity $\vec{v}$, or $\vec{p} = m\vec{v}$.

The quantity $(m\vec{v}_f - m\vec{v}_i)$ is thus the change in momentum of the object, $\Delta\vec{p} = \vec{p}_f - \vec{p}_i$, which is another way of describing the right side of our recast version of Newton's second law.

This quantity of motion, which we call *momentum*, depends both on the size (and direction) of the velocity and on the mass of the object. Momentum is a vector quantity, like velocity, and has the same direction as the velocity vector. Two different objects can have very different masses and velocities, but still have the same momentum. A 7-kg bowling ball, for example, moving with the relatively slow velocity of 2 m/s would have a momentum of the following magnitude:

$$p = mv = (7 \text{ kg})(2 \text{ m/s}) = 14 \text{ kg·m/s}.$$

A tennis ball, on the other hand, with a mass of just 0.070 kg (70 grams) but moving in the same direction as the bowling ball with the much larger velocity of 200 m/s, would have the same momentum as the bowling ball (see fig. 6.4).

Using these definitions of impulse and momentum, we can express Newton's second law in the following form:

$$\overrightarrow{\text{impulse}} = \Delta\vec{p},$$

where we have substituted the impulse for $\vec{F}\Delta t$ and the change in momentum, $\Delta\vec{p}$, for $m\Delta\vec{v}$. This statement of the second law is sometimes referred to as the *impulse/momentum theorem*. In words, it can be stated as follows:

The impulse acting upon an object produces a change in momentum of the object that is equal in both size and direction to the impulse.

It is important to bear in mind, however, that this is merely another way of expressing Newton's second law of motion.

Applications of the Impulse/Momentum Theorem

As an example of the application of these ideas, we might consider a golf ball, at rest initially, being struck by a golf club. If we assume that the golf club exerts an average force of 50 N on the ball, but that the club is in contact with the ball for only 0.1 s, then the magnitude of the impulse is just the force times the time of action, or

$$\text{impulse} = F\Delta t = (50 \text{ N})(0.10 \text{ s}) = 5.0 \text{ N·s}.$$

Before the ball is hit, its momentum, $m\vec{v}$, is zero because it is not moving, and its velocity is zero. Immediately after being hit, the ball takes off with a large velocity, $\vec{v}$, and its momentum after being hit is $m\vec{v}$. The change in momentum is just the difference between the final momentum $\vec{p}_f$ and the initial momentum, or in symbols,

$$\Delta\vec{p} = \vec{p}_f - \vec{p}_i.$$

In this case, then, $\Delta\vec{p}$ is equal to the final momentum because the initial momentum, $\vec{p}_i$ is zero. Since the magnitude of the change in momentum, Δp, must equal the impulse (according to the impulse/momentum theorem), Δp is equal to 5.0 N·s because that was the value of the impulse (fig. 6.5). The momentum of the ball immediately after the impact, then, must also be 5.0 N·s (or 5.0 kg·m/s). (Impulse and momen-

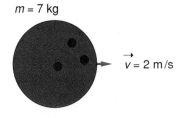

$m = 7$ kg

$\vec{v} = 2$ m/s

$m = .070$ kg

$\vec{v} = 200$ m/s

Figure 6.4   A bowling ball and a tennis ball with the same momentum. The tennis ball with its smaller mass must have a much larger velocity.

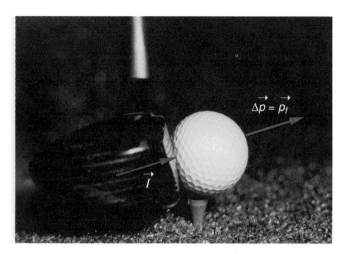

Figure 6.5 An impulse is delivered to the golf ball by the club head. If the initial momentum of the ball is zero, the final momentum is equal to the impulse delivered.

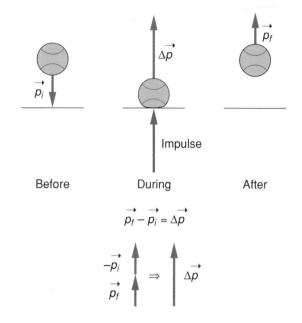

Figure 6.6 The impulse exerted by the floor on the tennis ball produces a change in its momentum.

tum must have the same units, and these can be expressed either as force times time, Newton-seconds (N·s), or mass times velocity, kilogram-meters per second (kg·m/s). These are equivalent units because 1 N = 1 kg·m/s^2.)

If the ball has a mass of 0.05 kg (50 grams), then we could also find its velocity after the impact by dividing the momentum by the mass.

$$p = mv,$$

so

$$v = \frac{p}{m} = \frac{5.0 \text{ kg·m/s}}{.05 \text{ kg}} = 100 \text{ m/s}.$$

Knowledge of the impulse, then, allows us to find both the change in momentum and the resulting velocity of the ball.

Returning to our earlier example of the ball bouncing off the floor, we can now redescribe this situation using these ideas. The floor exerts an impulse on the ball, which causes the momentum of the ball to change. The total change in momentum of the ball must be equal to the impulse. Since momentum is a vector quantity, the change in momentum is related to changes in both the magnitude *and direction* of the velocity vector. In this case, most of the change in momentum is associated with the change in direction of the ball as it hits the floor.

To illustrate this last point, we could assume that the ball bounces back with an upward velocity after hitting the floor that is equal in magnitude, but opposite in direction, to the velocity that it had just before hitting the floor. The change in momentum is again found by subtracting the initial momentum from the final momentum (those values just before and just after the collision with the floor). Suppose, for example, that the ball has a mass of 0.06 kg and a downward velocity of 10 m/s just before hitting the floor. Its initial momentum, *mv*, is then

$$p_i = mv = (0.06 \text{ kg})(-10 \text{ m/s}) = -0.60 \text{ kg·m/s},$$

where we have chosen the upward direction to be positive, so the minus sign indicates downward motion.

The momentum of the ball after hitting the floor would then be +0.60 kg m/s, since we are assuming that it bounces back with an equal but oppositely directed velocity. The change in momentum is therefore

$$\Delta p = p_f - p_i = 0.60 \text{ kg·m/s} - (-0.60 \text{ kg·m/s}) = 1.20 \text{ kg·m/s}.$$

Two minus signs are equivalent to a plus sign, so the magnitudes of the two momenta add. The change in momentum cannot be zero in this situation because momentum is a vector quantity, and its direction has changed. The magnitude of the momentum has not changed in this case, but its direction has. The change in momentum turns out to be just twice the magnitude of either the initial or final momentum for this situation.

An upwardly directed impulse is exerted on the ball by the floor to produce this change in momentum (fig. 6.6). According to Newton's second law in the form of the impulse/momentum theorem, that impulse must have a magnitude of 1.20 N·s in order to produce the change in momentum of 1.20 kg·m/s. If we were able to estimate the time of contact, we could also calculate the average force exerted by the floor on the ball from the definition of impulse, since the impulse is equal to the average force $\overline{F}$ times the time Δt. Since Δt is quite small, $\overline{F}$ will be relatively large. It must be considerably larger than the force of gravity on the ball, which continues to pull downward on the ball throughout this process.

The Second Law Revisited

It should be emphasized that all of this is just another way of working with Newton's second law of motion. In fact, when we write Newton's second law in terms of momentum, we are really writing it in a more generally valid form than when we write it as $\vec{F} = m\vec{a}$. Since $\overrightarrow{\text{impulse}} = \vec{F}\Delta t$, we could restate the second law as follows:

$$\overrightarrow{\text{impulse}} = \vec{F}\Delta t = \Delta\vec{p}.$$

Dividing both sides by the time interval, Δt, we have

$$\vec{F} = \frac{\Delta\vec{p}}{\Delta t} .$$

In words, Newton's second law in this form says that the net force acting upon an object is equal to the rate of change of momentum of the object. This reduces to $\vec{F} = m\vec{a}$ when the mass of the object in question is constant, as it often is. Then the change in momentum is just the mass times the change in velocity, which is related directly to the acceleration.

In cases where the mass is changing, however, the change in mass will also affect the momentum. For example, as a rocket burns fuel, its mass decreases while its velocity increases; both of these changes affect its momentum. It is the rate of change in momentum with time that must be used in the second law in such cases. In fact, this was the way in which Newton originally stated the second law; he said, in effect, that the net force is proportional to the rate of change of the quantity of motion. Newton's "quantity of motion" is what we now call *momentum*.

6.2 CONSERVATION OF MOMENTUM

The concept of momentum is particularly useful in analyzing collisions such as that between the fullback and the defensive back in our opening example. It is the principle of conservation of momentum, which also follows from Newton's laws, that makes momentum itself so useful. A look at this principle and its conditions of validity will illustrate why we can often predict many features of collisions without needing a detailed knowledge of the forces of impact.

The Conservation Principle

Suppose that we take a closer look at a head-on collision, such as that between our two football players, using the concepts developed in the previous section. We can readily see that each player exerts a strong impulse on the other during the collision. By Newton's third law, however, the forces involved in these impulses must be equal in magnitude but opposite in direction (fig. 6.7). Since the time interval, Δt, is the same for both impulses, the impulses themselves, $\vec{F}\Delta t$, must also be equal in magnitude but opposite in direction.

From Newton's second law in the form of the impulse/momentum theorem, however, we know that the

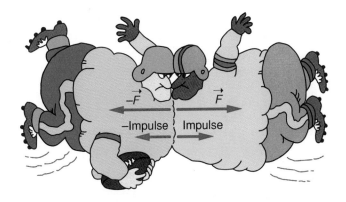

Figure 6.7 **Two football players colliding. The impulses acting upon the two players are equal in magnitude but opposite in direction.**

change in momentum of each player must equal the impulse acting upon that player ($\Delta p = \text{impulse} = \vec{F}\Delta t$). Since the impulses acting on each player are equal in magnitude but oppositely directed, so are the momentum changes for each player. In symbols this can be expressed as follows:

$$\Delta p_1 = -\Delta p_2.$$

The total change in momentum of the system (consisting of both players) is therefore zero because the changes cancel one another. In symbols,

$$\Delta p_T = \Delta p_1 + \Delta p_2 = -\Delta p_2 + \Delta p_2 = 0.$$

What this says is that the total momentum of the system does not change; in other words, it is conserved. The increase in momentum of one player equals the decrease in momentum of the other, so that the total change in momentum of the *system* is zero.

This, then, is an illustration of the principle of conservation of momentum. When we say that a physical quantity is *conserved,* we mean that it is constant or unchanging. In this case, we have assumed that the only important forces involved are the impact forces exerted by each player on the other. If this is so, then the total momentum of the system, *found by adding together (as vectors) the momenta of each object in the system,* is unchanged, or conserved. Both Newton's second law and his third law were involved in arriving at this conclusion.

A concise statement of the law of conservation of momentum would take the following form:

> **If the net external force acting on a system of objects is zero, then the total momentum of the system is conserved.**

The forces of interaction between the objects in the system are internal forces whose effects on the total momentum cancel because of Newton's third law of motion. Different portions of the system can exchange momentum without affecting the total momentum of the system. If there is a net

Before

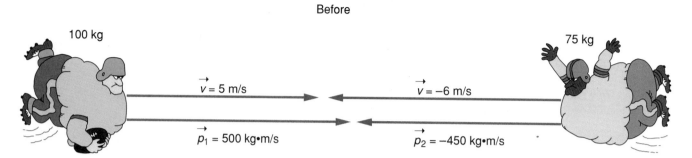

100 kg

$\vec{v}$ = 5 m/s

$\vec{v}$ = −6 m/s

75 kg

$\vec{p}_1$ = 500 kg•m/s

$\vec{p}_2$ = −450 kg•m/s

Figure 6.8 **The two football players before the collision, with velocity and momentum vectors for each.**

external force acting upon the system, however, then the entire system will be accelerated, and its momentum will change.

An Application of Conservation of Momentum

What can the principle of conservation of momentum tell us about our colliding football players? If we know the masses of the players and their initial velocities, we can find in what direction they will move (and with what velocity) after they collide. We do not need to know anything at all regarding the details of the strong interaction forces involved in the collision.

Suppose that the fullback has a mass of 100 kg (equivalent to a weight of about 220 lb) and is moving straight down field with a velocity of 5.0 m/s through the hole created by his linemen. His momentum before the collision is then

$$p_1 = m_1 v_1 = (100 \text{ kg})(5.0 \text{ m/s}) = 500 \text{ kg·m/s}.$$

The defensive back charges up to meet him with a velocity in the opposite direction of −6.0 m/s. (The minus sign indicates the direction; we are choosing the direction of the original motion of the fullback to be positive.) If the defensive back has a mass of 75 kg, then his momentum is

$$p_2 = m_2 v_2 = (75 \text{ kg})(-6.0 \text{ m/s}) = -450 \text{ kg·m/s}.$$

The total momentum of the system prior to the collision is then found by adding these two values, taking into account the difference in sign that is associated with the difference in direction (fig. 6.8). This yields

$$p_T = p_1 + p_2 = 500 \text{ kg·m/s} + (-450 \text{ kg·m/s}) = 50 \text{ kg·m/s}.$$

If we assume that both players have left their feet just before the collision, so that there are no frictional forces between their feet and the ground, then conservation of momentum should hold. The total momentum of the two players, now joined in the embrace of the tackle, should be 50 kg·m/s right after the collision also (fig. 6.9).

This conclusion allows us to compute the velocity with which the two players are traveling immediately after the

After

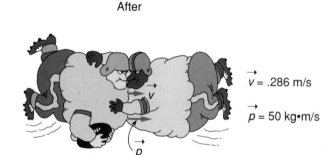

$\vec{v}$ = .286 m/s

$\vec{p}$ = 50 kg•m/s

Figure 6.9 **The two players immediately after the collision, with velocity and momentum vectors indicated.**

collision (and before hitting the ground). The total mass of the two players is the sum of their individual masses, 100 kg + 75 kg = 175 kg. Their common velocity is therefore:

$$v = \frac{p_T}{m_T} = \frac{50 \text{ kg·m/s}}{175 \text{ kg}} = 0.286 \text{ m/s}.$$

The fact that this value is positive indicates that the two players are traveling in the direction of the fullback's initial motion, which we had chosen to be the positive direction at the outset. The greater initial momentum of the fullback dictates that the defensive back will be carried backwards briefly before the two players hit the turf.

This, then, is a real example of momentum at work in the world of sports. The principles involved, however, apply to all sorts of situations involving collisions or explosions or other forms of brief but forceful interaction between objects. The next few sections of this chapter introduce other examples in which conservation of momentum is a powerful predictor of what will happen.

6.3 RECOIL

The phenomenon of recoil is a common part of everyday experience, and yet it is frequently misunderstood. It is involved in the firing of a gun, the throwing of a ball, or in the propulsion of a rocket, as well as in any situation where two objects push off against one another. Conservation of momentum provides the key to understanding it.

The Push-off

Imagine two ice skaters facing one another and pushing against one another with their hands (fig. 6.10). The frictional forces between their skates and the ice are presumably very small, so that we can neglect them. The upward normal force and the downward force of gravity cancel one another, since we know that there is no acceleration in the vertical direction. The net external force acting on the system of the two skaters is therefore essentially zero, and the law of conservation of momentum should apply.

Prior to the push-off, neither skater is moving, so the initial momentum of the system is zero. Since momentum is conserved, this means that after the push-off, the total momentum of the system should still be zero. But now we have both skaters moving in opposite directions. How fast does each one travel?

As pictured, one of the skaters is obviously somewhat larger than the other. Let us suppose that her mass is 100 kg and that the smaller skater has a mass of 50 kg. If the total momentum of the system is to remain zero after the push-off, the momenta of the two skaters must be equal in magnitude but opposite in direction. This implies that the smaller skater will end up with the larger velocity.

We can use symbols to present this argument more concisely. The total momentum of the system before and after the push-off is

$$p_T = p_1 + p_2 = 0,$$

which implies that

$$p_2 = -p_1.$$

The minus sign, again, indicates that the momentum of the second skater, p_2, is in the opposite direction to the momentum of the first skater, p_1. As expected, then, the two skaters will move away from one another after the push-off (fig. 6.11).

If we rewrite this second relationship in terms of the masses and velocities of the two skaters, we have

$$m_2 v_2 = -m_1 v_1.$$

The conclusion that we stated earlier is easily seen from this equality; in order for the equality to hold, the skater with the larger mass must have the smaller velocity and vice versa. In fact, by dividing both sides of the equality by mass m_2, we see that

$$v_2 = -\left(\frac{m_1}{m_2}\right)v_1.$$

It is the ratio of the two masses that determines the relative sizes of the velocities. If the big skater has a mass twice that of the smaller skater, as is the case with the numbers given here, then the velocity of the second (and smaller) skater would be twice that of the larger skater.

This example involving ice skaters illustrates the basic principles of any recoil situation. A brief force between two objects causes the objects to move in opposite directions, with their velocities being determined by the ratio of their masses. The lighter object gains the larger velocity, so that

Figure 6.10 Two skaters of different sizes prepare to push off against one another. Which one will acquire the larger velocity?

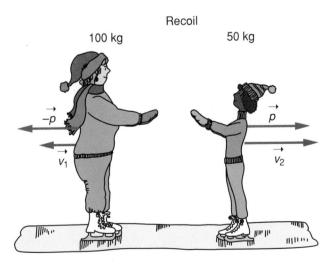

Figure 6.11 The two skaters after pushing off with velocity and momentum vectors indicated.

Figure 6.12 **The shot and the shotgun have equal but oppositely directed momentum after the gun is fired.**

Figure 6.13 **If a short blast is fired, a rocket and its exhaust gases gain equal momenta in opposite directions.**

the momenta of the two objects can be equal in magnitude. This yields a total momentum for the system of zero, which is what we started with.

Firing a Shotgun

If you have ever fired a twelve-gauge shotgun without holding it firmly against your shoulder, you have probably had painful firsthand experience with recoil. What is happening here? The explosion of the powder in a shotgun or rifle causes the shot or bullet to move very rapidly in the direction in which the gun is aimed. If the gun is somewhat free to move, it will move in the opposite direction with a momentum that is equal in magnitude to that of the shot or bullet (fig. 6.12).

Even though the mass of the shot is considerably less than that of the shotgun, its momentum can be quite large as a result of its large velocity. If the external forces acting on the system can be neglected, the shotgun recoils with a momentum equal in magnitude to that of the shot. The recoil velocity of the shotgun will be smaller than the shot velocity because of the larger mass of the gun, but it can still be very significant. As the gun comes slamming back against your shoulder, you will know that it has recoiled!

How can you avoid a bruised shoulder? The trick is to hold the gun firmly against your shoulder. Then your own mass becomes part of the system; the shotgun and the shooter can be regarded as one object. This increased mass will produce a smaller recoil velocity, even if you happen to be standing on ice so that there are no frictional forces between your feet and the earth. More important, there is essentially no motion of the shotgun relative to your shoulder. If we consider the shotgun as a separate object, your shoulder is exerting an external force on the shotgun, thus removing the condition necessary for conservation of momentum.

Rockets

The firing of a rocket provides another example of recoil at work. The exhaust gases that are shot out of the tail end of

the rocket have both mass and velocity and therefore momentum. If we can ignore external forces, then the momentum gained by the rocket in the forward direction will equal the momentum acquired by the exhaust gases in the opposite direction (fig. 6.13). Conservation of momentum holds, and the rocket and exhaust gases are recoiling from one another. Newton's third law is also at work; the exhaust gases and the rocket push against one another.

The difference between what happens with a rocket and what happens in our earlier examples of the skaters and the shotgun is that the firing of the exhaust gases in a rocket is usually a continuous process. The rocket gains momentum gradually rather than in a single short blast. The mass of the rocket is also changing as fuel is consumed and exhaust gases are fired out of the back end. This makes computing the final velocity more difficult than it was for the skaters. If a brief blast of the rocket occurs, however, the same type of analysis could be employed.

It is important to recognize that this recoil process will work in empty space; there is no need for two objects to push against anything else but each other, just as with the skaters and the gun. Rocket engines can therefore be used in space travel, unlike the propeller engines or jet engines used on airplanes. Airplane engines depend upon the presence of the atmosphere, both as a source of oxygen used in burning the fuel and as something to push against. A propeller, for example, pushes against the air, and the air, by Newton's third law, pushes against the propeller. It is this interaction that provides the force that accelerates the airplane. A rocket, on the other hand, is self-contained; it pushes against its own exhaust gases.

6.4 ELASTIC AND INELASTIC COLLISIONS

Collisions provide one of the most fruitful areas for applying conservation of momentum. As the earlier example involving football players illustrated, collisions generally involve large forces of interaction acting for very brief times and producing dramatic changes in the momenta of the colliding objects. Because the forces of interaction are so large, any external forces acting on the system are often unimportant by comparison, and conservation of momentum can be assumed to hold.

Inelastic Collisions

The easiest type of collision to analyze using conservation of momentum is one in which two objects collide head-on and then stick together after the collision, as happened with the two football players. The fact that they stuck together and moved as one object after the collision meant that we had just one final velocity to contend with.

Such a collision, in which the objects stick together after collision, is referred to as a *perfectly inelastic* collision. The objects do not bounce at all. If the total momentum of the system before the collision is known, and external forces can be ignored, than we can easily compute the final momentum and velocity of the stuck-together objects.

Coupling railroad cars provide a simple example of this type of collision. The sample exercise in box 6.1 uses conservation of momentum to predict the final momentum and velocity of railroad cars after they have coupled from knowledge of the momentum of the system prior to the collision. The process is essentially the same as the one we used in section 6.2 to predict the final velocity of the coupled football players.

The total mass of the coupled cars after the collision is five times that of the initial car, so the final velocity of the coupled cars must be ⅕ that of the initial car in order to yield a constant momentum. The final velocity that we computed in the sample exercise is that of the coupled cars immediately after they have collided; frictional forces will gradually decelerate the cars thereafter. These external frictional forces are small, however, in comparison to the very large force of interaction between the cars when they collide, so we are justified in assuming conservation of momentum for the collision itself.

Box 6.1

Sample Exercise

Four railroad cars, all with the same mass (20 000 kg) are sitting on a track as pictured in the drawing. A fifth car of identical mass approaches them with a velocity of 15 m/s. This car collides and couples with the other four cars. What is the velocity of the five coupled cars after the collision?

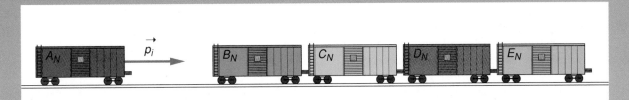

A railroad car approaches four others that are at rest on the track. What is the velocity of the cars after they couple?

$m_1 = 20\ 000$ kg.

$v_1 = 15$ m/s.

The initial momentum of the system is:

$$p_i = m_1 v_1 = (20\ 000\ \text{kg})(15\ \text{m/s})$$
$$= 300\ 000\ \text{kg} \cdot \text{m/s}.$$

$m_T = 5\ m_1$.

$m_T = 100\ 000$ kg.

$v_f = ?$

If momentum is conserved, then the final momentum is

$$p_f = m_T v_f = p_i.$$

The final velocity, then, is

$$v_f = \frac{p_i}{m_T} = \frac{(300\ 000\ \text{kg m/s})}{100\ 000\ \text{kg}}$$

$$= 3.0\ \text{m/s}.$$

If, instead of sticking together, the colliding objects bounce off of one another, then we must deal with two final velocities. We can easily compute the total momentum of the system after the collision from knowledge of the initial momentum, but this total would be the sum of the individual momenta of the two objects. We would need more information in order to determine the individual velocities of the objects. This is why the case in which the objects stick together in a perfectly inelastic collision is particularly easy to analyze.

Elastic Collisions

When the objects bounce, the collision is either *elastic* or only *partially inelastic* rather than perfectly inelastic. The distinction between these cases is based upon energy considerations. An elastic collision is one in which no kinetic energy is lost. A partially inelastic collision is one in which some energy is lost, but the objects do not stick together. The concepts of work and energy will be developed in chapter 8; these statements will be more meaningful after you understand those ideas.

For our purposes here, however, some sense of the meaning of kinetic energy may be sufficient. Kinetic energy is the energy that an object has by virtue of its motion. It is related to the mass and velocity of the object by the formula $KE = (\frac{1}{2})mv^2$. If the total kinetic energy of the objects after a collision is equal to the total kinetic energy of the objects before the collision, then the collision is elastic; no kinetic energy has been lost. (Energy is not a vector quantity, so the kinetic energies of different objects in the system can be added directly; they are always positive.)

In most collisions, some kinetic energy is lost; the collisions are not perfectly elastic. Heat is generated, the objects may be deformed, and sound waves are created, all of which represent the conversion of kinetic energy to other forms of energy. Even if the objects bounce, then, we cannot assume that the collision is elastic. It is more likely that the collision will be partially inelastic, implying that some, but not all, of the initial kinetic energy has been lost.

A ball bouncing against a floor or wall with no decrease in the magnitude of its velocity, as discussed in section 6.1, is an example of an elastic collision. Since the magnitude of the velocity does not change (only the direction of the velocity changes), the kinetic energy does not decrease, and no energy has been lost. It is more likely, of course, that some energy will be lost in such a collision, and that the magnitude of the ball's velocity after the collision will be a little smaller than it was before the collision.

The opposite of an elastic collision of a ball with the wall would be a perfectly inelastic collision in which the ball sticks to the wall. In this case, the velocity of the ball is zero after the collision and so is its kinetic energy; all of the kinetic energy is lost (fig. 6.14). Momentum can still be regarded as being conserved in this collision, however, if

we include the wall and the earth as part of our system. The ball imparts its momentum to the wall, which in turn may be firmly attached to the surface of the earth. The mass of the wall and the earth are so large in comparison to that of the ball that only a tiny velocity is required to yield a momentum equal to the initial momentum of the ball.

Playing pool provides examples of collisions that are essentially elastic; very little energy is lost when pool balls collide with one another. (Time spent playing pool can therefore be justified as being a form of experimental physics. Your intuition regarding elastic collisions can be improved in the process.) The simplest case is that of one ball striking another in a head-on collision in which the second ball is not moving before the collision (fig. 6.15).

What happens in this situation? If spin is not a major factor in the collision, the first ball stops dead at the point of impact, and the second ball moves forward with a velocity equal to that which the first ball had before the collision. This result can be readily explained. Both balls have the same mass, so that if the second ball acquires the same velocity that the first ball had before the collision, it then also has the same momentum, $m\vec{v}$, as the initial momentum of the first ball. Momentum is therefore conserved.

The kinetic energy is also conserved given this result. The first ball had a kinetic energy of $(\frac{1}{2})mv^2$ before the collision. After the collision its velocity and kinetic energy are zero, but the second ball now has a kinetic energy of $(\frac{1}{2})mv^2$ since its mass and velocity are the same as that of

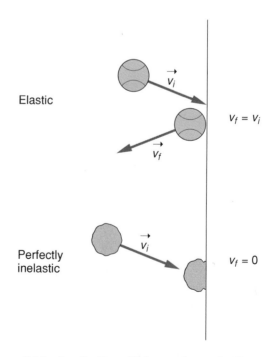

Figure 6.14   An elastic collision and a perfectly inelastic collision of a ball with a wall. The ball sticks to the wall in the perfectly inelastic collision.

Figure 6.15 A photo of a head-on collision between a cue ball and a second ball initially at rest. The cue ball (white) stops and the 11 ball moves forward.

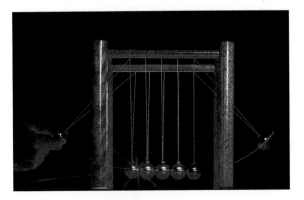

Figure 6.16 The swinging ball apparatus provides an example of collisions that are approximately elastic.

the first ball before the collision. Because of the equal masses of the two balls, it turns out that the only way that both momentum and kinetic energy can be conserved is for the first ball to stop and the second ball to move forward with the same momentum and kinetic energy originally associated with the first ball. This effect is familiar to any pool player.

The same phenomenon is involved in the familiar swinging ball demonstration, which is often seen as a decorative toy on mantles or desk tops (see fig. 6.16). This apparatus consists of a row of steel balls suspended with threads from a metal or wooden frame. If one ball is pulled back and released, the collision with the other balls results in a single ball from the other end of the row flying off with the same velocity that the first ball had just before the collision. Both momentum and kinetic energy are thus conserved.

If two balls are pulled back on one side and then released, then two balls fly off from the opposite side of the row after the collision. Again, both momentum and kinetic energy are conserved by this result. You can explore a variety of other combinations for your entertainment and edification. The process can be somewhat hypnotic.

Collisions between hard spheres, such as the steel balls in the swinging-ball apparatus or pool balls, will generally be approximately elastic. It is important to recognize, however, that most collisions involving everyday objects are inelastic to some degree; some kinetic energy is usually lost. Momentum will be conserved in any case, as long as our concern is with the values of momentum and velocity immediately before and immediately after the collision. External forces can usually be ignored for the brief time interval in which the collision forces themselves are dominant, and the law of conservation of momentum therefore applies.

6.5 COLLISIONS INVOLVING MOTION IN TWO DIMENSIONS

Some of the most interesting applications of conservation of momentum arise when the motion is not confined to a straight line, as it has been in all of the examples up to this point. The vector nature of momentum and its involvement in conservation of momentum are more obvious in cases where the objects are free to move in two dimensions. A pool table provides that freedom, as does any situation where objects move on a two-dimensional plane; cars colliding at an intersection are another common example.

An Inelastic Two-Dimensional Collision

What happens, for example, if two objects, traveling originally at right angles to one another, collide and stick together? In what direction will they move after the collision? The football players in the first example of this chapter can be our model.

Suppose that the fullback and the defensive back have the same masses and initial speeds as in the earlier example, but now they are traveling at right angles to each other prior to their collision, as shown in figure 6.17. Their momenta have the same magnitudes as before, but the direction of the defensive back's momentum is now directed across the field. What is the total momentum of the system prior to the collision in this case?

Since momentum is a vector, we need to add the individual momenta of the fullback and the defensive back as vectors to get the total momentum of the system. This can be done most easily by using a vector diagram in which the vectors are drawn to scale, and adding the vectors head-to-toe, as we have done before. (See chapter 2 and appendix C.) This process is illustrated in the diagram of figure 6.18.

(Top view)

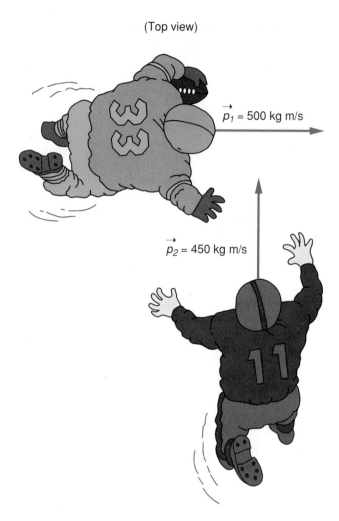

$\vec{p_1} = 500$ kg m/s

$\vec{p_2} = 450$ kg m/s

Figure 6.17 The fullback and the defensive back approaching each other at a right angle (viewed from above).

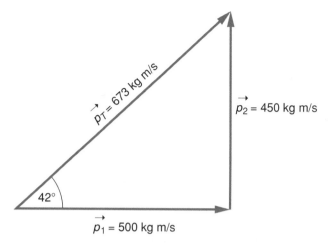

$\vec{p_T} = 673$ kg m/s

$\vec{p_2} = 450$ kg m/s

42°

$\vec{p_1} = 500$ kg m/s

Figure 6.18 The total momentum of the two football players prior to the collision is the vector sum of their individual momenta.

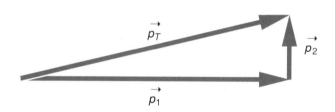

$\vec{p_T}$

$\vec{p_2}$

$\vec{p_1}$

Figure 6.19 If the defensive back has a smaller initial momentum $\vec{P_2}$, then the total momentum is closer in direction to the initial momentum of the fullback, $\vec{P_1}$.

We can then find the direction and the magnitude of the total momentum of the system by measuring the direction and length of the arrow that represents the vector sum of the two original momentum vectors. This sum is the initial momentum of the system prior to the collision. If we assume that conservation of momentum holds during the collision, then the total momentum of the system after the collision must have this same direction and magnitude. We can use this fact to find the direction in which the two players will be moving together after contact.

Using the masses and speeds given earlier, the momenta of the fullback and safety are $p_1 = 500$ kg·m/s and $p_2 = 450$·kg m/s. The total momentum of the system, p_T on the diagram, is found to be approximately 673 kg·m/s. (This number can also be found by using the Pythagorean formula for the hypotenuse of a right triangle.) Since the total momentum $p_T = m_T v$, dividing this value by the total mass of the two players yields

$$v = \frac{p_T}{m_T} = \frac{(673 \text{ kg·m/s})}{175 \text{ kg}} = 3.85 \text{ m/s}.$$

This is a considerably larger post-collision velocity than that obtained earlier for the head-on collision.

The direction of this velocity immediately after the collision is the same as the direction of the momentum. From the diagram, we see that the two players will be moving at an angle of approximately 42° from the original direction of the fullback. Obviously this direction depends on how fast each player is moving prior to the collision as well as on their masses and original directions. If the fullback is moving much faster than the safety prior to the collision, then the direction of motion immediately after the collision will be much closer to the original direction of the fullback, as shown in figure 6.19.

The example in box 6.2 involves a similar situation. In this case, we have two cars approaching an intersection at right angles to one another, colliding, and sticking together. Working backwards from information on the final direction of travel, the investigator can draw conclusions about the initial velocities. Conservation of momentum is an extremely important tool in accident analysis.

Box 6.2

Everyday Phenomenon:
Impact at the Crossroads: An Automobile Collision

The Situation. Officer Jones is investigating an automobile collision that has occurred at the intersection of Main Street and 19th Avenue. Driver A was traveling east on 19th Avenue when she was struck from the side by driver B, who was traveling north on Main. The two cars stuck together after the collision and ended up against the lamp post on the northeast corner of the intersection.

Both drivers claim to have just started up as the light in their direction changed to green and to have collided with the other driver who was running a red light and speeding. There are no other witnesses. Which driver is telling the truth?

The Analysis. Officer Jones, having taken a physics course during college and being trained in the art of accident investigation, makes the following observations.

1. The point of impact is well marked by the shards of glass from the headlights of B's car and other debris. He indicates this point on the diagram in his accident report form.
2. The direction of travel of the two cars after impact should be indicated by a line drawn from the point of impact to their final resting spot, which is also indicated on his diagram.

3. Both cars have about the same mass (both are compacts of roughly the same vintage and size).
4. Conservation of momentum should determine the direction of the momentum vector after the collision.

After sketching the diagram and noting the direction of the final momentum vector, Officer Jones concludes that B is lying. Why? The final momentum vector must be equal to the sum of the initial momentum vectors of the two cars prior to the collision. Since the two cars were traveling at right angles to one another, the two initial momenta form the two sides of a triangle whose hypotenuse is the total momentum of the system. The diagram shows clearly that the momentum of B's car must have been considerably larger than the momentum of A's car.

Since both drivers had claimed to be just starting from a complete stop, following a change in the traffic signal, the driver with the larger velocity before the collision is presumably not telling it like it was. B is the one who had the larger velocity, so B was presumably speeding through the red light and is appropriately cited by Officer Jones. Case closed.

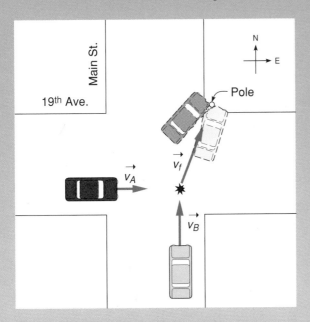

The collision at Main Street and 19th Avenue.

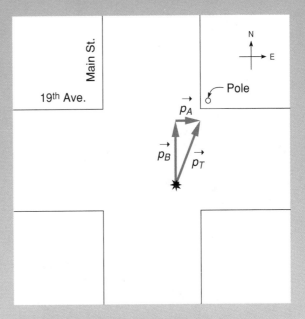

Officer Jones' accident report contains a vector diagram obtained from conservation of momentum.

Elastic Two-Dimensional Collisions

If the objects do not stick together after the collision, then again we have two final velocities to contend with, and these can be in two different directions. Although many real collisions are like this, their analysis is more complicated than the example we have just discussed, and we need more information in order to predict the final velocities. If we know that the collision is elastic, however, then conservation of kinetic energy can provide that additional information.

Experimental physics on the pool table again provides an interesting example (and one of practical importance to any pool player). Suppose that a pool ball strikes a stationary second ball at an angle (off-center), as shown in figure 6.20. What happens to the two balls after the collision? Conservation of momentum together with conservation of kinetic energy provide a unique result that is well-known to serious pool players.

The initial momentum of the system is simply that of the first ball, which is the only one moving. It is directed towards the top of the drawing in figure 6.21. The force of interaction (and the impulse) between the two balls is along a line joining the centers of the balls at the point of impact. The second ball will move off along this line because it is pushed in that direction by the contact force.

The total momentum of the system, however, must still be toward the top of the drawing since momentum should be conserved in the collision. The first ball must therefore move in such a direction that its momentum vector, when added to the momentum vector of the second ball, produces a total system momentum that is in the direction of the original momentum of the cue ball. This is illustrated in figure 6.21. The momenta of the two balls after the collision are

added here to give the total momentum of the system, p_T, which is equal to the initial momentum of the system.

Since the collision is elastic, though, the total kinetic energy after the collision must be equal to the initial kinetic energy of the first ball, $(\frac{1}{2})mv^2$. Putting this statement in the form of an equation, we have

$$\left(\frac{1}{2}\right)mv^2 = \left(\frac{1}{2}\right)m(v_1)^2 + \left(\frac{1}{2}\right)m(v_2)^2,$$

where v is the speed of the first ball before the collision, and v_1 and v_2 are the speeds of the two balls after the collision. Since the factor $(\frac{1}{2})m$ is the same for all three terms in this equation, however, the only way that the equation can be true is if

$$v^2 = (v_1)^2 + (v_2)^2.$$

This equation expresses a Pythagorean relationship between the three velocity vectors. In other words, in order for the sum of the squares of the two velocities after the collision to yield the square of the velocity of the first ball before the collision, the velocity vectors must form a right triangle. In this case, where the balls have equal mass, conservation of momentum requires that the velocity vectors add to form a triangle, but it is conservation of kinetic energy that dictates that the triangle be a right triangle.

This result tells us that the first ball will move off at a right angle (90°) to the direction of motion of the second ball after the collision. In pool, this is an important piece of

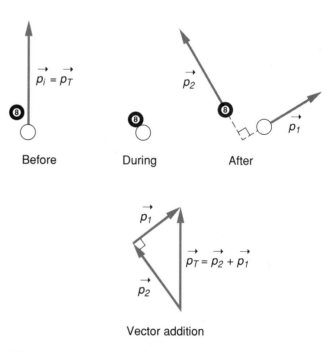

Vector addition

Figure 6.21 **The momenta of the two balls after the collision add to give the total (initial) momentum of the system. The paths of the two balls are approximately at right angles to one another after the collision.**

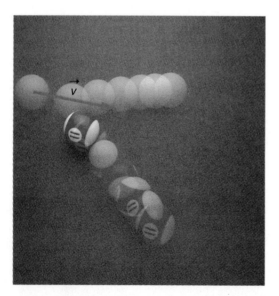

Figure 6.20 **The cue ball is aimed at a point off-center on the second ball to produce an angular collision.**

intelligence if you are planning ahead to your next shot. Conservation of momentum, combined with conservation of kinetic energy in this case, determine the result.

If you have a pool table handy, you can verify these conclusions using a variety of impact angles in your experiments. (Marbles or steel balls can provide a suitable substitute.) You may not get perfect 90° angles after the collision, because the collision is not perfectly elastic, and spin can sometimes be a factor. The angle between the two final velocities will usually be within a few degrees of being a right angle, though. Seeing is believing; give it a try!

SUMMARY

In this chapter Newton's second law has been recast, introducing the concepts of impulse and momentum, in order to describe interactions between objects, such as collisions, that involve strong forces acting for brief time intervals. The principle of conservation of momentum, which follows from Newton's second and third laws, plays an important role in these descriptions. The following key concepts were discussed.

Impulse. Impulse is defined as the average force acting upon an object times the time interval through which the force acts.

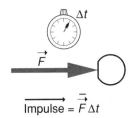

$$\overrightarrow{\text{Impulse}} = \vec{F}\,\Delta t$$

Momentum. Momentum is defined as the mass of an object times its velocity. It is Newton's "quantity of motion."

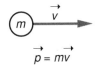

$$\vec{p} = m\vec{v}$$

Impulse/Momentum Theorem. With these new definitions, Newton's second law can be recast in terms of impulse and momentum, yielding the following statement: The impulse acting on an object equals the change in momentum of the object.

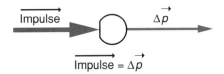

$$\overrightarrow{\text{Impulse}} = \Delta \vec{p}$$

Conservation of Momentum. Newton's second and third laws combine to yield the principle of conservation of momentum, which states that if the net external force acting on a system is zero, the total momentum of the system is a constant.

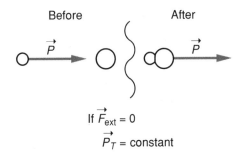

$$\text{If } \vec{F}_{\text{ext}} = 0$$
$$\vec{P}_T = \text{constant}$$

Recoil. If an explosion or push occurs between two objects initially at rest, conservation of momentum dictates that the total momentum after the event must still be zero. Therefore, the momenta of the two objects are equal but opposite in direction after the explosion.

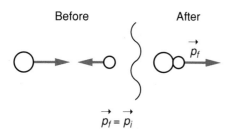

$$\vec{p}_2 = -\vec{p}_1$$

Perfectly Inelastic Collision. A perfectly inelastic collision is one in which the objects stick together after the collision. The momentum of the combined objects after the collision must equal the total momentum of the system before the collision.

Before   After

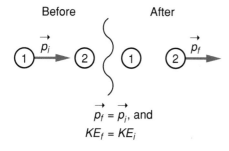

$$\vec{p}_f = \vec{p}_i$$

Elastic Collision. An elastic collision is one in which the total kinetic energy, $KE = (\frac{1}{2})mv^2$ is also conserved. Collisions between hard spheres, such as pool balls, are often elastic.

Before   After

$$\vec{p}_f = \vec{p}_i, \text{ and}$$
$$KE_f = KE_i$$

Two-Dimensional Collisions. Conservation of momentum holds for collisions in two dimensions. The total momentum of the system before and after the collision is found by adding the momentum vectors by the head-to-toe method.

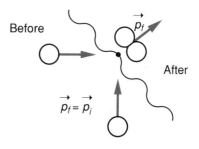

$$\vec{p}_f = \vec{p}_i$$

Study Hint

Except for the examples involving impulse, most of the situations described in this chapter have involved the principle of conservation of momentum. The logic used in applying conservation of momentum has taken the following form.

1. External forces are assumed to be much smaller than the very strong forces of interaction involved in a collision or other brief event. If external forces acting on the system can be ignored, then momentum is conserved.
2. The total momentum of the system before the collision or other brief interaction, $\vec{p}_i$ is set equal to the momentum after the event, $\vec{p}_f$. This represents the fact that momentum is conserved; it does not change.
3. This equality of momentum before and after the event is used to obtain other information about the motion of the objects after the event from knowledge of what was happening before the event.

For review, it would be a good idea to look back at how the logic just outlined was used in each of the examples treated earlier in the chapter. The total momentum of the system before and after the event is always found by adding the momenta of the individual objects *as vectors*. You should be able to describe the magnitude and direction of this total momentum for each of the examples.

QUESTIONS

Q6.1 Does the length of time during which a force acts have any bearing on the strength of the impulse produced? Explain.

Q6.2 Is it possible for a baseball to have as large a momentum as a much more massive bowling ball? Explain.

Q6.3 If a ball bounces off a wall so that its velocity coming back has the same magnitude that it had prior to bouncing, is there a change in the momentum of the ball? Explain.

Q6.4 Is there an impulse acting upon the ball of question 6.3 during its collision with the wall? Explain.

Q6.5 Is there an advantage to following through when hitting a baseball and thereby maintaining a longer time of contact between the bat and the ball? Explain.

Q6.6 Is the law of conservation of momentum always valid, or are there special conditions necessary for it to be valid? Explain.

Q6.7 A ball is accelerated down a fixed inclined plane under the influence of the force of gravity. Is the momentum of the ball conserved in this process? Explain.

Q6.8 In a collision between two objects for which conservation of momentum applies, is the momentum of each object involved in the collision conserved? Explain.

Q6.9 A compact car and an eighteen-wheel truck have a head-on collision. During the collision, which vehicle, if either, experiences

    a. the greater force of impact? Explain.

    b. the greater impulse? Explain.

    c. the greater change in momentum? Explain.

    d. the greater acceleration? Explain.

Q6.10 Which of Newton's laws of motion are involved in justifying the principle of conservation of momentum? Explain.

Q6.11 Two ice skaters, initially at rest, push against one another. What is the total momentum of the system after the push-off? Explain.

Q6.12 Two shotguns are identical in every respect (including the size of shell fired), except that one has twice the mass of the other. Which gun will tend to recoil with the greater recoil velocity when fired? Explain.

Q6.13 Is it possible for a rocket to work in empty space (in a vacuum) where there is nothing to push against except itself? Explain.

Q6.14 A railroad car collides and couples with a second railroad car of equal mass, producing a combined mass twice that of the original car. If external forces acting on the system can be ignored, is the total momentum of the system after the collision equal to, greater than, or less than that of the first car before the collision? Explain.

Q6.15 Is the collision in question 6.14 elastic, partially inelastic, or perfectly inelastic? Explain.

Q6.16 If momentum is conserved in a collision, does this indicate conclusively that the collision is elastic? Explain.

Q6.17 A ball bounces off of a wall with a velocity whose magnitude is less than that before hitting the wall. Is the collision elastic? Explain.

Q6.18 A pool ball strikes a second pool ball of equal mass, initially at rest. The first ball stops, and the second ball moves forward with a velocity equal to the initial velocity of the first ball. Is the collision elastic? Explain.

Q6.19 Two lumps of clay traveling through the air in opposite directions collide and stick together. Their momentum vectors are shown in the diagram. Sketch the momentum vector of the combined lump of clay after the collision, making the length and direction appropriate to the situation. Explain your result.

Q6.20 Two lumps of clay, of equal mass, are traveling through the air at right angles to each other with velocities of equal magnitude. They collide and stick together. Is it possible that their velocity vector after the collision is in the direction shown in the diagram? Explain.

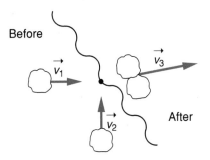

Q6.21 Two cars of equal mass collide at right angles to one another. Their direction of motion after the collision is as shown in the diagram. Which car had the greater velocity before the collision? Explain.

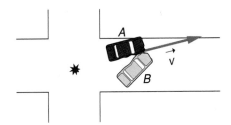

Q6.22 A car and a small truck, traveling at right angles to one another with the same speed, collide and stick together. The truck has a mass roughly twice that of the car. Sketch the direction of their momentum vector immediately after the collision. Explain your result.

EXERCISES

E6.1 An average force of 50 N acts for a time interval of 0.25 s on a golf ball.

   a. What is the impulse acting on the golf ball?

   b. What is the change in momentum of the golf ball?

E6.2 A bowling ball has a mass of 7 kg and a velocity of 0.4 m/s. A tennis ball has a mass of 0.06 kg and a velocity of 40 m/s.

   a. What is the momentum of each ball?

   b. Which ball has the larger momentum?

E6.3 An impulse of 12 N·s acts on a ball that is initially at rest. What is the final momentum of the ball?

E6.4 A ball traveling with an initial momentum of 3.0 kg·m/s bounces off of a wall and comes back in the opposite direction with a momentum of –3.0 kg·m/s.

   a. What is the change in momentum of the ball?

   b. What impulse would be required to produce this change?

E6.5 A ball traveling with an initial momentum of 5.0 kg·m/s bounces off of a wall and comes back in the opposite direction with a momentum of 4.0 kg·m/s.

   a. What is the change in momentum of the ball?

   b. What impulse would be required to produce this change?

E6.6 A ball hits a wall and experiences a change in momentum of 6.0 kg·m/s.

   a. What is the impulse acting on the ball?

   b. If the time of interaction is 0.15 s, what is the magnitude of the average force acting on the ball?

E6.7 A fullback with a mass of 90 kg and a velocity of 4.0 m/s due east collides head-on with a defensive back with a mass of 80 kg and a velocity of 6.0 m/s in the opposite direction (due west).

   a. What is the initial momentum of each player?

   b. What is the total momentum of the system before the collision?

   c. If they stick together and external forces can be ignored, in what direction will they be traveling immediately after they collide?

E6.8 An ice skater with a mass of 80 kg pushes against a second skater with a mass of 20 kg. Both skaters are initially at rest.

   a. What is the initial momentum of the system?

   b. What is the total momentum of the system after they push off?

   c. If the larger skater moves off with a speed of 3.0 m/s, what is the corresponding speed of the smaller skater?

E6.9 A rifle with a mass of 2.5 kg fires a bullet with a mass of 5.0 grams (0.005 kg). The bullet moves with a muzzle velocity of 350 m/s after the rifle is fired.

   a. What is the momentum of the bullet after the rifle is fired?

   b. If external forces acting on the rifle can be ignored, what is the momentum of the rifle after it is fired?

   c. What is the recoil velocity of the rifle under these conditions?

E6.10 A rocket ship traveling in space gives a short blast of its engine, firing 20 kg of exhaust gas out the back end with an average velocity of 150 m/s. What is the change in momentum of the rocket during this blast?

E6.11 A railroad car with a mass of 15 000 kg collides and couples with a second car of mass 20 000 kg, which is initially at rest. The first car is moving with a speed of 10 m/s prior to the collision.

   a. What is the initial momentum of the first car?

   b. If external forces can be ignored, what is the final velocity of the two cars after they couple?

E6.12 A truck with a mass of 6000 kg traveling with speed of 15 m/s collides head-on with a 1500-kg car that has an initial velocity of 30 m/s in the opposite direction.

   a. What is the momentum of the truck prior to the collision?

   b. What is the momentum of the car prior to the collision?

   c. What is the total momentum of the system prior to the collision?

   d. In what direction will the two vehicles move after the collision if they stick together?

E6.13

   a. Sketch to scale the momentum vectors of the two vehicles in exercise 6.12 prior to the collision.

   b. Add the two vectors on your sketch, using the head-to-toe method.

E6.14 A truck with a mass of 5000 kg traveling with a speed of 10 m/s collides at right angles with a car with a mass of 2000 kg traveling with a speed of 20 m/s.

   a. Sketch to proper scale and direction the momentum vectors of each vehicle prior to the collision.

   b. Using the head-to-toe method on your sketch, add the momentum vectors to get the total momentum of the system prior to the collision.

   c. In what direction will the two vehicles travel after the collision if they stick together?

CHALLENGE PROBLEMS

CP6.1 A fast ball thrown with a velocity of 40 m/s (approximately 90 MPH) is struck by a baseball bat, and a line drive comes back towards the pitcher with a velocity of 60 m/s. The ball is in contact with the bat for a time of just 0.01 s. The baseball has a mass of 150 g (0.150 kg).

   a. What is the change in momentum of the baseball during this process?

   b. Is the change in momentum greater (in magnitude) than the final momentum? Explain.

   c. What is the impulse required to produce this change in momentum?

   d. What is the average force that acts on the baseball to produce this impulse?

CP6.2 A bullet is fired into a block of wood sitting on a block of ice. The bullet has an initial velocity of 600 m/s and a mass of 0.005 kg. The wooden block has a mass of 2.0 kg and is initially at rest. The bullet remains embedded in the block of wood after the collision.

   a. What is the velocity of the block of wood (and bullet) after the bullet has embedded itself?

   b. What is the impulse that acts upon the block of wood in this process?

   c. Does the change in momentum of the bullet equal that of the block of wood? Explain.

CP6.3 A car traveling with a speed of 30 m/s (approximately 67 MPH) crashes into a solid concrete wall. The driver has a mass of 80 kg and is brought to a stop in a time of approximately 0.25 s.

   a. What is the change in momentum of the driver?

   b. What impulse and average force are required in order to produce this change in momentum?

   c. How does the application and magnitude of this force differ in two cases: one in which the driver is wearing a seat belt, and the other in which he is not and instead is stopped by contact with the windshield and steering column? Will the time of application differ? Explain.

CP6.4 Consider two cases in which the same ball is thrown against a wall with the same initial velocity. In case *A*, the ball sticks to the wall and does not bounce. In case *B*, the ball bounces back with the same speed that it came in with.

   a. In which of these two cases is the change in momentum the largest?

   b. Assuming that the time during which the change in momentum takes place is approximately the same for these two cases, in which case is the larger average force involved?

   c. Is momentum conserved in this collision? Explain.

CP6.5 Suppose that on a perfectly still day a sailboat enthusiast decides to bring along a powerful battery-operated fan in order to provide an air current against his sail, as shown in the diagram.

   a. What are the directions of the change in momentum of the air at the fan and at the sail?

   b. What are the directions of the forces acting on the fan and on the sail due to these changes in momentum?

   c. Would the sailor be better off with the sail up or down (furled) in this picture? Explain.

HOME EXPERIMENTS AND OBSERVATIONS

The following suggested experiments are not intended to yield results that can be accurately measured. They are designed, rather, to provide some intuitive feeling for how impulse and conservation of momentum are involved in situations that can be easily produced on a desk top or floor at home.

HE6.1 *Bouncing ball.* Take a small rubber ball or tennis ball and drop it from shoulder height to a hard floor. Try to estimate the following quantities.

a. How high does the ball bounce compared to its original starting height? (This is a measure of the inelasticity of the bounce; if it came back to the same height, the bounce would be perfectly elastic.) If more than one ball is available, which has the most elastic bounce?

b. How long is the ball in contact with the floor? Is it less than a quarter of a second? How can you make this estimate?

c. Is it possible to observe the deformation of the ball (change in shape) when it is in contact with the floor? Try pushing the ball against the floor and then releasing it. Is it possible to cause the ball to reach the same height in this manner as it did when it was released from shoulder height?

HE6.2 *Marble collisions.* Take two marbles or steel balls of the same size and practice shooting one into the other. Make the following observations.

a. If you produce a head-on collision with the second marble initially at rest, does the first marble come to a complete stop after the collision?

b. If the collision with a second marble initially at rest occurs at an angle, is the angle between the paths of the two marbles after the collision a right angle (90°)?

c. If marbles of different sizes and masses are used, how do the results in (*a*) and (*b*) differ from those obtained with marbles of the same masses?

HE6.3 *Pool ball collisions.* If you have access to a pool table, try (*a*) and (*b*) of experiment 6.2 on the pool table. What effect does putting spin on the first ball have on the collisions?

7

Rotational Motion of Solid Objects

In the park next door to my house there is a simple, child-propelled merry-go-round (fig. 7.1). It consists of a circular steel platform mounted upon an excellent bearing so that it will rotate without much frictional resistance. Once set in rotation, it will continue, so that a child can start it moving and then jump on (and sometimes fall off). Along with the swings, the slide, and the little animals mounted on heavy-duty springs, it is a popular center of activity in the park.

The motion of this merry-go-round bears both similarities and differences to the motions we have already considered. A child sitting on the merry-go-round undergoes circular motion, and some of the ideas discussed in chapter 5 come into play. What about the merry-go-round itself, though? It certainly moves, but it goes nowhere. How would we describe its motion?

Rotational motion of solid objects such as the merry-go-round is very common; the rotating earth, a spinning skater, a top, or a turning wheel all exhibit this type of motion. If Newton's theory of motion is to be broadly useful, it should be able to explain what is happening in these cases as well as in cases where an object moves from one point to another, which we might call *linear* motion. How is rotational motion caused? Can Newton's second law be used to explain such motion?

We will find that there is a very useful analogy between the linear motion of objects that we have considered in the preceding chapters and rotational motion. The questions that we have just posed are best answered by making full use of this analogy. By carefully noting the similarities of rotational motion to linear motion at the outset, we save valuable memory space in our mental computers!

Figure 7.1 The merry-go-round in the park provides an example of rotational motion. How might we describe and explain this motion?

Chapter Objectives

Starting with the example of the merry-go-round, and making use of the analogy between linear and rotational motion, we first consider how rotational motion can be described. We then turn to the causes of rotational motion, which involve a modified form of Newton's second law. The concepts of torque, rotational inertia, and angular momentum will be introduced as we proceed.

The objective is to develop a clear picture of both the description and causes of rotational motion. When you have studied this chapter, you should be able to predict what will happen in many common and simple examples of rotating objects, including yourself or other people. The world of sports is rich in examples of rotational motion.

Chapter Outline

❶ *Description of rotational motion.* What are the concepts needed to describe rotational motion? How are they related to similar concepts used to describe linear motion?

❷ *Torque and balance.* What is torque, and how is it involved in causing an object to rotate? When is a balance beam balanced? What condition determines whether or not objects will rotate?

❸ *Rotational inertia and Newton's second law.* How can Newton's second law be modified so that it can be used to explain the motion or lack of motion of rotating objects? What quantity describes the rotational inertia of an object—its resistance to changes in rotational motion?

❹ *Conservation of angular momentum.* What is angular momentum, and how is the principle of conservation of angular momentum useful in describing and predicting how the rotational motion of objects will change? How does a spinning skater or diver change her rotational velocity?

❺ *Riding a bicycle and other amazing feats.* Why does a bicycle remain upright when it is moving but not when it is stationary? How can rotational velocity and angular momentum be regarded as vector quantities?

7.1 DESCRIPTION OF ROTATIONAL MOTION

Suppose that a child begins to rotate the merry-go-round described earlier. She does so by holding on to one of the bars on the edge of the merry-go-round (fig. 7.1) and, starting from a standstill, begins to accelerate. Eventually the child is running and the merry-go-round is rotating quite rapidly.

How would we describe this motion of the merry-go-round in the park, or that of a record on a turntable or the second hand on a clock? What quantity would we use to describe how fast they are rotating, or how would we describe how far they have rotated? These are simple questions with some familiar answers, but a careful look at these ideas is necessary before we can discuss the causes of their motion.

Rotational Displacement and Velocity

The record on the turntable is a good example for introducing the concept of rotational velocity, although records in stereo systems have been mostly replaced now by tape cassettes and compact discs, and may no longer be quite so familiar. We talk about 45 RPM records, for example, meaning that the rate of rotation for proper sound reproduction is 45 revolutions per minute. This is a common way of describing a rotational velocity or a rate of rotation.

Revolutions per minute (RPM) is a rate, which we use to describe how fast the record is turning. It is analogous to speed or velocity, the quantities that we use to describe how fast an object is moving in the case of linear motion. Since velocity involves dividing a displacement by a time, we are evidently using revolutions to describe a rotational displacement. That is a common and familiar unit for this quantity. If we choose a point on the record or the merry-go-round, it is easy to count revolutions by noting when the chosen point has returned to its initial position. One complete revolution, or cycle, occurs each time the point returns to its starting position.

Another common unit of rotational displacement is the degree (°), which is often used to describe angles or changes in rotational position that are less than one complete cycle. If we draw a line from the axis of rotation to the point we are using to mark the motion, that line will rotate through an angle that can be used to describe how much the object has rotated. Since there are 360° in one complete rotation, or circle, it is easy to convert revolutions to degrees by simply multiplying by 360°/rev.

There is, however, a more natural measure of angular displacement that is used in physics, the radian, which may be somewhat less familiar. The *radian* is defined by dividing the arc length through which the point travels by the radius of the circle upon which it is moving. Thus, as illustrated in figure 7.2, if the point on the merry-go-round moves along

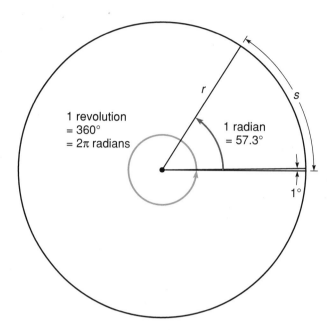

Figure 7.2 **Revolutions, radians, and degrees are different possible units for describing the rotational displacement of the merry-go-round.**

the arc length a distance *s*, the number of radians involved is *s* / *r*, where *r* is the radius. Since we are dividing one distance by another, the radian itself has no dimensions. Also, since the arc length, *s*, is proportional to the radius, *r*, it does not matter how large a radius we choose; the ratio of *s* to *r* will be the same for a given angle (fig. 7.3).

One complete revolution causes the point on the merry-go-round to travel a distance equal to the circumference of the circle upon which it is moving, $2\pi r$. (As you are probably aware, the symbol π represents the dimensionless mathematical constant having the value 3.1416.) One revolution, therefore represents an angle of

$$\frac{s}{r} = \frac{2\pi r}{r} = 2\pi \text{ radians,}$$

so there are 2π radians in one revolution. To compare the three commonly used measures of angle or *rotational displacement*, then, we have

$$1 \text{ rev} = 360° = 2\pi \text{ radians.}$$

Dividing 360° by 2π radians shows that one radian equals approximately 57.3° or a little less than ⅙ of a revolution.

In describing rotational motion, we usually use either revolutions or radians as the measure of rotational displacement; the degree is less commonly used for this purpose. The rate of rotation, or *rotational velocity,* is then naturally defined as follows:

The rotational velocity is the rate of change of rotational displacement, or the change in rotational dis-

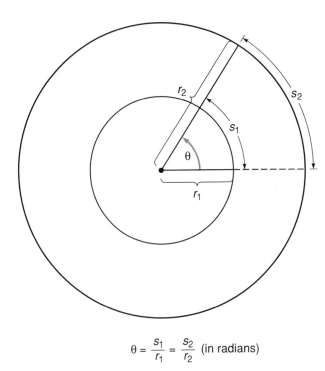

$$\theta = \frac{s_1}{r_1} = \frac{s_2}{r_2} \text{ (in radians)}$$

Figure 7.3 **The radian is defined as the arc length associated with a given angle divided by the radius of the circle. This ratio is the same regardless of the size of the circle.**

placement divided by the time taken for this displacement to occur.

There is an obvious parallel here to the definition of linear velocity in chapter 2. Linear velocity is the rate of change of linear displacement, $\vec{D}$, with time.

In symbolic shorthand, the definition of rotational velocity is often written as follows:

$$\omega = \frac{\Delta\theta}{\Delta t}.$$

The Greek letter ω (omega) is commonly used to represent rotational velocities, and the Greek letter θ (theta) is commonly used to represent angles or rotational displacements. (Greek letters are used to avoid confusion with other quantities represented by letters of the ordinary alphabet. Lack of familiarity with these symbols may be a problem for some students, but it should not if you have already become accustomed to using letters to represent numerical quantities. These are the symbols that physicists traditionally use for these quantities.)

Because angle measures are used to describe rotational displacement, rotational velocities are traditionally referred to as *angular velocities*. The terms *angular velocity* and *rotational velocity* mean the same thing; the terms *rotational velocity* or *rate of rotation* have become more commonly used recently in physics texts like this one. The commonly

used units of rotational velocity are revolutions per second (rev/s) or radians per second (rad/s), or sometimes revolutions per minute (rev/min = RPM). The latter is probably the most familiar. What units would you choose to describe the rotational velocity of the second hand on the clock? (Consider also the rotational velocity of the earth on its axis; what units might you use for that?)

Rotational Acceleration

In the initial example of the merry-go-round, the rate of rotation increases as the child runs alongside pushing. This involves a change in the rotational velocity, which suggests the concept of acceleration. *Angular* or *rotational acceleration* is the appropriate concept, and by analogy to the definition of linear acceleration in chapter 2, it can be defined as follows:

Rotational acceleration is the rate of change in rotational velocity or the change in rotational velocity divided by the time interval required for the change to occur.

In symbols, this definition can be written as follows:

$$\alpha = \frac{\Delta\omega}{\Delta t}.$$

The Greek letter α (alpha) is generally used to represent rotational acceleration. Alpha is the first letter in the Greek alphabet and therefore corresponds to a, which we use to represent linear acceleration. Rotational acceleration would have units of rev/s^2 or rad/s^2.

Actually, the definitions given for rotational velocity and rotational acceleration yield the average values of these quantities. Instantaneous values would require that Δt become very small, as in the definitions in chapter 2 of instantaneous velocity and instantaneous acceleration for linear motion. In any case, it is the rate of change that is important, just as with the linear quantities.

These definitions of rotational displacement, velocity, and acceleration are easy to remember if you keep in mind the complete analogy that exists between these quantities and those already used to describe linear motions. This analogy is summarized in figure 7.4. For linear motion in one dimension, the coordinate x is often used to indicate position; Δx is then the displacement or change in position, which corresponds to $\Delta\theta$, the change in rotational position. Average velocity and acceleration for linear motion are defined as before, with the corresponding definitions of rotational velocity and acceleration shown on the right in the figure.

Since the analogy between the linear and rotational quantities is so complete, it is also easy to write equations that describe what happens in the case of constant rotational acceleration. We simply substitute the rotational quantities for the corresponding linear quantities in the equations developed in chapter 3 for linear motion. Table 7.1 shows the

Linear motion

Rotational motion

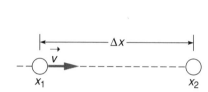

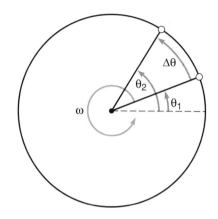

Displacement = Δx

Velocity = $\dfrac{\Delta x}{\Delta t}$ = v

Acceleration = $\dfrac{\Delta v}{\Delta t}$ = a

Displacement = $\Delta \theta$

Velocity = $\dfrac{\Delta \theta}{\Delta t}$ = ω

Acceleration = $\dfrac{\Delta \omega}{\Delta t}$ = α

Figure 7.4   There is a close analogy between the quantities used to describe linear motion and those used to describe rotational motion.

Table 7.1	
Constant Acceleration Equations for Linear and Rotational Motion	
Linear Motion	**Rotational Motion**
$v = v_0 + at.$	$\omega = \omega_0 + \alpha t.$
$\Delta x = v_0 t + \left(\dfrac{1}{2}\right) at^2.$	$\Delta \theta = \omega_0 t + \left(\dfrac{1}{2}\right) \alpha t^2.$
$v^2 = v_0^2 + 2a\Delta x.$	$\omega^2 = \omega_0^2 + 2\alpha\Delta\theta.$

resulting equations alongside of the corresponding equations for linear motion.

How might these equations be applied to the merry-go-round? Suppose that the merry-go-round is accelerated at a constant rate of 0.005 rev/s^2 for one minute. If it starts from rest, what is its rotational velocity at the end of that minute? Using the first rotational equation in table 7.1, we have

$$\omega = \omega_0 + \alpha t = 0 + (0.005 \text{ rev/s}^2)(60 \text{ s}) = 0.30 \text{ rev/s.}$$

This is a little less than a third of a revolution per second and is a fairly fast rotational velocity for such a merry-go-round.

How many revolutions does the merry-go-round make in one minute? Using the second equation, we have

$$\Delta\theta = \omega_0 t + \frac{1}{2}\alpha t^2 = 0 + \left(\frac{1}{2}\right)(0.005 \text{ rev/s}^2)(60 \text{ s})^2$$

$$= 9.0 \text{ rev.}$$

To start from a complete stop and get up to nine revolutions in the time of one minute represents a pretty good effort. Try it yourself if you have a park with a merry-go-round nearby.

Relationship Between Linear and Rotational Velocity

How fast is the rider going when the merry-go-round is rotating with a rotational velocity of 0.30 rev/s? The answer to this question depends upon where the rider is sitting; she will move faster when seated near the edge of the merry-go-round than near the center. The question involves a relationship between the linear speed of the rider and the rotational speed of the merry-go-round.

Figure 7.5 shows two circles on the merry-go-round with different radii representing different positions of a seated rider. Obviously, the rider seated at the greater distance from the center travels a larger distance in one revolution than the rider near the center because the circumference of her circle is greater. The outside rider must then be moving with a greater linear speed than the one near the inside.

If the merry-go-round were rotating with a rotational velocity of 1 rev/s, in 1 second the rider would cover a distance equal to the circumference of the circle that he or she is making as the merry-go-round turns. If the rider is sitting at distance r from the center, this circumference is just 2π times the radius of the circle, or $2\pi r$. Her speed is then just the distance covered in one revolution divided by the time taken for one revolution, which we have taken to be 1 second in order to simplify the discussion. This process is equivalent to mul-

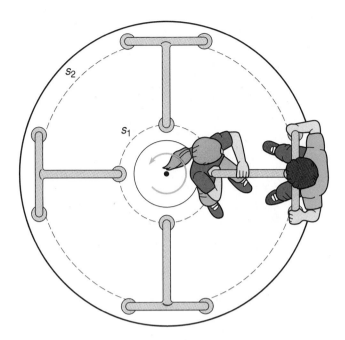

Figure 7.5 **The child near the edge of the merry-go-round travels a greater distance in one revolution than the child near the center.**

tiplying the distance covered in one revolution by the rotational velocity in revolutions per second. If the rider is sitting 2 m from the axis of rotation, this would give us

$$v = (2\pi r)\omega = 2\ (3.14)(2\ \text{m})(0.30\ \text{rev/s}) = 3.77\ \text{m/s}.$$

Done this way, the computation has a logic that is easy to follow, but it turns out that the relationship actually takes a simpler form if we choose to express rotational velocity in radians per second instead of revolutions per second. Since the radian is defined as the distance traveled on the circle divided by the radius of the circle, $\theta = s\,/\,r$, the distance s on the circle is just the radius of the circle times the angle in radians, or $s = r\theta$. In other words, the distance traveled, s, is directly proportional to the rotational displacement.

The linear speed is then also directly proportional to the rotational velocity, with the radius of the circle being the multiplier. In symbols,

$$v = r\omega.$$

In this expression for the speed, however, we would have to express ω in radians per second rather than in revolutions per second. Since there are 2π radians in one revolution, $0.30\ \text{rev/s} = 2\pi\ (0.30) = 1.885\ \text{rad/s}$. The speed could then be found as follows:

$$v = r\omega = (2\ \text{m})\ (1.885\ \text{rad/s}) = 3.77\ \text{m/s}.$$

This is the same result that we obtained above, but the equation has a simpler appearance.

The form of these two computations is really identical; the difference lies only in where we choose to multiply by 2π.

Because of the simpler form of the equation $v = r\omega$, however, we usually choose to express rotational velocity in radians per second if we want to find the linear speed or other linear quantities. Then we have simple relationships between the linear displacement, or speed, and the corresponding rotational quantities. In any case, it is important to note carefully the appropriate units for the expression you are using.

It is easy enough to see that the rate at which the merry-go-round or other object is turning affects the linear speed of a point on the rotating object. It is also easy to see that the distance from the axis of rotation affects the linear speed. These are the basic ideas here. A few of the exercises at the end of this chapter provide an opportunity for practice in using these ideas.

7.2 TORQUE AND BALANCE

To rotate or not to rotate; that is the question. What determines whether or not an object such as the merry-go-round will indeed rotate? Forces must be involved in some manner, but how must they be applied in order to be effective in producing rotation? The concept of torque provides the key to answering this question.

A Matter of Balance

A simple balance provides an example in which the concept of torque is easily visualized. Suppose that we consider a balance made of a very light rod supported by a fulcrum or pivot point, and upon which we can place weights, as shown in figure 7.6. If the rod itself is balanced before we place weights on it, and if we put equal weights at equal distances from the fulcrum, we would expect the rod to remain balanced. By this, of course, we mean that it will *not* tend to rotate about the fulcrum.

Suppose, though, that we wish to balance unequal weights on the rod. If, for example, a weight twice as large as the two initial weights is to be balanced by one of the initial weights, we would have to place the weights at unequal distances from the fulcrum. Where would we place the larger weight? Intuition might suggest that the larger weight must be placed closer to the fulcrum in order for the system to be balanced, but it may not tell you how much closer (fig. 7.7).

Trial and error with a simple balance will show that the appropriate distance for the larger weight is just half the distance of the smaller weight from the fulcrum, if the larger weight is twice the weight of the smaller. Try it yourself, using a ruler for the rod and a pencil for the fulcrum. Quarters or other coins can be used as the weights. A child's teeter-totter can provide a larger-scale test.

As trial and error will show, both the weight and the distance from the fulcrum are important. The farther that a weight is located from the fulcrum, the more effective it will be in balancing larger weights on the other side of the ful-

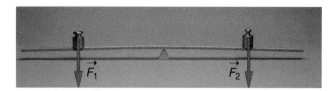

Figure 7.6   A simple balance with equal weights placed at equal distances from the fulcrum.

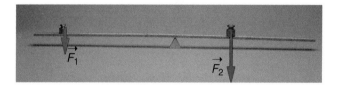

Figure 7.7   A simple balance showing unequal weights placed at different distances from the fulcrum. What determines whether the system will be balanced?

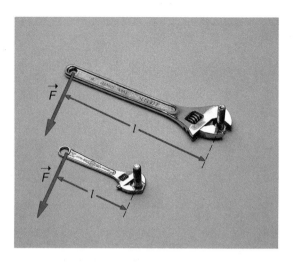

Figure 7.8   A wrench with a long handle is more effective than one with a short handle because of the longer lever arm for the longer wrench.

crum. The product of the weight (which is a force) and the distance from the fulcrum turns out to be the quantity that tells us whether or not the system will be balanced.

Definition of Torque

The product of the force and the distance from the fulcrum, which describes the effectiveness of the weight in tending to produce a rotation, is called the *torque*. More generally, we can define torque in words as follows:

> **The torque about a given axis or fulcrum is equal to the product of the applied force and the lever arm. The *lever arm* is the perpendicular distance from the axis of rotation to the line of action of the applied force.**

In symbols this definition takes the following form:

$$\tau = Fl,$$

where τ is the Greek letter tau, which is commonly used to represent torque. The distance l is the distance between the point of application of the force and the fulcrum and must be measured in a direction perpendicular to the line of action of the force. This distance is often called the *lever arm,* or *moment arm,* of the force in question. The strength of the torque, then, depends directly on both the size of the force and the length of its lever arm.

A wrench is convenient for illustrating the concept of torque, since most of us have tried to turn a nut with a wrench. You exert the force at the end of the wrench, in a direction perpendicular to the handle, as shown in figure 7.8. The handle is the lever, and its length determines the lever arm. A longer handle is more effective than a shorter one because the resulting torque is greater. If you make the

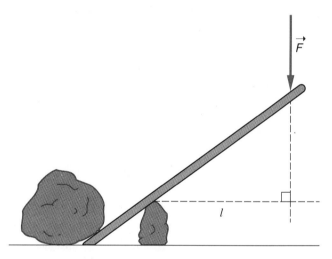

Figure 7.9   When the applied force is not perpendicular to the crowbar, the lever arm is found by drawing a perpendicular line from the fulcrum to the line of action of the force.

handle too long, though, you have trouble fitting the wrench in the tool kit! It might also have a tendency to bend or break.

As the term suggests, the concept of a lever arm has emerged from our use of levers to move objects. Moving a large rock with a crowbar, for example, involves leverage. The applied force is most effective if it is applied at the end of the bar *and* perpendicular to the bar. The lever arm, l, is then just the length of the bar from the fulcrum to the end of the bar at which the force is being applied. If the force is applied in some other direction, as is shown in figure 7.9, then the lever arm is shorter than it is if the force is applied in the perpendicular direction. It is found by drawing the perpendicular line from the fulcrum to the line of action of the force, as indicated in figure 7.9.

The direction of the torque is also important. Some torques tend to produce clockwise rotations and others counterclockwise rotations. The torque due to the heavier weight in figure 7.7, for example, which is on the right-hand side of the fulcrum, will tend to produce a clockwise rotation about the fulcrum if acting by itself. This is countered, however, by the equal torque of the weight on the left-hand side of the fulcrum which tends to produce a counterclockwise rotation. The two torques cancel one another when the system is balanced.

The condition that determines whether or not something will rotate is staring us in the face then. If the torque tending to produce a clockwise rotation equals that tending to produce a counterclockwise rotation, then there is no rotation. If one of these torques is larger than the other, the system will tend to rotate. If we remove the weight from one side of the beam, for example, the beam will rotate in the direction dictated by the remaining weight.

Adding Torques

Since torques can have opposing effects, it is natural to assign opposite signs to torques that tend to produce rotations in opposite directions. If, for example, we choose to call torques that produce a counterclockwise rotation positive, then torques producing clockwise rotations would be negative. (This is the conventional choice, but it is unimportant which direction is chosen to be positive as long as you are consistent with your choices for a given situation.) By choosing signs in this manner, we can add torques (subtracting, of course, if the sign is negative).

In the case of the balance beam, for example, the total (or net) torque is zero when the beam is balanced, because the two torques are equal but have opposite signs. The condition for balance, or equilibrium, then, is that the net torque acting on the system be zero, in symbols,

$$\Sigma\tau = 0.$$

where the Greek letter sigma (Σ) represents a summation as before. In other words, the total positive torque equals the total negative torque, and they cancel one another by adding to zero.

To get a feel for how this works, consider a different balance-beam situation. Suppose we want to balance a 0.3-kg mass against a 0.5-kg mass on a beam that is balanced when no masses are in place. If we place the 0.5-kg mass 20 cm from the fulcrum, how far would we have to place the 0.3-kg mass from the fulcrum in order to balance the system? (See figure 7.10.)

We can compute the torque due to the 0.5-kg mass from the information provided. By Newton's third law, the force that this mass exerts on the beam is simply the weight of the mass:

$$F = W = mg = (0.5 \text{ kg})(9.8 \text{ m/s}^2) = 4.9 \text{ N}.$$

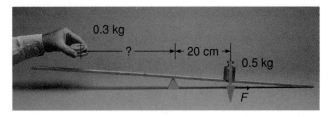

Figure 7.10   Where should the 0.3-kg mass be placed on the beam in order to balance the system?

The torque is then

$$\tau = Fl = (4.9 \text{ N})(0.20 \text{ m}) = 0.98 \text{ N·m}.$$

Notice that the units of torque are those of force times distance, or newtons times meters in the metric units that we are using. Since we have shown the 0.5-kg mass on the left-hand side of the fulcrum, where it would tend to produce a counterclockwise rotation, its sign would be positive by the conventional choice of sign.

A negative torque of the same magnitude would be required to balance the system. The weight of the 0.3-kg mass is 2.94 N (found by multiplying by g), so the torque due to this mass would be

$$\tau = -Fl = -0.98 \text{ N·m} = -(2.94 \text{ N})(l).$$

The unknown distance l (the lever arm) can be found then by dividing both sides of this equation by the force of 2.94 N.

$$l = \frac{-\tau}{F} = \frac{+0.98 \text{ N·m}}{2.94 \text{ N}} = 0.333 \text{ m}.$$

In other words, the 0.3-kg mass should be placed 33.3 cm from the fulcrum in order to balance the beam.

Center of Gravity

For a system that is in equilibrium, it is often important to know what torques and corresponding forces or lever arms are required to keep the system from rotating. How far, for example, could the child in figure 7.11 walk out on the plank without the plank tipping?

The weight of the plank itself is important in this case, and the concept of *center of gravity* is useful here. The center of gravity is the point in or near an object about which the weight of the object itself exerts no net torque. If we suspended an object from its center of gravity, there is no net torque about the suspension point, and the object is balanced. We can therefore locate the center of gravity of a planar object experimentally by suspending the object from two different points, seeing how it hangs, drawing a line straight downward from each point of suspension, and then locating the point of intersection of the two lines, as illustrated in figure 7.12. This works because the center of gravity will always lie directly below the point of suspension when the object is hanging freely.

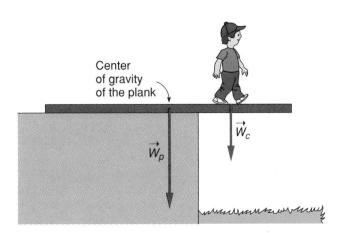

Figure 7.11   How far out can the child walk without tipping the plank? The entire weight of the plank can be treated as though it is located at the center of gravity.

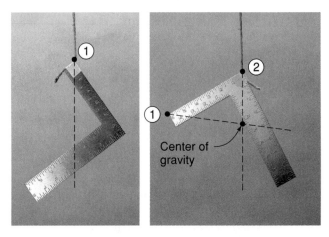

Figure 7.12   Locating the center of gravity of a planar object. The center of gravity does not necessarily lie within the object.

In the case of the plank, the center of gravity is at its geometric center if the plank is uniform in density and cut. The pivot point will be the edge of the supporting platform, so this is the point to consider when computing torques. The plank will not tip as long as the counterclockwise torque produced by the weight of the plank about this point is larger than the torque produced by the weight of the child. The weight of the plank is treated as though it is concentrated at the center of gravity of the plank.

The plank will be just on the verge of tipping when the torque of the child about the edge equals the torque of the plank in magnitude. This condition determines how far the child can walk on the plank before it will tip. If the torque of the plank about the edge of the platform is larger than that of the child, the child is safe. (The platform constrains the plank from rotating in the counterclockwise direction; what actually happens in that case is that the effective pivot point shifts to the left, so that the net torque is still zero.)

The idea of treating the weight of the plank as though it all acted through the center of gravity runs against intuition sometimes. Students often prefer to break the plank into two portions in problems like this, one on either side of the pivot point. This will work, but it requires the computation of three torques rather than two. Each portion of the plank will then have its own weight and center of gravity. It is easier to treat the entire plank as having a single weight and center of gravity, and therefore a single torque.

The location of the center of gravity of an object is a very important consideration in any balancing effort. If the center of gravity lies below the pivot point, as in the balancing toy pictured in figure 7.13, then the object will automatically return to a balanced condition when disturbed, because the center of gravity will come back to the position directly below the pivot point. This is where it produces no torque; the lever arm for the weight is then zero.

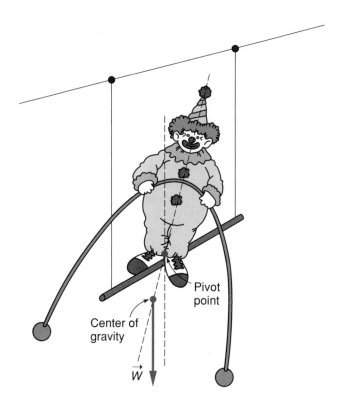

Figure 7.13   The clown automatically returns to an upright position because the center of gravity is below the pivot point.

The location of your own center of gravity is similarly an important consideration in performing various maneuvers, athletic or otherwise. Try, for example, touching your toes with your back and heels against a wall. Why is this apparently simple trick impossible for most people to do? Where is your center of gravity relative to the pivot point determined by your feet? Center of gravity and torque are at work here!

7.3 ROTATIONAL INERTIA AND NEWTON'S SECOND LAW

When a child runs alongside of a merry-go-round, starting it in rotational motion, the force exerted by the child must be producing a torque about the axle. The previous section established that the net torque acting upon an object determines whether or not it will begin to rotate. But what about the rate of rotation that is produced? Can we predict that from knowledge of the torque?

In the case of linear motion, the net force determines the acceleration of an object. It turns out that we can adapt Newton's second law, which predicts the rate of acceleration for linear motion, to cases of rotational motion. When we do so, it is now the rotational acceleration that is determined by the net torque. A new quantity, the rotational inertia, replaces the mass in our modified second law.

Newton's Second Law for Rotational Motion

Suppose that we return to the merry-go-round example. The propulsion system (one energetic child or tired parent) applies a force at the rim of the merry-go-round. The torque about the axle is found by multiplying this force by the lever arm, which in this case is just the radius of the merry-go-round (fig. 7.14). If the frictional torque at the axle is small enough to be ignored, then the torque produced by the child is the only one acting on the system.

If we wanted to find the linear acceleration produced by a force acting on an object, we would use Newton's second law, $\sum \vec{F} = m\vec{a}$. By analogy, then, we might seek a similar expression for rotational motion, where the net torque replaces the net force and the rotational acceleration, α, replaces the linear acceleration. What quantity, however, should we use in place of the mass?

In the case of linear motion, the mass represents the inertia, or resistance to a change in motion. A new concept is needed for rotational motion, which we call the *rotational inertia*. This is often also referred to as the *moment of inertia*, and the symbol I is used to represent this quantity. The rotational inertia (or moment of inertia) represents the resistance of an object to change in its rotational motion. It is related to the mass of the object, but it turns out to be related also to how that mass is distributed about the axis of rotation being considered. By analogy to Newton's second law, then, we can state the second law for rotational motion as follows:

Newton's second law for rotational motion states that the net torque acting upon an object about a given axis is equal to the rotational inertia of the object about that axis times the rotational acceleration of the object, or

$$\sum \tau = I\alpha.$$

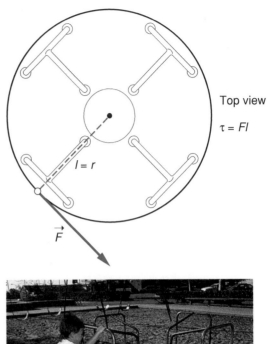

Top view

$\tau = Fl$

$l = r$

$\vec{F}$

Figure 7.14   The child exerts a force at the rim of the merry-go-round that produces a torque about the axle.

To put it differently, the rotational acceleration that is produced is equal to the net torque divided by the rotational inertia, $\alpha = \sum \tau / I$. The larger the net torque, the larger the rotational acceleration, but the larger the rotational inertia, the smaller the rotational acceleration. The key to understanding this relationship, however, lies in acquiring an understanding of the concept of rotational inertia and how it depends upon the distribution of mass.

In order to gain a feeling for how rotational inertia depends upon the distribution of mass, physicists often use the trick of considering the simplest possible situation. For rotational motion, this case would be that of a single, concentrated mass at the end of a very light rod, as pictured in figure 7.15. If a force is applied to this mass in a direction perpendicular to the rod, the rod and mass will begin to rotate about the fixed axle at the other end of the rod.

The component of the linear acceleration of the mass that is in the direction of the applied force can be found directly from Newton's second law, $F = ma$, in this case. We can also write this equation in terms of the rotational quantities, however, by using the definition of torque and the relationship

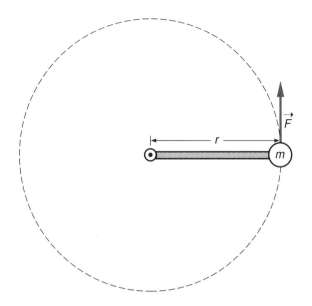

Figure 7.15 A single concentrated mass at the end of a light rod is set into rotation by the applied force *F*. Newton's second law can be used to find the acceleration.

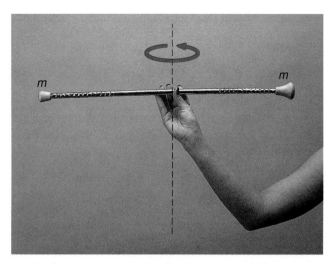

Figure 7.16 The rotational inertia of a baton is determined largely by the masses at either end.

between the linear acceleration and the rotational acceleration, $a = r\alpha$. We simply multiply both sides of Newton's second law by r:

$$Fr = mar = m(r\alpha)r = mr^2\alpha.$$

The far left-hand side of this equation is the torque acting on the mass m, so this equation has the following form:

$$\tau = (mr^2)\alpha = I\alpha.$$

Evidently the rotational inertia, I, of the concentrated mass at the end of the rod is equal to mr^2, the mass times the square of its distance from the axle. If this is true for any mass, then we can find the total rotational inertia of an object like our merry-go-round by adding the contributions of different masses at different distances from the axle:

$$I = \Sigma(mr^2).$$

All we have really done here is to recast Newton's law in a form suitable for discussing the rotational motion of a solid object. In the process, the appropriate expression for finding the rotational inertia of the object emerges. The rotational inertia dictates how difficult it will be to change the rotational velocity of the object, because the rotational acceleration is inversely proportional to the rotational inertia. This rotational inertia, I, depends both on the masses of different parts of the object and on how far these masses are from the axis of rotation; the greater the distance, the greater the contribution to the rotational inertia.

To get a feel for these ideas, it is helpful to consider a simple object such as a baton, which consists of two roundish masses at the end of a rod (fig. 7.16). If the rod itself is light, most of the rotational inertia of the baton is due to the masses

at either end. If you hold the baton at the center, you can apply a torque with your hand, thus producing a rotational acceleration and making the baton rotate.

Suppose that we can move these masses along the rod. If we move the masses in toward the center of the rod, so that their distance from the center is half the original distance, what happens to the rotational inertia? It decreases to one-fourth of its initial value, ignoring the contribution of the rod. This is because the dependence upon distance from the axis involves the *square* of the distance: doubling the distance produces a quadrupling of the rotational inertia; halving the distance decreases the rotational inertia by a factor of 4.

The baton will be four times harder to set into rotation when the masses are at the ends than when they are at half the distance from the ends. In other words, the torque needed to produce a given rotational acceleration will be four times larger when the masses are at the ends than when they are at the intermediate positions. If you had a rod upon which you could move concentrated masses, you could feel this difference in the amount of torque that you would need to apply to start it rotating. A pencil, with lumps of clay serving as the masses, might provide a reasonable substitute.

Rotational Inertia of Solid Objects

Finding the rotational inertia of an object like our merry-go-round is a more difficult task. This is because the mass of the merry-go-round is distributed at different distances from the axis in a continuous manner. Each portion of the total mass makes a contribution to the rotational inertia depending upon its distance from the axis. The mathematical tool for summing these contributions involves calculus, so we will not get into the details. The results of such computations for a few common shapes are shown in figure 7.17. The results do depend upon the location of the axis, as indicated by the two different values shown for different axes through the rod. When the axis is through the end of the rod (fig. 7.17a) we get a larger value than when it is through the center (fig. 7.17b) because the mass of the rod is, on the average, farther from the axis when the axis is at the end.

The merry-go-round could be considered to be a disk if we ignore the handles and other variations in the distribution of its mass. It is interesting to note that the rotational inertia of a disk (fig. 7.17c) is just half of that which would result if all of the mass were concentrated at the rim, $I = MR^2$ (fig. 7.17d). If the mass were all concentrated at the rim, of course, it would have a greater average distance from the axis, so we would expect a larger rotational inertia.

A child sitting on the merry-go-round will also affect the rotational inertia. If several children are all sitting near the rim of the merry-go-round, their contributions to the moment of inertia will make it considerably more difficult to set the merry-go-round into rotation with a given rotational acceleration. On the other hand, if the children are near the center, they provide less additional rotational inertia. If you are feeling tired, have them sit near the middle; you will save some effort!

Some numbers might help to place all this in perspective. Suppose that the merry-go-round has a mass of 400 kg and a radius of 2 m. If we treat it as a disk, its rotational inertia is

$$I = \left(\frac{1}{2}\right)MR^2 = \left(\frac{1}{2}\right)(400 \text{ kg})(2 \text{ m})^2 = 800 \text{ kg·m}^2.$$

A child with a mass of 40 kg sitting near the edge would add an additional contribution to the rotational inertia:

$$I_C = mr^2 = (40 \text{ kg})(2 \text{ m})^2 = 160 \text{ kg·m}^2.$$

The total rotational inertia of the merry-go-round plus the child would be the sum of these two terms, or 960 kg·m^2.

If we want to produce the rotational acceleration of 0.005 rev/s^2 that we used in section 7.1, the rotational form of Newton's second law can be used to find the required torque. First, however, we need to convert the rotational acceleration to rad/s^2, because we have used relationships between linear and rotational quantities in developing Newton's second law for rotational motion, and the expressions used are only valid when the rotational quantities are expressed in radians. The conversion is simple; we merely need to multiply by 2π radians per revolution:

$$\alpha = (0.005 \text{ rev/s}^2)(2\pi \text{ rad/rev}) = 0.0314 \text{ rad/s}^2.$$

The required torque is then found by multiplying the rotational acceleration by the rotational inertia, according to Newton's second law:

$$\tau = I\alpha = (960 \text{ kg·m}^2)(0.0314 \text{ rad/s}^2) = 30.1 \text{ N·m}.$$

(The units do yield newton-meters in the metric system because a newton is just 1 kg·m/s^2, and the radian is a dimensionless quantity.) Since a torque is equal to a force times a

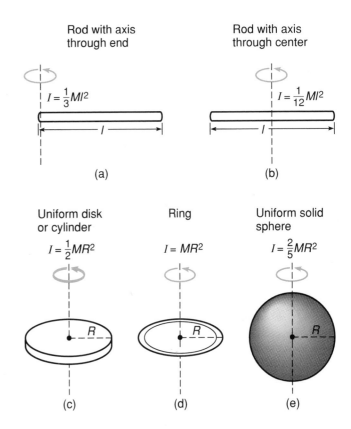

Figure 7.17 **Expressions for the rotational inertia of several objects, each with a uniform distribution of mass over its shape. A capital *M* is used to represent the total mass of the object.**

lever arm, to find the force required to produce this torque, we would divide the torque by the lever arm, which in this case would be the radius of the merry-go-round:

$$F = \frac{\tau}{l} = \frac{30.1 \text{ N·m}}{2 \text{ m}} = 15 \text{ N}.$$

If more children attempt to freeload on the ride, all of them sitting near the edge, the torque and force required to produce the same rotational acceleration become larger and the task of the pusher more difficult. Once accelerated, however, the merry-go-round will keep rotating for a longer time, assuming that the frictional torque on the axle is not affected by the greater number of children. This is because the greater rotational inertia will result in smaller negative rotational acceleration due to the frictional torque ($\alpha = \tau / I$). The greater the rotational inertia, the greater the tendency for the object to maintain its motion.

Resistance to change in rotational motion is the essence of rotational inertia. The rotational form of Newton's second law, $\tau = I\alpha$, provides the quantitative relationship. Torque takes the place of force, rotational inertia replaces mass, and rotational acceleration replaces linear acceleration in the ordinary form of Newton's second law for linear motion.

Pushing a merry-go-round or twirling a baton can provide an intuitive feeling for these ideas. Such activities are highly recommended for a study break.

7.4 CONSERVATION OF ANGULAR MOMENTUM

Have you ever watched an ice skater go into a spin? She starts the spin with her arms extended and then brings them in towards her body. As she brings her arms in, the rate of the spin increases; as she extends her arms again, her rotational velocity decreases (fig. 7.18). How is this accomplished; how might we explain this phenomenon?

The concept of angular or rotational momentum is very useful in analyzing situations like this. The principle of conservation of angular momentum explains a whole host of phenomena in sports and other areas that are similar to the ice-skater example. The analogy between linear and rotational motion is helpful in understanding these ideas.

Definition of Angular Momentum

Using the analogy between linear motion and rotational motion, we can quickly see the likely form of the concept of rotational, or angular, momentum. Since linear momentum is defined as the mass (inertia) times the velocity, $\vec{p} = m\vec{v}$, angular momentum should be defined as follows:

Angular momentum is the product of the rotational inertia times the rotational velocity; in symbols, $L = I\omega$.

The symbol L is commonly used to represent angular momentum. The term *angular momentum* is more com-

monly used than *rotational momentum,* but either can be used to describe this concept.

Just as with linear momentum, then, angular momentum depends upon two quantities, an inertia and a velocity. A bowling ball spinning slowly might have the same angular momentum as a baseball spinning much more rapidly, because of the larger rotational inertia of the bowling ball. Because of its enormous rotational inertia, the earth has a very large angular momentum associated with its daily turn about its axis, even though the angular velocity is small.

It is sometimes useful to express the angular momentum of an individual mass about some axis in terms of the mass and its linear momentum. For example, the angular momentum of the planets about the sun is involved in understanding Kepler's second law, as we mentioned in chapter 5. To compute this angular momentum, we could use the definition $L = I\omega$, but we can also express this definition in terms of linear quantities. This is done by using the definition of rotational inertia and the relationship between rotational velocity and the resulting linear velocity, $v = r\omega$. Substituting these expressions in our definition for angular momentum yields

$$L = I\omega = (mr^2)\left(\frac{v}{r}\right) = mvr.$$

We see, then, that the angular momentum of a concentrated mass about some center of rotation is just the linear momentum, mv, times the distance r from the center of rotation or any other axis that we might choose. (Actually, the velocity v must be perpendicular to r in order for the expression mvr to be valid.) The distance r is essentially a lever arm for the linear momentum about the axis of interest (fig. 7.19).

Figure 7.18 The rotational velocity of the skater increases as she pulls her arms in towards her body.

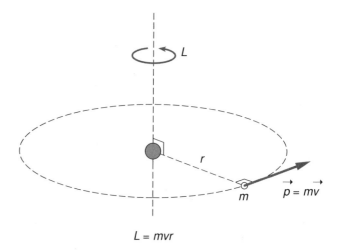

$$L = mvr$$

Figure 7.19 A concentrated mass such as a planet can be regarded as having an angular momentum about an appropriate axis.

The Conservation Principle

If there is no net torque acting on a system, and the rotational inertia, I, is not changing, then the rotational velocity ω and the angular momentum will be constant. This follows from Newton's second law, since the rotational acceleration is zero if there is no net torque.

It turns out, however, that (more generally) the angular momentum is constant when there is no net external torque, even if the rotational inertia is changing. In other words, the torque in the rotational form of Newton's second law is really related to the rate of change of angular momentum; if there is no torque, the angular momentum does not change. This is analogous to the situation for linear motion; if the net external force is zero, then the total momentum of the system does not change, even in cases (such as the rocket) where the mass may be changing.

What we have just expressed is the principle of conservation of angular momentum, which we can state more formally as follows:

If the net torque acting upon a system is zero, then the total angular momentum of the system is conserved.

This is a very similar statement to that for the conservation of linear momentum; net torque replaces net force, and angular momentum replaces ordinary, or linear, momentum.

We are now in a position to explain what happens to our spinning ice skater. The torque acting upon the skater about her axis of rotation is very small, so that we have approximately the condition necessary for conservation of angular momentum. When her arms are extended, they contribute a relatively large portion to her total rotational inertia because their average distance from her axis of rotation is much larger than that of other portions of her body. The rotational inertia depends upon the square of the distance from the axis, so the effect of this distance is substantial, even though the hands and arms make up a small portion of the total mass of the skater.

When the skater pulls her arms in towards her body, their contribution to her rotational inertia decreases and, therefore, her total rotational inertia decreases. Conservation of angular momentum requires that her angular momentum remain constant, so her angular velocity must increase if her rotational inertia is decreasing. Expressed in symbols, this argument takes the following form:

$$L = I_1\omega_1 = I_2\omega_2.$$

If I decreases, ω must increase, and vice versa in order for the product of I times ω, the angular momentum, to remain constant. These ideas are illustrated in the numerical example in box 7.1.

This same phenomenon can be demonstrated if we have available a rotating platform or stool with good bearings, so

---

Box 7.1

Sample Exercise

Suppose that an ice skater has a rotational inertia of 1.20 kg·m^2 when her arms are extended and a rotational inertia of 0.50 kg·m^2 when her arms are pulled in close to her body. If she goes into a spin with her arms extended and has a rotational velocity of 1.0 rev/s, what is her rotational velocity when she pulls her arms in close to her body?

$I_1 = 1.20$ kg·m^2 Since angular momentum is conserved,

$I_2 = 0.50$ kg·m^2 $I_2\omega_2 = I_1\omega_1$.

$\omega_1 = 1.0$ rev/s.   Dividing both sides by I_2, we have

$\omega_2 = ?$ $\omega_2 = \left(\dfrac{I_1}{I_2}\right)\omega_1$

$$= \left(\frac{1.20 \text{ kg·m}^2}{0.50 \text{ kg·m}^2}\right)(1.0 \text{ rev/s})$$

$$= (2.4)\,(1.0 \text{ rev/s}) = \textbf{2.4 rev/s.}$$

---

that the frictional torques are small. In these demonstrations, we often have the subjects hold masses in their hands, which has the effect of increasing the change in rotational inertia that occurs when the arms are drawn in toward the body (fig. 7.20). A striking increase in rotational velocity can then be achieved!

A similar effect is at work when a diver pulls into a tuck position to produce a spin. In this case, the dive is started with the body extended and a slow rate of rotation about an axis through the center of gravity of the diver's body, as shown in figure 7.21. As the diver pulls into a tuck position, the rotational inertia about this axis is reduced, and the rotational velocity increases. Near the end of the dive, the diver comes out of the tuck, thus increasing the rotational inertia and decreasing the rotational velocity. The net torque about the center of gravity due to the gravitational attraction of the earth on the diver is zero because of the way in which center of gravity is defined.

There are many other examples like these, in which the rotational velocity can be altered by changing the rotational inertia. It is much easier to produce a change in the rotational inertia of a body than it is to change the mass of the body itself. We simply need to change the distance of various portions of the mass from the axis of rotation. To consider the

Figure 7.20 **A student holding masses in each hand while sitting upon a rotating stool can achieve a large increase in rotational velocity by bringing his arms in towards his body.**

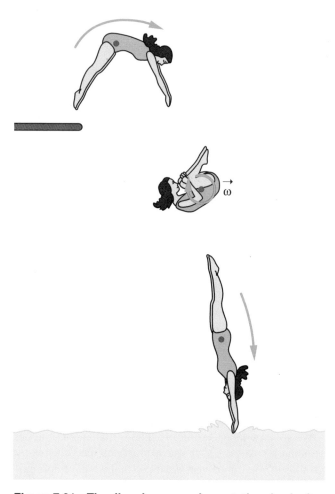

Figure 7.21 **The diver increases her rotational velocity by pulling into a tuck position, thus reducing her rotational inertia about her center of gravity.**

internal forces and torques involved in this process would be a complex undertaking, but conservation of angular momentum provides a convenient explanation for these phenomena.

Kepler's Second Law

In the case of a planet orbiting the sun, conservation of angular momentum also plays a role. The gravitational force acting upon the planet produces no torque about the sun because its line of action passes directly through the sun. Thus the lever arm for this force is zero, and the resulting torque, which equals the force times the lever arm, must also be zero (fig. 7.22). Angular momentum is conserved, therefore.

Using the form of the definition of angular momentum that we developed for a concentrated mass, $L = mvr$, we can now explain Kepler's second law of planetary motion. Since the angular momentum is constant, the velocity v must get larger when the distance r gets smaller in order for the product mvr to remain the same. It is not difficult to show that this requirement results in equal areas being swept out by the radius line in equal times, although we will skip the details here. The velocity must be larger when the radius gets smaller in order to keep the area that is being swept out the same.

A related effect can be observed in a simple experiment with a ball on a string. If you allow the string to wrap around your finger as it rotates, thus producing a smaller radius of rotation, the ball will increase its rotational velocity about

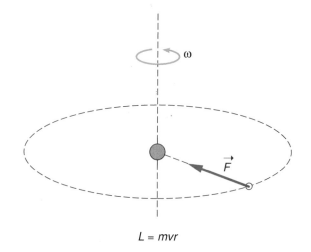

$$L = mvr$$

Figure 7.22 **The gravitational force acting on the planet produces no torque about an axis through the sun because the lever arm is zero for this force.**

your finger The product *mvr* remains constant, since there is no net torque about your finger. The speed *v* increases as the radius *r* decreases. Alternatively, the rotational velocity ω increases as the rotational inertia *I* decreases due to the decreased radius. Either way, angular momentum is conserved. Try it!

7.5 RIDING A BICYCLE AND OTHER AMAZING FEATS

Have you ever wondered why a bicycle remains upright when it is moving but promptly falls over when it is not moving? It turns out that conservation of angular momentum is involved here also, but an additional wrinkle is needed to provide the explanation. We must consider the directional aspects of angular momentum in order to get to the bottom of things.

The Vector Nature of Angular Momentum

Does angular momentum have a direction associated with it? Linear momentum is a vector, with the direction of the momentum $\vec{p}$ being the same as that of the velocity $\vec{v}$ of the object. Perhaps we should ask, then, whether rotational velocity has a direction. It turns out that rotational velocity and acceleration can be treated as vectors, and since angular momentum is related to rotational velocity, angular momentum can also be treated as a vector. Let's begin by returning to the merry-go-round example and exploring how we might define the direction of rotational velocity.

If the merry-go-round is rotating in a counterclockwise direction, as pictured in figure 7.23, how might we indicate that direction with an arrow? The term counterclockwise indicates the direction of rotation as seen from a given perspective, but not a unique direction in space; we must also specify the axis of rotation and how we are looking at the object. An object that is seen as rotating in a counterclockwise manner when viewed from above is seen as rotating in

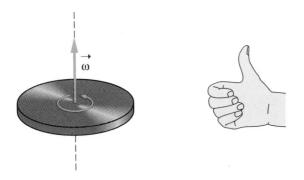

Figure 7.23 The direction of the rotational velocity vector for the counterclockwise rotation is defined to be upwards along the axis of rotation, as is indicated by the thumb on the right hand with the fingers curled in the direction of rotation.

a clockwise manner when viewed from below. We could draw an axis of rotation and a curved arrow around it, as we often do, but it would be more desirable to be able to specify direction with a simple arrow.

The normal solution to this problem is to define the direction of the rotational velocity vector as being along the axis of rotation and in the upward direction for the counterclockwise rotation pictured in figure 7.23. A simple rule for deciding which direction the vector should have along the axis of rotation, up or down, can be stated with the help of your right hand. If you place your right hand with the fingers curling around the axis of rotation in the direction of the rotation, then your thumb points in the direction of the rotational velocity vector. Thus, if the merry-go-round were rotating in the clockwise direction, instead of counterclockwise, your thumb would point downwards, and that would be the direction of the rotational velocity vector.

By analogy to our definition of linear momentum, the direction of the angular momentum vector is the same as that of the rotational velocity. In fact, we can write the definition of angular momentum as a vector expression, $\vec{L} = I\vec{\omega}$. Since the rotational inertia, *I*, is a scalar quantity that has no direction associated with it, this definition implies that $\vec{L}$ is in the same direction as $\vec{\omega}$. Conservation of angular momentum then requires that the direction of the angular momentum vector remain constant as well as its magnitude, because a vector quantity changes if either its direction or magnitude change.

The Bicycle

This is where the bicycle comes in. The wheels of a bicycle acquire angular momentum when the bicycle is moving. (Torque is applied to the rear wheel via the pedals and chain in order to produce a rotational acceleration.) If the bicycle is moving in a straight path, the direction of the angular momentum vector is the same for both wheels and is perpendicular to the vertical plane, as shown in figure 7.24.

A torque is required to change the direction of the rotational velocity or angular momentum, just as a force is required to change the direction of a linear velocity. The larger the angular momentum, however, the larger the torque that is required to produce an appreciable change in this direction, as is indicated in figure 7.25. A change of a given size in angular momentum, $\Delta \vec{L}$, produces only a small change in the direction of the angular momentum, $\vec{L}$, when $\vec{L}$ is large (fig. 7.25a) but a much greater change in the direction of $\vec{L}$ when the angular momentum is small (fig. 7.25b). The size of $\Delta \vec{L}$ is proportional to the torque.

This torque would normally come from the gravitational force acting upon the rider and the bicycle through their center of gravity. When the bicycle is exactly upright, though, this force acts straight downward and passes through the axis of rotation that would be involved if the bicycle were to start to fall, namely the line along which the tires contact

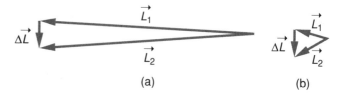

Figure 7.25 The same change in angular momentum $\Delta\vec{L}$, produces a greater change in the direction of $\vec{L}$ when $\vec{L}$ is small than when $\vec{L}$ is large.

Figure 7.24 The angular momentum vector for each wheel is in the horizontal direction when the bicycle is upright.

the road. The torque will therefore be zero because the line of action of the force passes through this axis of rotation, and the lever arm is zero. Small deviations from the upright position produce small torques, but their effect on the direction of the angular momentum is usually small and can be easily compensated for by the rider shifting his weight.

What is the difference, then, between the situation when the bicycle is moving and when it is stationary? When the bike is stationary, there is no angular momentum associated with the wheels. A small gravitational torque that results from a slight deviation from a perfectly upright position will cause the bike to begin to rotate quickly about the axis joining the tires with the ground and the bike falls.

When the bike is moving, however, this torque must also change the direction of the angular momentum vectors associated with the wheels. The direction of the rotational velocity associated with the bike tipping over is actually, by our definition of the rotational velocity vector, horizontal (parallel to the ground). This must also be the direction, then, of the change in angular momentum, $\Delta\vec{L}$, that we have pictured in figure 7.25. When $\Delta\vec{L}$ is added to $\vec{L}$, then, its effect is to turn the wheel as the bike begins to lean over. This is an intuitively surprising result, and yet all of us who have ridden bicycles take advantage of it routinely. By leaning into a curve, we actually cause the bicycle wheels to turn, thus helping us to make the curve!

This effect of the torque required to change the direction of an angular momentum vector can be observed by holding a

bicycle upright on its rear wheel and having a friend spin the front wheel. You can feel that it is harder to change the direction of this wheel when it is spinning rapidly than when it is spinning slowly or not at all. You will also get the feeling that the wheel has a mind of its own; as you try to tilt the wheel, it tends to turn in a direction perpendicular to that tilt!

A bicycle tire mounted on a hand-held axle is even more effective for sensing the torque that is required. This is a common piece of demonstration apparatus; usually the tire is filled with steel cable rather than air, which gives it a larger rotational inertia and thus a larger angular momentum for a given rate of spin. Holding the axle on either side while the wheel is spinning in a vertical plane and then trying to tip the wheel will give you an appreciation for what happens when you are riding a bicycle. It also demonstrates how difficult it is to change the direction of the angular momentum of a rapidly spinning wheel.

Other demonstrations that point out the vector nature of angular momentum can be performed with this hand-held bicycle wheel. If, for example, we have a student hold the wheel with its axle in the vertical direction while standing or sitting on a rotating platform or stool, conservation of angular momentum produces striking results. The idea is to start the tire rotating while holding the stool so that it does not rotate initially. With the tire rotating, then, we have the student flip the tire, thus reversing the direction of the angular momentum vector (fig. 7.26).

Can you imagine what happens then? In order for angular momentum to be conserved, the original direction of the angular momentum vector must be maintained. The only way that this can occur is for the stool, with the student volunteer, to begin to rotate in the same direction in which the tire was rotating initially. The sum of the two vectors, that of the tire and that of the student and stool (including the tire) must add to yield the original angular momentum (see fig. 7.27). This means that the angular momentum gained by the student and stool will be exactly twice the original angular momentum of the wheel. The student can stop the rotation of the stool by flipping the wheel axis back to its original direction.

Directional effects involving angular momentum and its conservation are important in many other situations. The angular momentum associated with the rotors of a helicopter becomes an extremely important factor in helicopter design,

Figure 7.26 A student holds a spinning bicycle wheel while sitting upon a stool that is free to rotate. What happens if the wheel is turned upside down?

for example. The motion of a top also provides some fascinating effects. If you have access to a top, try observing what happens to the direction of the angular momentum vector as the top slows down. As the top begins to tip, the change in direction of the angular momentum vector causes the rotation axis of the top to rotate (*precess*) about a vertical line. Can you see an analogy here to what happens with a bicycle wheel?

Angular momentum and its directional aspects are also extremely important factors in atomic and nuclear physics. The particles that make up atoms have spins associated with them, and these spins, of course, imply angular momentum.

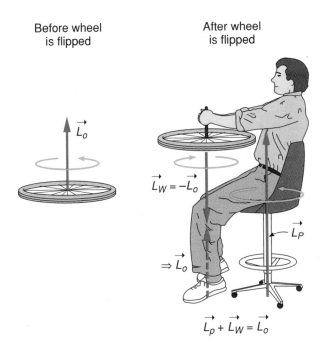

Figure 7.27 The angular momentum of the person and stool, $\vec{L}_p$, adds to that of the flipped wheel, $\vec{L}_w$, to yield the original angular momentum of the wheel, $\vec{L}_o$.

The ways in which these angular momentum vectors add can be used to explain a variety of atomic phenomena. These are not exactly everyday phenomena, of course, but it is useful to recognize that atoms and nuclei have things in common with bicycle wheels and the solar system.

Box 7.2

Everyday Phenomenon:
Achieving the State of Yo

The Situation. A physics professor noticed that one of his students, Tom Struthers, often carried a yo-yo with him to class and was quite proficient at putting it through its paces. The professor challenged Struthers to explain the behavior of the yo-yo using the principles that they had just studied in class involving torque and angular momentum.

In particular, the professor asked Struthers to explain why the yo-yo sometimes comes back and sometimes can be made to "sleep," or continue to rotate at the end of the string. What are the differences in these two situations?

Continued

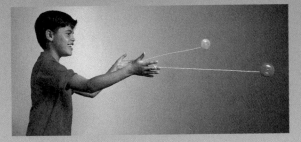

A yo-yo will come back to your hand or, with sufficient skill, you can make it "sleep" at the end of its string.

Box 7.2 Continued

The Analysis. Struthers examined carefully how his yo-yo was constructed and how the string was attached. He noticed that the string was not tied tightly to the axle of the yo-yo, but instead ended in a loose loop around the axle. When the yo-yo is at the end of its string, then, the string can slip on the axle. When the string is wound around the axle on the other hand, it is less likely to slip.

The yo-yo is generally started with the string wound around the axle and looped around the middle finger. As the yo-yo is released from the hand, the string unwinds, and the yo-yo gains rotational velocity and angular momentum. A torque must be at work, Struthers reasoned. He drew a force diagram for the yo-yo that looked like the one shown here. Only two forces are acting on the yo-yo: its weight acting downwards and the tension in the string acting upwards.

Since the yo-yo is accelerated downwards, the weight must be greater than the tension in order to produce a downward net force. The weight does not produce a torque about the center of gravity of the yo-yo, however, because its line of action passes through the center of gravity, and the lever arm is therefore zero. The tension, on the other hand, acts along a line that is off-center and produces a torque that will cause a counter-clockwise rotation about the center of gravity, as shown in the drawing.

The torque due to the tension in the string produces a rotational acceleration, and the yo-yo gains rotational velocity and angular momentum as it falls. The torque gets smaller as the string unwinds because the lever arm, which is the distance of the line of action of the tension from the center of the yo-yo, gets smaller. The yo-yo has a substantial angular momentum, however, when it reaches the bottom of the string, and in the absence of external torques that might change this angular momentum, it will be conserved. This is what happens when the yo-yo sleeps at the bottom of the string; the only torque acting is the frictional torque of the string slipping on the axle, and this will be small if the axle is smooth.

What happens, however, when the yo-yo returns to the hand? The yo-yo can be caused to return by jerking lightly on the string at the instant that it reaches the bottom of the string. This jerk provides a brief impulse and upward acceleration of the yo-yo. Since it is already spinning, it continues to spin in the same direction, and the string rewinds itself around the axle. The line of action of the tension in the string is now on

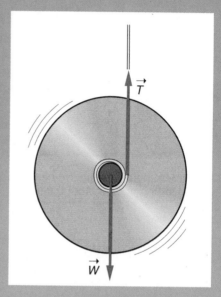

A cut-away diagram showing the forces acting upon the yo-yo when it is falling. Its weight and the tension in the string are the only significant forces.

the opposite side of the axle, though, and its torque now causes the rotational velocity and angular momentum to decrease. The rotation should stop when the yo-yo has returned to the hand.

As the yo-yo is rising, the net force acting on it is still downwards, and its linear velocity decreases along with its rotational velocity. The only time that there is a net force acting upwards is when the upward impulse is delivered by jerking on the string. The situation is similar in this respect to that of a ball bouncing on the floor; the net force is downward except during the very brief time of contact with the floor. It is our ability to affect the nature and timing of the impulse imparted by the string at the bottom of the yo-yo's path that causes the yo-yo to either sleep or return. This is what the art of yo is all about.

Struther's analysis provided him with a renewed appreciation of his art. (Additional insight can be gained by bringing in the energy ideas found in chapter 8.) Whether or not Struther's understanding was useful in improving his skills is debatable; a lot of yo-yoing skill is due to practice. The subtle pulls on the string occur too rapidly to allow detailed analysis. Struthers was still a lot better at it than his physics professor!

SUMMARY

We have considered how to describe the rotational motion of a solid object and also what causes rotational motion. An analogy to quantities used to describe linear motion has been useful in getting an understanding of the rotational concepts. The rotational form of Newton's second law, $\sum\tau = I\alpha$, was central to our discussion. The key concepts were the following.

Rotational Displacement, Velocity, and Acceleration. Rotational displacement is described by an angle; rotational velocity is the rate of change of that angle with time; and rotational acceleration is the rate of change of rotational velocity with time.

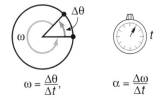

$$\omega = \frac{\Delta\theta}{\Delta t}, \qquad \alpha = \frac{\Delta\omega}{\Delta t}$$

Torque. A torque is what causes an object to rotate. It is defined as a force times the lever arm of the force, which is the perpendicular distance from the line of action of the force to the axis of rotation. If the net torque acting on an object is zero, the object will not change its state of rotation.

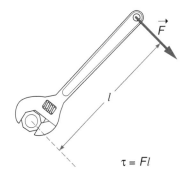

$$\tau = Fl$$

Newton's Second Law and Rotational Inertia. In the rotational form of Newton's second law, the torque takes the place of the force, rotational acceleration replaces the ordinary linear acceleration, and rotational inertia replaces mass. The rotational inertia depends upon the distribution of mass about the axis of rotation.

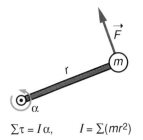

$$\sum\tau = I\alpha, \qquad I = \sum(mr^2)$$

Conservation of Angular Momentum. By analogy to linear momentum, angular momentum is defined as the rotational inertia times the rotational velocity. It is conserved when there is no net external torque acting on a system.

$$L = I\omega$$
If $\sum\tau_{ext} = 0$, $L = $ constant

Riding a Bicycle. The vector nature of angular momentum can be used to explain why a bicycle stays upright when moving, as well as many other phenomena.

QUESTIONS

Q7.1 Which of the following units would not be appropriate for describing a rotational velocity: rad/min, rad/m, rev/hr, m/s? Explain.

Q7.2 Which of the following units would not be appropriate for describing a rotational acceleration: rad/s, rev/hr², rad/m², degrees/s²? Explain.

Q7.3 Consider a rotating object that is gradually slowing down in its rate of rotation. Does this object have a rotational acceleration? Explain.

Q7.4 Will the expression $v = r\omega$ be valid for finding the linear velocity of a point on a rotating object if ω is expressed in rev/s ? Explain.

Q7.5 Which of the forces pictured as acting upon the rod shown here will produce a torque about the axis at the left end of the rod? Explain.

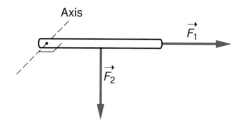

Q7.6 Which of the two orientations of the equal-magnitude forces shown in the diagram below will produce the greater torque on the wheel? Explain.

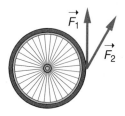

Q7.7 Is it possible to balance two masses of different weights on a simple balance beam resting upon a fulcrum? Explain.

Q7.8 Is it possible for the net force acting on an object to be zero, but the net torque to be greater than zero? Explain. (Hint: The forces contributing to the net force might not lie along the same line.)

Q7.9 You are attempting to move a large rock, using a steel rod as a lever. Will it be more effective to place the fulcrum nearer to your hands or nearer to the rock? Explain.

Q7.10 Two forces, equal in magnitude but opposite in direction, act at the same point on an object. Is it possible for the net torque about any axis on the object to be greater than zero in this situation? Explain.

Q7.11 An object is rotating with a constant rotational velocity. Can there be a net torque acting on the object? Explain.

Q7.12 Two objects have the same total mass, but object *A* has its mass concentrated closer to the axis than object *B*. Which object will be easier to set into rotational motion? Explain.

Q7.13 Is it possible for two objects with the same mass to have different rotational inertias? Explain.

Q7.14 Can you change your rotational inertia about a vertical axis through the center of your body without changing your total weight? Explain.

Q7.15 Is angular momentum always conserved? Explain.

Q7.16 Is it possible for two objects with the same mass and the same rotational velocities to have different angular momenta? Explain.

Q7.17 A child on a freely rotating merry-go-round moves from near the center towards the edge. Will the rotational velocity of the merry-go-round increase or decrease? Explain.

Q7.18 Is it possible for an ice skater to change her rotational velocity without the involvement of any external torque? Explain.

Q7.19 A top falls over quickly if it is not spinning, but will remain approximately upright for some time when it is spinning. Explain why this is so.

Q7.20 Does the direction of the angular momentum vector of the wheels change when a bicycle goes around a corner? Explain.

Q7.21 Can a yo-yo be made to sleep if the string is tied tightly to the axle? Explain.

EXERCISES

E7.1 Suppose that a merry-go-round is rotating at the rate of 10 rev/min.

    a. Express this rotational velocity in rev/s.

    b. Express this rotational velocity in rad/s.

E7.2 Suppose that a disk rotates through half a revolution in 1 second.

    a. What is its displacement in radians in this 1 second?

    b. What is its rotational velocity in rad/s, assuming that it is rotating with constant rotational velocity?

E7.3 A standard long-playing record rotates at a rate of 33.3 rev/min.

    a. Express this rotational velocity in rev/s.

    b. Express this rotational velocity in rad/s.

    c. Through how many revolutions does the record turn in a time of 10 s?

E7.4 A small disk is rotationally accelerated at the constant rate of 4 rev/s^2.

    a. If it starts from rest, what is its rotational velocity after 5 s?

    b. Through how many revolutions does it turn in this time?

E7.5 The rotational velocity of a merry-go-round increases at a constant rate from 1.0 rad/s to 1.5 rad/s in a time of 10 s.

    a. What is the rotational acceleration of the merry-go-round?

    b. Through how many radians does the merry-go-round turn during this time?

E7.6 A force of 10 N is applied at the end of a wrench handle that is 30 cm long. The force is applied in a direction perpendicular to the handle in order to turn a nut at the opposite end of the wrench.

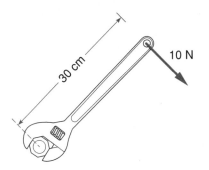

E7.7 Two forces are applied to a merry-go-round with a radius of 2.5 m, as shown in the diagram. One force has a magnitude of 80 N, and the other a magnitude of 50 N.

    a. What is the torque applied to the nut by the wrench?

    b. What would the torque be if the force were applied halfway up the handle instead of at the end?

    a. What is the torque about the axle of the merry-go-round due to the 80-N force?

    b. What is the torque about the axle due to the 50-N force?

    c. Assuming that these two torques are the only ones involved, what is the net torque acting on the merry-go-round, given the force directions shown?

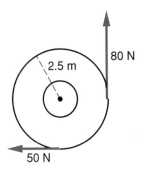

E7.8 A weight of 15 N is located at a distance of 20 cm from the fulcrum of a simple balance beam. At what distance from the fulcrum should a weight of 20 N be placed on the opposite side in order to balance the system?

E7.9 A weight of 5 N is located 10 cm from the fulcrum of a simple balance beam. What weight should be placed at a point 4 cm from the fulcrum on the opposite side in order to balance the system?

E7.10 A net torque of 10 N·m is applied to a disk with a rotational inertia of 2.0 kg·m^2. What is the rotational acceleration of the disk?

E7.11 A torque of 40 N·m tending to produce a counterclockwise rotation is applied to a wheel about its axle. A frictional torque of 10 N·m also is applied at the axle.

    a. What is the net torque applied to the wheel?

    b. If the wheel is observed to accelerate at the rate of 2.0 rad/s^2 under the influence of these torques, what is the rotational inertia of the wheel?

E7.12 Two 2-kg masses are located at either end of a rod that is 1 m long, very light, and rigid, as shown in the diagram.

a. What is the rotational inertia of this system about an axis through the center of the rod?

b. What is the rotational inertia of this system about an axis through one of the masses at the end of the rod?

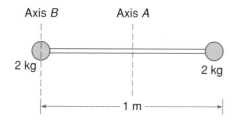

E7.13 A uniform disk with a mass of 3 kg and a radius of 0.15 m is rotating with a rotational velocity of 12 rad/s.

a. What is the rotational inertia of the disk? $(I = \frac{1}{2}MR^2)$

b. What is the angular momentum of the disk?

E7.14 A mass of 3 kg is located at the end of a very light, rigid rod 25 cm in length. The rod is rotating about an axis at its opposite end with a rotational velocity of 5.0 rad/s.

a. What is the rotational inertia of the system?

b. What is the angular momentum of the system?

E7.15 A very light rod with two equal masses at either end has an initial rotational inertia of 4.0 kg·m² and is rotating about its center with a rotational velocity of 2.0 rev/s. The masses are suddenly pulled in towards the center of the rod by a spring to a distance just half of the initial distance, reducing the rotational inertia of the system to 1.0 kg·m². If there are no external torques, what is the new rotational velocity of the system?

CHALLENGE PROBLEMS

CP7.1 A merry-go-round in the park has a radius of 2.2 m and a rotational inertia of 1500 kg·m². A child pushes the merry-go-round with a constant force of 80 N applied at the rim and parallel to the rim. A frictional torque of 16 N·m acts at the axle of the merry-go-round.

a. What is the net torque acting on the merry-go-round about its axle?

b. What is the rotational acceleration of the merry-go-round?

c. At this rate, what will the rotational velocity of the merry-go-round be after 30 s if it starts from rest?

d. If the child is running alongside the merry-go-round as he pushes, what is his approximate speed after 30 s? Could he go much faster than this?

e. What is the rotational acceleration of the merry-go-round if the child stops pushing after 30 s? How long will it take for the merry-go-round to stop turning?

CP7.2 A 4-m plank with a weight of 100 N is placed on a dock with 1 m of its length extended over the water, as shown in the diagram. The plank is uniform in density, so that the center of gravity of the plank is located at the center of the plank. A boy with a weight of 40 N is standing on the plank and moving out slowly from the edge of the dock.

a. What is the torque exerted by the weight of the plank about the pivot point at the edge of the dock? (Treat the weight as all acting through the center of gravity of the plank.)

b. How far from the edge of the dock can the boy move until the plank is just on the verge of tipping?

c. Can the boy test this conclusion without falling in the water? How might he go about this?

d. If the boy is nearer to the dock than this tipping point, why does the plank not tend to rotate in the opposite direction? What happens to the location of the effective pivot point?

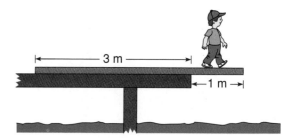

CP7.3 Suppose that several children with a total mass of 300 kg are riding on the merry-go-round in the park, which has a rotational inertia of 1500 kg·m² and a radius of 2.2 m. The average distance of the children from the axle of the merry-go-round is 2.0 m initially, since they are all riding near the rim.

a. What is the rotational inertia of the children about the axle of the merry-go-round? What is the total rotational inertia of the children and the merry-go-round?

b. The children now move inwards so that their average distance from the axle is 0.5 m. What is the new rotational inertia for the system?

c. If the initial rotational velocity of the merry-go-round was 1.2 rad/s, what is the rotational velocity after the children move in towards the center, assuming that the frictional torque can be ignored? (Use conservation of angular momentum.)

d. Is the merry-go-round rotationally accelerated during this process? If so, where does the accelerating torque come from?

CP7.4 A student sitting on a stool that is free to rotate, but is initially at rest, holds a bicycle wheel that is rotating with a rotational velocity of 2.0 rev/s about a vertical axis, as shown in the diagram. The rotational inertia of the wheel is 1.25 kg·m² about its center, and the rotational inertia of the student and wheel and stool about the rotational axis of the stool is 6.0 kg·m².

a. What are the magnitude and direction of the initial angular momentum of the system?

b. If the student flips the axis of the wheel, thus reversing the direction of its angular momentum vector, but not changing its magnitude, what is the rotational velocity (magnitude and direction of the student and the stool about their axis) after the wheel is flipped? (See figure 7.27.)

c. Where does the torque come from that accelerates the student and the stool? Explain.

HOME EXPERIMENTS AND OBSERVATIONS

HE7.1 If there is a park nearby containing a freely rotating child's merry-go-round, take some time with a friend to observe some of the phenomena that we have been discussing throughout this chapter. In particular, the following observations might be made.

a. What is a typical rotational velocity that can be achieved with the merry-go-round? How would you go about measuring this?

b. Applying a more or less constant force, how long does it take to reach this typical rotational velocity? Is it difficult to apply a constant force as the merry-go-round accelerates?

c. How long does it take for the merry-go-round to come to rest after you stop pushing? Could you estimate the frictional torque from this information? What other information would you need?

d. If you or your friend are riding on the merry-go-round, what happens to the rotational velocity when you move inward or outward from the axis of the merry-go-round? How is this effect explained?

HE7.2 Create a simple balance, using a ruler as the balance beam and a pencil as the fulcrum. (A pencil or pen with a hexagonal cross section is easier to use than one with a round cross section.) Make the following observations.

a. Does the ruler balance at exactly its midpoint? What does this imply about the ruler?

b. Using a nickel as your standard, what are the ratios of the weights of pennies, dimes, and quarters to that of the nickel? What procedure do you use to find these ratios?

c. Do different quarters or different nickels have exactly the same weight? Try using coins from different years.

d. Is the distance from the fulcrum necessary to balance two nickels exactly half that of a single nickel on the opposite side? How would you account for any discrepancy?

HE7.3 Tape or staple a small rubber ball to a string 50 cm or more in length. Start the ball rotating in a vertical circle and then let the string wrap around your finger.

a. What happens to the rotational velocity of the ball about your finger as the string wraps around your finger? Can conservation of angular momentum explain this phenomenon?

b. Can you estimate a value for the rotational velocity before the string wraps around your finger?

c. Using conservation of angular momentum, by how much would you expect the rotational velocity to increase when the ball is near your finger? Do your observations seem to confirm this expectation?

HE7.4 A simple top can be made by cutting out a circular piece of cardboard, poking a hole through the center, and using a short, dull pencil for the post. A short wooden dowel with a rounded end can be substituted for the pencil for better effect.

   a. Try building such a top and testing it. How far up on the pencil should the cardboard disk be placed for best stability?

   b. Observe what happens to the axis of rotation as the top slows down. What is the direction of the angular momentum vector, and how does it change?

   c. Try spinning the top in the air some distance above the floor or other hard surface and then dropping it. Is angular momentum conserved as the top falls?

   d. If pennies or other coins are taped to the cardboard near its rim, how does this affect the behavior of the top? How does this change the rotational inertia of the top?

   e. Putting black and white markings on the top (lines, dots, bars, etc.) can produce some interesting visual effects when the top is spinning. Do you see colors when the top is spinning, even though none were present initially?

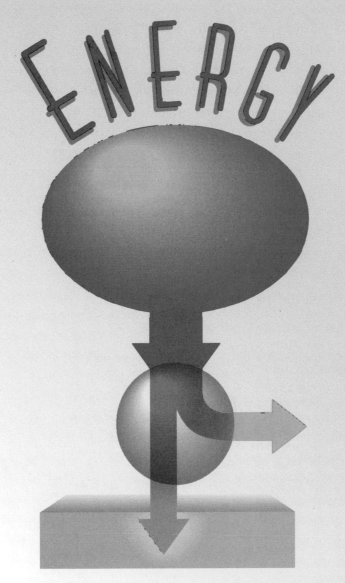

Part 2

ENERGY

and Thermodynamics

8 *Energy and Oscillations*

Have you ever watched a ball on the end of a string swinging back and forth in a repeating pattern? A pendant on the end of a chain, a swing in the park, or the pendulum on a grandfather's clock all exhibit the same hypnotic motion (fig. 8.1). Galileo is said to have amused himself during boring sermons in church by contemplating the motion of the chandeliers swaying slowly back and forth at the ends of their chains. Whether or not it kept him awake is not reported.

The thing that intrigued Galileo and others who have observed such oscillations is that the object at the end of the pendulum always seems to return to its initial position in each swing. It may fall a little short of the prior position in each successive swing, but the motion can go on for a long time before coming to a complete stop. The velocity, on the other hand, is continually changing, from zero at the end points of the swing to a maximum at the low point in the path. How is it that it can go through such changes in velocity and yet always return to its starting point?

Evidently something is being conserved in this process, but it is obviously not momentum. The momentum of the swinging object is changing continuously. The quantity that is conserved in such processes turns out to be what we now call *energy*. Although some of the fundamental ideas were recognized in Newton's day, the concept of energy did not play a large role in Newton's theory of mechanics. It was not until the nineteenth century that energy and energy transformations were elevated to the central position that they now hold in our understanding of the physical world.

The motion of a pendulum and other types of oscillation can be understood by appealing to the principle of conservation of mechanical energy. The potential energy that the pendulum has at its end points is converted to kinetic energy at the low point and then back to potential energy. But for the moment, these are just words; what is energy and how does it get into the system in the first place? Why is it that the concept of energy now plays such a central role in physics and all of science? These questions are explored in this chapter as well as in the following two chapters and, to some extent, throughout the rest of the book.

Since energy plays such a central role in economic and environmental issues as well as in physics and science more generally, it is important to have a thorough understanding of the fundamental ideas relating to energy. Energy is the basic currency of the physical world; in order to spend it wisely we must be familiar with its nature.

Figure 8.1　A pendant swinging at the end of a chain. Why does it return to approximately the same point after each swing?

Chapter Objectives

We usually approach the concept of energy by first considering how energy is added to a system. In mechanics, this involves the idea of work, which has a specialized meaning to physicists. If a force does work on a system, the energy of the system increases. We will begin, then, by defining work and then considering how, in different circumstances, work done on a system can result in either an increase in kinetic energy or in potential energy. Finally, we will put it all together by introducing the principle of conservation of energy and considering its application to a number of practical situations.

Chapter Outline

1 *Work and power.* How is the concept of work defined, and how is work involved in adding energy to a system? What are the units of work and of power, the rate of doing work?

2 *Kinetic energy.* If the work done on a system is used to increase the speed of an object, what type of energy results, and how does it depend upon the speed? Does rotational velocity also involve energy?

3 *Potential energy.* If the work done on a system is used to lift an object, what type of energy results, and how does it depend upon the height through which the object has been lifted? Are there other types of potential energy than that involving gravity?

4 *Conservation of energy and oscillations.* How do we define the total mechanical energy of a system, and under what conditions is it conserved? How can the principle of conservation of energy be used to explain oscillations and other phenomena?

5 *Sleds, hills, and friction.* What is the effect of frictional forces on the energy of a system? How can we handle situations in which the mechanical energy of a system is not conserved?

8.1 WORK AND POWER

Suppose that we have a simple pendulum consisting of a ball at the end of a string. What do you have to do in order to start it swinging? If the ball is hanging from a fixed support, you might start by pulling the ball away from its equilibrium position, (which lies directly below the support). In order to pull the ball from this position, you must apply a force to the ball with your hand or by some other means (fig. 8.2).

To a physicist, this simple process of applying a force to move an object through some distance involves doing work, even though the actual exertion required may be slight. Doing work on a system such as the pendulum increases the energy of the system, and this energy can be used to account for the motion of the pendulum. The quantity of work done is therefore important. How do we go about defining this quantity?

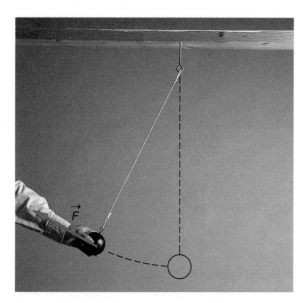

Figure 8.2   The force applied does work to move the pendulum bob from its original position located directly below the point of suspension.

Definition of Work

Although the term *work* is used in everyday language, the meaning that physicists give to this term is much more specific. Consider, for example, a simple situation in which a constant horizontal force is applied to a heavy crate to move it across a concrete floor, as illustrated in figure 8.3. The crate will move in the direction of the applied force, and we can define the work done in this process in a straightforward manner. The work is defined to be the product of the applied force and the distance that the crate is moved under the influence of this force.

In symbols, this definition is expressed as follows:

$$W = Fd,$$

where d is the distance that the crate has moved. If, for example, a force of 50 N is used to move the crate a distance of 4 m, the work done is

$$W = Fd = (50 \text{ N})(4 \text{ m}) = 200 \text{ N·m} = 200 \text{ joules}.$$

The *joule* (J) is the metric unit of energy and is defined as 1 N·m, the product of the units of force and of distance in the metric unit system that we are using.

In doing this work, we have transferred 200 J of energy from the pusher to the crate or other portions of the system. Exactly how this affects the system depends upon what other forces are acting upon the crate. If, for example, the floor is greased, so that the frictional forces are negligible, then the applied force has the effect of accelerating the crate. The velocity of the crate increases, and we say that its *kinetic energy* has increased. (The expression for kinetic energy was introduced in chapter 6, and it will be explored more fully in the next section of this chapter.) In any case, the work done by the applied force has had the effect of transferring energy from one object or system to another, and this energy transfer is an important aspect of the definition of work.

Work computations are not always as simple as the one that we have just done for the crate. Sometimes the force involved in doing work is not in the same direction as the displacement of the object. In this case the work done is not found by multiplying the total force times the distance moved. Instead, we use only that portion of the force which is in the direction of the motion.

Figure 8.3   A crate is moved a distance, *d,* across a concrete floor under the influence of a constant horizontal force, *F.*

Suppose, for example, that we are pulling a box across the floor with a rope, as illustrated in figure 8.4. The rope is at an angle to the floor, so that the force is pointing partly upwards rather than parallel to the floor. The box moves along the floor, however, not in the direction of the force. As we have done before, we can picture the force as consisting of two components, one parallel to the floor and the other perpendicular to the floor. Added together as vectors, these two components equal the total force applied by the rope.

Only the portion of the force that is in the direction of motion is used in computing the work. This portion, or component, can be found from a vector diagram like that in figure 8.4. If the total applied force is 50 N, but the component in the direction of the motion is 30 N, then the work done is found by multiplying 30 N times the distance moved. If this distance is again 4 m, then the work done is

$$W = Fd = (30 \text{ N})(4 \text{ m}) = 120 \text{ J}.$$

The component of the force that is perpendicular to the floor (and to the direction of motion) does no work. This is true also for some of the other forces acting on the box. The weight of the box, for example, acts straight downward and is therefore perpendicular to the direction of motion. It has no component in the direction of the motion and therefore does no work. The same is true for the normal force pushing upwards on the box as a result of its contact with the floor.

A more complicated situation arises when the force of interest is not constant. In this case we either have to find some appropriate means of averaging the force or break the motion down into small increments and use different values of the force for each small increment. These two choices are really the same process, and there are mathematical techniques for accomplishing these tasks. We will consider one such case in our discussion of potential energy.

A general definition of work, then, might be stated as follows:

The work done by a given force is the product of the component of the force that is along the line of motion of the object times the distance that the object moves under the influence of the force. If the force varies during the motion, then the motion must be broken down into small increments and the work for each increment added to that of the others to get the total work.

As noted, the definitions of various forms of energy follow from this definition of work. If the work is used to accelerate the object, kinetic energy is gained by the object, as mentioned earlier. The energy gained is equal to the work done, and this energy transfer, which often occurs when objects interact, is an important aspect of the work concept.

Kinetic energy is not the only type of energy that can result when work is done. If, in the more usual situation, frictional forces are also acting upon the crate, then the applied force may just be used to overcome or balance the frictional forces. The net force acting upon the crate would then be zero. The crate will not accelerate in this situation, but will move with constant velocity. In this case, the energy added to the system has the effect of increasing the temperature of the crate and the floor, thereby increasing the *internal energy* of the system. The concept of internal energy is explored more thoroughly in the next chapter on thermodynamics.

If the crate or box is being pushed up a ramp, so that its height above the floor is increased, the *potential energy* of the crate is increased. This situation is considered in more detail in section 8.3. In all of these cases, however, there has been a transfer of energy to the crate or other portions of the system from the person or entity applying the force. The work done describes the amount of energy transferred; this is the basic effect of doing work.

Simple Machines

One of the early applications of the concept of work was in the analysis of devices such as levers, pulley systems, or inclined planes, which we sometimes refer to collectively as *simple machines*. A simple machine is any mechanical device that multiplies the effect of an applied force; perhaps the simplest example is a lever.

Figure 8.5 illustrates the use of a lever. The force applied at one end of the lever by the person results in a larger force being applied to the rock at the other end. There is a *mechanical advantage* to this arrangement, since the use of the lever has a multiplying effect upon the input force. The mechanical advantage is defined quantitatively as the ratio of the

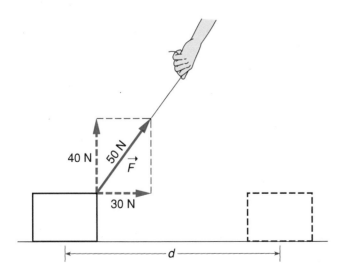

Figure 8.4 A rope is used to pull a box across the floor. Only the component of the force $\vec{F}$ that is in the direction of motion is used in computing the work.

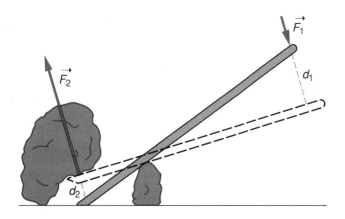

$$F_1 d_1 = F_2 d_2 = W$$

Figure 8.5 A lever is used to lift a rock. The small force $\vec{F}_1$ acts through a larger distance, d_1 than the distance d_2 that the rock moves, so that the work done on the lever equals the work done on the rock.

Figure 8.6 An inclined plane, or ramp, is also a simple machine. A smaller force acts through a larger distance to do the same amount of work as would be involved in lifting the crate straight up into the truck.

output force to the input force, or F_o / F_i. In the case of the lever this ratio depends upon the lengths of the lever arms of the two forces on either side of the fulcrum. When there is no rotational acceleration, the torques produced by these two forces, that of the rock on the lever and that of the person on the lever, will cancel each other.

It turns out, however, that the work done on or by the lever by the two forces is the same. As figure 8.5 shows, the smaller force applied by the person must move that end of the lever a greater distance than the distance through which the rock is moved at the other end. The product of the force and the distance moved (the work) ends up being the same; the work done by the person on the lever equals the work done by the lever on the rock. In general this turns out to be the case in all simple machines; the work output is less than or equal to the work input. There is no free lunch! The price we pay for being able to use a smaller force is the need to apply that force through a greater distance.

An inclined plane provides another simple example of this principle. In pushing a crate up an inclined plane, we can get away with a smaller force than we would need to lift the crate straight up. The distance moved along the plane, however, is greater than the height to which the crate is lifted by this process (fig. 8.6). The mechanical advantage is greater than 1, but the work done in pushing the crate is at least as large as the effective work output.

Power

The rate at which work is done or energy transferred is often of obvious interest. The rate at which a car can be accelerated, for example, which involves transfer of energy from the

fuel in the engine to the motion of the car, depends upon the *power* of the engine. Power can be defined as follows:

Power is the time rate of doing work; it is found by dividing the amount of work done by the time required. In symbols this takes the following form:

$$P = \frac{W}{t}.$$

Suppose, for example, that it takes 5 seconds for the puller to move the box in figure 8.4 the 4 m across the floor. We have already computed a work value of 120 J for this situation, so the power exerted is found by dividing this value by the time:

$$P = \frac{W}{t} = \frac{(120 \text{ J})}{5 \text{ s}} = 24 \text{ J/s} = 24 \text{ watts}.$$

The watt is the metric unit of power and is defined as 1 J/s . It is familiar because we use it commonly in discussing electrical power, but it is used much more generally for any situation involving the rate of transfer of energy.

The faster the box is moved, the greater the rate of energy transfer and the larger the power required. In fact, for situations like the moving box, the power is directly proportional to the speed of the box. This can be seen by writing the power definition in terms of the work definition:

$$P = \frac{W}{t} = \frac{Fd}{t} = Fv,$$

since d / t is the rate at which distance is covered, or the speed of the box. If the box is moved slowly, a low power is involved, even though the same amount of work may be accomplished as for a faster moving box. The force must be applied for a longer time, however, to accomplish the same amount of work for the slower moving box as for the faster moving box.

Another unit of power still used in this country for discussing the power of automobile engines is the horsepower (hp). One horsepower is equal to 746 watts, or 0.746 kilowatts. The day may come when we compare the power of different engines in kilowatts rather than in horsepower, but for the time being we still rate engines in terms of horsepower. The relationship of this unit to the power that a horse can generate is dubious at best, but the idea of comparing the iron horse to the flesh and blood variety still has a certain appeal.

8.2 KINETIC ENERGY

What happens if the force involved in doing work on an object has the effect of accelerating the object? Obviously, the velocity of the object increases, and we say that its *kinetic energy* increases in the process. Since the amount of energy transferred is equal to the amount of work done, we must define kinetic energy so that this is indeed the case. Our definition of work thus serves as the starting point.

Definition of Kinetic Energy

Suppose that we return to the crate being pushed across a well-greased floor (fig. 8.3). If the frictional forces are small enough to be ignored, then the net force in the direction of motion of the crate is just that applied by the pusher. The acceleration of the crate is related to this net force and the mass of the crate by Newton's second law, $F = ma$.

The work done by this force is

$$W = Fd = (ma)d = m(ad),$$

where d is the distance that the crate has moved. If the force and the resulting acceleration are constant, however, this distance is related to the crate's rate of acceleration and its initial and final velocities by one of the constant acceleration equations developed in chapter 3, $v^2 = v_0^2 + 2aD$. For straight-line motion, the displacement D will equal the distance traveled d and the product of acceleration times distance, ad, that appears in our expression for the work is therefore directly related to the difference between the *squares* of the initial and final velocities of the crate, $v^2 - v_0^2$. Since $v^2 - v_0^2 = 2aD$,

$$aD = \left(\frac{1}{2}\right)(v^2 - v_0^2).$$

The work done on the crate is then

$$W = m(aD) = m\left(\frac{1}{2}\right)(v^2 - v_0^2) = \left(\frac{1}{2}\right)mv^2 - \left(\frac{1}{2}\right)mv_0^2.$$

Evidently the quantity that has been *increased* by doing work on the crate is equal to one-half the mass times the square of the velocity, or

$$KE = \left(\frac{1}{2}\right)mv^2,$$

which is called the *kinetic energy.*

Kinetic energy is the energy that an object possesses as a result of its motion and is equal to one-half the mass of the object times the square of its speed (magnitude of the instantaneous velocity): $KE = (\frac{1}{2})mv^2$.

The abbreviation KE is commonly used to represent kinetic energy (fig. 8.7).

Sometimes numbers help to clarify the picture. Suppose, for example, that we move the crate a distance of 4 m using a net force of 50 N, as before. The work done is then $W = Fd = 200$ J. If the crate started from rest, its initial kinetic energy would be zero. The final kinetic energy could be found by computing the acceleration of the crate from Newton's second law and then finding the final velocity from the constant acceleration equations. Suppose the crate has a mass of 100 kg. From Newton's second law, its acceleration would then be

$$a = \frac{F}{m} = \frac{50 \text{ N}}{(100 \text{ kg})} = 0.50 \text{ m/s}^2.$$

The final velocity can be found from the constant acceleration equation stated earlier,

$$v^2 = v_0^2 + 2aD = 0 + 2(0.50 \text{ m/s}^2)(4 \text{ m}) = 4 \text{ m}^2/\text{s}^2.$$

Taking the square root to find v, we find that $v = 2$ m/s. Substituting this value into our definition of kinetic energy yields

$$KE = \left(\frac{1}{2}\right)mv^2 = \left(\frac{1}{2}\right)(100 \text{ kg})(2 \text{ m/s})^2 = 200 \text{ J}.$$

In other words, doing 200 J of work to accelerate the crate results in an increase in kinetic energy of exactly 200 J. This

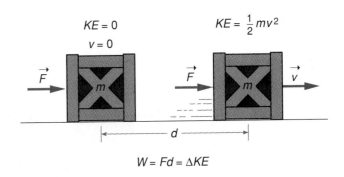

$$W = Fd = \Delta KE$$

Figure 8.7 **The work done on an object by the net force acting on that object results in an increase in the kinetic energy of the object.**

is no accident; we have defined kinetic energy so that this will be true.

Notice that the value of kinetic energy is not proportional to the speed itself, but rather to the square of the speed. Thus if we double the speed, the kinetic energy quadruples. Four times as much work must be done to reach the doubled speed than to reach the original speed.

A practical application of this fact can be found in the stopping distances required for cars traveling at different speeds. When we apply the brakes to stop a car, the frictional force between the car's tires and the road does negative work on the car as it produces a negative acceleration (fig. 8.8). Negative work results when the force in question is in the opposite direction to the motion of the object.

If we assume that the frictional force does not depend on the speed of the car, then the negative work is proportional to the distance through which the frictional force must act; $W = -fd$, where f is the frictional force, and d is the stopping distance. The amount of negative work required to stop the car is equal to the kinetic energy of the car before the brakes are applied, because this is the amount of energy that must be removed from the system. Since kinetic energy is proportional to the square of the speed, the work required (and the stopping distance) is four times as large for a car traveling at 60 mph than for one traveling at 30 mph.

A more careful analysis of this situation would have to account for the fact that the frictional force does vary somewhat with the speed of the car. The stopping distances quoted in driver-training manuals do indeed increase rapidly with speed, although perhaps not exactly in proportion to v^2. The more kinetic energy present initially, the more negative work is required to reduce this energy to zero, and the greater the required stopping distance.

Rotational Kinetic Energy

Other examples involving kinetic energy arise in the discussion of conservation of energy later in this chapter. As a final note on the definition of kinetic energy, however, we should point out that a rotating object also possesses kinetic energy. The expression for this form of kinetic energy can be written simply by using the analogy between rotational motion

and linear motion introduced in the previous chapter. This yields

$$KE = \frac{1}{2} I\omega^2,$$

since mass is replaced by rotational inertia, I, and linear velocity, v, is replaced by rotational velocity, ω.

In fact, we can extend the analogy to the definition of work also. A child who pushes a merry-go-round, for example, does work on the merry-go-round in the process of accelerating it. The work can be found by multiplying the torque times the rotational displacement ($W = \tau \Delta\theta$), since torque replaces force in our analogy, and rotational displacement replaces linear displacement, or distance. The work done by the net torque exerted by the child equals the increase in rotational kinetic energy of the merry-go-round (fig. 8.9).

Figure 8.9 **A child exerts a torque on the merry-go-round, thus doing work and increasing the rotational kinetic energy of the merry-go-round.**

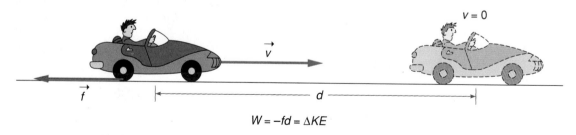

$$W = -fd = \Delta KE$$

Figure 8.8 **The frictional forces exerted upon a car's tires by the road surface do negative work in stopping the car, resulting in a decrease in kinetic energy.**

A rolling object has both linear and rotational kinetic energy; it is both rotating and moving from one place to another. The total kinetic energy in this case is the sum of linear ($\frac{1}{2}mv^2$) and rotational ($\frac{1}{2}I\omega^2$) kinetic energy terms. If an object is moving, it has kinetic energy, and that motion can be either linear or rotational (or a combination of the two).

8.3 POTENTIAL ENERGY

Suppose that instead of accelerating a crate or other object, a force does work to lift an object to a higher position. Have we increased the energy of the object in such a process? If so, what is the nature of the energy that the object or system gains? If the velocity of the object has not increased in the process, there is no increase in kinetic energy. What happens to the energy that is transferred via the work?

In order to lift the crate, we need to apply a force that pulls or pushes upwards on the crate. As shown in figure 8.10, this will not be the only force acting on the crate; the force of gravity (the weight of the crate) pulls downward. If we lift the crate with a force that is exactly equal to the force of gravity, but opposite in direction, then the net force acting on the crate will be zero, and the crate will not accelerate.

Actually, of course, we need to accelerate the crate a little bit at the start of the motion and then decelerate it at the end of the motion, so that the net effect of the operation is to start with the crate at rest on the ground and to end with it at rest at the higher position on the loading dock. There is no gain in kinetic energy. While the crate is moving, it travels with constant velocity, since the net force acting on the crate is zero. The force applied by the rope does work, however, and this work can be readily calculated.

Gravitational Potential Energy

By our definition of work, the work done by the force applied via the rope is equal to the force times the distance moved, because the force and the displacement of the crate are in the same direction. Since we want the net force acting on the crate to be zero, the applied force must be equal to the weight of the crate, *mg*. If the crate is lifted through a distance equal to the increase in height *h*, then the work done is simply

$$W = Fd = (mg)h = mgh.$$

The energy of the system must have increased by this amount.

We say that the crate, or more accurately the system consisting of the crate and the earth, has gained *potential energy* in this process. In this situation, where the applied force has been used to move the object against the force of gravity, the potential energy gained is equal to the work we have just computed. An expression for the gravitational potential energy can then be written as follows:

$$PE_g = W = mgh.$$

The symbol *PE* is often used to represent potential energy.

If the crate has a mass of 100 kg, then the potential energy that it gains in being lifted through a height of 2 m to the dock would be

$$PE_g = mgh = (100 \text{ kg})(9.8 \text{ m/s}^2)(2 \text{ m}) = 1960 \text{ J}.$$

The notion of a *reference level* is necessarily involved in such computations. Here we have chosen the original position of the crate on the ground to be our reference level. If we raised the crate an additional 2 m above the level of the loading dock, it would have a potential energy of 1960 J relative to the dock and an amount twice this, 3920 J, relative to the ground. Usually we choose the lowest point in the probable motion of the object as our reference level in order to avoid negative values of potential energy.

The term *potential energy* implies a storage of energy that could be used later on for other purposes, and certainly this feature is present in the situation just described. The crate, for example, could be left indefinitely at the higher height on the loading dock. If we push it off the dock, however, it would rapidly gain kinetic energy as it fell. This, in turn, could be used to compress objects lying underneath, drive pilings into the ground, or perform other such useful violence (fig. 8.11). Kinetic energy could also be said to have this feature, however, so this is not really the distinguishing characteristic of potential energy.

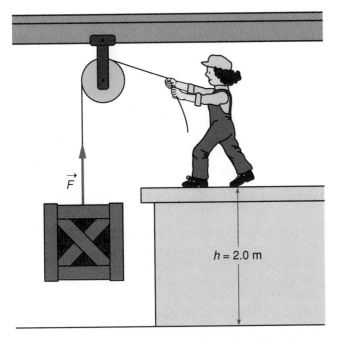

$\vec{F}$

$h = 2.0$ m

Figure 8.10 **A rope and pulley are used to lift a crate to a higher position on the loading dock, resulting in an increase in potential energy.**

Figure 8.11 **The potential energy of the raised crate can be converted to kinetic energy and used for other purposes.**

The essence of the idea of potential energy involves the fact that we have changed the *position* of the object along the line of action of some force. The source of that force is then an important part of our system. In the case of gravitational potential energy, that force is the gravitational attraction between the object and the earth. The farther we move the object from the center of the earth, the greater the gravitational potential energy.

The Potential Energy of a Spring

The essential characteristic of potential energy can be best illustrated by choosing a different system and force that result in a different kind of potential energy. The force that a spring exerts on an object is convenient for this purpose. Imagine a spring attached to a wall, as pictured in figure 8.12, with a wooden block or similar object attached to the other end. If we pull the block back from the original position, in which the spring is not stretched, the system gains potential energy .

Obviously, it requires work to pull the block against the force exerted by the spring. Most springs exert a force that is proportional to the distance that the spring is stretched:

$$F_s = (-)kx,$$

where x is the distance that the spring is stretched, and k is a constant called, not surprisingly, the *spring constant*. The spring constant is a property of the spring itself; the larger the value of k, the stiffer the spring, and the larger the force required to stretch the spring a given distance. A minus sign is often included in this expression to indicate that the force

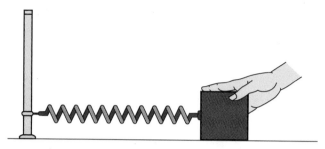

Figure 8.12 **A wooden block is attached to a spring that is attached to a fixed support at the opposite end. Stretching the spring increases the elastic potential energy of the system.**

exerted by the spring is in the opposite direction to the displacement of the end of the spring.

How do we find the potential energy of such a system? Just as before, we need to compute the work done by the force used to move the object against the force that we are making part of our system. In the first example, this was the gravitational force; in this case, it is the force of the spring. The applied force must be adjusted so that it is always equal in magnitude, but opposite in direction, to the force of the spring. The net force on the block will then again be zero, so that there is no acceleration and no kinetic energy gained. The work required to stretch the spring is then equal to the potential energy gained by the system.

Unfortunately, the computation of the work done by the force used to stretch the spring is a little more complicated than that involved in the gravitational case. The reason for this is that the force in question is not constant; it varies with position. The force must get larger, according to the relationship $F = kx$, as x gets larger, in order to balance the increasing force of the spring. As shown in figure 8.13, the force starts at zero and grows to a maximum value of kx_f, where x_f is the total distance stretched when the mass has reached its final position.

It turns out that in this situation an average value of the force, $(1/2)kx_f$, can be used to compute the work. The work done by the force in stretching the spring is then

$$W = Fd = \left(\frac{1}{2} kx_f\right)(x_f) = \left(\frac{1}{2}\right)kx_f^2 = PE_s.$$

The potential energy of the stretched-spring system, PE_s, is therefore one-half the spring constant times the square of the distance stretched. We can also store energy by compressing the spring; the same expression for the potential energy results, but x would then be the distance the spring is compressed from its initial relaxed position. This potential energy that a system gains when it is stretched or compressed is often called *elastic* potential energy.

Once again, just as with gravitational potential energy, the potential energy stored in the spring can be converted to

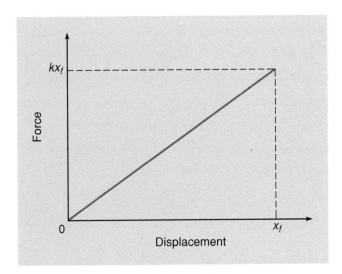

Figure 8.13 **The applied force used to stretch a spring varies with displacement, going from initial value of zero to a final value of kx_f.**

other forms or put to various uses. Obviously, if we let go of the block when the spring is either stretched or compressed, the block will gain kinetic energy as it is accelerated. Our intuitive sense of stored energy is probably even stronger for this situation than for the gravitational case. The essential feature, however, is again the change of position of the object relative to a specific force whose source is part of our system.

Potential energy can result from work done against a variety of different forces in addition to those of gravity and springs. Not all forces lend themselves to this description, however. Work done against frictional forces, for example, does not result in an increase in the potential energy of the system. Heat is generated in this case, which either transfers energy out of the system or increases the microscopic internal energy of the system.

Forces such as gravity or the spring force for which we can develop potential energy expressions are referred to as *conservative forces*. When work is done against conservative forces, the energy gained by the system is completely recoverable for use in other forms. Friction is a nonconservative force that does not result in recoverable mechanical energy. We are finally in a position, then, to put forward a general definition of potential energy.

> **Potential energy is the energy that an object has by virtue of its position along the line of action of some conservative force (such as gravity or the spring force) that has been incorporated into the system. It represents stored energy that does not involve motion of the object.**

Such a definition is intelligible only in the context of the examples that we have already considered. It is important to

recognize, once again, the role that work plays in increasing the potential energy of a system and in leading to the specific expressions for gravitational potential energy and the potential energy of a spring. It is easy enough to sense intuitively the idea that energy is stored in an object that has been lifted to a table top or in a spring that has been stretched or compressed. The system is poised to release that energy, converting it to kinetic energy or other forms of energy.

8.4 CONSERVATION OF ENERGY AND OSCILLATIONS

What are these energy ideas good for? You now have definitions and some understanding of the concepts of work, kinetic energy, and potential energy. How can they be used to explain phenomena in ways that are more convenient than the direct application of Newton's laws? The usefulness of energy concepts becomes much more apparent when we bring them all together in the principle of conservation of energy and examine its applications.

The Pendulum Revisited

Let us return to the pendulum and look at its motion from an energy perspective. Imagine a pendulum that consists of a ball that is initially hanging motionless on the end of a string. You pull the ball aside and then release it to start it swinging. What happens to the energy of the system in this process?

In the first step of this process, work is done on the ball by your hand. The net effect of this work is to increase the potential energy of the ball since the height of the ball above the ground increases as the ball is pulled aside. The work done transfers energy from the person doing the pulling to the system consisting of the pendulum and the earth. It shows up as gravitational potential energy, $PE_g = mgh$.

When the ball is released, this potential energy begins a conversion to kinetic energy as the ball begins its swing. At the bottom of the swing (the initial position of the ball when it was just hanging), the potential energy is at its minimum and the kinetic energy has its maximum value. The ball does not stop here, however; its momentum carries it on to a point opposite the release point where the kinetic energy is again zero and the potential energy has approximately its initial value before release. The ball then swings back, repeating the transformation of potential energy to kinetic energy and back to potential energy (fig. 8.14).

This process is an example of the principle of conservation of energy. If we define the total mechanical energy of the system as the sum of the potential plus kinetic energies ($PE + KE$), then this quantity remains constant during the motion of the pendulum. It remains constant because there is no work being done by external or internal forces that would have the effect of increasing or decreasing the

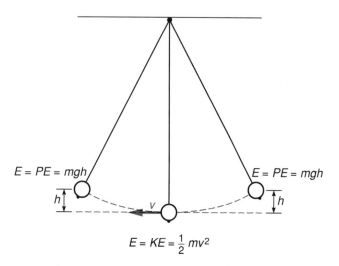

$E = PE = mgh$ $E = PE = mgh$

h v h

$E = KE = \frac{1}{2}mv^2$

Figure 8.14 **Potential energy is converted to kinetic energy and back to potential energy as the pendulum swings back and forth.**

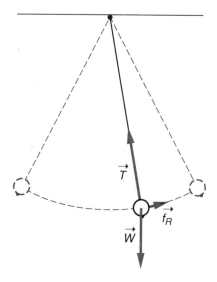

Figure 8.15 **Of the three forces acting upon the pendulum bob, only the force of air resistance represents an external force that does work on the system to change its total energy. The tension does no work, and the work done by gravity is already taken into account in the potential energy term.**

mechanical energy. The principle of conservation of mechanical energy can be stated as follows:

If there are no nonconservative forces doing work on a system, then the total mechanical energy of the system remains constant. The total mechanical energy is the sum of the kinetic energy and the potential energy of the system.

Since we have emphasized the role that work plays in increasing or decreasing the energy of a system, this principle should make sense. If no energy is added or removed by forces doing either positive or negative work, then the total energy should not change. In symbols, this statement takes the following form:

If $W = 0$, then $E = PE + KE = $ constant,

where E is the symbol generally used to represent the total energy.

The statement of this principle seems simple enough, and almost self-evident given the way in which we have used the concept of work. Some points, however, deserve close attention. Why, for example, do we not include the work done by gravity on the pendulum in our considerations? The answer is that we have made the gravitational force of the earth on the ball part of our system by including the gravitational potential energy of the ball in our description. Gravity is a conservative force whose work has already been taken into account.

What about other forces acting on the ball? The tension of the string acts in a direction that is always perpendicular to the motion of the ball. This force does no work, therefore, because it has no component in the direction of the motion. The only other force that concerns us is the frictional force of air resistance. This does do negative work on the ball and has

the effect of slowly decreasing the total mechanical energy of the system (fig. 8.15). The total energy is not really completely constant in this situation, therefore; it would be so only if air resistance were totally negligible. The frictional effects are often small, though, and can be ignored in an approximate treatment of the problem.

What are the advantages of using this energy description and the principle of conservation of energy? Energy considerations do permit us to make certain predictions about the behavior of the system. To the extent that we can ignore frictional effects, we can predict, for example, that the ball will reach the same height at either end of its swing. This is because the kinetic energy is zero at the end points of the swing (the ball is momentarily at rest there) and the total energy is just potential energy at these points. If no energy has been lost, the potential energy must have the same value that it had at the point of release. Since $PE_g = mgh$, this implies that the same height, h, is involved.

A demonstration sometimes performed in physics lecture rooms illustrates this idea in a dramatic fashion, using a bowling ball as the pendulum bob. The bowling ball is suspended from a support near the ceiling so that when pulled to one side, the ball is at about chin height on the physics instructor. The instructor pulls the ball to this position, releases it and lets it swing across the room, and then stands without flinching as the ball returns and stops just a few inches from his or her chin (fig. 8.16). The act of not flinching requires some faith in the principle of conservation of energy! (The success of this demonstration, however,

Figure 8.16 A bowling ball at the end of a cable suspended from the ceiling is released and allowed to swing across the room and back, stopping just in time.

requires that the ball not be given any initial velocity when it is released. What happens if it is pushed?)

We can also use the principle of conservation of energy to predict what the speed will be at any point in the swing. We know, of course, that the speed is zero at the end points and has its maximum value at the low point of the swing. If we place our reference level for measuring potential energy at this low point, then the potential energy will be zero at the low point because the height is zero. At this point, then, all of the initial potential energy has been converted to kinetic energy. If the initial potential energy was mgh_0, then the kinetic energy at the low point must be equal to this value, or

$$E = KE = \left(\frac{1}{2}\right) mv^2 = mgh_0.$$

This expression can be solved to yield a value for v at the low point, $v = \sqrt{2gh_0}$, as is shown in the example exercise in box 8.1.

We can, in fact, find the velocity at any other point in the swing by setting the total energy, $E = PE + KE$, at any point equal to the initial energy, mgh_0 (fig. 8.17). Different values of the height h above the low point yield different values of the kinetic energy and velocity. This idea is sometimes clearer when expressed in symbolic shorthand:

$$KE = E - PE = mgh_0 - mgh.$$

The kinetic energy is the difference between the initial potential energy and the remaining potential energy at any other height, h. The system has just so much energy. It can show up as either potential or kinetic energy or some of both, but it cannot exceed the initial value, mgh_0.

Box 8.1

Sample Exercise

Suppose that a pendulum bob with a mass of 0.50 kg is released from a position at which the bob is 12 cm above the low point in its swing. What is the velocity of the bob as it passes through the low point in its swing?

$m = 0.50$ kg.   The initial energy is:

$h_0 = 12$ cm. $E = PE = mgh_0.$

$v = ?$ at the low point. $= (0.50 \text{ kg})(9.8 \text{ m/s}^2)(0.12 \text{ m})$

$= 0.588$ J.

At the low point,

$$PE = 0, \text{ so } E = PE + KE = KE = \left(\frac{1}{2}\right) mv^2 = 0.588 \text{ J}.$$

Dividing both sides by $\left(\frac{1}{2}\right) m$, we obtain

$$v^2 = \frac{(0.588 \text{ J})}{\left(\frac{1}{2}\right) m}$$

$$= \frac{(0.588 \text{ J})}{\left(\frac{1}{2}\right)(0.50 \text{ kg})}$$

$$= 2.35 \text{ m}^2/\text{s}^2.$$

Taking the square root of both sides yields

$$v = \textbf{1.53 m/s.}$$

(Notice that we multiplied by the mass to get the potential energy and then divided again by the mass to get the speed, so that the value of the speed does not depend upon the mass.)

Springs and Simple Harmonic Motion

In the case of the pendulum, the energy transformations result in a repetitive oscillation. We often characterize an oscillation by stating its *period, T,* which is the time taken for one complete cycle. The *frequency, f,* which is the number of cycles per unit time is also used. The frequency is just the reciprocal of the period, $f = 1 / T$, given these definitions. The common unit used for frequency is the *hertz,* which is equal to one cycle per second. The frequency of oscillation of a pendulum depends primarily upon its length. Some simple experiments with a ball on a string will give you some

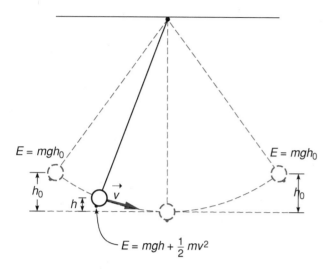

Figure 8.17 **The kinetic energy and velocity of the pendulum bob can be found at any height *h* from knowledge of the initial energy, *E = mgh*$_o$.**

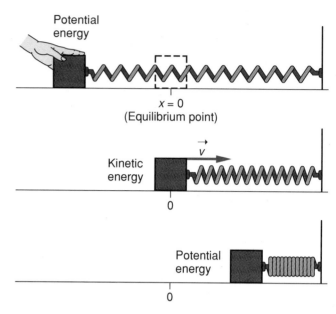

Figure 8.18 **Energy added by doing work to stretch the spring is then transformed back and forth between the potential energy of the spring and kinetic energy of the mass.**

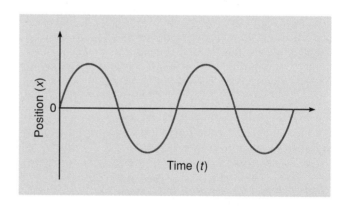

Figure 8.19 **The horizontal position *x* of the mass on the spring is plotted against time as it moves back and forth. The resulting curve is a sine function.**

idea of how the frequency varies with length; try it and see if you can find a trend.

Another simple system that undergoes an oscillation much like that of the pendulum is the mass on a spring that we described in the previous section. The situation is really quite similar to that of the pendulum. We add energy by doing work in pulling the mass to one side of its equilibrium position (where the spring is unstretched). This increases the potential energy of the spring/mass system, resulting in an initial energy of the form: $PE_s = (1/2)kx_m^2$ where x_m is the maximum distance that the spring is stretched from the equilibrium position.

Once the mass is released, the potential energy is converted to kinetic energy as the mass approaches its equilibrium position. Again, as with the pendulum, its momentum carries it beyond the equilibrium position, and the spring is compressed, gaining potential energy again (fig. 8.18). When the kinetic energy is converted completely to potential energy, the mass stops and returns, and the whole process repeats. If frictional effects can be ignored, the total energy of the system remains constant while the mass oscillates back and forth.

What determines the frequency of this system? Intuitively we expect that a loose spring will yield a low frequency of oscillation and a stiff spring a high frequency, and this is indeed the case. The mass that is oscillating also has an effect; larger masses provide greater resistance to a change in motion and therefore yield lower frequencies. An analysis using Newton's second law yields an expression for this frequency of the following form:

$$f = \frac{1}{2\pi} \sqrt{\frac{k}{m}}$$

where k is the spring constant that describes the stiffness of the spring.

Oscillations such as that of the mass on the spring or of the pendulum are referred to as *simple harmonic motion*. This terminology follows from the fact that if we plot the position of the mass against time, the resulting curve takes the form shown in figure 8.19. The mathematical functions that describe such curves are the sine or cosine functions from trigonometry, which are collectively referred to as *harmonic* functions. The variation of velocity with time can also be described by a harmonic function that is similar to the curve shown for position but reaches its maximum value at a different point in the oscillation. The velocity has its maximum value when the distance from the equilibrium point is zero.

In any oscillatory motion, there must be a force tending to pull the object back towards its equilibrium position when the object is displaced from that position. In the case of the pendulum, that force is gravity, whereas in the case of the mass on the spring, the force is that of the spring itself, $F = -kx$. Simple harmonic motion occurs when such a force is directly proportional to the displacement from equilibrium, as is clearly the case with the spring. If the force varies in some more complicated fashion with distance, the result is not simple harmonic motion, but there may still be an oscillation.

Look around to find systems that vibrate. There are numerous examples, ranging from a springy piece of metal to a ball rolling in a depression of some kind. What force provides the pull back towards the equilibrium position in each case? Is the motion likely to be simple harmonic motion, or is it a more complicated oscillation? Most important, given the theme of this chapter, what kind of potential energy is involved? The analysis of vibrations such as these forms an important subfield of physics that plays a role in music, communications, and many other areas.

8.5 SLEDS, HILLS, AND FRICTION

A sled on a hill or a roller coaster are popular examples often used to illustrate the principle of conservation of energy. The reason is that frictional forces can be assumed to be small in these cases, and conservation of energy can be used to make predictions that would be very difficult to make using Newton's laws directly. A simple energy accounting can provide a good overview of the phenomena.

A Sled Ride

Let's look in some detail at the energy transformations involved in a typical ride down a snowy hill on a sled. (For those of you unfamiliar with that particular type of thrill, substitute a water slide or a roller coaster; the description is essentially the same.) Where does the energy come from, where does it go, and what insights can an energy analysis provide?

We will assume that the sled ride begins at the bottom of the hill and that a larger person pulls the sled and rider up the hill and gives them a push to send them on their way. The first stage of the process thus involves work being done by the puller in dragging the sled and rider up the hill. Ignoring frictional effects, the work done equals the gain in potential energy. If the hill has a height of 20 m, and the combined mass of the sled and rider is 80 kg, then the energy added is easily found from our expression for gravitational potential energy:

$$W = PE_g = mgh_0 = (80 \text{ kg})(9.8 \text{ m/s}^2)(20 \text{ m}) = 15\ 680 \text{ J}.$$

What is the source of this energy? The energy is provided via the work done by the puller, so it must have been stored in the body of the person doing the pulling. Muscle groups were activated and chemical changes took place that released potential energy stored in chemical compounds in the body. Ultimately this energy had to come from food consumed by the person doing the pulling, which in turn involves plant life that received energy from the sun. A puller who did not have a good breakfast or attempts too many trips up the hill may not have sufficient energy to get to the top.

Once at the top, the puller may give the sled and rider a push to get them started down the hill. This involves additional work done by the pusher, which has the effect of accelerating the sled and increasing its kinetic energy (fig. 8.20). Suppose that an initial velocity of 4 m/s is provided at the top of the hill by the pusher. This represents a kinetic energy of

$$KE = \left(\frac{1}{2}\right)mv^2 = \left(\frac{1}{2}\right)(80 \text{ kg})(4 \text{ m/s})^2 = 640 \text{ J}.$$

The total energy at the top of the hill after the push is then 15 680 J of potential energy plus 640 J of kinetic energy, or approximately 16 300 J. This represents the amount of energy deposited in the account, to use a financial analogy.

As the sled and rider head down the hill, the potential energy decreases, and the kinetic energy increases. If there were no friction at all, the total energy would remain constant; there would be no withdrawals from the account. We could use conservation of energy to compute the kinetic energy and velocity for any height on the hill. At the bottom of the hill, for example, the potential energy would be zero, and the kinetic energy would have to be 16 300 J, since the total energy remains constant. This value of kinetic energy represents a velocity of about 20 m/s, as you can readily confirm by plugging this value into the kinetic energy expression. (Since 20 m/s is approximately 45 MPH, the rider may wish to drag her feet on the way down!)

Similar computations could be made for any other point on the hill. At the hump with a height of 15 m pictured in figure 8.20, for example, the potential energy would be

$$PE_g = mgh = (80 \text{ kg})(9.8 \text{ m/s}^2)(15 \text{ m}) = 11\ 800 \text{ J}.$$

Since the total energy is 16 300 J, the kinetic energy must be

$$KE = E - PE_g = 16\ 300 \text{ J} - 11\ 800 \text{ J} = 4500 \text{ J}.$$

This represents a velocity of 10.6 m/s, or roughly 24 MPH. The sled could become airborne at this point if the gravitational force is not large enough to provide the centripetal force necessary to keep it on the ground. This would depend on the radius of curvature of the hill at that point, since the radius and velocity will determine the necessary centripetal acceleration, v^2/r.

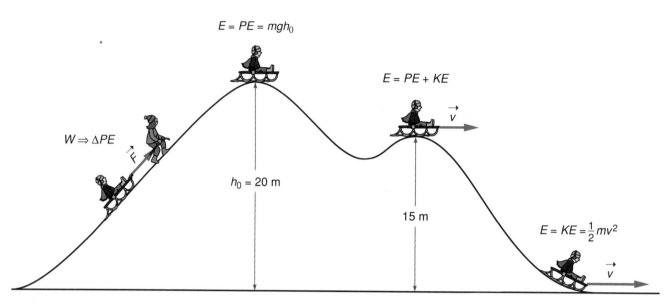

$$E = PE = mgh_0$$

$$E = PE + KE$$

$W \Rightarrow \Delta PE$

$h_0 = 20$ m

15 m

$$E = KE = \frac{1}{2}mv^2$$

Figure 8.20 **Work done in pulling the sled up the hill produces an increase in the potential energy of the sled and rider. This initial energy is then converted to kinetic energy as they slide down the hill.**

The basic idea in all of these computations, as the financial analogy is intended to show, is simply an energy accounting. The initial energy of 16 300 J must show up either as potential or kinetic energy (or some of both) at other points on the hill. This initial energy sets limits upon how high the sled may go and how fast it may go without the addition of more energy. Once that energy has been expended against friction or by dragging your feet, more energy must be added by work done in pulling the sled in order to get it going again. There is no free lunch!

Effects of Friction

The velocities that we have computed are really upper limits because we have ignored frictional effects. The actual velocities achieved in a sled ride would be somewhat lower, since air resistance and some friction, however small, between the runners of the sled and the snow are always present. These frictional effects always result in energy being removed from the system; they represent withdrawals from the account, since negative work is involved. How might we incorporate these effects into our energy accounting?

A precise analysis would be difficult because the frictional forces are likely to vary for different velocities and angles on the hill. In general terms, it is easy enough to state what happens, however. The frictional forces are always in the opposite direction to the motion, and so they do negative work on the system. This removes mechanical energy from the system, thus yielding smaller values of kinetic energy than those that we have computed (fig. 8.21). The farther the sled has traveled, the larger is the negative work done by frictional forces, and the smaller the total energy of the system.

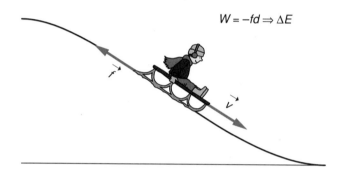

$$W = -fd \Rightarrow \Delta E$$

Figure 8.21 **The work done by frictional forces is negative; it removes mechanical energy from the system.**

If we knew how much work had been done by frictional forces at some particular point on the hill, we could easily adjust our energy account to reflect this change. If, for example, −2000 J of work had been done by frictional forces when the sled reached the hump pictured in figure 8.20, then instead of 16 300 J of total energy, we would have 14 300 J. Subtracting the 11 800 J of potential energy represented by the height of the hump would leave us with just 2500 J of kinetic energy, 2000 J less than in our earlier computation. This, of course, represents a smaller velocity than we computed earlier; the sled would be moving at a rate of 7.9 m/s instead of 10.6 m/s.

What happens to the energy that is lost? As we have indicated earlier, there is a heating effect whenever frictional forces are involved. The mechanical energy that is lost by our system shows up as thermal energy added to the air, the sled, and the snow. This has the effect of raising slightly the temperature of the sled runners and the snow. The elevated pressure underneath the sled runners also lowers the melting point of ice, and this effect, together with the heat generated by friction, causes a small amount of snow that is in contact with the runners of the sled to melt. In fact, the small amount of water produced by melting is an important factor in lubricating the contact between the runners and the snow, thus reducing the frictional forces (fig. 8.22).

When heat (thermal energy) is added to a system, the internal energy of the system increases. This means that the kinetic and potential energies of the atoms and molecules making up the system are increased. The effect of this increase may show up as an increase in temperature, or it may take other forms, such as the melting of ice to produce water. In any case, however, it is not convenient to convert this thermal energy back to mechanical work; in fact it is impossible to do so completely. For all practical purposes, then, this energy is lost from our mechanical system. What can and cannot be accomplished with thermal energy is the subject matter of thermodynamics, which the next two chapters discuss much more thoroughly.

In many situations, then, the mechanical energy of a system is not conserved, largely as a result of frictional or resistive forces. When a ball is dropped and allowed to bounce from the floor or other hard surface, for example, it usually does not return to its original height. Energy is lost through air resistance and also through the internal frictional forces involved in the distortion of the ball and the floor when they are in contact.

A lot of energy is lost in the collision of two cars or other vehicles. Metal is distorted, sound waves are produced, and the cars also get warmer. For this reason, conservation of momentum is a much more useful tool than conservation of energy in analyzing collisions. As already indicated, most real collisions are *inelastic,* meaning that some mechanical energy is lost. Perfectly elastic collisions are

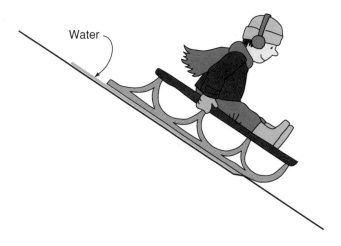

Figure 8.22 Pressure effects and the thermal energy generated from frictional forces cause a small amount of snow to melt, thus reducing the friction between the sled's runners and the snow.

hard to find; collisions between hard spheres such as pool balls come the closest of those that we encounter in our everyday experience.

If few systems actually completely meet the conditions of conservation of mechanical energy, then why make so much of this principle? It is still useful to set limits on what is possible, even when mechanical energy is not conserved. In addition, if we include the internal energy that is generated via the frictional forces in our definition of total energy, then the overall energy is always conserved. This is indeed what we do in thermodynamics and in other more general uses of energy concepts. Energy becomes the currency of the physical world; an understanding, therefore, of energy accounting is crucial to both science and economics. Our utilization of energy resources will continue to be a critical issue for our nation and the world at large.

Box 8.2

Everyday Phenomenon:
Energy and the Pole Vault

The Situation. Ben Lopez is out for track. He specializes in the pole vault and helps out sometimes with the sprint relays, where he can use his speed to good advantage. His coach, being aware that Ben is also taking an introductory physics course, suggests that Ben try to understand the physics of the pole

vault. What factors determine the height reached? How can we optimize these factors?

The coach knows that energy considerations are important in the pole vault. What types of energy transformations are involved? Could a good under-

Continued

Box 8.2 Continued

A pole vaulter on the way up. What energy transformations are taking place?

standing of these effects help an athlete's performance in the vault?

The Analysis. It was not difficult for Ben to provide a general description of the energy transformations that take place in the pole vault. The vaulter begins by running down a path leading to the vaulting standard and pit. During this phase he is accelerating and increasing his kinetic energy at the expense of chemical energy stored in his muscles. When he reaches the standard, he plants the end of the pole in a notch in the ground. At this point, some of his kinetic energy is stored in the potential energy of the bent pole, which acts like a spring. The rest is converted to gravitational potential energy as he begins to rise over the standard.

Near the top of his vault, the potential energy in the bent pole is also converted to gravitational potential energy as the pole straightens out. In addition, the vaulter does some additional work with his arm and upper-body muscles to provide an additional boost in height. At the very top of his flight, his kinetic energy should be essentially zero, with only a minimal horizontal velocity left to carry him over the standard. Too large a kinetic energy at this point would indicate that he had not optimized his jump by converting as much energy as possible into gravitational potential energy.

What insights could Ben gain from this analysis?

The most obvious is the importance of speed. The more kinetic energy he generates in his approach, the more energy is available for conversion to gravitational potential energy (mgh), which will largely determine the height of his vault. This is why successful pole vaulters are usually good sprinters.

The characteristics of the pole and where Ben grips it are also important factors. If the pole is too stiff, or if he has gripped it too close to the bottom end, he will experience a jarring impact in which little useful potential energy is stored in the pole, and some of his initial kinetic energy is lost in the collision. On the other hand, if the pole is too limber, or if it is gripped too far from the bottom end, it will not spring back soon enough to provide useful energy at the top of his vault.

Finally, upper-body strength is important in the final phase of clearing the standard. Good upper-body conditioning should improve Ben's vaulting success. Obviously, though, timing and technique are critical, and these can be improved only through practice. As far as his coach was concerned, that may have been the most important message.

The flexibility of the pole and the point at which the vaulter grasps the pole are important to the success of the vault.

SUMMARY

The concept of work played a central role in this chapter. Energy is transferred into a system by doing work on the system, which results in an increase in either the kinetic energy or the potential energy of the system. If no additional work is done on the system, the total energy of the system remains constant. This principle of conservation of energy allows us to explain many features of the behavior of the system. The following concepts are important for understanding this type of analysis.

Work and Power.   Work is defined as the force times the distance involved in moving an object, provided that only the portion of the force that is in the direction of the motion is used. Power is the rate of doing work.

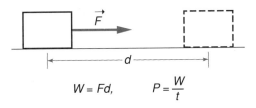

$$W = Fd, \qquad P = \frac{W}{t}$$

Kinetic Energy.   The work done by the net force acting on an object is used to accelerate the object, and the object then gains kinetic energy. The kinetic energy is equal to one-half the mass of the object times the square of its speed.

$$KE = \left(\frac{1}{2}\right)mv^2$$

Potential Energy.   If work done on an object has the effect of moving the object without acceleration against a conservative force acting in the opposite direction, then the potential energy of the object is increased. Two types of potential energy were considered: gravitational potential energy and the potential energy of a spring.

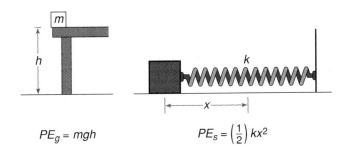

$$PE_g = mgh \qquad\qquad PE_s = \left(\frac{1}{2}\right)kx^2$$

Conservation of Energy.   If no work is done on a system by nonconservative forces, then the total mechanical energy (kinetic plus potential) remains constant. This is the principle of conservation of mechanical energy.

$$E = KE + PE = \text{constant}$$

Simple Harmonic Motion.   The motion of a simple pendulum or that of a mass on a spring both illustrate the principle of conservation of energy, but involve different types of potential energy. These are both examples of simple harmonic motion, which results whenever the restoring force is proportional to the displacement.

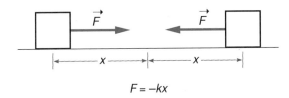

$$F = -kx$$

Effects of Friction.   When frictional forces are acting on an object, the work done against friction removes energy from the system, and the mechanical energy is not conserved. Knowing how much energy has been removed still allows the use of energy principles to predict how the system will behave.

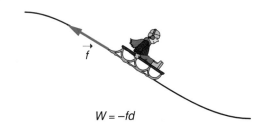

$$W = -fd$$

QUESTIONS

Q8.1 A string is used to pull a wooden block across the floor without accelerating the block. The string makes an angle to the horizontal, as shown in the diagram.

    a. Does the force applied via the string do work on the block? Explain.

    b. Is the total force involved in doing work, or just a portion of the force? Explain.

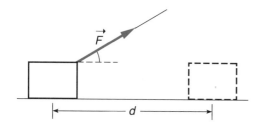

Q8.2 In the situation pictured in question 8.1, if there is a frictional force opposing the motion of the block, does this frictional force do work on the block? Explain.

Q8.3 In the situation pictured in question 8.1, does the normal force of the floor pushing upwards on the block do any work? Explain.

Q8.4 A ball is being twirled in a circle at the end of a string; the string provides the centripetal force necessary to keep the ball moving in the circle at constant speed. Does the force applied via the string do work on the ball in this situation? Explain.

Q8.5 A lever is used to lift a rock in the manner shown in the diagram. If the rock is not accelerated, will the work done by the person on the lever be greater than, equal to, or less than the work done by the lever on the rock? Explain.

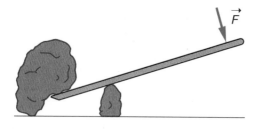

Q8.6 A block is being pulled across the floor with a force applied by a horizontally held string. A smaller frictional force also acts upon the block, which yields a net force that is smaller than the force applied by the string. Does the work done by the force applied to the block by the string equal the change in kinetic energy? Explain.

Q8.7 If there is just one force acting on an object, does its work necessarily result in an increase in kinetic energy? Explain.

Q8.8 Two blocks of the same mass are accelerated by different net forces. One block gains a velocity twice that of the other block in the process. Is the work done by the net force on the faster moving block just twice that done on the slower block? Explain.

Q8.9 A block is moved from the floor up to a tabletop, but gains no velocity in the process. Is there work done on the block? If so, what has happened to the energy added to the system?

Q8.10 When work is done to increase the potential energy of an object without increasing the kinetic energy of the object, what is the work done by the *net* force acting on the object? Explain.

Q8.11 Suppose that work is done upon a large crate to tilt it so that it is balanced on one edge, as shown in the diagram, rather than sitting squarely on the floor as it was initially. Has the potential energy of the crate increased in this process? Explain.

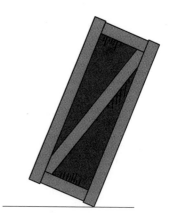

Q8.12 A child's toy consists of a rubber ball connected by an elastic string to a wooden paddle. A force is applied to the ball to pull it back from the paddle, thus stretching the string but not increasing either the height or the velocity of the ball. Has the energy of the system increased? Explain.

Q8.13 When a bow and arrow are cocked, a force is applied to the string in order to pull it back. Is the energy of the system increased in this process? Explain.

Q8.14 A pendulum is pulled back from its equilibrium (center) position and then released.

    a. What form of energy is added to the system prior to its release? Explain.

b. At what points in the motion of the pendulum after release is its kinetic energy the greatest? Explain.

c. At what points is the potential energy the greatest? Explain.

Q8.15 Is the total mechanical energy conserved in the motion of a pendulum? Will it keep swinging forever? Explain.

Q8.16 A mass attached to a spring, which in turn is attached to a wall, is free to move upon a friction-free, horizontal surface. The mass is pulled back and then released.

a. What form of energy is added to the system prior to the release of the mass? Explain.

b. At what points in the motion of the mass after its release is its potential energy the greatest? Explain.

c. At what points is the kinetic energy the greatest? Explain.

Q8.17 Suppose that a mass is hanging vertically at the end of a spring. The mass is pulled downward and released to set it into oscillation. Is the potential energy of the system increased or decreased when the mass is lowered? Explain.

Q8.18 A sled is given a push at the top of a hill. Is it possible for the sled to cross a hump in the hill that is higher than its starting point under these circumstances? Explain.

Q8.19 Can work done by a frictional force ever increase the total mechanical energy of a system? Explain.

Q8.20 Is gravitational potential energy the only type of potential energy that is involved in pole vaulting? Explain.

Q8.21 If two pole vaulters run with the same speed during their approach, will they necessarily reach the same height in their vaults? Explain.

EXERCISES

E8.1 A horizontally directed force of 15 N is used to pull a box a distance of 2 m across a tabletop. How much work is done by the 15-N force?

E8.2 A force of 50 N is used to drag a crate 4 m across a floor. The force is directed at an angle upward from the crate such that the vertical component of the force is 30 N and the horizontal component is 40 N.

a. What is the work done by the horizontal component of the force?

b. What is the work done by the vertical component of the force?

c. What is the total work done by the 50-N force?

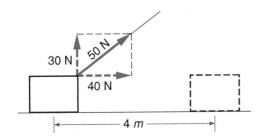

E8.3 A net force of 30 N accelerates a mass of 4 kg through a distance of 5 m.

a. What is the work done by this net force?

b. What is the increase in kinetic energy of the mass?

E8.4 A 0.5-kg ball has a velocity of 20 m/s. What is the kinetic energy of the ball?

E8.5 A box with a mass of 4.5 kg is lifted (without acceleration) through a height of 2 m in order to place it on a shelf.

a. What is the increase in potential energy of the box?

b. How much work was required to lift the box to this position?

E8.6 A spring with a spring constant, k, of 80 N/m is stretched a distance of 20 cm (0.20 m) from its original unstretched position. What is the increase in potential energy of the spring?

E8.7 Suppose that 20 J of work is done to stretch a spring a distance of 0.10 m.

a. What is the increase in potential energy of the spring?

b. What is the value of the spring constant of the spring?

E8.8 At the low point in its swing, a pendulum bob with a mass of 1 kg has a velocity of 2 m/s.

a. What is its kinetic energy at the low point?

b. Ignoring air resistance, how high will it swing above the low point?

E8.9 A 0.20-kg mass attached to a spring is pulled back horizontally across a table so that the potential energy of the system is increased from zero to 150 J.

a. Ignoring friction, what is the kinetic energy of the system after the mass is released and the potential energy has decreased to 60 J?

b. What is the velocity of the mass at this point?

E8.10 A sled and rider with a combined mass of 60 kg are at the top of a hill that is 15 m above the level ground below. The sled is given a push, providing an initial kinetic energy at the top of the hill of 1800 J.

a. Choosing a reference level at the bottom of the hill, what is the potential energy of the sled and rider at the top of the hill?

b. After the push, what is the total mechanical energy of the sled and rider at the top of the hill?

c. If friction can be ignored, what will be the kinetic energy of the sled and rider at the bottom of the hill?

E8.11 A roller coaster has a potential energy of 500 000 J and a kinetic energy of 100 000 J at point *A* in its travel. At the low point of the ride, the potential energy is zero, and 50 000 J of work have been done against friction since it left point *A*. What is the kinetic energy of the roller coaster at this low point?

E8.12 A roller coaster car with a mass of 1200 kg starts at rest from a point 20 m above the ground. At point *B*, it is 10 m above the ground.

a. What is the initial potential energy of the car?

b. What is the potential energy at point *B*?

c. If the initial kinetic energy was zero, and the work done against friction between the starting point and point *B* is 20 000 J, what is the kinetic energy of the car at point *B*?

E8.13 A child with a mass of 30 kg starts sliding from the top of a slide that is 3 m higher than the bottom of the slide. As the child slides down, 282 J of work are done against friction.

a. What is the initial potential energy of the child relative to the bottom of the slide?

b. What is the kinetic energy of the child at the bottom of the slide?

c. What is the velocity of the child at the bottom?

CHALLENGE PROBLEMS

CP8.1 Suppose that two horizontal forces are acting upon a 0.5-kg wooden block as it moves across a laboratory table: an 8-N force pulling the block and a 3-N frictional force opposing the motion. The block moves a distance of 2 m across the table.

a. What is the work done by the 8-N force?

b. What is the work done by the net force acting upon the block?

c. Which of these two values should you use to find the increase in kinetic energy of the block? Explain.

d. What happens to the energy added to the system via the work done by the 8-N force? Can it all be accounted for? Explain.

e. If the block started from rest, what is its velocity at the end of the 2-m motion?

CP8.2 Suppose that we want to move a 120-kg crate to a loading platform that is 1.5 m above the ground. We have available a 30° inclined plane, or ramp, and can move the crate on rollers with minimal friction. With a 30° ramp, the distance moved along the ramp is exactly twice the height of the ramp, and the force required to counter the weight of the crate is just half the weight of the crate.

a. What is the increase in potential energy of the crate?

b. How much work would be done to lift the crate into position without using the ramp?

c. How much work must be done to push the crate up the ramp on the rollers?

d. Which of these two processes would you choose to use? Explain.

e. How is the use of the ramp in this situation similar to the use of a lever to lift a rock? Explain.

CP8.3 A slingshot consists of a rubber strap attached to a Y-shaped frame, with a small pouch at the center of the strap to hold a small rock or other projectile. The rubber strap behaves much like a spring. Suppose that for a particular slingshot a spring constant of 1200 N/m is measured for the rubber strap as mounted in the frame. The strap is pulled back approximately 30 cm (0.30 m) prior to being released.

a. What is the potential energy of the system prior to release?

b. What is the maximum possible kinetic energy that can be gained by the rock after release?

c. If the rock has a mass of 50 g (0.050 kg), what is its maximum possible velocity after release?

d. Will the rock actually reach these maximum values of kinetic energy and velocity? What are the possible avenues of energy loss in this system? Explain.

CP8.4 Suppose that a 150-g mass (0.150 kg) is oscillating at the end of a spring upon a horizontal surface that is essentially friction-free. The spring can be both stretched and compressed and has a spring constant of 250 N/m. It was originally stretched a distance of 10 cm (0.10 m) from its equilibrium (unstretched) position prior to release.

a. What is its initial potential energy?

b. What is the maximum velocity that the mass will reach in its oscillation? Where in the motion is this maximum reached?

c. Ignoring friction, what are the values of the potential energy, kinetic energy, and velocity of the mass when it is 5 cm from the equilibrium position?

d. How does the value of the velocity computed in (*c*) compare to that computed in (*b*)? (What is the ratio of the values?)

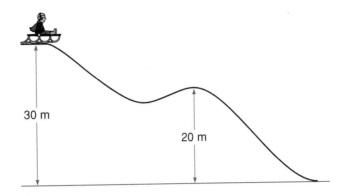

CP8.5 A sled and rider with a total mass of 60 kg are perched at the top of a hill, as pictured in the diagram. The top of this hill is 30 m above the low point in the path of the sled. A second hill is 20 m above this low point. Suppose that approximately 5000 J of work is done against friction as the sled travels between these two high points.

30 m

20 m

a. Will the sled make it to the top of the second hill if no kinetic energy is given to the sled at the start of its motion? Explain.

b. What is the maximum height that the second hill could be in order for the sled to reach the top, assuming that the same work against friction will be involved and that no initial push is provided? Explain.

HOME EXPERIMENTS AND OBSERVATIONS

HE8.1 Construct a simple pendulum by attaching a ball to a string (with tape or a staple) and fixing the other end of the string to a rigid support. (A pencil taped firmly to the end of a desk or table will do nicely.) Make the following observations.

a. Measure the frequency of oscillation by timing the swings. The usual method is to use a watch to measure the time required for 10 or more complete swings. The period, *T*, (the time required for one swing) is then the total time divided by the number of swings counted, and the frequency, *f*, is just 1 / *T*.

b. How does the frequency change if you vary the length of your pendulum?

c. Approximately how many swings will the pendulum make before coming to a stop? What fraction of the original energy is lost in each swing? Does this fraction change if the pendulum is started with a larger initial displacement?

HE8.2 Make a ramp for a marble or small steel ball by bending a long strip of cardboard into a V-shaped groove. Two such ramps can be placed end to end, as pictured in the diagram, to produce a track, in which the marble will oscillate.

a. Can you measure a frequency of oscillation for this system? Does this frequency depend upon how high up the ramp the marble starts?

b. How high up the second ramp does the marble go? Is more energy lost per swing in this system than for a pendulum?

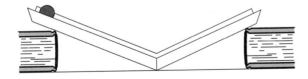

HE8.3 Make a marble launcher by stapling a rubber band across a cardboard support, as shown. Staple the rubber band so that it meets the marble halfway up when the marble is sitting on the base of the launcher. You can then launch marbles horizontally from a table or desk top, and measure the distance they travel before they hit the floor.

a. How does the distance traveled depend upon the distance that the rubber band is pulled back before release?

b. Does this dependence correspond to what you would expect on the basis of your knowledge of projectile motion and of the energy transformations involved with the launcher? Explain.

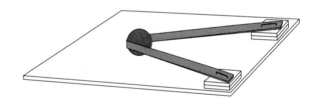

HE8.4 The height to which a ball bounces after being dropped provides a measure of how much energy is lost in the collision with the floor or other surface. A small portion of the energy is lost to air resistance as the ball is moving, but most is lost in the collision.

a. Using a number of different balls that you may have available, test the height of the bounce using the same height of release for all of the balls. Which loses the most energy, and which the least?

b. What characteristics of the balls tested provide the best bounce? Does a different floor surface (carpet, tile, etc.) make a difference in your results?

9

Temperature and Heat

Have you ever touched a drill bit right after drilling a hole in a piece of wood or metal? Chances are that you removed your hand quickly since the bit was probably hot, particularly if you had been drilling in metal. The same phenomenon is at work when brakes get hot on your bicycle or car, or in any process where one surface is rubbing on another.

We could also make the drill bit hot by placing it in a pot of boiling water or in the flame of a blow torch. In either case, when the bit gets hotter, we say that its temperature has increased. But what is temperature, and how do we go about comparing one temperature to another? Does it make any difference at all to the final condition of the drill bit whether it has been warmed by the drilling process (fig. 9.1) or by being placed in contact with hot water or flame gases?

Questions such as these lie in the domain of thermodynamics, the study of heat and its effects upon matter. In the last chapter you saw that mechanical energy is sometimes lost to a system in the form of heat generated by frictional processes. The subject of thermodynamics, then, is ultimately about energy, but in a broader context than that of mechanical energy as discussed in chapter 8. The first law of thermodynamics, which is introduced in this chapter, extends the principle of conservation of energy to include the effects of heat upon matter.

What is the difference between the concepts of heat and temperature? Our everyday language often confuses these two ideas. A true appreciation for this distinction did not really emerge until about the middle of the last century when the laws of thermodynamics were developed. These ideas are critical, however, to understanding why things get hot or cold and why some remain that way longer than others. A cool night will often follow a very hot day in the desert; why does this not happen in Miami? An understanding of the distinction between heat and temperature will help to clarify such questions.

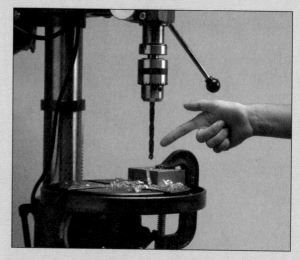

Figure 9.1   The drill bit feels hot after drilling a hole in a piece of metal. What causes its increase in temperature?

Chapter Objectives

The first two sections of this chapter are concerned with the definitions of temperature and heat respectively, as well as with the relationship between these two concepts. We then introduce the first law of thermodynamics, which helps us to explain why the drill bit gets hot. Providing a basis for an understanding of the distinction between heat and temperature and of the meaning of the first law of thermodynamics are the primary objectives of this chapter.

Chapter Outline

❶ *Temperature and its measurement.* What is temperature, and how do we go about measuring temperature differences? Where should the zero of a temperature scale lie?

❷ *Heat, specific heat capacity, and latent heat.* What is heat, and how does it affect the temperature of objects? Does adding heat always change the temperature of a substance, or are other effects possible? What do the terms *specific heat* and *latent heat* mean? Why do some foods stay hot longer than others?

❸ *Joule's experiment and the first law of thermodynamics.* Are there other ways to change the temperature of an object besides adding heat? What does the first law of thermodynamics say and mean?

❹ *Gases, balloons, and the first law.* How can the behavior of gases be explained with the help of the first law of thermodynamics? What is an ideal gas?

❺ *The flow of heat.* What are the different ways in which heat can be transferred from one object or system to another? How do these ideas apply to keeping a house warm in the winter and cool in the summer?

9.1 TEMPERATURE AND ITS MEASUREMENT

Suppose that you are not feeling well and are looking for an excuse to skip class. You think you might have a fever, and your roommate, by placing a hand on your forehead, agrees that you feel a little hot. In the interest of making that judgment somewhat more precise, however, you pull a thermometer out of the medicine cabinet, or perhaps go to the school nurse, who takes your temperature (fig. 9.2). The thermometer reads 101.3°.

What exactly does this tell you? It indicates that your temperature is above 98.6° Fahrenheit, which is generally considered normal, and that you therefore have a fever. You are probably justified in spending the day in bed. The thermometer provides a quantitative measure of how hot or cold things are and therefore a basis for comparison of your current temperature to its level when you do not have a fever. This quantitative measure generally carries more credibility with the boss or a professor than the vaguer statement that you felt like you had a fever.

Making a Temperature Measurement

The process of taking a temperature or reading a thermometer is certainly a common experience. Besides the obvious function of providing a basis for comparing the hotness or coldness of different objects, however, do the numbers have any more fundamental meaning? To ask an even more basic question, how do hot objects differ from cold objects? What is this thing we call *temperature?*

Although we all have an intuitive sense of what *hot* and *cold* mean, putting that sense into words can be difficult. (Try it before you read on; how would you define the term *hot*?) Even our senses can give us misleading impressions; the pain that we feel when we touch something that is very hot can be very similar to, and hard to distinguish from, the pain that we feel when we touch a very cold object. Two objects that are at the same temperature but are made of different materials often seem to be at different temperatures if we rely only on our sense of touch.

What, then, are we doing in making a temperature measurement? The process is ultimately one of comparison, and the relative terms *hotter* and *colder* actually have more meaning than *hot* or *cold* themselves. A closer look at what

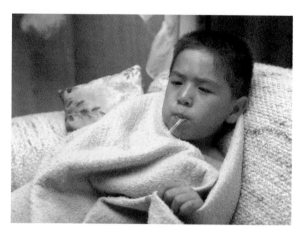

Figure 9.2   Taking a temperature. What does the thermometer tell us?

happens when we use a thermometer can help clarify these issues.

The traditional clinical thermometer consists of a sealed glass tube partially filled with some fluid, usually mercury. The inside diameter of the tube is very small, but it widens at the bottom into a reservoir that contains most of the mercury as shown in figure 9.3. You usually place the thermometer in your mouth under your tongue and then wait for a few minutes until the thermometer reaches the same temperature as the inside of your mouth. Initially the mercury rises in the thin tube, and you assume that it has reached the same temperature as your mouth when the mercury level in the tube is no longer changing.

Why does the mercury rise? Most materials expand as they get warmer. It happens that mercury (as well as many other liquids) expands at a greater rate than glass. The mercury in the reservoir must go somewhere as it expands, so it rises in the narrow tube. We are using the physical property of the thermal expansion of mercury to give us an indication of temperature; by placing marks along the tube, we can create a temperature scale. Any other physical property that changes with temperature could, in principle, be used.

Implicit in this whole process is the idea that if two objects are in contact with one another long enough so that the physical properties of the objects are no longer changing, then the two objects have the same temperature. When we take a temperature with the clinical thermometer, we wait until the mercury stops rising before reading the thermometer. This idea is really a part of the definition of temperature; it defines when two or more objects have the *same* temperature. When the physical properties (such as volume) are no longer changing, the objects are said to be in *thermal equilibrium.* Two or more objects that are in thermal equilibrium with one another have the same temperature. This assumption is sometimes referred to as the *zeroth law of thermodynamics* (with respect to the first law), because it underlies the definition of temperature.

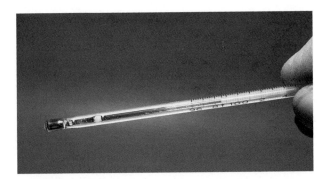

Figure 9.3 The traditional clinical thermometer contains mercury in a thin glass tube with a wider reservoir of mercury at the bottom.

Temperature Scales

What significance can we attribute to the numbers on a thermometer? When the first crude thermometer was constructed, the numbers associated with the divisions on the scale were completely arbitrary. They were useful for comparing temperatures only if you used the same thermometer. If a temperature measured in Germany with one thermometer was to be compared to another one measured in England with a different thermometer, some means of standardizing a temperature scale was needed.

The first widely used temperature scale was devised by Gabriel Fahrenheit (1686–1736) in the early 1700s. The Celsius scale came somewhat later, in 1743. Both of these scales used the fixed points of the freezing point and boiling point of water to anchor the scales. (The modern Celsius scale uses the triple point of water, where ice, water, and water vapor are all in equilibrium as its primary reference temperature. This point varies only slightly from the freezing point at normal atmospheric pressure.) Fahrenheit set the freezing point of water at 32° on his scale and the boiling point at 212°, whereas these two points are set at 0° and 100° respectively on the Celsius scale.

As figure 9.4 shows, the Celsius degree is larger than the Fahrenheit degree, since only 100 Celsius degrees are required to span the temperature range between the two fixed points, whereas 212 − 32 = 180 Fahrenheit degrees are needed to span the same range. The ratio of Fahrenheit degrees to Celsius degrees needed to span the same temperature range is evidently 180 / 100 = 9 / 5. The Fahrenheit degree must therefore be only 5 / 9 the size of the Celsius degree. This fact is used in converting temperatures from one scale to the other.

Do the zero points on either scale have any special significance? Beside the fact that the zero point on the Celsius scale is the freezing point of water, there is no fundamental significance to these points; they were selected arbitrarily. It is possible, of course, for the temperature to go below zero on either scale, as happens frequently in places like Minnesota or Alaska in the winter. Fahrenheit's 0° was based originally on the temperature of a mixture of salt and ice in a salt solution, and his 100° point may have been intended

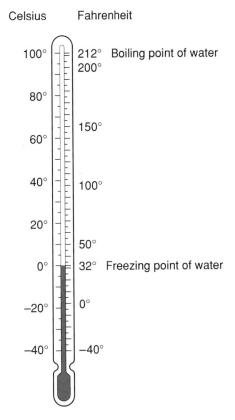

Figure 9.4 The Fahrenheit and Celsius scales use different numerical values for the freezing and boiling points of water. The Celsius degree is larger than the Fahrenheit degree.

to be approximately the temperature of the human body. (His test person may have had a fever!)

Since the Celsius scale is used in science as well as in most of the world, it is useful to be able to convert the Fahrenheit temperatures still commonly used in this country to Celsius temperatures. Since the zero points differ as well as the size of the degree, we must take both factors into account in the conversion. For example, an ordinary room temperature of 72° Fahrenheit is 72° − 32°, or 40° above the freezing point of water. Since it takes only 5 / 9 as many Celsius degrees to span the same range, this temperature would be (5 / 9)(40°) = 22 Celsius degrees above the freezing point. The Celsius scale has its zero at the freezing point of water, so this temperature is then 22° C.

What we have done here is to find how many Fahrenheit degrees (40) this temperature is above the freezing point of water by subtraction, and then multiplied by a factor (5 / 9) that represents the ratio of the size of the Fahrenheit degree to the Celsius degree, since we know it will take fewer Celsius degrees to span this range. We can express this process in an equation of the following form:

$$T_c = \left(\frac{5}{9}\right)(T_f - 32).$$

Using the same logic to perform the reverse conversion from Celsius to Fahrenheit, you should be able to confirm that the equation becomes

$$T_f = \left(\frac{9}{5}\right)T_c + 32.$$

Multiplying by 9 / 5 tells us how many Fahrenheit degrees this temperature is above the freezing point; this value must be added to 32° F, the freezing point on the Fahrenheit scale. Using this relationship, for example, you can easily show that the boiling point of water ($T_c = 100°$ C) equals 212° F, which must be true from the definitions of the two scales. Normal body temperature (98.6° F) is equivalent to $T_c = 37.0°$ C, and so on. A temperature of 38° C would be enough to keep you home from school or work.

Absolute Zero

Although the zero points of both the Fahrenheit and Celsius scales are arbitrarily selected, it is possible to define an absolute zero, which has a more fundamental significance. This was not recognized until one hundred years or so after the original development of these scales. The first suggestion that we might be able to define an absolute zero for our temperature measurements came from studying changes in pressure and volume that take place in gases when the temperature is changed.

If we hold the volume of a gas constant while increasing the temperature, the pressure of the gas will increase. Pressure is the force per unit area (F / A) exerted by the gas upon its surroundings and is often measured by noting the height of a column of mercury that a given pressure will support. (A more extensive discussion of the concept of pressure appears in chapter 19 on the behavior of fluids.) Because the pressure of a gas held at constant volume varies consistently with temperature, we can use this property as a means of measuring temperature (fig. 9.5). It is another physical property that changes with temperature, just like the volume of mercury in a mercury thermometer.

If we plot the pressure of a gas as a function of the temperature (on the Celsius scale, for example), we get a graph like that shown in figure 9.6. If the temperatures and pressures involved are not too high, an interesting feature emerges from plots of this nature. The curves are all straight lines, and when extrapolated backwards to zero pressure, they all intersect the temperature axis at the same point. This is true regardless of the nature of the gas used (oxygen, nitrogen, helium, etc.) or the quantity of gas involved.

The point at which the curves all intersect the axis lies at −273.2° C. Since a negative pressure would have no meaning, this suggested that the temperature could not get any lower than −273.2° C. It is important to recognize, however, that we cannot approach this point and have the gases remain gases; most gases will condense to liquids and then solidify well before this point is reached.

We now refer to this temperature (−273.2° C) as absolute zero. The Kelvin, or absolute, temperature scale has its zero at this point and uses the same size degree as the Celsius scale. To convert Celsius temperature, then, to kelvins, we simply add 273.2 to the Celsius temperature, or

$$T_K = T_c + 273.2.$$

The room temperature of 22° C is therefore approximately 295 K. (The term *degree* or the ° symbol are not usually used in expressing absolute temperatures; they are simply referred to as kelvins.)

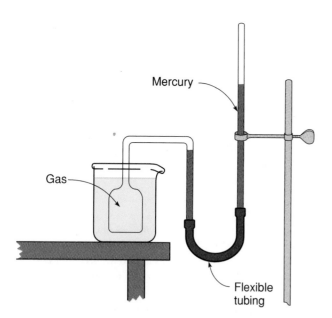

Figure 9.5 A constant-volume gas thermometer allows the pressure to vary with temperature while the volume is held constant. The difference in height of the two mercury columns is proportional to the pressure.

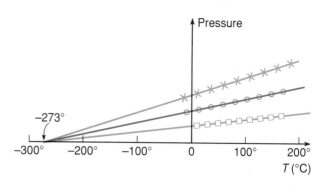

Figure 9.6 The pressure plotted against temperature for different amounts of different gases. When extrapolated backwards, the lines all intersect the temperature axis at the same point, −273° C.

We now have even more fundamental means of defining absolute zero, but the absolute temperature scale currently in use is essentially the same as that originally suggested by observing the behavior of gases. Absolute zero is a temperature that we can approach, but never really reach. It represents a limiting value, beyond which it is not meaningful to speak of anything getting any colder.

If you know how to measure temperature and you know what it means to say that two objects are at the same temperature, you are well on the way to understanding the concept of temperature. The next section of this chapter considers the missing piece of the definition of temperature, namely, what is involved when objects are at different temperatures, and the related distinction between the concepts of heat and temperature.

9.2 HEAT, SPECIFIC HEAT CAPACITY, AND LATENT HEAT

Suppose that your morning cup of coffee is too hot to drink, but you are already late for class or work. What can you do to cool it to the point where it is safe to drink without scalding your tongue? You can blow on it, but this is not a very efficient way of cooling it quickly. If you take milk with your coffee, you can pour a little cold milk into the cup, and the temperature of the coffee will be lowered (fig. 9.7).

What happens when objects or fluids at different temperatures come in contact with one another? The temperature of the colder object increases and the temperature of the warmer one decreases until both objects ultimately come to the same intermediate temperature. This familiar event suggests that something is flowing from the hotter to the colder body (or vice versa). But what is it that is flowing?

Early attempts at explaining such phenomena involved the idea that an invisible fluid called *caloric* flowed from the hotter to the cooler object. The amount of caloric that was transferred dictated the extent of the temperature change. Different materials were regarded as being capable of storing different amounts of caloric in the same amount of mass, which helped to explain some of the experimental observations.

Although the caloric model was successful in explaining many simple phenomena involving temperature changes, there were problems with this idea that we discuss in more detail in the next section. We now use the term *heat* to represent the quantity that flows from one object to another when two objects of different temperatures come in contact. As we will see in the next section, heat flow represents a form of energy transfer between objects. For now, however, it plays the same role as caloric did in the earlier models; it is the invisible something that flows between objects to produce temperature changes.

Figure 9.7 Pouring cold milk into a hot cup of coffee lowers the temperature of the coffee.

Specific Heat Capacity

A large variety of phenomena can be explained using the simple idea of transfer of heat. The varying effects of different materials, for example, in producing temperature changes can be neatly handled. If you drop a 50-gram steel ball at room temperature into your cup of coffee, you will find that it is less effective than 50 grams of milk or water, originally at room temperature, in cooling your coffee. This is explained by the fact that steel has a lower *specific heat capacity* than milk or water. (Milk, of course, is mostly water, as is coffee.) Less heat is needed to change the temperature of 50 grams of steel than to change the temperature of an equal mass of water by the same amount.

The specific heat capacity of a material is the amount of heat needed to change a given mass of the material by a specified amount in temperature (for example, to change the temperature of 1 gram by 1° C). It is a property of the material in question.

It turns out that water has an unusually large specific heat capacity compared to most other materials. In order to make quantitative predictions, we can use water as a point of reference for measuring the specific heat capacities of other materials. We define a unit of heat called the *calorie* as the amount of heat required to raise the temperature of 1 gram of water 1° C in temperature. The specific heat capacity of water is then, by definition of the calorie, 1.0 cal/g·C°. If one calorie of heat is added to a gram of water, its temperature increases by 1 C°; if one calorie of heat is removed, its temperature decreases by 1 C°. The specific heat capacities for a few common substances are shown in table 9.1.

Table 9.1
Specific Heat Capacities of a Few Common Substances

Specific Heat

Substance	Capacity in cal/g·C°
Water	1.0
Ice	0.49
Ethyl alcohol	0.58
Glass	0.20
Granite	0.19
Steel	0.11
Aluminum	0.215
Lead	0.0305

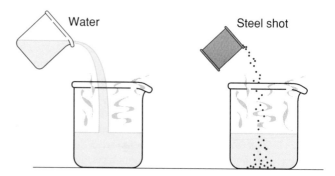

Figure 9.8 **100 grams of room-temperature (20° C) water is more effective than 100 grams of steel shot at 20° C in cooling a beaker of water initially at 100° C.**

From table 9.1, we see that the specific heat capacity of steel is approximately 0.11 cal/g·C°, much less than that of water. If we dropped a 100-g steel ball initially at 20° C into an insulated cup containing 100 g of hot water initially at 100° C, the final temperature of both the water and steel will be approximately 92° C. (If there are no heat losses to the environment, a computation of this final temperature would yield a value of 92.1° C.) The water has decreased in temperature by 7.9 C°. It has given up an amount of heat equal to

$$Q = m_w c_w \Delta T = (100 \text{ g})(1.0 \text{ cal/g·C°})(7.9 \text{ C°}) = 790 \text{ cal,}$$

where Q is the standard symbol for a quantity of heat, and c_w is the symbol for the specific heat capacity of water.

The steel, on the other hand, has increased in temperature by 92.1° C – 20° C, or 72.1 C°. It has gained an amount of heat equal to

$$Q = m_s c_s \Delta T = (100 \text{ g})(0.11 \text{ cal/g·C°})(72.1 \text{ C°}) = 790 \text{ cal,}$$

where c_s is the specific heat capacity of steel. In other words, 790 cal of heat has flowed from the water to the steel ball in order to produce the temperature changes observed. In spite of the fact that the masses of the steel and water were the same, the final temperature is much closer to the initial temperature of the water than of the steel. This is because water has a much larger specific heat capacity than steel; larger quantities of heat are needed to change the temperature of water a given amount than to change the temperature of steel by the same amount (fig. 9.8).

If instead of using the 100-g steel ball, we poured 100 g of water initially at 20° C into the cup containing 100 g of water at 100° C, the final temperature of the 200 g of water now in the cup will be 60° C, halfway between the two initial temperatures. (This fact results from having equal quantities of water (100 g each), as well as identical materials (water) with the same specific heat capacity.) The temperature changes for the two initial 100-g portions of water are

each 40 C°, and the amount of heat that has flowed from one to the other is

$$Q = m_w c_w \Delta T = (100 \text{ g})(1.0 \text{ cal/g·C°})(40 \text{ C°}) = 4000 \text{ cal.}$$

Because of its greater specific heat capacity, water is obviously more effective for cooling than steel.

This same phenomenon is responsible for the fact that variations in temperature are less extreme near the coast of an ocean or large lake than they are inland. A large body of water with its large specific heat capacity requires larger quantities of heat to change its temperature than does the nearby land. (Note from table 9.1, for example, that the specific heat capacity of granite is less than 1/5 that of water.) The water will therefore have a moderating influence on the air temperature in its vicinity; nights will be warmer and days cooler than at some distance inland because it is harder to change the temperature of the water.

Distinction between Heat and Temperature

We are now in a position to complete our definition of temperature and to make a clear distinction between the concepts of heat and temperature, which are often confused. When two objects at different temperatures are placed in contact, heat will flow from the object with the higher temperature to the one with the lower temperature. Heat added increases the temperature; heat removed decreases the temperature, but heat and temperature are not the same thing. The amount of heat that must be added or removed to produce a given change in temperature depends upon the quantity of material involved (its mass) and the specific heat capacity of the material (fig. 9.9).

Temperature is a quantity that tells us in which direction heat will flow. If the two objects are at the same temperature, no heat will flow; if they are at different temperatures, the direction of heat flow is from the higher to the lower temperature. The quantity of heat that is transferred depends upon the temperature difference between the two materials as well as upon their masses and specific heat capacities, as

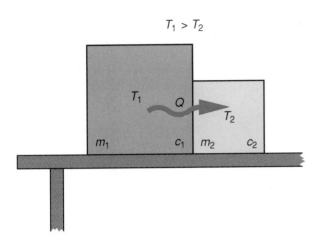

$T_1 > T_2$

Figure 9.9 Heat flows from the hotter to the cooler object when two objects at different temperatures are placed in contact. The changes in temperature that result depend upon the quantities of material and the specific heat capacity of each object.

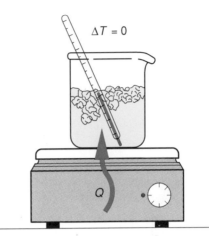

$\Delta T = 0$

Figure 9.10 Adding heat to a mixture of water and ice at 0° C melts the ice without changing the temperature of the mixture.

can be seen in the relationship $Q = mc\Delta T$. Heat and temperature are therefore closely related concepts, but play different roles in our explanation of heating and cooling processes.

Heat is thermal energy that flows from one object to another when there is a difference in temperature between the objects.

Temperature is the quantity that indicates whether or not, and in which direction, heat will flow. Objects at the same temperature are in thermal equilibrium, and no heat will flow from one to the other.

Latent Heat and Changes in State

Are there instances in which heat can be added to or removed from a substance without changing the temperature at all? It turns out that this does indeed occur when substances go through a change of phase or state. The melting of ice or the boiling of water are the most familiar examples of changes in phase; they are involved whenever we cool a drink using ice or cook an egg in boiling water. What is happening in these cases?

Ice, of course, is just solid (frozen) water. If water is cooled to 0° C and we continue to remove heat, the water will freeze into ice. Likewise, if we warm ice to 0° C and continue to add heat, the ice will melt into water. In either case, the temperature of the ice and water remains at 0° C while the freezing or melting is taking place (fig. 9.10). Heat is being removed or added, but no change in temperature is occurring! Apparently, adding or removing heat can produce changes other than just temperature changes.

Careful measurements have shown that it takes approximately 80 cal of heat to melt 1 gram of ice. This quantity, 80 cal/g, is referred to as the *latent heat of fusion, L_f,* of water. It is the quantity of heat needed to change the *phase* of water without changing its temperature. The ice/water mixture remains at 0° C during this process until all the ice has been melted. Likewise, approximately 540 cal of heat are required to vaporize 1 gram of water at 100° C into steam. This quantity, 540 cal/g, is referred to as the *latent heat of vaporization, L_v.* These values are valid only for water; other substances have their own particular latent heats of fusion and vaporization.

What happens, then, when we cool a glass of water by adding ice? Initially, the ice and water are at different temperatures, the ice somewhere below 0° C and the water somewhere above this temperature. Heat flows from the water to the ice until the ice has reached a temperature of 0° C. At this point, the ice begins to melt, and more heat flows from the water to the ice. Assuming that enough ice is present, this flow of heat will continue until the water has also reached a temperature of 0° C. Heat is therefore flowing into the ice without changing the temperature of the ice. (If an insufficient amount of ice was present initially, the ice may melt completely, and the final temperature of the water will then be somewhat above 0° C.)

If the glass is well insulated, so that heat does not continue to flow into the glass from the warmer surrounding air, then the ice and water mixture will remain at 0° C after both the water and the ice have reached this temperature, and no more ice would melt. In the usual situation, however, heat flows into the system from the surroundings and the ice continues to melt more slowly. Once the ice is gone, the drink begins to warm and will ultimately reach room temperature as heat flows in from the surrounding air.

Once it has reached thermal equilibrium, an ice/water mixture, well stirred, will remain at the temperature of 0° C

Box 9.1

Sample Exercise

If the specific heat capacity of ice is 0.50 cal/g·C°, how much heat would have to be added to 200 g of ice, initially at a temperature of −10° C, in order to raise the ice to the melting point and then completely melt the ice?

m = 200 g.

c_i = 0.50 cal/g·C°.

T_i = −10° C.

L_f = 80 cal/g.

Q = ? to completely melt.

Heat required to raise the temperature to 0° C (the melting point):

$Q_1 = mc_i \Delta T_i$

$\quad$ = (200 g)(0.50 cal/g·C°)(10° C)

$\quad$ = 1000 cal.

Heat required to melt the ice:

$Q_2 = mL_f$

$\quad$ = (200g)(80 cal/g) = 16 000 cal.

Total heat required to raise the ice to 0° C and to melt it:

$Q = Q_1 + Q_2$ = 1000 cal + 16 000 cal = **17 000 cal**

$\quad$ = **17 kcal.**

even though small quantities of heat may be flowing into or out of the mixture from the surroundings, and small quantities of ice may be melting or freezing. This is one reason why this temperature is useful as a reference point for our temperature scales. The mixture provides a stable and reproducible temperature.

When we cook food by boiling, we take advantage of the fact that water boils at 100° C and that we can continue to add heat without changing the temperature. Adding heat from the burner simply causes the liquid water to change to water vapor while the temperature remains at 100° C. Since the temperature is constant, the cooking time required to boil an egg or a potato is then also fairly constant, although it does depend upon the size of the egg or potato. If the water is not at the boiling point, cooking times will vary widely depending on the temperature.

Cooks should be aware, however, of the effects of altitude upon the boiling temperature. In cities like Denver or Albuquerque, where the altitude is in the neighborhood of 5000 ft (1500 m), water boils at a temperature of about 96° C because of the lower atmospheric pressure at this altitude. It takes longer to boil an egg or potato in this case than at altitudes near sea level, where the boiling point is 100° C. (When I moved from the east coast to Albuquerque, it took some time to adjust to this fact, despite my knowledge of physics. My potatoes were crunchy in the center for a while.)

9.3 JOULE'S EXPERIMENT AND THE FIRST LAW OF THERMODYNAMICS

Can the temperature of an object be increased without placing the object in contact with a warmer object? What is going on when things get warmer due to rubbing? These questions were a subject of scientific debate during the first half of the last century. They were resolved by the experimental work of James Prescott Joule (1818–89), which was followed by the statement of the first law of thermodynamics around 1850.

One of the first people to raise the question forcefully was the colorful, American-born scientist and adventurer, Benjamin Thompson (1753–1814), also known as Count Rumford. Thompson, after finding himself as a Tory on the losing side of the American Revolutionary War, migrated to Europe, where he served the King of Bavaria as a consultant on weaponry, including cannons. It was here that he gained the title, Count Rumford. From his experience in supervising the boring of cannon barrels, Rumford knew firsthand that the barrels and drill bits became quite hot in the drilling process (fig. 9.11). He once demonstrated that water could be boiled by placing it in contact with the cannon barrel as it was being drilled!

The question, of course, was what was the source of the caloric (heat) that caused the increase in temperature? No hotter body was present from which heat might flow. Horses were used to turn the drill, but their temperature, although warm, was certainly below the boiling point of water. The horses performed mechanical work on the drilling engine; perhaps this work was capable of generating caloric?

Joule's Experiments

Although Rumford's demonstration occurred in 1798, and the question was discussed by many scientists during the early part of the nineteenth century, quantitative experiments on this heating effect were not performed until the 1840s. It was then that Joule did a famous series of experiments in which he showed that various ways of performing mechanical work upon a system had a consistent and predictable effect in raising the temperature of the system.

Although Joule demonstrated this idea in a variety of ways, the one that was most dramatic (and most often presented) was one in which he turned a simple paddle wheel in an insulated beaker of water and measured the increase in temperature. The arrangement is illustrated schematically in figure 9.12. A system of weights and pulleys was used to turn the paddle wheel, thus transferring energy from the

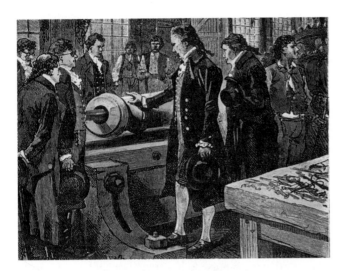

Figure 9.11 Rumford's cannon-drilling apparatus. What is the source of heat that raises the temperature of the drill and the cannon barrel?

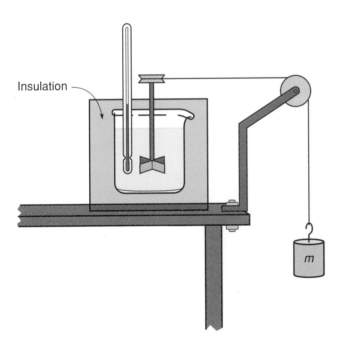

Figure 9.12 A falling mass turns a paddle in an insulated beaker of water in this schematic representation of Joule's apparatus for measuring the temperature increase produced by doing mechanical work on a system.

weights to the water. As the weights fell, losing gravitational potential energy, the paddle wheel did work against the drag forces of the water, equal in amount to the loss in potential energy of the weights.

A thermometer placed in the water allowed the measurement of the increase in temperature. Joule found that 4.19 J of work were required to raise the temperature of 1 gram of

water by 1C°. Since 1 calorie of heat is, by definition, the amount of heat required to raise the temperature of 1 gram of water 1C° this implied that 4.19 J of work is equivalent to 1 cal of heat in its effect upon the temperature of water. (The use of the joule (J) as an energy unit, of course, did not come about until well after Joule's work; his own results were stated in terms of more archaic units.)

Joule's experiments involved a number of different ways of performing work upon a system, as well as variations on the type of system used. The results were always the same, however: the effect of 4.19 J of work was the same as that of 1 cal of heat in increasing the temperature of the system. The final state of the system provides no clue as to whether the temperature was increased by adding heat or by doing mechanical work.

The First Law of Thermodynamics

At the time of Joule's work, several people had already suggested that the flow of heat represented the transfer of kinetic energy between the atoms or molecules of the systems between which the heat was flowing. Joule's experiments bolstered this view and led directly to the statement of the first law of thermodynamics by Lord Kelvin (William Thomson, 1824–1907). The idea underlying the first law is that both work and heat flow represent transfers of energy into or out of a system. The internal energy of the system increases or decreases accordingly; the change in internal energy is equal to the net amount of heat and work transferred into the system. An increase in temperature is one way that an increase in internal energy might affect the system.

To summarize these ideas in words, we can state the first law as follows:

The increase in the internal energy of a system is equal to the amount of heat added to a system minus the amount of work done on the system.

In symbols, we often write this statement as follows:

$$\Delta U = Q - W,$$

where U represents the internal energy of the system, Q represents the amount of heat added to the system, and W represents the amount of work done *by* the system (fig. 9.13). The minus sign in this statement is a direct result of the convention of choosing work done *by* the system to be positive rather than work done *on* the system. Work done by the system on the surroundings removes energy from the system and thus reduces the internal energy. This is the sign convention used in most introductory physics texts, and it is convenient for discussing heat engines, which are the primary focus of chapter 10.

On its surface, the first law seems like a simple statement of conservation of energy, and indeed it is. Its apparent simplicity, though, obscures the important insights that it con-

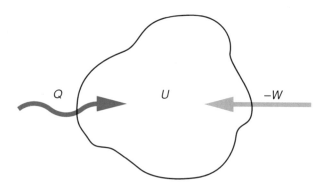

Figure 9.13   The internal energy, *U*, of a system is increased by the transfer of either heat or work into the system.

tains. First and foremost among these insights is the recognition that heat flow represents a transfer of energy, an idea that was strongly reinforced by Joule's experiments. Prior to 1850, the use of the energy concept had been restricted to mechanics; in fact it was just coming into its own within mechanics. It did not play an important role in Newton's original theory of mechanics. The first law of thermodynamics thus extended the energy concept into a new area of application.

The first law also introduced the concept of the internal energy of a system. An increase in internal energy can show up in a variety of ways, depending upon the state of the system. One way is as an increase in temperature, as in Joule's experiment with the paddle wheel. A temperature increase is related to an increase in the average kinetic energy of the atoms or molecules making up the system.

A change in phase, such as melting or vaporization, is another way in which an increase in the internal energy of a system can affect the system. In this case the average *potential* energy of the atoms and molecules is increased as the atoms and molecules are pulled farther away from each other. No temperature change occurs during this process, but there has been an increase in internal energy. Thus an increase in internal energy can show up as either an increase in the kinetic energy or the potential energy (or both) of the atoms or molecules making up the system:

The internal energy of a system is the sum of the kinetic and potential energies of the atoms and molecules making up the system.

Internal energy turns out to be a property of a system that is uniquely determined by the state of the system. Thus, if we know that a system is in a certain phase, and the temperature, pressure, and quantity of material are specified, then there is only one possible value for the internal energy. The amount of heat or work that have been transferred to the system, on the other hand, are not unique to the state of the system. As Joule's experiment showed, either heat or work

could be transferred to the beaker of water to produce the same increase in temperature.

The system in question can be anything that we wish it to be: a beaker of water, a steam engine, or an orangutan. It is best, however, to consider a system that has distinct boundaries, so that we can easily define where it begins and ends. Choosing the system is similar to isolating an object in order to apply Newton's second law in mechanics. In mechanics, forces define the interaction of the object chosen with other objects; in thermodynamics the transfer of heat or work defines the interaction of a system with other systems or the surroundings.

A Simple Application of the First Law   Consider a glass containing equal amounts of water and ice as an example of a system that we can use to illustrate some of these ideas. If we initially have 100 g of ice and 100 g of water in equilibrium in the glass, then the temperature must be 0° C. It requires 80 cal per gram of heat or work to melt the ice, so if we placed the glass on a hot plate and added an amount of heat,

$$Q = mL_f = (100 \text{ g})(80 \text{ cal/g}) = 8000 \text{ cal},$$

to the mixture, all of the ice would melt. The internal energy of the system will have increased by 8000 cal, according to the first law of thermodynamics:

$$\Delta U = Q - W = 8000 \text{ cal} - 0 = 8000 \text{ cal}.$$

(This computation ignores the small amount of work that is done on the system as it decreases slightly in volume.)

Since there are approximately 4.19 J/cal, this increase in internal energy could also be expressed as follows:

$$\Delta U = 8000 \text{ cal } (4.19 \text{ J/cal}) = 33\ 500 \text{ J}.$$

The Joule and the calorie are just two different units of energy; either can be used to express an increase in internal energy.

If, on the other hand, we insulated the glass so that no heat could flow from our hand or the surroundings, we could also melt the ice by stirring vigorously (fig. 9.14). Evidently, −33 500 J of stirring work would do the job, and the internal energy would again increase by 33 500 J:

$$\Delta U = Q - W = 0 - (-33\ 500 \text{ J}) = +33\ 500 \text{ J}.$$

The minus sign is used for the stirring work because this is work done on the system by the stirrer; given our sign convention, this is negative work. Quite obviously, it should have the effect of increasing the internal energy of the system, and it does, given this choice of sign.

In fact, any combination of heat or work equal to 33 500 J added to the system would melt all of the ice. During this process, the temperature of the system remains at 0° C; we have been increasing the potential energy of the atoms and molecules, but not the average kinetic energy. Once all of the ice has melted, however, additional stirring has the effect of increasing the average kinetic energy of the water molecules. This shows up as an increase in the temperature of the water in the glass.

Counting Calories   A final note on the units is worth some mention. When you eat, you add energy to your system by adding material that can release potential energy via chemical reactions. We often measure that energy in a unit called a *calorie*, but this unit is actually a *kilocalorie*, or 1000 cal as we have defined the calorie here. The calorie counting that is done for food, then, uses the following unit:

$$1 \text{ Cal} = 1 \text{ kilocalorie} = 1000 \text{ cal.}$$

The capital *C* is used in the abbreviation to indicate that this is a larger unit than the ordinary calorie.

The Calorie values given for different types of food, then, represent the amount of energy that these foods can release when they are digested and metabolized. This energy may be stored in various ways within the body, including the formation of fat cells. Our bodies are constantly converting energy stored in muscles to other forms of energy. When we do physical work, for example, we can transfer mechanical energy to other systems, and we also warm the surroundings by heat flow from our bodies. Thus when you count calories, you are counting energy intake. If your output is less than your intake, you should expect to get larger.

Figure 9.14   The ice in the beaker can be melted either by adding heat from a hot plate or by doing work in stirring the ice and water mixture.

9.4 GASES, BALLOONS, AND THE FIRST LAW

What happens when we compress air in a cylinder? How does a hot-air balloon work? The behavior of gases in these and other situations can be treated using the first law of thermodynamics with the help of some additional facts and laws of gas behavior. The atmosphere, which is made up of gases, provides another interesting system for exploring the laws of thermodynamics.

Compressing a Gas

To start with a simple case, consider a gas contained in a cylinder with a movable piston (fig. 9.15). If the piston is pushed inward by an external force, work is done *on* the gas in the cylinder by the piston. From the definition of work given in chapter 8, this work is equal to the force exerted on the piston times the distance that the piston moves. By the sign convention that we are using here, however, this is negative work, since it is done on the system (the gas in the cylinder) rather than by the system.

Since pressure is the force per unit area (F / A), the force exerted on the piston by the gas will equal the pressure of the gas times the area of the piston ($F = PA$). If we move the piston without accelerating it, the external force applied to the piston must be equal to the force exerted by the gas on the piston and is therefore directly related to the pressure of the gas. The motion of the piston produces a change in the volume of the gas within the cylinder, and this change in volume is clearly related to the distance that the piston moves. The work done on the piston by the external force is then expressed as follows:

$$W = Fd = (PA)d = P\Delta V$$

where ΔV is the change in volume and is equal to the area of the piston times the distance that it moves (Ad).

If we think of ΔV, the change in volume, as being positive when the gas is expanding and negative when the gas is being compressed (a natural choice), then this expression

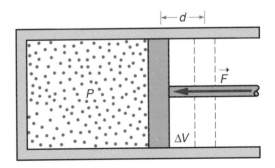

Figure 9.15   A movable piston compresses a gas in a cylinder. The work done is $W = Fd = P\Delta V$.

yields negative work for the case we are considering. This is consistent with our sign convention for work in the first law of thermodynamics ($\Delta U = Q - W$); work is negative when it is done on the system.

When work is done on the gas in the cylinder to compress it, the internal energy increases according to the first law of thermodynamics. If the heat flow into or out of the gas, Q, is zero, the first law tells us that the change in internal energy must be equal to the work done on the gas:

$$\Delta U = Q - W = -W = -P\Delta V.$$

ΔU ends up being positive, representing an increase in internal energy, because ΔV is negative when the gas is compressed. This makes perfect sense, of course; if we do work on the gas, its energy should increase.

In order for Q, the heat flow into or out of the system, to be zero, we must either insulate the cylinder from its surroundings or do the compression quickly enough so that the heat flow is negligible. Such a process (for which there is no heat flow) is called an *adiabatic* process. The first law tells us, then, that the internal energy of a gas will increase in an adiabatic compression by an amount equal to the work done on the gas.

What happens when we increase the internal energy of a gas? An interesting feature of gases involves the relationship between internal energy and temperature. In general, internal energy is made up of both the kinetic and potential energies of the atoms and molecules in the system. In a gas, however, the internal energy is almost exclusively kinetic energy, because the molecules are so far apart, on the average, that the potential energy of interaction between them is negligible. Since the internal energy is essentially kinetic energy and the temperature of a substance is related to the average kinetic energy of the atoms and molecules in a system, the internal energy of a gas turns out to be directly proportional to its temperature.

An *ideal* gas is one for which the forces between atoms (and the associated potential energy) are small enough that we can completely ignore them. This is a good approximation, provided that the temperature of the gas is not too low, or the pressure too high, so that the gas is not on the verge of condensing into a liquid. For an ideal gas, then, the internal energy is completely in the form of kinetic energy, and we have a direct relationship between temperature and internal energy. If the internal energy increases, the temperature must also increase in direct proportion. A plot of internal energy as a function of temperature takes the form of the graph in figure 9.16.

We conclude that the temperature of a gas will increase in an adiabatic compression. We have already seen that the internal energy of the gas will increase by an amount equal to the work done on it. Since the temperature of an ideal gas is directly related to the internal energy, the temperature increases in direct proportion to the amount of work done on the gas in an adiabatic compression.

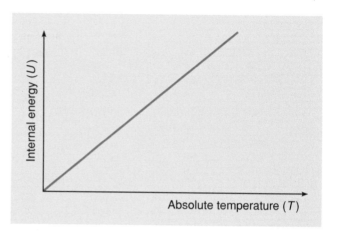

Figure 9.16 **Internal energy is plotted against absolute temperature for an ideal gas. The internal energy increases in direct proportion to the temperature.**

The reverse is also true. If the gas expands against the piston, thus doing work on its surroundings, the work term in the first law becomes positive, and the internal energy decreases. Energy is being removed from the system in the form of work done on the surroundings, and the temperature of the gas decreases if no heat is permitted to flow into it. This is essentially what occurs in a refrigerator; a pressurized gas is allowed to expand, which lowers its temperature. The cool gas is then circulated through coils inside the refrigerator, removing heat in the process.

Isothermal Processes

What happens if we allow heat to flow into or out of the gas? Is it possible for the temperature to remain constant even if heat is added to the system? Again, the first law of thermodynamics can help us analyze this situation. If the temperature is to remain constant, the internal energy must not change, because of the direct relationship between internal energy and temperature in an ideal gas. Thus by the first law,

$$\Delta U = Q - W = 0,$$

and

$$W = Q.$$

In other words, if an amount of heat, Q, is added to the gas, an equal amount of work must be done by the gas upon its surroundings if the temperature and the internal energy are to remain constant.

The gas does work upon its surroundings by expanding. A process in which temperature remains constant (does not change) is called an *isothermal* process. We see, then, that if we heat a gas isothermally, the gas must expand, doing work upon its surroundings, so that the internal energy and the temperature remain the same. Heat can be added without changing the temperature as long as an equal amount of energy is removed in the form of work done by the gas.

Here again, the reverse is also true. In order to compress a gas while holding the temperature constant, heat must be removed from the gas according to the first law of thermodynamics. If energy is added to the gas in the form of work, an equal amount of energy in the form of heat must be removed in order for the internal energy and the temperature to remain constant.

Constant Pressure Processes and Hot-Air Balloons

A somewhat more complicated process is involved in a hot-air balloon. When the gas is heated in a hot-air balloon, it is the pressure, not the temperature, that remains constant (fig. 9.17). The temperature increases, and so does the internal energy. The gas does expand, however, in this process as well as in the isothermal case discussed above. A process in which pressure remains constant is called an *isobaric* process.

The gas expands in the isobaric heating process due to another property of ideal gases that was discovered experimentally near the beginning of the nineteenth century. A series of experiments showed that the pressure, volume, and absolute temperature of an ideal gas are related by the equation

$$PV = NkT,$$

where N is the number of molecules in the gas, and k is a constant of nature called *Boltzmann's constant*. (Although the concept of absolute temperature did not emerge until after these experiments were done, we must use absolute temperature in this relationship in order to obtain the simple form shown here. The introduction of absolute temperature in the first section of this chapter illustrates this relationship.)

Figure 9.17   Heat is added to the air in a hot-air balloon from the propane burner in the gondola. The heated air expands making it less dense than the surrounding air.

This equation ($PV = NkT$) is referred to as the *equation of state* of an ideal gas. It merely states an experimentally determined relationship between the pressure, volume, temperature, and quantity of the gas for any given equilibrium state of the gas. Using this relationship, we can see that if the temperature increases while the pressure and number of molecules are held constant, the volume of the gas must also increase. In other words, the gas must expand.

In this situation, the first law of thermodynamics tells us that the heat added to the gas must be greater than the amount of work done by the gas in expanding. This follows from the fact that the internal energy, U, increases as the temperature increases. Thus,

$$\Delta U = Q - W > 0.$$

Since W is positive, because work is being done by the gas in the expansion, the amount of heat, Q, that is added must be larger than the amount of work done by the expanding gas in order for ΔU to be positive.

Since the gas inside the balloon has expanded, the same number of molecules now occupy a larger volume, and the density of the gas (mass per unit volume m / V) has therefore decreased. In other words, the density of the gas inside the balloon is less than the density of the air outside the balloon, which causes the balloon to rise. (The buoyant force acting upon the balloon, which arises from this difference in densities, is described by Archimedes' principle, discussed in chapter 19.)

The same phenomenon occurs on a much larger scale in the atmosphere. A warm air mass will rise relative to a cooler air mass for the same reason that a hot-air balloon rises. Air that is heated near the ground on a warm summer day expands, becoming less dense, and rises in the atmosphere. This rising air carries water vapor with it, which provides the moisture for the formation of clouds at higher elevations. The rising columns of warm air also create air turbulence, which can become a problem for aircraft. These *thermals* are a real boon, however, to hang-glider enthusiasts or soaring hawks.

Although we have barely scratched the surface here, the behavior of gases and the implications of the first law of thermodynamics play important roles in our understanding of weather phenomena. The concept of latent heat is also important; it is involved in the evaporation and condensation of water as well as in the formation of ice crystals in snow or sleet. The source of energy for all of these processes is, of course, the sun.

Gas behavior and the first law also come into play in our understanding of heat engines (such as an automobile engine), which we will discuss in the next chapter. Expanding gases do work on moving pistons, and this work is transmitted to the drive shaft of the automobile. The atmosphere itself can in some ways be regarded as an enormous heat engine.

9.5 THE FLOW OF HEAT

In most climates, we need to heat our homes during the winter. We do this by burning wood or fossil fuels such as coal, oil, or natural gas in a stove or furnace, or by electric heating. In any case, we pay for the fuel that we use and are therefore concerned with how to keep the heat it generates from escaping. How does heat flow away from a heated house or any other warm object? What mechanisms are involved, and how can we affect the loss of heat by our awareness of these mechanisms?

Conduction, convection, and radiation are the three basic mechanisms by which heat flows from one object to another. All three of these are important in heat loss (or gain) from a house as well as in other common phenomena, such as the freezing of road surfaces. An understanding of each of these mechanisms can help you achieve better comfort, as well as fuel savings, in a house, a dormitory room, or workplace.

Conduction

Conduction is the mechanism that is at work when objects at different temperatures are placed in direct contact with one another. Heat flows directly through the material from the hotter body to the cooler body at a rate that depends upon the temperature difference between the bodies and a property of their materials called *thermal conductivity* (fig. 9.18).

Suppose, for example, that you pick up a block of metal that is at room temperature. Your body is typically 10 to 15 C° warmer than room temperature (37° C verses 20 to 25° C), so that heat flows from your hand to the metal block. The block feels cool to the touch because it is removing heat from your hand.

Since heat flow represents a flow of energy, it is actually kinetic energy of the atoms and electrons in your hand and the metal that is being transferred. The average kinetic energy of the atoms in your hand is larger than that of the atoms and electrons in the metal because of the higher temperature of your hand. As the atoms bump into one another, kinetic energy is transferred, and the average kinetic energy of the atoms and electrons in the metal increases at the expense of that of the atoms in your hand. The free electrons

in the metal play an important role in this process; metals are therefore good conductors of heat as well as of electrical charge.

Thermal conductivity is a property of materials that can vary considerably from one material to another. If, for example, you pick up a block of wood at room temperature in one hand and a block of metal of similar size and also at room temperature in the other hand, the metal block will feel cooler than the wood block (fig. 9.19). Since the two blocks are actually at the same temperature, the difference lies not in the temperature but in the thermal conductivity of the two materials. Metal conducts heat better than wood, so the metal removes heat from your hand at a greater rate than does the wood and therefore feels colder.

In discussing home insulation, we often use the concept of thermal resistance, or *R values,* to describe the relative effectiveness of different materials as insulators. The *R* value is related both to the thickness of the material and to its thermal conductivity. The greater the conductivity, the lower the *R* value, since good thermal conductors are poor thermal insulators. The greater the thickness of an insulating material, however, the greater its insulating effect, so the *R* values are proportional to the thickness of the material.

Since still air is a good thermal insulator, porous materials with trapped pockets of air make excellent insulators and therefore have high *R* values. Rock wool or fiberglass insulation take advantage of this fact: their low thermal conductivities result from the air pockets trapped by these materials, rather than from the material itself. When we use such materials to insulate the walls, ceilings, and floors of a home, we

Figure 9.19 When a metal block and a wooden block, both at room temperature, are picked up, the metal block feels cooler.

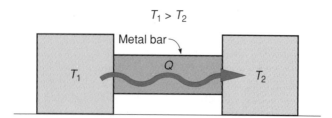

$T_1 > T_2$

Metal bar

T_1 Q T_2

Figure 9.18 Thermal conduction is the process by which thermal energy flows through a material when there is a temperature difference across the material.

reduce their ability to conduct heat and therefore reduce the heat loss from the house.

Convection

Convection is another process or mechanism of heat transfer. If we heat a volume of air to a higher temperature and then move the air via blowers or natural convection, we are transferring heat by convection. In heating a home we often move air, water, or steam to convey heat from a central furnace to different rooms. These are all fluids that can be directed to different parts of the house through pipes or ducts. Convection, then, is the process in which heat is transferred by motion of a fluid containing thermal energy; it is therefore the primary process by which heat is transferred within a house.

Natural convection, for example, is involved within a room. Radiators are usually placed near floor level. As noted in the previous section, warm air is less dense than cooler air, and so it tends to rise toward the ceiling. In so doing, it sets up air currents within the room that distribute the energy around the room. Warm air rises along the wall containing the heat source, and cooler air falls along the opposite wall, as shown in figure 9.20.

Convection is also involved in heat loss from buildings. When warm air leaks out of a building, or cooler air leaks in from outside, we say that we are losing heat to infiltration, which is a form of convection. This can be prevented, to some extent, by weather-stripping doors and windows and sealing other cracks or holes that might be involved. It is not desirable, however, to completely eliminate infiltration; a certain amount of turnover of the air within a house is desirable to keep the air fresh and free of odors.

Homes that are tightly constructed for purposes of energy conservation also run the risk of buildup of the radioactive gas, radon, which comes from natural sources of radioactivity in the earth and building materials. This problem has only recently been recognized and has caused us to rethink the desirability of designing ultratight, energy-efficient houses. Some infiltration is essential!

Radiation

The third mechanism of heat flow, that of radiation, is a little more difficult to explain given the background provided thus far. Radiation involves the flow of energy via electromagnetic waves. Wave motion and electromagnetic waves in particular are discussed in chapter 14. Light and radio waves are both forms of electromagnetic waves, but those that are primarily involved in the transfer of heat lie in the infrared portion of the electromagnetic wave spectrum. The infrared region includes wavelengths that are between the longer radio waves and the shorter wavelengths of visible light.

Unlike conduction or convection, which require a medium to travel through (conduction) or to be carried along with (convection), radiation can take place across a vacuum. Radiation is the only process that can be involved, then, in the flow of heat across an evacuated barrier such as that in a thermos bottle (fig. 9.21). The radiation can be reduced to a

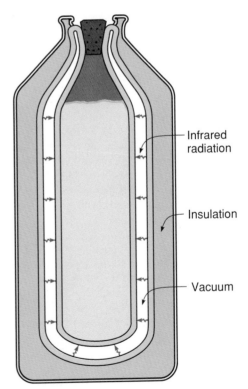

Infrared radiation

Insulation

Vacuum

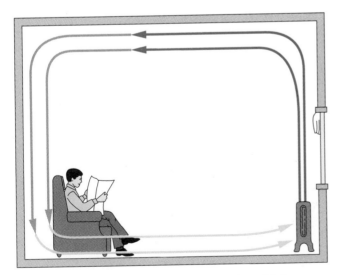

Figure 9.20 **Convection is the process by which thermal energy is transferred by motion of a heated fluid. This occurs in a room when heated air flows from a radiator or vent.**

Figure 9.21 **Radiation is the only process that can carry thermal energy across the evacuated space in a thermos bottle. Silvering the walls of the bottle reduces the radiation flow.**

minimum by silvering the facing walls of the evacuated space. Silvering causes the electromagnetic waves to be reflected rather than absorbed and therefore reduces the energy flow.

The same principle is involved in the use of foil-backed insulation. Some of the heat flow across the dead-air space within a wall is due to radiation and some to conduction through the air. Using foil-backed insulation can reduce the portion due to radiation. Thin sheets of foil-backed insulation have *R* values that are equivalent to those of somewhat thicker sheets of insulation without the foil backing.

Radiation is also involved in the loss of energy from roofs or from road surfaces to the dark (and cold) night sky. Black-top road surfaces in particular (and black surfaces in general) are very effective radiators of electromagnetic waves. The road surface therefore radiates energy to the sky, thus cooling the surface more rapidly than the surroundings. Because of this effect, the surface of a blacktop road can drop below the freezing point, creating icy conditions even when the air temperature is still above freezing.

Good absorbers of electromagnetic radiation are also good emitters, and black materials in general absorb more solar

Box 9.2

Everyday Phenomenon:
Solar Collectors and the Greenhouse Effect

The Situation. Flat-plate solar collectors are frequently used to collect solar energy for use in heating water or a home. The flat-plate collector consists of a metal plate with tubes bonded to its surface that carry water. The plate and tubes are painted black and are contained in a frame that is insulated below the plate and has a glass or transparent plastic cover on top. The challenge is to explain how the flat-plate collector works. What heat flow processes are involved? Secondly, what relationship does the solar collector have, if any, to the much-discussed greenhouse effect. This effect is thought to be causing a slow warming of the earth, which could eventually lead to the partial melting of the ice caps and coastal flooding.

The Analysis. The flat-plate collector receives energy from the sun in the form of electromagnetic radiation, which is the only form of heat flow that can take place across the vacuum of space. The electromagnetic waves that are emitted by the sun are primarily in the visible part of the electromagnetic spectrum, the wavelengths to which our eyes are sensitive. These waves pass easily through the transparent cover of the collector.

The black surface of the collector plate absorbs much of the visible light, reflecting very little. This absorbed energy raises the temperature of the plate. Water traveling through the tubes bonded to the plate is, in turn, heated by conduction; the plate must be at a higher temperature than the water in order for this heat flow to take place. The insulation below the plate reduces heat flow to the surroundings via conduction.

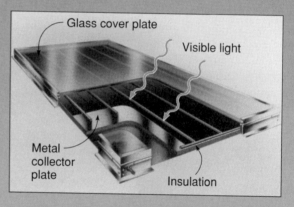

The black-surfaced metal collecting plate in a flat-plate solar collector is insulated below and covered with a glass or transparent plastic plate above.

Because the metal plate is warmer than its surroundings, it radiates heat in the form of infrared electromagnetic waves. It turns out, however, that glass is opaque to infrared radiation; the infrared waves cannot pass through the glass cover. The cover therefore reduces heat loss from radiation, as well as from convection, since it keeps air currents from passing directly over the metal plate. The glass enclosure of a greenhouse has essentially the same purpose. In addition to reducing convection losses, the glass allows visible sunlight in, but prevents heat loss from radiation at longer wavelengths; it is an energy trap! This one-way transmission of radiation at visible energy in the summer and lose more heat in the winter. A black roof may look nice, but it is not the best color from the perspective of energy conservation; a lighter color is more effective both in keeping your home cool in the summer and warm in the winter.

Knowledge of the basic mechanisms of heat transfer is very useful in designing a house. Glass is a good conductor of heat, so large windows represent areas of high heat loss, even when double-pane glass is used. If the window is on the south side of the house, however, this heat loss may be somewhat offset by the heat gain from the sun. Convection and radiation also play important roles, as already indicated. All three of these processes are important in understanding how to make the best use of solar energy. Some of these ideas are discussed in box 9.2.

wavelengths but not at the longer infrared wavelengths is therefore known as the *greenhouse effect.*

The earth itself can also be thought of as a large greenhouse. In this case, the substance that is transparent to visible radiation but not to infrared radiation is carbon dioxide gas (CO_2) as well as some other gases present in the atmosphere in smaller quantities. Carbon dioxide is produced as a natural by-product in the combustion (burning) of any carbon-based fuel such as oil, natural gas, coal, or wood. It is taken up by plant life and converted to other carbon compounds, so that the global plant cover is also a factor in determining the amount of carbon dioxide in the atmosphere.

The recent concern of scientists has been that our heavy use of fossil fuels such as oil, natural gas, and coal may be increasing the amount of carbon dioxide in the upper atmosphere. If this is so, and some measurements seem to support this concern, then the greenhouse effect attributable to the carbon dioxide should be letting less heat escape by radiation, thus slowly increasing the earth's temperature. Reduction in global forest cover and other forms of plant life may also contribute to the carbon dioxide buildup.

The flat-plate collector and the greenhouse (or your car on a summer day) are small-scale models, therefore, of what happens on a much larger scale in the upper atmosphere of the earth. If the earth's temperature is indeed slowly increasing, the polar ice caps will begin to recede. This in turn will put more water in the oceans, with the possible consequence that low-lying coastal regions may be flooded. Changes in global weather phenomena could also effect crop production.

The only solutions to this problem would be to reduce our dependence upon the burning of fossil fuels or to somehow increase the plant cover on the earth's surface. It obviously becomes important for us to be able to predict the potential size of the effect and its consequences in order to be able to develop a rational environmental policy. Any changes that we decide are necessary will surely take time to implement and take effect.

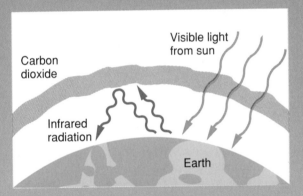

Carbon dioxide and other gases in the atmosphere of the earth play the same role as the glass cover on a solar collector; visible light passes through, but the infrared radiation does not.

SUMMARY

The ideas in chapter 9 are centered upon the nature of heat, which is described, in part, by the first law of thermodynamics. We also considered the effects of heat in changing the temperature or phase of a substance and the mechanisms of heat flow between objects. The distinction between the concepts of heat and temperature is very important to a good understanding of these ideas. The following concepts were discussed.

Temperature. Temperature scales have been devised to provide a consistent means of describing how hot or cold an object is. When objects are at the same temperature, no heat flows between them, and their physical properties remain constant.

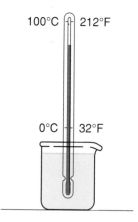

Heat and Heat Capacity. When objects at different temperatures are placed in contact, energy will flow from the warmer to the cooler object. This flow of energy is called *heat*, and the amount of heat required to change the temperature of one unit of mass by one degree is called the *specific heat capacity*.

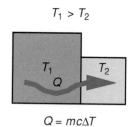

$$Q = mc\Delta T$$

Latent Heat. Latent heat is the amount of heat per unit mass that is required to change the phase of a substance without changing the temperature. Adding heat will cause the substance to melt or vaporize; removing heat will produce freezing or condensation.

$$Q = mL$$

First Law of Thermodynamics. The internal energy of a substance or system can be increased either by adding heat or by doing work on the system. An increase in internal energy may show up as an increase in temperature, a change of phase, or as changes in other properties of the system.

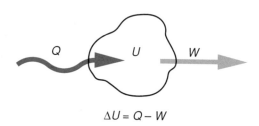

$$\Delta U = Q - W$$

Gas Behavior. For an ideal gas (one at high temperatures and low pressures), the internal energy varies directly with the temperature of the gas. Work done by the gas is proportional to the change in volume. The first law can therefore be used to predict many features of gas behavior.

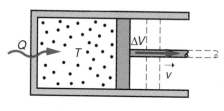

$$\Delta U = Q - P\Delta V \Rightarrow \Delta T$$

Heat Flow. The three mechanisms by which heat flows from one system to another are conduction, the direct movement of heat between objects in contact; convection, the flow of heat carried by a moving fluid; and radiation, the flow of heat via electromagnetic waves. All three are important in discussing heat loss from a house.

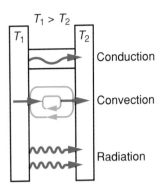

QUESTIONS

Q9.1 Is an object that has a temperature of 0° C hotter than, colder than, or at the same temperature as one that has a temperature of 0° F? Explain.

Q9.2 Which represents a greater change in temperature, a change of 10 Fahrenheit degrees or a change of 10 Celsius degrees? Explain.

Q9.3 Is it possible for a temperature to be lower than 0° C? Explain.

Q9.4 Is it possible for a temperature to be lower than 0 K on the Kelvin temperature scale? Explain.

Q9.5 Two objects at different temperatures are placed in contact with one another but are insulated from the surroundings. Will the temperature of either object change? Explain.

Q9.6 Is it possible for the final temperature of either of the two objects discussed in question 9.5 to be greater than the higher of the two initial temperatures? Explain.

Q9.7 Two objects of the same mass but of different materials are initially at the same temperature. Equal amounts of heat are added to each object. Will the final temperature of the two objects necessarily be the same? Explain.

Q9.8 Is it possible to add heat to a substance without changing its temperature? Explain.

Q9.9 What happens if we add heat to water that is at the temperature of 100° C? Does the temperature change while the water is boiling? Explain.

Q9.10 What happens if we remove heat from water at 0° C? Does the temperature change as the water freezes? Explain.

Q9.11 Is it possible to change the temperature of a glass of water by stirring the water, even though the glass is insulated from its surroundings? Explain.

Q9.12 Is it possible to change the temperature of a system that is insulated from its surroundings so that no heat can flow into or out of the system? Explain.

Q9.13 Suppose that the internal energy of a system has been increased, resulting in an increase in its temperature. Is it possible to tell from the final state of the system whether the change in internal energy was due to the addition of heat or of work to the system? Explain.

Q9.14 Heat is added to an ideal gas, and the gas expands in the process. Is it possible for the temperature to remain constant in this situation? Explain.

Q9.15 Heat is added to an ideal gas at constant volume. Is it possible for the temperature of the gas to remain constant in this process? Explain.

Q9.16 Heat is added to ice, causing the ice to melt but producing no change in temperature. The volume of the water obtained from the melting ice is less than the initial volume of ice. Does the internal energy of the ice/water system change in this process? Explain.

Q9.17 A block of wood and a block of metal have been sitting on a table for a long time. The block of metal feels colder to the touch than the block of wood. Does this mean that the metal is actually at a lower temperature than the wood? Explain.

Q9.18 Heat is sometimes lost from a house through cracks around windows and doors. What mechanism of heat transfer (conduction, convection, or radiation) is involved in this process? Explain.

Q9.19 Is it possible for water on the surface of a road to freeze even though the temperature of the air just above the road is above 0° C? Explain.

Q9.20 What heat-transfer mechanism (conduction, convection, or radiation) is involved when heat flows through a glass window pane? Explain.

Q9.21 Is it possible for heat to flow across a vacuum? Explain.

Q9.22 What property does glass share with carbon dioxide gas that makes them both effective in producing the greenhouse effect? Explain.

Q9.23 What mechanism of heat flow (conduction, convection, or radiation) is involved in transporting useful thermal energy from a flat-plate solar collector to its point of application? Explain.

Q9.24 Will a solar power plant (one that generates electricity from solar energy) have the same tendency to increase the greenhouse effect in the atmosphere as a coal-fired power plant? Explain.

EXERCISES

E9.1 An object has a temperature of 50° C. What is its temperature in degrees Fahrenheit?

E9.2 The temperature on a winter day is 20° F. What is the temperature in degrees Celsius?

E9.3 The temperature in a dormitory room is 20° C. What is the temperature of the room on the absolute (Kelvin) temperature scale?

E9.4 The temperature on a hot summer day is 95° F. What is this temperature:

   a. in ° Celsius?

   b. on the absolute (Kelvin) scale?

E9.5 How much heat is required to raise the temperature of 50 grams of water from 20° C to 70° C?

E9.6 How much heat must be removed from a 200-g block of copper to lower its temperature from 150° C to 20° C? The specific heat capacity of copper is 0.093 cal/g·C°.

E9.7 How much heat must be added to 200 g of ice at 0° C in order to completely melt the ice producing 200 g of water at 0° C?

E9.8 How much heat must be added to 200 g of water initially at 60° C to

   a. heat it to the boiling point?

   b. completely convert it to steam?

E9.9 If 600 cal of heat are added to a system, how much energy has been added in joules?

E9.10 Suppose that 1200 J of heat are added to 200 g of water initially at 20° C.

   a. How much energy does this represent in calories?

   b. What is the final temperature of the water?

E9.11 The volume of an ideal gas is increased from 1 m^3 to 3 m^3 while maintaining a constant pressure of 1000 N/m^2.

   a. How much work is done by the gas in this expansion process?

   b. If no heat has been added, what is the change in the internal energy of the gas?

E9.12 Suppose that 800 J of heat is added to an ideal gas while the gas does 1400 J of work on its surroundings.

   a. What is the change in internal energy of the gas?

   b. Does the temperature of the gas increase or decrease?

E9.13 Work of 1200 J is done on an ideal gas, but the internal energy increases by only 500 J. What is the amount and direction of heat flow into or out of the system?

E9.14 Work of 600 J is done by stirring an insulated beaker containing 300 g of water.

   a. What is the change in internal energy of the system?

   b. What is the change in temperature of the water?

CHALLENGE PROBLEMS

CP9.1 Heat is added to an object initially at 25° C increasing its temperature to 70° C.

   a. What is the temperature change of the object in Celsius degrees?

   b. What is the temperature change of the object in Fahrenheit degrees?

   c. What is the temperature change of the object in kelvins?

   d. Is there any difference in the numerical value of a heat capacity expressed in cal/g·C° from one expressed in cal/g·K? Explain.

CP9.2 A student constructs a thermometer and invents her own temperature scale such that the ice point of water is 0° S (S = Student) and the boiling point of water is 20° S. She measures the temperature of a beaker of water with her thermometer and finds it to be 15° S.

a. What is the temperature of the water in ° C?

b. What is the temperature of the water in ° F?

c. What is the temperature of the water in kelvins?

d. Is the temperature range spanned by the one Student degree larger or smaller than that spanned by one Celsius degree? Explain.

CP9.3 50 grams of ice are initially at a temperature of –10° C. The specific heat capacity of ice is 0.5 cal/g·C° and that of water is 1.0 cal/g·C°. The latent heat of fusion of water is 80 cal/g.

a. How much heat is required to raise the ice to 0° C and completely melt the ice?

b. How much additional heat is required to heat the water (obtained by melting the ice) to 25° C?

c. What is the total heat that must be added to convert ice at –10° C to water at +25° C?

CP9.4 Suppose that 200 g of a certain metal, initially at 120° C are dropped into an insulated beaker containing 100 g of water initially at 20° C. The final temperature of the metal and water in the beaker is found to be 35° C. Assume that the heat capacity of the beaker can be ignored.

a. How much heat has been transferred from the metal to the water?

b. Given the temperature change and mass of the metal, what must be the specific heat capacity of the metal?

c. How much of this same metal at the same initial temperature would we need in order for the final temperature of the 100 g of water and the metal to be 70° C (equal temperature changes for the water and metal)?

CP9.5 A beaker containing 300 g of water has 900 J of work done upon it by stirring and 200 cal of heat added to it from a hot plate.

a. What is the change in internal energy of the water in joules?

b. What is the change in internal energy of the water in calories?

c. What is the temperature change of the water?

d. Would your answers to the first three parts differ if there had been 200 J of work done and 900 cal of heat added? Explain.

HOME EXPERIMENTS AND OBSERVATIONS

HE9.1 Look around your home or dormitory to see what kinds of thermometers you can find; indoor, outdoor, clinical, or cooking thermometers can be found in most homes. For each thermometer that you find, note the following:

a. What temperature scale is used?

b. What is the temperature range of the thermometer?

c. What is the smallest temperature change that can be read on each thermometer?

d. What changing physical property is used to indicate change in temperature for each thermometer?

HE9.2 Take two styrofoam cups, partially fill them with equal amounts of cold water, and then drop equal amounts of ice into each cup so that each cup contains approximately the same mixture of ice and water. Stir each cup lightly for a minute, so that both come to thermal equilibrium.

a. After thermal equilibrium has been reached, stir the water and ice mixture in one of the cups vigorously with a nonmetallic stirrer until all of the ice has been melted. Note the time that it takes to melt the ice.

b. Set the second cup aside and observe it every fifteen minutes or so until all the ice has melted. Note the time that it takes for the ice to melt.

c. Where is the energy coming from to melt the ice in each cup? Can you estimate your stirring power?

HE9.3 Fill a styrofoam cup with very hot water (or coffee). Take objects made of different materials, such as a metal spoon, a wooden pencil, a plastic pen, or a glass rod. Put one end of each object into the water and make the following observations:

a. How close to the surface of the water do you need to grasp the object in order for it to feel noticeably warm to the touch?

b. From your observations, which material would you judge to be the best conductor of heat and which the worst? Can you order the materials according to their ability to conduct heat?

HE9.4 Light a candle and move it to different parts of your room, observing the direction that the candle flame moves. Make observations near the floor as well as near the ceiling. These observations are most interesting in the winter when the heat is on.

a. Can you make a rough map of the air flow patterns in the room?

b. What are the heat sources in the room, if any? What are the heat convection patterns related to these sources?

10 Heat Engines and the Second Law of Thermodynamics

Many of us have spent a good portion of our lives either driving or riding in automobiles. We speak knowledgeably about how many miles per gallon we get and, in some cases, even about such things as fuel injection or compression ratios. We regularly drive into gas stations to buy unleaded gasoline, and we generally have a good idea of what this fuel costs at different places (fig. 10.1).

How many of us, though, as we turn the ignition switch and depress the gas pedal, have any real understanding of what is going on inside the engine? The engine roars into action, consumes fuel, and powers the car at our command without requiring a detailed knowledge of its principles of operation. Ignorance may be bliss until something goes wrong—and then some understanding may or may not be useful. What does an automobile engine do, and how does it work?

The internal combustion engine used in all modern automobiles is a heat engine. Steam engines were the earliest practical versions of heat engines, and it was an attempt to better understand the operating principles of steam engines that led to the development of the science of thermodynamics during the nineteenth century. Building more efficient engines was the primary objective. Although steam engines are usually external combustion engines, in that the fuel is burned outside of the engine rather than internally, the fundamental principles of operation of steam engines and automobile engines are the same; they are both heat engines.

What, then, are the principles of operation of heat engines? What factors determine the efficiency of a heat engine, and how can this efficiency be maximized? The second law of thermodynamics plays a central role in any discussion of these questions. An examination of the nature of heat engines will lead us to an exploration of the second law and the related concept of entropy.

Figure 10.1   Filling up the family chariot. How does the engine use this fuel to produce motion?

Chapter Objectives

After an initial discussion of the nature of heat engines, this chapter introduces the second law of thermodynamics. Obtaining an understanding of the second law is the primary objective of the chapter. The second law is central to an understanding of the efficiency of heat engines as well as to a general appreciation of the nature of heat energy. Using the second law and the first law, introduced in chapter 9, we can consider questions related to the use of energy resources in today's global economy.

The questions raised in this chapter and the explanations provided have a fundamental significance with regard to the problems facing modern society. The quality of our environment and the health of our economy depend upon intelligent choices on energy issues. Obtaining a general understanding of the laws of thermodynamics and the nature of heat is a critical first step in that process.

Chapter Outline

1. *Heat engines.* What is a heat engine? How does the first law of thermodynamics govern the quantities of heat and work, that flow into or out of a heat engine?

2. *The second law of thermodynamics.* How would an ideal heat engine operate if we could build one? How is the concept of an ideal engine related to the second law of thermodynamics, and what does the second law say and mean?

3. *Refrigerators, heat pumps, and entropy.* What does a refrigerator or heat pump do? What are the reasons for the limitations placed upon the use of heat by the second law of thermodynamics? What is entropy, and how are these ideas related to the concept of entropy?

4. *Thermal power plants and energy resources.* What are the implications of the second law of thermodynamics for our utilization of resources such as coal, oil, natural gas, uranium, geothermal resources, and solar energy? What do we really mean when we talk about conserving energy resources?

5. *Perpetual motion and energy frauds.* Is perpetual motion possible within the framework of the laws of thermodynamics? How can we judge the claims of engine inventors?

10.1 HEAT ENGINES

Place yourself in the driver's seat, once again, and think about what the internal combustion engine in your car does. We have said that both steam engines and gasoline engines are heat engines. What does this mean? What does a heat engine do? Can we develop a model of what a heat engine accomplishes that can be applied to any heat engine?

It is not difficult to produce a basic description of what our gasoline engine does. Fuel in the form of gasoline is mixed with air and introduced into a chamber of the engine which, because of its shape, we call a cylinder. A spark produced by a spark plug ignites the gas/air mixture and it burns explosively. Heat is released from the fuel as it burns, and some of this heat energy causes the gases in the cylinder to expand, thus doing work on the piston, as discussed in chapter 9. Through various mechanical connections, this work is transferred to the drive shaft and ultimately to the drive wheels of the car (fig. 10.2). By pushing against the road surface, the wheels convert it to work done to move the car.

We see, then, that the gasoline engine partially converts heat obtained from burning the fuel to work done to move the car. If you held your hand near the tail pipe, you can also feel that some heat is released to the environment in the form of warm exhaust gases. In other words, not all of the heat obtained from burning the fuel is converted to work.

This simple description includes features that are common to any heat engine. Thermal energy (heat) is introduced into the engine in some manner; some of this energy is converted to mechanical work, and some heat is released to the environment at a temperature lower than that of the input heat. Figure 10.3 presents these ideas in a schematic form. The circle represents the heat engine, the box at the top represents the high-temperature source of heat, and the box at the bottom represents the lower-temperature environment to which the waste heat is released.

Efficiency of an Engine

The main question, of course, is how much useful mechanical work can such an engine produce for a given amount of energy input in the form of heat? The concept of efficiency is introduced to address this question. *Efficiency* is defined as the ratio of the work done by the engine to the amount of heat that must be supplied to accomplish this work; in symbols,

$$e = \frac{W}{Q_H},$$

where e is the efficiency, W is the work done by the engine, and Q_H is the amount of heat taken in by the engine from the high-temperature source, or *heat reservoir*. By the sign convention introduced in the last chapter, W is positive, since it represents work done *by* the engine on the surroundings.

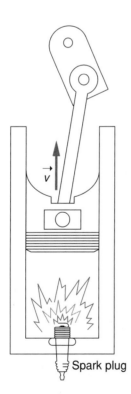

Figure 10.2 Thermal energy released by burning gasoline in the cylinder of an automobile engine causes the piston to move, converting some of the thermal energy to work.

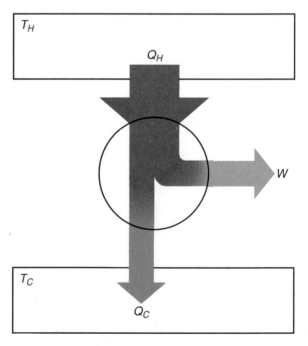

Figure 10.3 A schematic representation of a heat engine. Heat is taken in at a high temperature, T_H, some is converted to work, and the remainder is rejected at a lower temperature, T_C.

Suppose, for example, that we have an engine that takes in 1200 joules of heat from the high-temperature heat source in each cycle and does 400 joules of work per cycle. The efficiency of this engine would be

$$e = \frac{W}{Q_H} = \frac{400 \text{ J}}{1200 \text{ J}} = \frac{1}{3}.$$

We would usually state this efficiency as the decimal fraction 0.333 or 33.3%. Such an efficiency would be greater than that of most automobile engines, but less than that achieved by the steam turbines powered by coal or oil that are used in many electric-power plants.

Heat Engines and the First Law

Notice that we have used heat and work values for one complete cycle in computing this efficiency. An engine usually functions in repetitive cycles: it regularly returns to its initial state and then repeats the same process over and over. It is necessary to use one complete cycle or an average over several complete cycles in computing efficiency because heat and work input and output occur at different points within the cycle. Since the engine returns to its initial state at the end of one cycle, its internal energy at the end of the cycle has the same value that it did at the beginning. The change in the internal energy of the engine for one complete cycle is therefore zero.

The first law of thermodynamics can therefore tell us something about the behavior of heat engines. Since the change in internal energy is zero, the first law becomes

$$\Delta U = Q - W = 0,$$

or

$$Q = W.$$

This says that the net heat flowing into or out of the engine per cycle must equal the work done by the engine. The net heat, Q, represents the difference between the heat input from the high-temperature source, Q_H, and that released, to the lower-temperature environment, Q_C. (The subscripts H and C stand for *hot* and *cold*.)

$$Q = Q_H - Q_C = W.$$

The first law thus tells us how much heat is emitted in the form of exhaust in each cycle. In our earlier example, 400 J of work was done from a heat input of 1200 J from the high-temperature source. There must then be 800 J of work released to the environment in each cycle, so that the net heat, 1200 J – 800 J = 400 J, is equal to the amount of work done by the engine. Energy is conserved, in other words; the net heat input equals the work done.

The schematic representation of the engine in figure 10.3 uses the width of the arrows to represent the quantities of heat and work involved. This helps us to form a mental picture of what is happening. Figure 10.4 shows the arrows

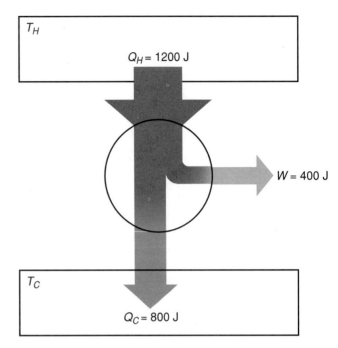

Figure 10.4 **The arrow widths in this diagram represent the quantities of energy involved in the example discussed in the text.**

drawn to represent the quantities in our example; 1200 J of heat are taken in, 400 J ($\frac{1}{3}$ of the total) are converted to work, and 800 J ($\frac{2}{3}$ of the total) of heat are released to the lower-temperature reservoir. The arrow representing the emitted heat is therefore twice as wide as that representing the work, and the combined width of these two arrows equals the width of the arrow representing the heat taken in by the engine.

Automobile engines, diesel engines, jet engines, and the steam turbines used in power plants represent different types of heat engines. You could create a simple steam turbine by placing a pinwheel in front of the steam spout of a tea kettle, as shown in figure 10.5. Although the engine boundaries are not well defined in this case, the heat input is clearly in the form of the hot steam emerging from the spout, which ultimately comes from the power source of your stove or hot plate. Some of this heat is converted to work in turning the pinwheel. The rest of the heat is released to the room and is not utilized except, perhaps, to warm the room slightly. If the shaft of the pinwheel had a string or thread wrapped around it, the work done on the pinwheel could be used to lift a small weight.

This simple steam turbine would not be very powerful nor very efficient. Designing better engines with both high power (rate of doing work) and high efficiency (fraction of input heat converted to work) has been a goal of mechanical engineers for the last two hundred years. Gaining an understanding of the factors that affect efficiency has been an important part of this quest.

Figure 10.5 **Steam emerging from the tea kettle causes the pinwheel to turn in this simple steam turbine. Work could be done to lift a small weight with such an engine.**

10.2 THE SECOND LAW OF THERMODYNAMICS

Some heat engines, automobile or otherwise, are much more efficient than others. What are the basic factors that determine this efficiency? This question is as important for automotive engineers in Detroit, Japan, or Germany, or for designers of modern power plants, as it was for the early steam engine designers.

The Carnot Engine

One of the earliest engineers to be intrigued by these questions was a young French engineer named Sadi Carnot (1796–1832). Carnot was inspired, in part, by the work of his father, also an engineer, who had studied and written about the optimal design of water wheels, which were a competing source of mechanical power. Carnot made a rough analogy between the pressure difference, or *head,* that was important to the operation of a water wheel and the temperature difference that seemed to be important to the operation of steam engines.

Carnot attempted to define what he thought should be an ideal heat engine. He reasoned that the greatest efficiency could be obtained by taking in all of the input heat at a single high temperature and emitting all of the unused heat at a single low temperature. This would have the effect of maximizing the effective temperature difference within the limits provided by the temperatures of the heat source and the environment. It was not unlike the idea of bringing all of the water in at the highest point of a water wheel and releasing it at the lowest point to maximize the efficiency of the water wheel. The water-wheel analogy was a very important part of Carnot's thinking.

The other requirement that Carnot imposed upon his ideal engine was that all of the processes should occur without

undue turbulence or disequilibrium. This also paralleled his father's ideas on water wheels. In the case of the ideal heat engine, this meant that the working fluid of the engine (steam or whatever else might be used) should be approximately in thermal equilibrium at all points in the cycle. In practical terms, this condition requires that there be no rapid fluctuations in temperature or pressure within the engine. This also implied that the engine would be completely reversible; that is, it could be turned around and run the other way at any point in the cycle because it was always in a near-equilibrium state. This was the ideal; a real engine might depart considerably from these conditions.

At the time that Carnot published his paper on the ideal heat engine (1824), the energy aspects of heat and the first law of thermodynamics were not yet understood. Carnot pictured caloric as flowing through his engine as water flowed through a water wheel. It was not until after the development of the first law around 1850, that the full impact of Carnot's ideas became apparent. It was then clear that heat does not simply flow through the engine; some of it is converted to mechanical energy in the form of work done by the engine.

The cycle pictured by Carnot for his ideal engine was a very simple one, as illustrated in figure 10.6. Imagine a gas or some other fluid contained in a cylinder with a movable piston. In step 1 of the cycle heat is allowed to flow into the cylinder at a single high temperature T_H. The gas or fluid expands during this process doing work on the piston. This process is referred to as an *isothermal* (constant temperature) expansion and is the energy-input step in the cycle. In step 2, the fluid continues to expand, but no heat is allowed to flow between the cylinder and its surroundings. This process is called an *adiabatic* (no heat flow) expansion.

Step 3 is an isothermal compression in which work is done on the fluid by the piston to compress the fluid. This is the exhaust step, since heat, Q_C, flows out of the fluid at a single low temperature, T_C. The final step, step 4, returns the fluid to its initial condition by an additional compression process done adiabatically. All four of these steps must be done slowly, so that the fluid is in approximate equilibrium at all times. The complete process is then said to be *reversible,* which, as mentioned earlier, is a crucial feature of the Carnot engine.

When the fluid is expanding during steps 1 and 2, it is doing positive work on the piston that can be transmitted by appropriate mechanical linkages for external use. During steps 3 and 4, the fluid is being compressed, which requires that work be done on the engine by external forces. The amount of work that must be added in steps 3 and 4 is less than the amount done by the engine in steps 1 and 2, however, so that the engine does do a net amount of work on the surroundings.

Following the development of the first law of thermodynamics, it became possible to compute the efficiency of the Carnot cycle by assuming that the working fluid was an ideal gas. The process is straightforward; it involves computing

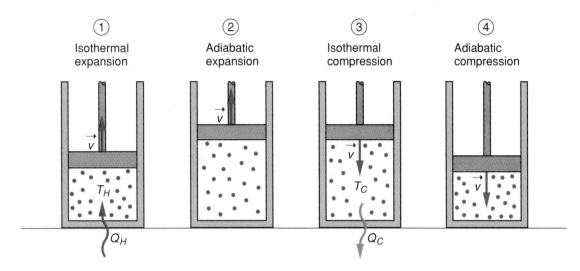

Figure 10.6   The Carnot cycle. Step 1: isothermal expansion; Step 2: adiabatic expansion; Step 3: isothermal compression; Step 4: adiabatic compression.

the work done on or by the gas in each step and the quantities of heat that must be added and removed in steps 1 and 3. Although the full analysis is beyond the scope of this book, a relatively simple expression emerges when we put the values of heat and work into the definition of efficiency provided in the previous section:

$$e_c = \frac{W}{Q_H} = \frac{T_H - T_C}{T_H}.$$

Here e_c is the Carnot efficiency, and T_H and T_C are the *absolute* temperatures of the high- and low-temperature reservoirs between which the engine is operating.

If, for example, a Carnot engine is able to take in heat at a temperature of 600 K (327° C) and rejects heat to the environment at 300 K (27° C), the efficiency of this engine would be

$$e_c = \frac{T_H - T_C}{T_H} = \frac{600 \text{ K} - 300 \text{ K}}{600 \text{ K}} = 0.5.$$

In other words, the efficiency of this engine would be 50%. According to Carnot's ideas, this would be the maximum efficiency possible for any engine operating between these two temperatures. See box 10.1 for another illustration of these ideas.

The Second Law of Thermodynamics

The absolute temperature scale was developed in the 1850s by William Thomson (Lord Kelvin) in England. As discussed in the previous chapter, the Kelvin scale emerged from considerations involving the behavior of an ideal gas. Kelvin was also aware of Joule's experiments and had a hand in the statement of the first law of thermodynamics. He was therefore in an excellent position to take a fresh look at Carnot's ideas on heat engines.

Using Carnot's ideas together with the new recognition that heat flow represented the transfer of energy, Kelvin stated what he took to be a general principle or law of nature. This principle, which we now call the *second law of thermodynamics,* is usually stated as follows:

No engine, working in a continuous cycle, can take heat from a reservoir at a single temperature and convert that heat completely to work.

In other words, it is not possible for any heat engine to have an efficiency of 1.0 or 100%.

Using this statement of the second law, it is possible to prove that no engine operating between two given temperatures can have a greater efficiency than a Carnot engine operating between the same two temperatures. This proof justifies Carnot's contention that the Carnot engine represents the best that can be achieved, and it involves the reversible feature of the Carnot cycle. If we assume that some engine could have a greater efficiency than a Carnot engine operating between the same two reservoirs, we could then use the Carnot engine run in reverse to pump the exhaust heat back to the original reservoir. The net effect would be to take heat from the high-temperature reservoir and convert it completely to work, which violates the Kelvin statement of the second law.

This argument is illustrated in figure 10.7. The engine whose efficiency is assumed to be greater than the Carnot engine would have to produce a greater amount of work than the Carnot engine for the same amount of heat, Q_H, taken from the high-temperature reservoir. Some of this work could then be used to run the Carnot engine in reverse, thus returning the heat rejected by the first engine, Q_C, to the high-temperature reservoir. The quantities of heat and work involved for the Carnot engine are the same as if it were being run in the forward direction; we merely reverse the directions of the arrows.

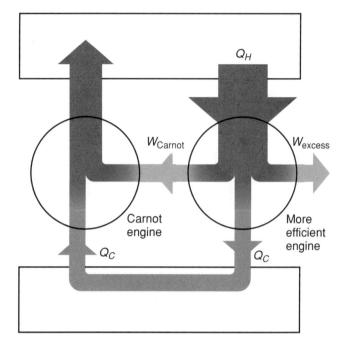

Figure 10.7 **A diagram showing that no engine can have a greater efficiency than a Carnot engine operating between the same two temperatures. If this were possible, a violation of the second law of thermodynamics would occur.**

Box 10.1

Sample Exercise

Suppose that a steam turbine takes in steam at a temperature of 400° C and rejects steam at a temperature of 120° C.

a. What is the Carnot efficiency for this engine?
b. If the turbine takes in 500 kJ of heat in each cycle, what is the maximum amount of work that could be generated by the turbine in each cycle?

$T_H = 400°$ C.
$T_C = 120°$ C.
$Q_H = 500$ kJ.

a. $e_c = ?$
Since the temperatures are given in ° C, they must be converted to absolute temperatures (in kelvin):

$$T_H = 400° \text{ C} + 273° \text{ C} = 673 \text{ K};$$
$$T_C = 120° \text{ C} + 273° \text{ C} = 393 \text{ K}.$$

$$e_c = \frac{T_H - T_C}{T_H}$$

$$= \frac{673 \text{ K} - 393 \text{ K}}{673 \text{ K}}$$

$$= \frac{280 \text{ K}}{673 \text{ K}} = \textbf{0.416} \text{ (41.6\%).}$$

b. Maximum possible $W = ?$

$$e = \frac{W}{Q_H}.$$

Multiplying both sides by Q_H, we obtain

$$W = eQ_H = (0.416)\,(500 \text{ kJ}) = \textbf{208 kJ.}$$

(Since e_c represents the maximum possible efficiency for these temperatures of operation, using e_c for the efficiency yields the maximum possible work for this engine.)

The Carnot efficiency for two given temperatures is therefore the maximum possible efficiency for any heat engine operating between these two temperatures, as stated earlier. Since the proof of this fact follows from the second law of thermodynamics, the Carnot efficiency is sometimes referred to as the *second-law efficiency.* It is the theoretical limit to the efficiency of any engine operating between the specified temperatures.

The second law of thermodynamics itself cannot be proved. It is a law of nature, which, as near as we know, cannot be violated. It sets a limit upon what can be achieved with energy in the form of heat; time has shown this to be an accurate statement of what is possible. The second law is consistent with everything that we know about heat transfer, heat engines, refrigerators, and many other phenomena.

The remaining work would be available for external use, and no heat ends up in the lower-temperature reservoir. The net effect of the two engines operating in tandem is to take a small quantity of heat from the high-temperature reservoir and convert it completely to work! This violates the Kelvin statement of the second law of thermodynamics. If you accept the Kelvin statement, then it also follows that no engine can have a greater efficiency than a Carnot engine operating between these two reservoirs.

10.3 REFRIGERATORS, HEAT PUMPS, AND ENTROPY

In the previous section, we talked about running a Carnot engine in reverse. What possible advantage could there be to running something as useful as a heat engine backwards? We ended up using work to pump heat from a colder to a hotter reservoir. Is this what a refrigerator does? What relationship, if any, exists between heat engines and refrigerators?

Refrigerators and Heat Pumps

Figure 10.8 shows a schematic diagram of a heat engine run in reverse. It is the same diagram used for heat engines, but the directions of the arrows representing flow of energy in the form of heat and work are reversed. Work W is done on the engine, heat Q_C is removed from the lower-temperature reservoir, and a greater quantity of heat, Q_H is released to the higher temperature reservoir. The first law of thermodynamics requires that the magnitude of the heat released at the higher temperature equal the energy that was put into the engine in the form of heat and work, or

$$|Q_H| = Q_C + |W| *$$

for one complete cycle. This assumes, again, that the engine returns to its initial condition at the end of each cycle and thus the internal energy of the engine does not change.

The net effect of this process is that heat is moved from a colder reservoir to a warmer reservoir by means of work from some external source that runs the engine in reverse. A device that does this is called a *heat pump* or a *refrigerator*. The two terms really mean the same thing, but they are used in different contexts depending upon the purpose.

The term *refrigerator* requires little explanation; we are all familiar with refrigerators, even if we do not always think much about what they are actually doing. A refrigerator keeps foods cold by pumping heat out of the colder interior into the warmer room (fig. 10.9). An electric motor or gas-powered engine is required to provide the necessary work. A refrigerator also warms the room, as you can easily verify by holding your hand near the coils on the back of the refrigerator when it is running.

The term *heat pump* is generally used in the context of heating a building by pumping heat from the colder outdoors to the warmer interior (fig. 10.10). The external source of work is usually an electric motor. Energy is provided in the form of work, but a greater amount of heat energy is available to heat the house because Q_H is equal in magnitude to the sum of work W and heat Q_C removed from the outside air. Thus, a heat pump can provide us with a greater amount of heat than we would obtain by directly converting the electrical energy to heat, as is done in an electric furnace.

Heat pumps are most effective for heating houses in climates where the difference between the outside temperature and inside temperature is not too large. Locations where the winters are mild, such as the Southeast or the Pacific Northwest, are therefore the most favorable. Two sets of coils are used for heat exchange; one sits outside and removes heat from the outside air, and the other is inside the building and gives up heat to the inside air (fig. 10.11). Many heat pumps are built to pump heat in either direction and can therefore be used as air conditioners in the summer.

* We have used absolute-value symbols here because Q_H and W are both negative quantities by our first-law sign convention.

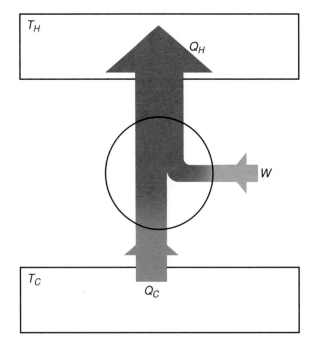

Figure 10.8 A schematic diagram of a heat engine run in reverse. Run in this manner, the engine becomes a heat pump or refrigerator.

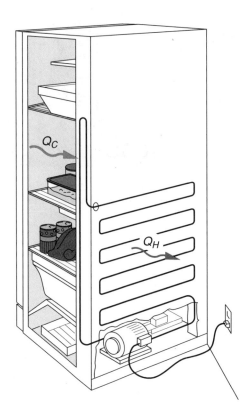

Figure 10.9 A refrigerator pumps heat from the colder interior of the refrigerator to the warmer room. The heat exchange coils that release heat to the room are usually on the back side of the refrigerator.

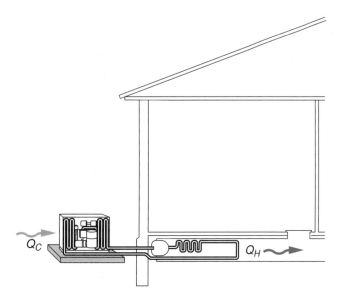

Figure 10.10 A heat pump removes heat from the outside air and pumps it into the warmer house.

Figure 10.11 The exterior heat-exchange coils for an air-to-air heat pump are usually located on a concrete pad behind the house or building.

A heat pump can usually deliver an amount of heat to the inside of a building that is two to three times the amount of electrical energy supplied as work. Again, the first law of thermodynamics is not violated, since the extra energy is supplied from the outside air. The work supplied to the heat pump allows us to move thermal energy in the direction opposite to its natural tendency, much as a water pump allows us to move water uphill.

The Clausius Statement of the Second Law

Heat normally flows, of course, from hotter to colder objects. This natural tendency of heat flow, in fact, is the basis for another statement of the second law of thermodynamics. This

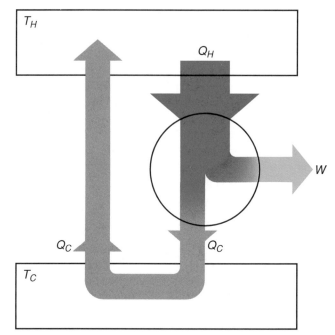

Figure 10.12 If we assume that heat can flow spontaneously from a colder to a hotter reservoir, the Kelvin statement of the second law of thermodynamics is violated.

statement, often called the *Clausius statement* after its originator, Rudolf Clausius (1822–1888), takes the following form:

> **Heat will not flow from a colder to a hotter body unless some other effect is also involved.**

In the case of a heat pump, the other effect is the introduction of work, which is used to pump the heat against its normal direction of flow.

Although this statement of the second law sounds very different from the Kelvin statement given in section 10.2, it turns out that these two statements express the same fundamental law of nature. They both place limits on what can be done with heat, and the limits stated, although they seem different, can be shown to be equivalent. The arguments used to show this equivalence are similar in form to that used in the previous section to show that no engine can have a greater efficiency than a Carnot engine operating between the same two temperatures.

Figure 10.12 illustrates the argument schematically. If it were possible for heat to flow from the colder to the hotter reservoir without requiring any work, then the heat released by the heat engine on the right side of the diagram could be allowed to flow back to the hotter reservoir. The net effect of the engine and the heat pipe back to the hotter reservoir would then be that an amount of heat, $Q_H - Q_C$, would be removed from the hotter reservoir and converted completely to work. No net heat would be added to the cooler reservoir. This result would violate the Kelvin statement of the second law of thermodynamics, which says that it is not possible to take heat from a reservoir at a single temperature and to convert it completely to work.

We see, then, that assuming a violation of the Clausius statement of the second law implies a violation of the Kelvin statement. A similar argument can be used to show that an assumed violation of the Kelvin statement implies a violation of the Clausius statement. (You may want to try your hand at developing this argument.) These two statements are therefore logically equivalent; they deal with the same fundamental law of nature.

Entropy

What is it about heat that is responsible for the limitations described in the different statements of the second law of thermodynamics? Both statements of the second law describe processes that would not violate the first law of thermodynamics (conservation of energy), but apparently are not possible. We cannot accomplish certain things with energy that is available in the form of heat.

The answer to this question can be seen partly by going back to Carnot's ideas. Suppose that a certain amount of heat is available in a high-temperature reservoir (perhaps just a container of hot water). We could imagine two different ways to remove heat from the hot source. First, we could simply allow the heat to flow through a good heat conduc-

tor to the cooler environment. This is illustrated in figure 10.13a, which shows a straight pipe or channel from the higher- to the lower-temperature reservoir.

Second, we could use this heat to run a heat engine, thereby doing some useful work (fig. 10.13b). If the engine is assumed to be a Carnot engine, then the process is completely reversible, and we have obtained the maximum possible work from the available heat. The first process, on the other hand, in which the heat simply flows through a conductor to the lower-temperature reservoir, is irreversible. The system is not in equilibrium while this flow is taking place, and the energy involved is not converted to any useful work.

In the irreversible process of heat flow, we have lost some ability to do useful work. The quantity that describes the extent of this loss we call *entropy*. For our purposes here, entropy can be defined as a measure of the disorder of the system, although this definition does not do full justice to the concept. The entropy of a system increases in any process in which the disorder or randomness of the system increases. Having our system organized into two reservoirs at two distinct temperatures is, in this sense, more ordered than having the energy all at a single intermediate temperature.

If we run a completely reversible Carnot engine with the heat taken from the higher-temperature reservoir, there is no increase in the entropy of the overall system. We have gotten the most that can be obtained in useful work from the available energy. The entropy of an isolated system (and of the universe) increases, however, in any irreversible process involving nonequilibrium conditions. It remains constant in reversible processes. The entropy of a system can decrease only if it interacts with some other system whose entropy is increased in the process. The entropy of the universe at large never decreases.

The statement that we have just made is, in fact, yet another statement of the second law of thermodynamics:

The entropy of the universe or of an isolated system can only increase or remain constant; it cannot decrease.

It turns out that the random nature of heat energy is responsible, then, for the limitations stated in the second law of thermodynamics. The entropy of the universe would decrease if heat could flow by itself from a colder to hotter body. The second law says that this cannot happen. Likewise, the entropy of the universe would decrease if heat at a single temperature could be converted completely to work.

You can begin to understand the reason for these limitations from the diagram of figure 10.14. The thermal energy present in a gas consists of the kinetic energy of its molecules. The velocities of these molecules are randomly directed, however. Only some of them have components in a direction appropriate to pushing the piston to produce work. If the molecules were all moving in the same direction, on the other hand, we could convert their kinetic energy completely to work done on the piston. This would be a

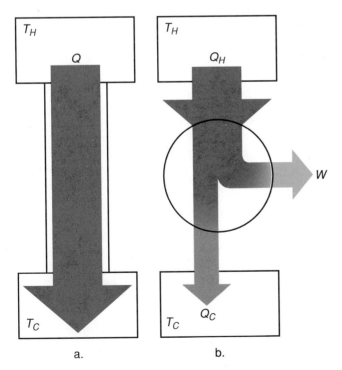

Figure 10.13 Heat can be removed from a high-temperature source either by direct flow through a conductor or by being used to run a heat engine.

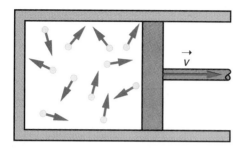

Figure 10.14 The random directions of the velocities of molecules in a gas prevent us from completely converting their kinetic energy to useful mechanical work.

lower-entropy (more organized) condition; it does not represent the normal condition for the thermal energy in a gas, which has a higher entropy (more disorganized). This disorganized nature of energy in the form of heat is responsible for the limitations on its use stated in the second law.

Like internal energy, entropy is a quantity that can be computed for any state of a given system. Changes in entropy can also be calculated, but the methods involved take us well beyond the scope of this discussion. The notion of disorder or randomness is at the heart of the matter, however. The state of my office or of a student's dormitory room tends naturally to become more disordered. This natural tendency of an increase in entropy can only be countered by the introduction of energy in the form of work to straighten things up again!

10.4 THERMAL POWER PLANTS AND ENERGY RESOURCES

We use electric power so routinely that most of us do not stop to think about where that energy comes from. The energy crisis was a major issue in the mid-seventies, but it faded into the background in the following years. Concerns about the greenhouse effect (see box 9.2) and acid rain have recently pushed questions about how we generate electric power back into public debate. How is electric power generated in your area? If hydroelectric power is not a major contributor in your area, most of your power must come from a thermal power source involving the use of heat engines.

The concepts of thermodynamics are extremely important in any discussion of the use of energy in our economy. What are the most efficient ways to use energy resources such as coal, oil, natural gas, nuclear energy, solar energy, or geothermal energy? How do the laws of thermodynamics and the efficiency of heat engines bear upon these questions? Most people are not aware of some of the fundamental limitations that underlie these issues.

A Thermal Power Plant

We produce electric power in this country in a variety of ways, but the most common method is via a coal- or oil-fired thermal power plant. As we have already indicated, the heart of such a plant is a heat engine. A fossil fuel (coal, oil, or natural gas) is burned, causing the temperature of the working fluid (usually water and steam) to increase. The steam is then run through a turbine (fig. 10.15) that turns a shaft connected to an electric generator. (The principle of operation of an electric generator is discussed in chapter 13.) Electric energy is then transmitted via the power lines to consumers (homes, offices, industries, etc.).

Because the steam turbine is a heat engine, its efficiency is inherently limited by the second law of thermodynamics to that given by the Carnot efficiency: $e_c = (T_H - T_C) / T_H$. (The temperatures used in this expression *must* be absolute temperatures.) The efficiency of a real heat engine is always less than this limit as a result of departures from the ideal conditions that define a Carnot engine; irreversible processes always take place in any real engine. A steam turbine generally comes closer to the ideal, however, than engines such as the internal combustion engine in an automobile. The explosions of the gasoline and air mixtures in an automobile engine are highly turbulent and irreversible processes.

Since the maximum possible efficiency is dictated by the temperature difference between the high- and low-temperature reservoirs, it is desirable to heat the steam to as high a temperature as the materials will permit. The materials from which the boiler and turbine are made do set an upper limit on the temperatures that can be tolerated; if these materials begin to soften or melt, the equipment will obviously deteriorate. For most steam turbines, this temperature limit is about 600° C, well below the melting point of steel. In practice most turbines operate at input temperatures below this limit; 550° C is a typical input temperature.

If we assume that a steam turbine is operating between an input temperature of 600° C (873 K) and an exhaust temperature near the boiling point of water (100° C or 373 K), where the steam condenses to water, it is a simple matter to compute the maximum possible efficiency for this turbine:

$$e_c = \frac{T_H - T_C}{T_H} = \frac{500}{873} = 0.572.$$

In other words, such a turbine would, at best, have an efficiency of about 57%. This is the ideal efficiency; the actual efficiency will be somewhat lower and usually runs between 40% and 50% for modern coal-fired or oil-fired power plants.

At best, then, we can convert about half of the thermal energy released in burning coal or oil to mechanical work or electrical energy. The rest must be released to the environment at temperatures too low to be useful for running heat engines or for most other functions except space heating. The exhaust

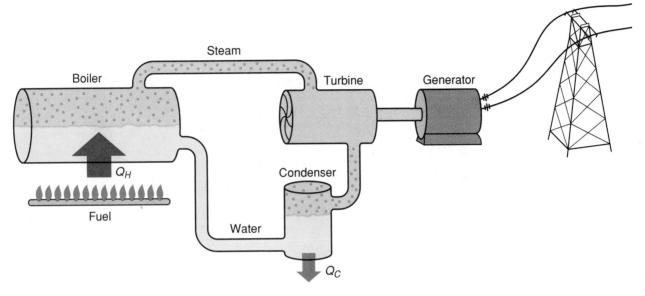

Figure 10.15 **A schematic diagram showing the basic components of a thermal-electric power plant. Heat from the boiler generates steams, which turns the steam turbine. The turbine, in turn, does work to generate electricity in the generator.**

Figure 10.16 **The cooling towers that are a common feature of many thermal power plants transfer heat from the cooling water to the atmosphere.**

side of the turbines must be cooled in order to achieve maximum efficiency, and the cooling water is either returned to some body of water or run through cooling towers, where the heat is dissipated into the atmosphere (fig. 10.16). This is the *waste heat* that we often refer to in discussions of the environmental impact of power plants. If the heat is dumped into a river, the temperature of the river will rise, causing undesirable effects for fish and other life-forms.

Other Energy Sources

Nuclear power plants, which are discussed in chapter 17, also generate heat to run steam turbines. Because of the effects of radiation upon materials, however, it is not feasible to run the turbines in a nuclear plant at temperatures as high as those used in coal- or oil-fired plants. The thermal efficiencies for nuclear plants are somewhat lower therefore, typically between 30% and 40%. The amount of heat that must be released to the environment is thus somewhat larger for a nuclear plant than for a fossil-fuel plant for the same amount of electrical energy generated. On the other hand, nuclear power plants do *not* release carbon dioxide and other exhaust gases to the atmosphere; they do not contribute, therefore, to the greenhouse effect.

Heat is also available from other sources. Geothermal energy is heat that comes from the interior of the earth, for example. Hot springs and geysers indicate the presence of hot water near the surface of the earth that might be utilized for power production. The temperatures at which this water is available are usually not much greater than 200° C, however. In places where steam is available in the form of geysers, low-temperature steam turbines can be run, as is done in northern California at the Geysers power plant (fig. 10.17). Their thermal efficiency is quite low, though, because of the relatively low input temperature.

If the water temperature is below 200° C, steam turbines are not very effective, and some other fluid with a lower boiling temperature than water is preferable to run a heat engine. Freon, which is used in most refrigerators, and isobutane have been studied as possible working fluids for low-temperature heat engines. The efficiency of such an engine

Figure 10.17 The Geysers power plant in California uses geothermal heat to run steam turbines.

Figure 10.18 High temperatures are achieved from solar energy by focusing sunlight onto a central boiler with an array of mirrors at a solar-thermal power plant at Barstow, California.

would be quite low, however. If, for example, water was available at a temperature of 150° C (423 K), and cooling water was available at 20° C (293 K), the ideal efficiency would be only

$$e_c = \frac{T_H - T_C}{T_H} = \frac{130}{423} = 0.307,$$

or around 31%. In practice, of course, the actual efficiency will be less than this, so it is common to assume efficiencies of only 20% to 25% for proposed geothermal power plants.

Warm ocean currents are yet another source of heat. Power plants that would operate by utilizing the temperature difference of warm water at the surface of an ocean and cooler water drawn from greater depths have been proposed and prototypes developed. Typically, however, this temperature difference is only 20 C° or so, and the thermal efficiency is very low. (For $T_H = 25°$ C = 298 K and $T_C = 5°$ C = 278 K, $e_c = 0.067 = 6.7\%$.) Although the efficiency is low, so is the cost of the warm water, so it still might be economically feasible to produce power in this manner.

The sun is the source of the heat that warms ocean currents and represents an energy source with enormous potential for development if the costs could become competitive. The temperatures that can be achieved with solar power depend upon the type of collection system that is used. The ordinary flat-plate collector is capable of achieving only relatively low temperatures of 50° to 100° C. On the other hand, concentrating collectors that use mirrors or lenses to focus the sunlight can achieve much higher temperatures. In a solar power plant near Barstow, California, an array of mirrors is used to focus sunlight upon a boiler that is located on a central tower (fig. 10.18). Temperatures comparable to those used in coal- or oil-fired plants can be obtained in this manner, so that similar steam turbines can be used.

Low-Grade and High-Grade Energy

We have seen that the temperature at which heat can be supplied makes a big difference in how much useful work can be

extracted from that heat. The second law of thermodynamics and the related Carnot efficiency set the limits upon what can be done. What does this say about our everyday use of energy? Or more broadly, what implications do these ideas have for our national policies on energy?

Clearly, heat at temperatures between 500° and 600° C (or even higher) is much more useful for running heat engines and producing mechanical work or electrical energy than heat available at lower temperatures. Such heat is sometimes referred to as *high-grade* heat because of its potential for producing work. Even then, as we have seen, only 50% or less of the heat can actually be converted to work in practice.

Heat available at lower temperatures can be used to produce work, but with considerably lower efficiency. Thus heat available at temperatures around 100° C or lower would generally be called *low-grade* heat. Low-grade heat is better used for purposes such as heating a home or building (space heating) or similar functions. This, then, is the optimal use for heat collected by flat-plate solar collectors, or even from geothermal sources if they are near enough to buildings that need heat. Geothermal heat is used for space heating in Klamath Falls, Oregon, as well as in some other parts of the world where conditions are favorable.

A lot of low-grade heat available from power plants or remote sources goes to waste simply because it is not economical to transport it to places where space heating may be needed. Nuclear power plants, for example, are not usually built in populated areas. There are other possible uses of low-grade heat, such as in agriculture or aquaculture, but we have not gone very far at this point in developing such uses.

The primary advantage of electrical energy is that it can be easily transported via power lines to users at locations far from the point of generation. Electrical energy can also be readily converted to mechanical work via electric motors. Electric motors can operate at efficiencies of 90% or greater because they are *not* heat engines. The efficiency involved in producing the electrical power in the first place may be considerably lower, of course.

Electrical energy can also be converted back to heat if desired. Electricity is used for space heating in locations like the Pacific Northwest, where electric power is relatively cheap because of the extensive hydroelectric resources on the Columbia River and its tributaries. The development of these resources has been subsidized by the federal government as well as by other governmental policies related to public power. Low prices encourage the use of a high-grade form of energy for purposes that might be just as well served by lower-grade sources of heat.

Energy is still relatively cheap in this country as well as in many other parts of the world. Continued economic development and depletion of fossil fuel resources will gradually change this picture. As this occurs, questions of optimal uses of energy resources will become more important. As with many issues, wise decisions will depend upon the participation of an informed citizenry. (That means you!) An understanding of the limitations imposed by the laws of thermodynamics, is a part of the scientific literacy needed.

10.5 PERPETUAL MOTION AND ENERGY FRAUDS

Inventors have long been fascinated with the idea of perpetual motion. The lure of finding a way to design an engine that could run on and on, using no fuel or somehow drawing energy from water or some other plentiful source, is not unlike that of finding gold. Indeed, the inventor of such an engine could become even richer than if he or she had discovered gold.

Is such an engine possible? The laws of thermodynamics certainly place limits upon what is possible in this regard. Since these laws are consistent with everything that physicists know about energy and engines, any claim that violates the laws of thermodynamics should be rejected. We tend to analyze claims of perpetual motion or similar phenomena, then, by testing to see if they violate either the first or second laws of thermodynamics.

Perpetual-Motion Machines of the First Kind
A proposed engine or machine that would violate the first law of thermodynamics is called a *perpetual-motion machine of the first kind.* Since the first law of thermodynamics is essentially a statement of the principle of conservation of energy, a perpetual-motion machine of the first kind would be one that puts out more energy in the form of work or heat than it takes in. If the machine or engine is operating in a continuous cycle of some sort, the internal energy must return to its initial value, and the energy output of the engine must equal its energy input, as we have already seen.

These ideas are illustrated in the schematic diagram of figure 10.19. The sum of the work output and the heat output is greater than the magnitude of the heat input, as represented by the width of the arrows. This could only happen if there were some source of energy within the engine itself, such as a battery. If this were the case, though, the energy in the battery would gradually be depleted, and the internal energy of the engine would be decreasing. The engine could not run indefinitely in this manner.

The fact that physicists would reject such an engine as impossible does not keep inventors from proposing them. From time to time we see claims reported in newspapers and other popular media of engines that can run indefinitely using only a gallon of water or a minuscule quantity of gasoline as fuel. Given the price of gasoline, such claims have an obvious appeal and often attract investors and other interest. With a little knowledge of physics, however, you could ask the inventor some simple questions. Where is the energy coming from? How can you get more energy out than you put in? Hang on to your wallet; it looks like a poor prospect.

Perpetual-Motion Machines of the Second Kind
Claims of engines that would violate the second law of thermodynamics without violating the first law are often a little more subtle. These inventors have learned how to answer the questions in the previous paragraph; they have a source of

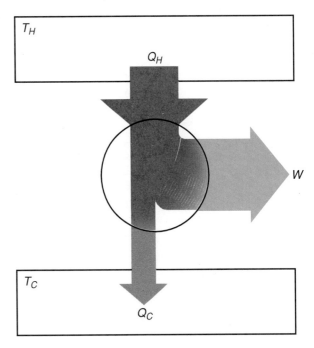

Figure 10.19 A perpetual-motion machine of the first kind. The energy output exceeds the input and therefore violates the first law of thermodynamics.

208

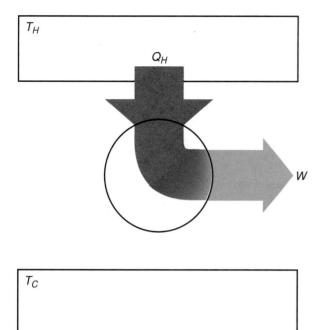

T_H

Q_H

W

T_C

Figure 10.20 A perpetual-motion machine of the second kind. Heat is extracted from a reservoir at a single temperature and converted completely to work, thus violating the second law of thermodynamics.

energy. Perhaps they plan to extract heat from the atmosphere or the ocean. These claims must be evaluated in terms of the second law of thermodynamics. If the second law is violated, the inventor has proposed a *perpetual-motion machine of the second kind* (fig. 10.20).

The second law states that it is not possible to take heat from a reservoir at a single temperature and to convert it completely to work. We must have a lower-temperature reservoir available, and some heat must be released to that reservoir. In addition, even if we have a lower-temperature reservoir available, the efficiency of any engine running between the two reservoirs will be quite low if the temperature difference between the reservoirs is not very great. Any claim of efficiency greater than the Carnot efficiency (for the available temperature difference) would also violate the second law of thermodynamics.

Theoretically it would be possible to take heat from the atmosphere in the summer and run a heat engine by rejecting heat to a river or some other colder body of water. The efficiency of such an engine would be quite low, however, and the project would probably fail on economic rather than physical grounds. The atmosphere is a low-grade source of heat even on a very warm day, and it is difficult to exploit such a source.

The laws of thermodynamics are useful for evaluating an inventor's claims as well as for guiding our attempts to build better engines and to use energy more efficiently. Better engines can and will be developed, and their inventors may indeed be enriched in the process. A host of possibilities exist that involve relatively low-grade energy sources or other special circumstances. The development of special materials and innovative designs for engines that could use even higher temperatures than are common in today's coal and oil-fired plants is also an active area of study.

Most physics departments in the country receive inquiries from local inventors seeking endorsement or help with their schemes for producing or using energy. The inventors are often sincere and sometimes quite well-informed. Occasionally their ideas have merit, although often they are based upon misunderstandings of the laws of thermodynamics (see box 10.2). Unfortunately, it is often difficult to convince them that their ideas will not work, even when they involve clear violations of the laws of thermodynamics.

Other cases are more clearly the work of charlatans. Inventors who begin with good intentions sometimes discover that their ideas attract money from eager investors. There have been cases in which inventors managed to raise several million dollars for the design and testing of prototype engines. Somehow the engines are never quite finished, or the tests are inconclusive and more money is needed to continue the work. The inventors, in the meantime, live quite well, finding that promoting their inventions is more lucrative than actually designing and testing them.

Investors beware. Under no known circumstances have the laws of thermodynamics been violated. The failure of repeated attempts to violate them reinforces our belief in their validity. They cannot be proved, but their standing as the basis of highly successful theory makes physicists confident that they will continue to be useful indicators of what is possible.

Box 10.2

Everyday Phenomenon:
A Productive Pond

The Situation. A local farmer consulted the author about an idea he had for generating electricity on his farm. The farmer had a pond that he thought he could use to run a water wheel, which in turn could power an electric generator. He had a drawing with him which looked something like the sketch shown here.

The idea was as follows. A drain pipe would be located at the bottom of the pond. The farmer was aware that water would flow down such a drain pipe with considerable velocity. His plan was to direct this water through a pipe to the side of the pond and then up above the water level, where it would flow out onto a water wheel or turbine, thus powering the generator. The water would return to the pond after leaving the water wheel, so there would be no need to replenish the water supply, except to replace the water lost to evaporation or leakage.

How would you advise the farmer? Will this plan work? Does it represent a perpetual motion machine? If so, of what kind?

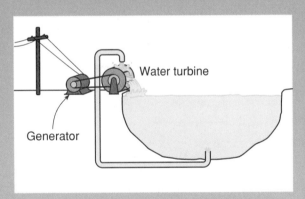

A sketch of the farmer's scheme for obtaining electrical power from his pond. What law of thermodynamics is violated here?

The Analysis. If we ignore the details of operation and just look at the overall result, we see that work is being obtained to turn the generator without the input of any energy in the form of heat or work. Since the pond returns to its initial state (with the same internal energy), the first law of thermodynamics, or the principle of conservation of energy, is violated by the proposal. It therefore represents a perpetual-motion machine of the first kind.

If we look at the mechanics in more detail, we can see that the water will indeed gain kinetic energy as it flows down the drain. This gain in kinetic energy comes at the expense of a loss in potential energy as the pond level is lowered. If we now direct this water upward, it regains potential energy at the expense of kinetic energy and slows down. If there are no frictional losses, its velocity should reach zero at the original level of the pond. Although in an initial surge it might overshoot and then oscillate around the original pond level, it cannot be raised above the original pond level in a steady-state, continuous-cycle process. In the steady state, the water in the vertical pipe will have the same level as the pond, and no water will flow. The proposal will not work.

Although the author explained these ideas to the farmer carefully, and the farmer was an educated and intelligent person, he still was not convinced that the idea would not work. Whether or not he actually tried out his plan was never reported to the author. The author did encourage him (if he was not convinced by the theoretical arguments) to try it out with a small scale model before investing larger sums of money in plumbing his pond. Models or prototypes are a good way to test ideas; demonstrations can be much more convincing than theoretical arguments!

SUMMARY

An interest in how to build a better steam engine led us to an analysis of the principles of operation of heat engines generally, and ultimately to a statement of the second law of thermodynamics. The second law is closely tied to an understanding of heat engines, refrigerators, and the concept of entropy. The prudent use of energy in our economy depends, in part, on our understanding and use of the laws of thermodynamics. The following basic ideas were covered in chapter 10.

Heat Engines.   A heat engine is any device that takes in heat from a high-temperature source and converts some of it to useful mechanical work. Some heat is always released at lower temperatures to the environment in this process. The efficiency is defined as the work output divided by the heat input.

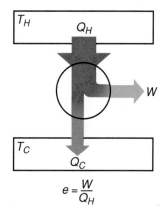

$$e = \frac{W}{Q_H}$$

The Second Law of Thermodynamics.   The Kelvin statement of the second law says that it is not possible for an engine working in a continuous cycle to take in heat at a single temperature and convert that heat completely to mechanical work. The maximum possible efficiency for any engine operating between two given temperatures is that of a Carnot engine, and it is always less than 1.

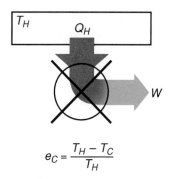

$$e_C = \frac{T_H - T_C}{T_H}$$

Heat Pumps and Refrigerators.   A heat pump, or refrigerator, is a device that moves heat from a reservoir at a low temperature to one at a higher temperature. Mechanical work must be supplied to the heat pump in order to accomplish this. The Clausius statement of the second law says that heat will not flow spontaneously from a colder to a hotter temperature.

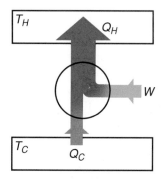

Entropy.   Entropy can be thought of as a measure of the disorder or randomness of a system. Irreversible processes such as the spontaneous flow of heat from a hotter to a colder body without the production of useful work always increase the entropy of an isolated system and of the universe.

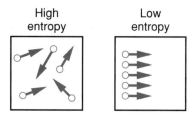

Thermal Power Plants.   Power plants that use heat generated from coal, oil, natural gas, nuclear fuels, geothermal resources, or the sun to produce electrical energy are all examples of thermal power plants. Their efficiency can not be greater than the Carnot efficiency, so it is desirable to make the input temperature as high as possible.

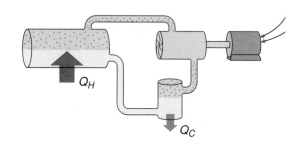

Perpetual-Motion Machines. A proposed device that would violate the first law of thermodynamics is a perpetual-motion machine of the first kind; one that would violate the second law, but not the first, is a perpetual-motion machine of the second kind. As far as we know, neither of these cases is possible, but that knowledge has not prevented inventors from trying.

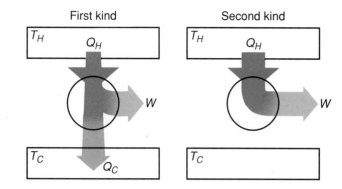

QUESTIONS

Q10.1 Which of the following are heat engines?

   a. An automobile engine

   b. An electric motor

   c. A steam turbine

Explain your reasons for classifying or not classifying each of these as a heat engine.

Q10.2 In applying the first law of thermodynamics to a heat engine, why can we take the change in internal energy of the engine to be zero? Explain.

Q10.3 Is the total amount of heat released to the lower-temperature reservoir in each cycle by a heat engine ever greater than the amount of heat introduced in each cycle? Explain.

Q10.4 From the perspective of the first law of thermodynamics, is it possible for a heat engine to have an efficiency greater than 1.0? Explain.

Q10.5 Is it the first or the second law of thermodynamics that requires the work output of a heat engine to be equal to the difference in the quantities of heat taken in and released by the engine? Explain.

Q10.6

   a. Is it possible for a heat engine to operate in the manner shown in diagram (*a*)? Explain.

   b. Is it possible for a heat engine to operate in the manner shown in diagram (*b*)? Explain.

   c. Is it possible for a heat engine to operate in the manner shown in diagram (*c*)? Explain.

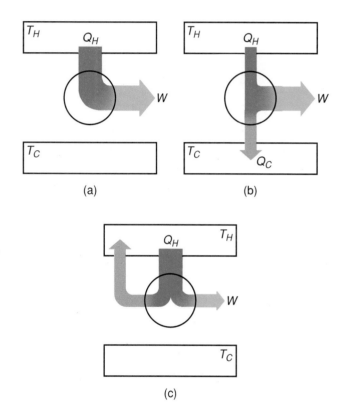

Q10.7 Is it possible for the efficiency of a heat engine to equal 1.0? Explain.

Q10.8 Is it possible for a Carnot engine to operate in an irreversible manner? Explain.

Q10.9 Which would have the greater efficiency, a Carnot engine operating between the temperatures of 400° C and 300° C, or one operating between the temperatures of 400 K and 300 K? Explain.

Q10.10 Is a heat pump essentially the same thing as a heat engine? Explain.

Q10.11 Is a heat pump essentially the same thing as a refrigerator? Explain.

Q10.12 Is it possible to cool a closed room by leaving open the door of a refrigerator in the room? Explain.

Q10.13 Is it ever possible to move heat from a cooler to a warmer temperature? Explain.

Q10.14

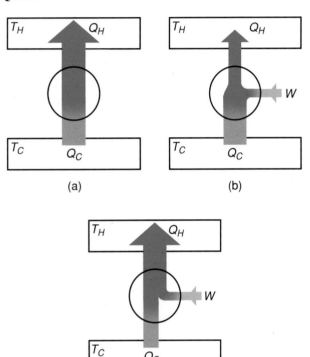

(a)　　　　　　　(b)

(c)

a. Is it possible for a heat pump to operate in the manner shown in diagram (*a*)? Explain.

b. Is it possible for a heat pump to operate in the manner shown in diagram (*b*)? Explain.

c. Is it possible for a heat pump to operate in the manner shown in diagram (*c*)? Explain.

Q10.15 Which would normally have the greater thermal efficiency, a coal-fired power plant or a geothermal power plant? Explain.

Q10.16 In what ways is a nuclear power plant similar to a coal-fired plant? Explain.

Q10.17 Is an automobile engine a perpetual-motion machine? Explain.

Q10.18 An engineer proposes a device that will extract heat from the atmosphere, convert some of it to work, and release the remaining heat back to the atmosphere. Is this a perpetual-motion machine? If so, of what kind? Explain.

EXERCISES

E10.1 In one cycle, a heat engine takes in 500 J of heat from a high-temperature reservoir, releases 300 J of heat to a lower-temperature reservoir, and does 200 J of work. What is its efficiency?

E10.2 In one cycle, a heat engine takes in 800 J of heat from a high-temperature reservoir and releases 500 J of heat to the lower-temperature reservoir.

a. How much work is done by the engine in each cycle?

b. What is its efficiency?

E10.3 In one cycle, a heat engine does 400 J of work and releases 500 J of heat to a lower-temperature reservoir.

a. How much heat does it take in from the higher-temperature reservoir?

b. What is the efficiency of the engine?

E10.4 A Carnot engine takes in heat at a temperature of 800 K and releases heat to a reservoir at a temperature of 300 K. What is its efficiency?

E10.5 A Carnot engine takes in heat from a reservoir at 500° C and releases heat to a lower-temperature reservoir at 150° C. What is its efficiency?

E10.6 A Carnot engine operates between temperatures of 500 K and 300 K and does 400 J of work in each cycle.

a. What is its efficiency?

b. How much heat does it take in from the higher-temperature reservoir in each cycle?

c. How much heat does it release to the lower-temperature reservoir in each cycle?

E10.7 A heat pump takes in 300 J of heat from a low-temperature reservoir in each cycle and uses 150 J of work per cycle to move the heat to a higher-temperature reservoir. How much heat is released to the higher-temperature reservoir in each cycle?

E10.8 In each cycle of its operation, a refrigerator removes 20 J of heat from the inside of the refrigerator and releases 30 J of heat to the room. How much work per cycle is required to operate this refrigerator?

E10.9 A typical electric refrigerator has a power rating of 400 watts, which is the rate in J/s at which electrical energy is supplied to do the work needed to move heat from the refrigerator. If the refrigerator releases heat to the room at a rate of 1200 watts, at what rate (in watts) does it remove heat from the inside of the refrigerator?

E10.10 A typical nuclear power plant delivers heat from the reactor to the turbines at a temperature of 540° C. If the turbines release heat at a temperature of 240° C, what is the maximum possible efficiency of this power plant?

E10.11 An ocean thermal-energy power plant takes in warm surface water at a temperature of 25° C and releases heat at 4° C to cooler water drawn from deeper in the ocean. Is it possible for this power plant to operate at an efficiency of 10%? (Show by computing the Carnot efficiency.)

CHALLENGE PROBLEMS

CP10.1 Suppose that a typical automobile engine operates at an efficiency of 25%. One gallon of gasoline releases approximately 150×10^6 J, or 150 MJ of heat when it is burned.

a. Of the energy available in one gallon of gas, how much energy is used to do useful work in moving the automobile and running its accessories?

b. How much heat per gallon is released to the environment in the exhaust gases and via the radiator?

c. If the car is moving at constant speed on a level road, how is the work output of the engine used?

CP10.2 Suppose that a certain Carnot engine operates between the temperatures of 400°C and 150°C and produces 30 J of work in each complete cycle.

a. What is the efficiency of this engine?

b. How much heat does it take in from the 400°C reservoir in each cycle?

c. How much heat is released to the 150°C reservoir in each cycle?

d. What is the change in internal energy of the engine, if any, in each cycle?

CP10.3 A Carnot engine operating in reverse as a heat pump moves heat from a cold reservoir at 10°C to a warmer one at 25°C.

a. What is the efficiency of a Carnot engine operating between these two temperatures?

b. If the Carnot heat pump releases 300 J of heat to the higher-temperature reservoir in each cycle, how much work must be provided in each cycle?

c. How much heat is removed from the 10°C reservoir in each cycle?

d. If the performance of a refrigerator or heat pump is described by a *coefficient of performance* defined as $K = Q_C / W$, what is the coefficient of performance for this Carnot heat pump?

e. Are the temperatures used in this example appropriate to the application of a heat pump for home heating? Explain.

CP10.4 Suppose that an oil-fired power plant is designed to produce 100 MW (megawatts) of electrical power. The turbine operates between temperatures of 600°C and 300°C and has an efficiency that is 80% of the ideal Carnot efficiency for these temperatures.

a. What is the Carnot efficiency for these temperatures?

b. What is the efficiency of the actual oil-fired turbines?

c. How many kilowatt-hours (kW·hr) of electrical energy does the plant generate in one hour? (The kilowatt-hour is an energy unit equal to one kilowatt of power times one hour.)

d. How many kilowatt-hours of heat must be obtained from the oil in each hour?

e. If one barrel of oil yields 1700 kW·hr of heat, how much oil is used by the plant each hour?

HOME EXPERIMENTS AND OBSERVATIONS

HE10.1 A simple, very low-power heat engine can be devised using a tea kettle as a heat source and building a simple steam turbine. The turbine can be just a pinwheel constructed from stiff plastic (from a milk bottle or beverage container, perhaps) and mounted on a nail or pencil or other suitable axle. If a thread or string is wrapped around the axle, the engine could be used to lift small weights, thus doing work on an external object.

a. Sketch a design of a simple steam turbine that could be built from materials you have at hand.

b. If time and materials permit, build a turbine according to your design.

c. Test your model to see how much weight it is capable of lifting. (Do not set your sights too high!)

d. You could estimate the power output of your engine by measuring the time required to raise the weight through a given height ($P = W / t = mgh / t$).

HE10.2 Study the construction of a refrigerator and use a thermometer to make the following observations.

a. What energy source (gas or electricity) is used to provide the work necessary to move the heat? Can you find a power rating somewhere on the refrigerator?

b. Where are the heat-exchange coils that are used to reject heat to the room? (They are usually on the back of the refrigerator, sometimes near the bottom.)

c. What are the temperatures inside the refrigerator, inside the freezer compartment, near the heat-exchange coils outside, and in the room at some distance from the refrigerator? These measurements might be repeated at different times to see how much these temperatures vary.

HE10.3 What types of power plants are used to generate electricity in your area? Do they employ heat engines? If so, what is the source of heat? Are cooling towers or other cooling structures used to release waste heat to the atmosphere? (Your local utility should be able to provide such information if requested.)

Part 3

ELECTRICITY

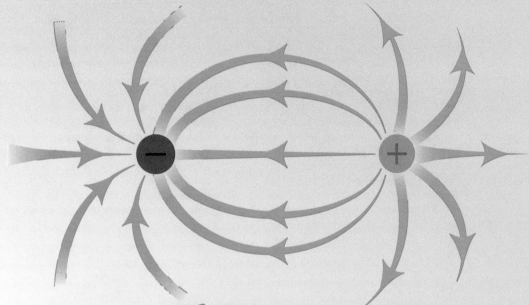

and Magnetism

11 *Electrostatic Phenomena*

Most of you have had the experience of running a comb through your hair on a dry winter day and hearing a crackling sound and perhaps even seeing sparks. Depending on the length and freedom of your hair, it may have stood on end, away from your head, as shown in figure 11.1. This phenomenon is intriguing, but also annoying. You must often wet the comb in order to get your hair to behave.

What is happening here? Some force appears to be causing the individual hairs to repel one another. Most of you would probably be able to identify that force as an electrostatic force, or at least you might say that static electricity is the cause of the misbehaving hair. To name a phenomenon is not the same thing as explaining it, however. Why does it occur under these conditions, and what is the basic nature of the force in question?

The hair-and-comb fireworks are just one of a long list of electrostatic phenomena that are part of our everyday experience. Why does a piece of plastic refuse to leave your hand after being peeled off of a package? Why do you get a slight shock after you walk across a carpeted floor and then touch a light switch? The phenomena range from static cling and similar annoyances to the much more dramatic displays of lightning in a good thunderstorm. They are familiar and yet, to many of us, mysterious.

Gaining an understanding of the nature of the electrostatic force and the related topics of electric current and magnetism was one of the major achievements of nineteenth-century physics. This understanding of electromagnetism has led to enormous technological advances that have revolutionized the performance of everyday tasks. Most of these changes have taken place within the current century.

Despite their importance to modern life, however, many of us are surprisingly ignorant of the basic nature of electrical phenomena. Since we cannot see much of what goes on when we turn a switch, there is an abstractness to the subject of electricity that often puts people off. Fear of electrical shock may also play a role. It need not be so, however; many of the phenomena can be explained with simple ideas that are understandable to all of us. We should beware of crediting the wizards of physics and electrical engineering with arcane powers that are beyond the rest of us!

Figure 11.1   Hair sometimes seems to have a mind of its own when combed on a dry winter day. What causes the hairs to repel one another?

Chapter Objectives

The primary purpose of this chapter is to describe and explain the basic nature of the electrostatic force, as well as concepts such as electric field and electric potential that physicists have developed as an aid to understanding electrical phenomena. The ideas are developed through the description and interpretation of simple experiments. The nature of electric charge and the charging process, the distinction between conductors and insulators, and many other ideas can be understood in this manner.

Chapter Outline

❶ *Electric charge: some simple experiments.* What is the nature of electric charge, and how is it involved in the electrostatic force? What is the process by which objects acquire charge?

❷ *More experiments: insulators and conductors.* What is the distinction between insulating materials and conducting materials, and how can this distinction be easily tested? How does the process of charging by induction work? Why are bits of paper or styrofoam attracted to charged objects?

❸ *Coulomb's law: the electrostatic force.* What is the nature of the electrostatic force described by Coulomb's law? How does it vary with charge and distance, and what are its similarities to (and differences from) the gravitational force?

❹ *The electric field.* How is the concept of electric field defined and why is it useful? How are field lines used to help us visualize electrostatic effects?

❺ *Electric potential.* How is the concept of electric potential defined, and how is it related to potential energy? What is voltage, and how are the concepts of electric field and potential related?

11.1 ELECTRIC CHARGE: SOME SIMPLE EXPERIMENTS

If we think about the hair-combing example and some of the other phenomena mentioned in the introduction, can we note any similarities among them? The comb passing through your hair, your shoes passing over the rug, or plastic being peeled from a box all involve different materials moving in contact with one another. A rubbing process between different materials occurs in all of these cases. This could be a clue to an initial understanding of what is happening.

How might we test such an idea? An obvious approach might be to collect some different materials, rub them together in various combinations, and see what you can generate in the way of sparks or observable effects. Some simple experiments of this nature could help to establish which combinations of materials are most effective in producing electrostatic crackles and pops. Before very long in this process, however, we would want to develop some consistent means of gauging the strength of the effects that we are observing.

Experiments with Pith Balls

The common method of demonstrating simple electrostatic effects is to rub plastic or glass rods with various furs or fabrics. If, for example, we take a plastic or hard rubber rod and rub it with a piece of cat hide with the fur attached, we observe the hairs on the cat fur standing out and snapping at one another. Like the hair-combing phenomenon, these effects are most striking on a dry winter day when the air contains very little moisture.

The other piece of equipment that is generally used in these demonstrations is a small stand to which two pith balls are attached by threads (fig. 11.2). Pith balls are small wads of dry, paperlike material. They are light enough to be strongly influenced by electrostatic forces.

An interesting sequence of events occurs when a plastic rod, having been vigorously rubbed with cat fur on a dry day, is brought near to the pith balls. Initially they are attracted to the rod, much as bits of iron are to a magnet. After contacting the rod and perhaps sticking to it for a few seconds, however, the pith balls dance away from the rod. Not only are they repelled by the rod at this point, but also by each other. The threads supporting the pith balls now make an angle to the vertical, as shown in figure 11.3, rather than hanging straight down.

How might we explain what has happened here? Obviously, a repulsive force of interaction exists between the two pith balls after they have been in contact with the rod. We might suppose that the balls have gained some substance (call it *electric charge*) from the rod that is responsible for the observed force. This charge, whatever it is, was somehow generated by rubbing the rod with the cat fur. It also causes the hairs in the fur to repel one another, making them stand on end.

Another interesting event occurs if we rub a glass rod with a piece of synthetic fabric such as nylon. If this glass rod is brought near to the pith balls that have been charged by the plastic rod, they are now attracted to the glass rod. They are still repelled, however, by the plastic rod. If we allow the pith balls to touch the glass rod, they repeat the sequence of events that we observed earlier with the plastic rod. At first they are attracted and stick briefly to the rod; then they dance away and are repelled by the rod and by each other. If we

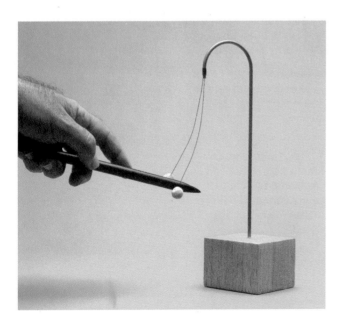

Figure 11.2 Pith balls suspended from a small stand are attracted to a charged plastic rod.

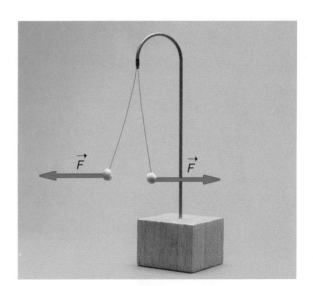

Figure 11.3 Once the pith balls have moved away from the rod, they also repel each other, indicating the presence of a repulsive force.

bring the plastic rod nearby, we now find that the balls are attracted to the plastic rod!

These observations complicate the picture somewhat. Apparently there are two types of charge; one generated by rubbing a plastic rod with fur, and the other generated by rubbing a glass rod with nylon. Perhaps there are more? Further experiments with different materials would indicate that we do not need to propose any more than two types. Other charged objects will cause either an attraction or repulsion, like the two types already identified, and can be placed in one or the other of these two categories.

The Electroscope

The observations just described can be reinforced and made slightly more quantitative by using a simple device called an *electroscope*. A simple electroscope consists of two leaves of a thin metallic foil suspended face-to-face from a metal hook (fig. 11.4). At one time, gold leaf was the preferred material for the foil, but aluminized mylar is now more commonly used. The metal hook is directly connected by a metal rod to a metal ball that protrudes from the top of the instrument. The hook and foil leaves are usually protected from air currents and other disturbances by a glass-walled container.

If the foil leaves are uncharged, they will hang straight down. If, however, we bring a charged rod or other charged object in contact with the metal ball on top, the leaves will immediately spread apart. They will stay spread apart after the rod is removed, presumably because they are now charged.

If any charged object is now brought near to the metal ball, the electroscope will indicate what type of charge is present, as well as roughly how much charge is present. If an object with the same type of charge as the original rod is brought near to the ball, the leaves will spread further apart; an object with the opposite charge will cause the leaves to come closer together.

Although somewhat less dramatic than the pith balls, the electroscope offers several advantages for detecting the type and strength of charge present. In the first place, the foil leaves do not dance around and become tangled together. The metal ball offers a stationary point for testing other objects. Also, the degree to which the leaves move apart or come together when a charged object is brought near to the ball provides a consistent indication of the strength of the charge.

One additional point about the charging process can be quickly established with the electroscope. When we charge the plastic rod by rubbing it with the cat fur, the fur becomes charged also, with a charge that is opposite to that of the rod. We can verify this by bringing the fur close to the ball on the electroscope after it has been charged by the rod. The leaves move closer together when the fur approaches the metal ball, whereas they move further apart when the rod

Figure 11.4   A simple electroscope consists of two metallic-foil leaves suspended from a metal post inside a glass-walled container.

approaches. Similar experiments show that the glass rod and the nylon fabric also acquire opposite types of charge when rubbed together.

The Single-Fluid Model

By the middle of the eighteenth century, the facts about the charging process that we have just been discussing had been well established. There were apparently two types of charge that, when separated, produced electrostatic forces between charged objects. The forces could be either attractive or repulsive according to a simple rule:

Like charges repel each other, and unlike charges attract each other.

There was less agreement, however, on what to call these two types of charge. *Charge-produced-on-a-rubber-rod-when-rubbed-by-cat's-fur* and *charge-produced-on-a-glass-rod-when-rubbed-with-silk* are unwieldy labels! The names that we now use were introduced by the American statesman and scientist Benjamin Franklin (1706–1790), around the year 1750. During the 1740s, Franklin performed a series of experiments on electrostatic charge similar to those we have discussed.

Franklin proposed that the facts, as they were known, could all be explained by the action of a single fluid that could be transferred from one object to another in the charging process. The charge acquired when a surplus of this fluid was gained could then be called *positive* (+) and that associated with a shortage of this fluid could be called *negative* (−) (fig. 11.5). Which of these was which was not clear, since the fluid itself was invisible and not detectable in any other

Like charges   Unlike charges

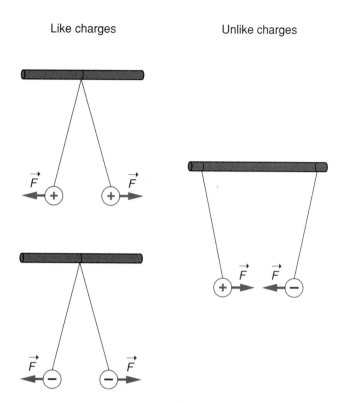

Figure 11.5   Like charges repel and unlike charges attract. The plus and minus signs were introduced in Franklin's single-fluid model.

would now regard electrons as being positively charged. Surpluses and shortages of electrons would then have more appropriate signs, and the lives of students would be somewhat easier! We cannot blame Ben Franklin, however; electrons were not discovered until almost 150 years after he developed his model.

11.2   MORE EXPERIMENTS: INSULATORS AND CONDUCTORS

The experiments with the pith balls and the electroscope described in the previous section can be used to establish some basic facts about the electrostatic force. Different properties of materials are also important, however, in understanding the range of electrostatic phenomena that are observed. Why, for example, were the pith balls initially attracted to the rod even when, presumably, they were not charged? Why are the leaves and the ball on the electroscope made of metal? The distinction between insulators and conductors is critical to many of these phenomena.

Insulators and Conductors

Suppose that you use one of the charged rods to charge the electroscope, causing the leaves to repel one another. Either the plastic or the glass rod will do; the sign of the charge is not important for this experiment. Once the electroscope is charged, touch the metal ball on top with your finger. What happens? The leaves of the electroscope immediately droop to a straight-down position; you have discharged the electroscope.

Suppose that you again charge the electroscope and now touch the metal ball on top with an uncharged rod made of plastic or glass. What happens in this case? There is essentially no effect upon the leaves of the electroscope. If, however, you now touch the ball with a hand-held metal rod, the leaves immediately droop to a vertical position again; the electroscope is discharged (fig. 11.6).

How can we explain these observations? Apparently both the metal rod and your fingers allow charge to flow from the leaves of the electroscope to your body. Your body represents a large neutral sink for charge; you can easily absorb the charge upon the electroscope without much affecting your overall charge. The plastic or glass rod, on the other hand, does not seem to permit the flow of charge from the electroscope to your body.

The metal rod and our bodies are examples of *conductors:* materials that readily permit the flow of charge. Plastic or glass, on the other hand, are examples of *insulators:* materials that do not ordinarily allow the flow of charge. By performing similar experiments with the charged electroscope, we could easily test many other materials. We would discover that all metals are good conductors, whereas glass,

manner. Franklin arbitrarily proposed that the charge obtained on a glass rod when rubbed with silk (synthetic fabrics were unavailable then) be called positive.

Besides simplifying the naming of the two types of charge, Franklin's model also provided a picture of what might be occurring in the charging process itself. The idea that there was some neutral, or stable, amount of this invisible fluid present in all objects, and that rubbing objects together could transfer some of this fluid from one object to the other explained why the two objects became oppositely charged in this process. This was a simpler model than that involving two different types of fluid for the two types of charge.

Franklin's model comes surprisingly close to our modern view of what takes place in the charging process. In this view, electrons are transferred between objects when they are rubbed together. Electrons are small, negatively charged particles that are present in all atoms and therefore in all materials. (See chapter 16 for a discussion of atomic structure.) A negatively charged object has a surplus of electrons, and a positively charged object has a shortage of electrons. The atomic or chemical properties of materials dictate which way the electrons will flow when objects are rubbed together.

The fact that electrons are negatively charged follows from Franklin's original choice of what to call *positive* and *negative*. If Franklin had made the opposite choice, we

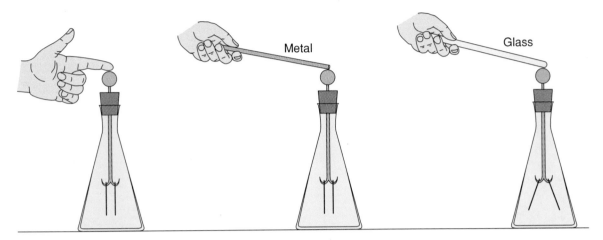

Figure 11.6 Touching the ball on top of a charged electroscope with either your finger or a metal rod causes the electroscope to discharge. Touching it with an uncharged glass rod produces no effect.

Table 11.1		
Some Common Conductors, *Insulators, and Semiconductors*		
Conductors	**Insulators**	**Semiconductors**
Copper	Glass	Carbon
Silver	Plastic	Silicon
Iron	Ceramics	Germanium
Gold	Paper	Damp wooden rod
Salt solution	Oil	
Acids		

plastic, and most nonmetallic materials are good insulators. Table 11.1 lists some examples of insulating and conducting materials.

The difference in the ability of good insulators and good conductors to conduct electric charge is amazingly large. Charge flows much more readily through several miles of copper wire than it does through just a few inches of the ceramic material that is used as an insulator for electrical transmission lines. There is simply no comparison between the two types of material.

Some materials, however, have properties that are intermediate between those of a good conductor and a good insulator. These materials are called *semiconductors;* carbon and silicon are probably the most familiar examples. A damp wooden rod serves as an example of a semiconductor for purposes of discharging the electroscope. With the appropriate amount of moisture, the rod will cause a slow discharge of the electroscope.

Although semiconductors are far less common than good conductors or good insulators, their importance to modern technology is enormous. The ability to control the level of conduction in these materials by adding small amounts of other substances to them has led to the development of miniaturized electronic devices such as transistors and integrated circuits. The entire computer revolution is based on the use of these materials (mostly silicon). Chapter 20 provides a closer look at some of these developments.

Charging by Induction

Is it possible to charge an object without actually touching it with another charged object? It turns out that this can be done. The process is called *charging by induction,* and it involves the conducting property of metals. It is easy enough to demonstrate.

Suppose that you charge a plastic rod with cat's fur, obtaining a negative charge, and then bring the rod near to a metal ball mounted upon an insulating post, as illustrated in figure 11.7. Since charges are free to flow within the metal ball, the free electrons in the ball are repelled by the negatively charged rod. This produces a collection of negative charge (excess electrons) on the side of the ball opposite the rod, and a net positive charge (shortage of electrons) on the side nearest the rod. The overall charge of the metal ball is still zero.

In order to charge the ball by induction, you now touch the ball on the side opposite the rod, still holding the rod near to, but not touching, the ball. The negative charge then flows from the ball to your body, since it is still being repelled by the negative charge on the rod. If you now remove your finger and then remove the rod, there will be a net positive charge left upon the ball (fig. 11.8). You can easily test this fact by bringing the metal ball near an electroscope that has been given a negative charge from the plastic rod. The leaves will come closer together when the metal ball charged by

Figure 11.7 The negatively charged rod is brought near to a metal ball mounted upon an insulating post, thus producing a separation of charge on the ball.

Figure 11.8 Touching a finger to the opposite side of the metal ball draws off the negative charge, leaving the ball with a net positive charge.

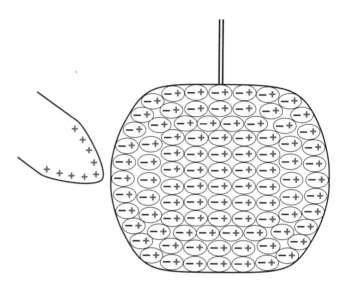

Figure 11.9 The negative charge in the atoms of the pith ball is attracted to the positively charged glass rod, while the positive charge in the atoms is repelled. This produces a polarization of charge in the atoms. The size of the atoms is grossly exaggerated.

This process of charging by induction illustrates the mobility of charges on a conducting object such as the metal ball. It is also important in the operation of machines used for generating electrostatic charge, as well as in explaining some of the phenomena that occur during lightning storms.

Polarization of Insulating Materials

In our initial experiments with the pith balls, we noticed that they were attracted to the charged rod before they had a chance to become charged themselves (see fig. 11.2). How can we explain this phenomenon? What happens inside an insulating object when it is brought near to a charged object?

Unlike the charges in the metal ball, the charges in the pith ball (or other insulating material) are not free to migrate through the material. Instead, they are tied to the atoms or molecules of which the material is made. Within an atom or molecule, however, charges have some freedom of movement; the distribution of charge within the atom or molecule can change.

Without worrying about details of atomic structure, we can develop a rough picture of what happens to the charge in atoms of an insulating material that is brought near to a charged object. The basic idea is illustrated in figure 11.9, which magnifies the pith ball and greatly exaggerates the relative size of the atoms. Within each atom, a small distortion of the charge distribution takes place: the negative charge in the atom is attracted to the positively charged rod, and the positive charge is repelled.

induction approaches, indicating the presence of positive charge on the metal ball.

The proper sequence of events is important if this experiment is to work. The charged rod must be held in place while you touch the ball on the opposite side and remove your finger. Only then can the charged rod be moved away. The ball ends up with a charge that is opposite to that of the rod. This would also be true if we had used a positively charged glass rod; the metal ball would then end up with a negative charge.

Each atom has now become an *electric dipole,* in which the center, or average location, of the negative charge is separated by a slight distance from the center of the positive charge. The atom now has positive and negative *poles,* hence the term *dipole,* and we say that the material has become *polarized.* The overall effect of these atomic dipoles within the insulating material is to produce a slight negative charge on the surface of the pith ball nearest to the positively charged rod and a slight positive charge on the opposite surface. The adjacent positive and negative charges within the material cancel one another.

The pith ball itself, then, becomes an electric dipole in the presence of the charged rod. Since the negatively charged surface is closer to the rod than the positively charged surface, it experiences a stronger electrostatic force, so that the pith ball is attracted to the charged rod. At this point, the overall charge on the pith ball is still essentially zero. Once the ball comes in contact with the charged rod, however, some of the charge on the rod can be transferred to the pith ball, which then becomes positively charged like the rod. The pith balls are then repelled by the rod, as we observed earlier.

This ability to become polarized is an important property of insulating materials. The extent of polarization can differ substantially for different materials, a fact that is characterized by the *dielectric constant* of the material. Materials with a large dielectric constant can be polarized to a greater extent than those with a small dielectric constant.

Polarization explains why small bits of paper or styrofoam are attracted to a charged object such as a synthetic-fabric sweater that has rubbed against some other material. Electrostatic precipitators, used to remove particles from smoke in industrial smoke stacks, make use of this property. Polarized particles are attracted to charged plates in the precipitator and removed from the emitted gases. There are many other applications.

11.3 COULOMB'S LAW: THE ELECTRO-STATIC FORCE

Although we cannot see electric charge, we can see the effects of the force that acts between charged objects. Pith balls and foil leaves refuse to hang straight downward when they have the same charge; a repulsive force pushes them apart. Can we describe this force in a quantitative way? How does it vary with distance and quantity of charge? Is it similar in some way to the gravitational force?

Questions such as these were actively explored by scientists in the latter part of the eighteenth century. Like the gravitational force, the electrostatic force was apparently present even when objects were not in contact; it acted at a distance. Although others had already speculated about the form of the electrostatic force law, it was the experiments of Charles Coulomb (1736–1806) during the 1780s that settled the matter, establishing what we now know as *Coulomb's law.*

Coulomb's Experiment

At first glance, measuring the strength of a force such as the electrostatic force might seem like a simple exercise. In reality, however, it is not an easy thing to do. Although much stronger than the gravitational force between ordinary objects, the electrostatic force is still relatively weak. Its measurement required the development of techniques for measuring small forces. In addition, Coulomb was faced with the problem of defining how much charge was present, which was far from a trivial matter.

Coulomb's answer to the problem of measuring weak forces was to develop what we now call a *torsion balance,* which is pictured schematically in figure 11.10. Two small metal balls are balanced upon an insulating rod that is suspended at the middle from a thin wire. A force applied upon either ball perpendicular to the rod produces a torque that causes the wire to twist. If we have previously measured how much force is required to produce a given angle of twist (a calibration process), then we have a method of measuring a weak force. The balls and wire are contained within a glass-walled enclosure to avoid disturbance from air currents.

To measure the electrostatic force, we must charge one of the balls in some manner. A third ball, also charged, can then be inserted on the end of an insulating rod, as shown in fig-

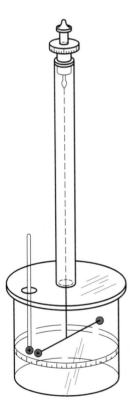

Figure 11.10 A schematic diagram of Coulomb's torsion balance. The degree of twist of the wire provides a measure of the repulsive force between the two charges.

ure 11.10. If it has the same charge as the ball on the end of the suspended rod, they will repel one another, and the resulting torque will twist the wire. By adjusting the distance between the two charged balls, we can measure the strength of the repulsive force at different distances.

The problem of determining the quantity of charge required even more ingenuity. An electroscope gave only a rough indication of the quantity of charge present and no consistent set of units or procedures for specifying quantity of charge existed at the time of Coulomb's work. His solution to this problem was to develop a system of charge division. He started with an unknown quantity of charge on a single metal ball mounted upon an insulating stand, as shown in figure 11.11. He then touched this ball to an identical ball mounted upon a similar stand. He reasoned that the two balls would now contain equal quantities of charge (half of the original amount on the first ball), and this could be verified by bringing the balls near to an electroscope.

If a third identical metal ball were touched to one of the first two, then the charge would again be divided equally, and these two balls would each have one-half the charge remaining upon the ball that did not participate in this second exchange. With several more identical balls, this process of splitting the charge can be continued. Although this process does not provide an absolute measure of charge, we can say with confidence that one ball contains twice as much charge, or perhaps four times as much charge, as another. We therefore know the relative quantities of charge that are involved.

Coulomb used this process to test the effects of different amounts of charge. If the balls used in the charge-division process were the same size as those used in the torsion balance, then touching one of the torsion-balance balls with one of the others represented one more charge division.

Using these procedures, Coulomb was able to determine how the strength of the electrostatic force varied with the amount of charge as well as with the distance between the two charged objects.

Coulomb's Law

The results of Coulomb's work can be simply stated:

The electrostatic force between two charged objects is proportional to the magnitudes of each of the charges and inversely proportional to the square of the distance between the charges.

If we let the letter q represent quantity of charge, then this statement can be expressed in symbols as follows:

$$F = \frac{kq_1q_2}{r^2},$$

where k is simply a constant (called *Coulomb's constant*) whose value depends upon the units used for charge and distance, and r is the distance between the centers of the two charges (see fig. 11.12).

Although Coulomb himself used different units, we now usually express charge in units called, appropriately enough, *coulombs* (C). If distance is measured in meters, then the value of Coulomb's constant turns out to be approximately

$$k = 9.0 \times 10^9 \text{ N·m}^2/\text{C}^2.$$

The definition of the coulomb itself is determined by measurements of electric current, which are discussed in chapter 12. Coulomb's constant is regarded as a constant of nature that is determined experimentally using the techniques pioneered by Coulomb. A force computation using Coulomb's law and Coulomb's constant appears in box 11.1.

Figure 11.12 presents a diagram to go with the words and formula associated with Coulomb's law. The two charges shown are both positive, so the force is repulsive since like charges repel. If one charge was negative and the other positive, the direction of both arrows would be reversed. The forces obey Newton's third law: the two charges experience equal but oppositely directed forces.

Comparisons to Newton's Law of Gravitation

The electrostatic force has the same inverse-square variation with distance as Newton's law of gravitation (see chapter 5). If we double the distance between the two charges, the force between them is only one-fourth of what it was before the distance was doubled. Tripling the distance produces a force that is one-ninth of that at the original distance, and so forth. The strength of the interaction between the two charges falls off rapidly as the distance between them increases.

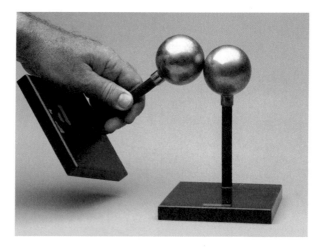

Figure 11.11   By bringing two identical metal balls into contact, one charged and the other one initially uncharged, equal quantities of charge are obtained upon the two balls.

Figure 11.12 Two positive charges exert equal but oppositely directed forces upon one another according to Coulomb's law and Newton's third law of motion. The force is inversely proportional to the square of the distance *r* between the two charges.

Box 11.1

Sample Exercise

Two positive charges, one 2.0 μC and the other 7.0 μC, are separated by a distance of 20 cm. What is the magnitude of the electrostatic force that each charge exerts upon the other?

$q_1 = 2.0 \times 10^{-6}$ C.   (1 μC = 1 microcoulomb = 10^{-6} C.)

$q_2 = 7.0 \times 10^{-6}$ C.

$r = 20$ cm = 0.20 m.

$$F = \frac{kq_1q_2}{r^2}$$

$$= \frac{(9.0 \times 10^9 \text{ N·m}^2/\text{C}^2)(2.0 \times 10^{-6} \text{ C})(7.0 \times 10^{-6} \text{ C})}{(0.20 \text{ m})^2}$$

$$= \frac{0.126 \text{ N·m}^2}{0.040 \text{ m}^2} = \textbf{3.15 N}.$$

(Each charge has a 3.15 N force acting upon it, but the two forces are in opposite directions. The forces are repulsive, since the two charges have the same sign; see figure 11.12.)

Since the gravitational force and the electrostatic force are two of the basic forces of nature, we should take a closer look at their similarities and differences. Putting their symbolic expressions side by side, we have

$$F_g = \frac{Gm_1 m_2}{r^2}; \qquad F_e = \frac{kq_1q_2}{r^2}.$$

An obvious difference, of course, is that the gravitational force depends upon the product of the masses of the two objects, and the electrostatic force depends upon the product of the charges of the two objects. Otherwise, however, the forms of these two force laws appear to be very similar.

A somewhat more subtle difference, however, has to do with direction. The gravitational force is always attractive; as far as we know, there is no such thing as negative mass. The electrostatic force, on the other hand, can be either attractive or repulsive according to the signs of the charges of the two objects. The rule that like charges repel and unlike charges attract determines the direction.

Another difference has to do with the relative strengths of these two forces. For objects of ordinary size and for subatomic particles, the gravitational force is much weaker than the electrostatic force for typical values of mass and charge. An enormous mass is required for at least one of the objects in order to produce a detectable gravitational force. For charged particles at the atomic or subatomic level, the electrostatic force totally dominates the gravitational force.

Although the basic forms of these two force laws have been known for over two hundred years, physicists are still trying to understand the underlying reasons for the relative strengths of these and other fundamental forces of nature. The search for a *unified field theory* that would explain the relationships between all of the fundamental forces is a major area of research within modern theoretical physics. Some of today's students will undoubtedly play important roles in this search.

11.4 THE ELECTRIC FIELD

Using Coulomb's law, we can readily compute the magnitude of the electrostatic force between any two charged objects, provided that the objects are small compared to the distance between them. Such small charged objects are often referred to as *point charges*. If there are more than two point charges, we can compute the net force on any one of them by adding as vectors (see appendix C) the forces due to each of the other charges. If there are several charges, however, or if the charge is distributed over a surface or throughout the volume of a large object, these computations can become very tedious indeed.

In order to describe the effect of such a distribution of charges on some other individual charge, we use the concept of *electric field*. The utility of the field idea goes well beyond this, however; it has become a central concept in modern theoretical physics.

To most of us, the word *field* suggests a field of wheat or a meadow filled with wild flowers. Unfortunately, the concept of electric field in physics is somewhat more abstract, which often causes problems for people who are trying to gain an initial understanding of the concept. Approaching it first in the context of a simple example involving point charges should be helpful.

A Force Computation

Suppose that we have two point charges, q_1 and q_2, separated by some distance, as shown in figure 11.13. These could be pith balls or small metal balls that have been charged from a charged rod or any other means. A third charge, q_0 is placed between the initial two charges, and we want to find the net electrostatic force acting upon this third charge.

The coulomb is a rather large unit of charge for quantities on the scale of small objects such as pith balls, so we have chosen the charge amounts to be of the order of micro-coulombs, or 10^{-6} coulombs. Computing the net force on q_0 is just a matter of finding the force on this charge due to each of the other two charges and then adding the results as vectors. Since the first charge has a magnitude of 3×10^{-6} C and is 10 cm from q_0, which has a magnitude of 4×10^{-6} C, the force exerted on q_0 due to q_1 is given by Coulomb's law:

$$F_1 = \frac{kq_1q_0}{r^2} = \frac{(9 \times 10^9 \text{ N·m}^2/\text{C}^2)(3 \times 10^{-6} \text{ C})(4 \times 10^{-6} \text{ C})}{(0.10 \text{ m})^2}$$

$$= 10.8 \text{ N}.$$

A similar computation for force F_2 due to charge q_2, which has a magnitude of 2×10^{-6} C and is 20 cm from q_0, yields a force of 1.8 N. This second charge is slightly weaker than the first, but it is also twice as far from q_0 as the first. Since the Coulomb force is inversely proportional to the square of the distance, this second charge exerts a consider-ably weaker force on the middle charge. All three charges are positive, so both of the forces acting upon q_0 are repulsive; they act in opposite directions, as shown in figure 11.14. The net force acting upon q_0 is thus

$$\Sigma F = F_1 - F_2 = 10.8 \text{ N} - 1.8 \text{ N} = 9.0 \text{ N}.$$

If the charges were not tied down somehow, this net force would have the effect of accelerating charge q_0 to the right towards q_2, which exerts the weaker repulsive force. The same general procedure used to compute this force could be used to compute the net force acting upon the other two charges. If the three charges did not all lie upon the same

$+3 \times 10^{-6}$ C $\qquad +4 \times 10^{-6}$ C $\qquad\qquad +2 \times 10^{-6}$ C

$q_1 \qquad\qquad\quad q_0 \qquad\qquad\qquad\qquad q_2$

$\longmapsto$ 10 cm $\longmapsto\longleftarrow$ 20 cm $\longrightarrow$

Figure 11.13 **A charge q_0 is placed between the charges q_1 and q_2 at the distances shown. What is the net force on q_0?**

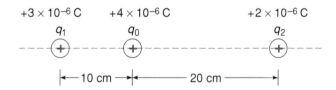

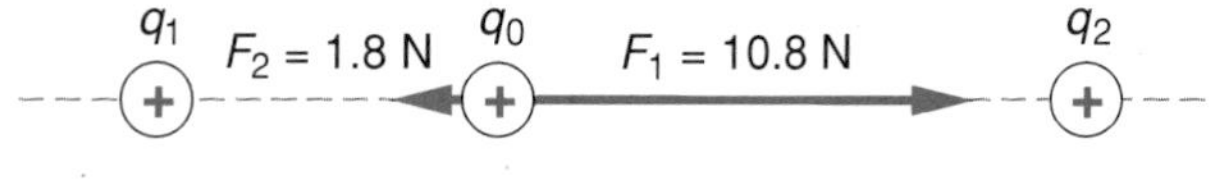

$q_1 \quad F_2 = 1.8 \text{ N} \quad q_0 \qquad F_1 = 10.8 \text{ N} \qquad\qquad q_2$

Figure 11.14 **The two forces acting upon q_0 are in opposite directions, yielding a net force of 9.0 N.**

straight line, however, the forces could not be simply added or subtracted; they would have to be added as vectors by using the head-to-toe techniques introduced in the chapters on mechanics and in appendix C. Obviously, several charges in a two- or three-dimensional array would provide us with a messy computation!

The Definition of Electric Field

Suppose that we wanted to know the force on some other charge at the same location as q_0 but with a different magnitude or sign. Would we have to go through this same computation over again? It could be done this way, but there is an easier way that involves the concept of electric field.

We could regard the charge q_0 as a test charge that had been inserted at this particular location in order to assess the strength of the electrostatic effect at that point. Since we obtained a force of 9.0 N (towards the right) on this charge, the force per unit positive charge would be

$$\frac{F}{q_0} = \frac{9.0 \text{ N}}{4 \times 10^{-6} \text{ C}} = 2.25 \times 10^6 \text{ N/C}.$$

Knowing the force per unit charge would permit us to compute the force on any other charge placed at that same point. For example, if we placed a charge of 5×10^{-6} C at that point, the force would be

$$F = \left(\frac{F}{q_0}\right)q = (2.25 \times 10^6 \text{ N/C})(5 \times 10^{-6} \text{ C}) = 11.25 \text{ N}.$$

This idea of using the force per unit charge as a measure of the strength of the electrostatic effect at some point in space lies at the heart of the electric-field concept. In fact, we define an electric field as follows:

The electric field at a given point in space is the electric force per unit positive charge that would be exerted on a charge if it were placed at that point. It is a vector having the same direction as the force that would be exerted on a positive charge at that point.

In symbols, this definition can be stated as follows:

$$\vec{E} = \frac{\vec{F}}{q_0},$$

where the standard symbol $\vec{E}$ represents the electric field.

In other words, the ratio F / q_0 that we computed in our example was the magnitude of the electric field at that point. We can use the field to find the force on any other charge placed at that point by performing the following simple operation:

$$\vec{F} = q\vec{E},$$

as illustrated in our computation of the force on the 5×10^{-6} C charge. If the charge happened to be negative, the minus sign in this computation would simply indicate that the direction of the force on a negative charge is opposite to the direction of the field. The direction of the field is that of the force that would exist on a *positive* charge placed at that point.

It is important to recognize, however, that the field and the force are not the same thing. We can talk about the field at a point in space with no charge located at that point. The field tells us the magnitude and direction of the force that would be exerted on any charge placed at that point. There must be a charge at this point if there is to be a force, but the field exists regardless of whether there is a charge there or not. The term *field* actually refers to the collection of values of the field vector $\vec{E}$ at *all* points in space in the vicinity of the charge distribution that produces the field.

The electric field is a property of space; it can exist even in a vacuum. When we use the field concept, we shift our focus from the interactions between particles or objects to the way in which a charged object affects the space surrounding it. The field concept is not restricted to electrostatics; we can also define and use the concepts of a gravitational field or a magnetic field, as well as others.

Field Lines

The concept of an electric field was formally introduced by the Scottish physicist James Clerk Maxwell (1831–1879) around 1865 as part of his highly successful theory of elec-

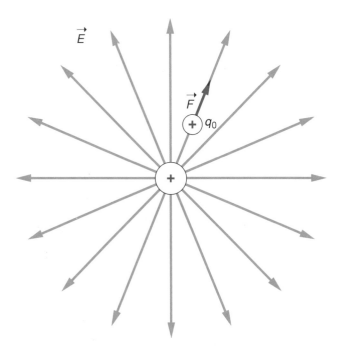

Figure 11.15   **The directions of the electric field lines around a positive charge can be found by imagining a positive test charge q_0 being placed at various points around the source charge. The field has the same direction as that of the force on a positive test charge.**

tromagnetism. The idea had been used informally, however, by Michael Faraday (1791–1867), who used the concept of what we now call *field lines* as an aid in visualizing electric and magnetic effects. Faraday was not adept at mathematics, but he was a brilliant experimentalist who made good use of mental pictures.

To illustrate the idea of field lines, suppose that we consider the electric field associated with just a single positive charge. If we used a positive test charge to assess the field direction and strength at various points, we would find that the test charge is repelled by the initial positive charge wherever we place it around this charge. If we draw lines to indicate the direction of the force on the test charge (which is also the direction of the field), we obtain a drawing like that shown in figure 11.15.

Actually, the diagram in figure 11.15 is just a two-dimensional slice of what should be a three-dimensional diagram. The electric field lines associated with a single positive charge radiate in all directions from the charge. If we also adopt the convention that the lines always start on positive charges and terminate on negative charges, then the lines can provide some additional information. The density of the field lines crossing a perpendicular area is proportional to the strength of the field.

In figure 11.15, this second property of the field lines is illustrated by the fact that the lines are much closer together (denser) near the charge than they are farther away. According to Coulomb's law, the force on a test charge (and the electric field) decreases proportionally to $1/r^2$ as its distance from the initial charge increases. The surface area of a sphere centered on the charge, on the other hand, increases proportionally to r^2. Thus the number of field lines per unit area that cross the surface of such a sphere decreases proportionally to $1/r^2$, as does the strength of the field; the area of the sphere increases with distance from the charge, but the number of field lines crossing the spherical surface remains constant.

The field lines, then, give us a means of visualizing both the direction and the strength of the field. A picture is worth a thousand words and maybe as many as a hundred equations! Figure 11.16 provides a two-dimensional slice of the electric field lines associated with a negative charge. Here the field lines terminate on the charge and must be directed inwards to indicate the proper direction for the force on a positive test charge.

As a final example, consider the field lines associated with an electric dipole in figure 11.17. An electric dipole is simply two charges of equal magnitude, but opposite sign, separated by a small distance. The field lines originate on the positive charge and terminate on the negative charge. Try to imagine a positive test charge placed at various points around the dipole. Do the field lines agree with your sense of what the direction of the force on the test charge should be? This is the acid test for any field-line diagram.

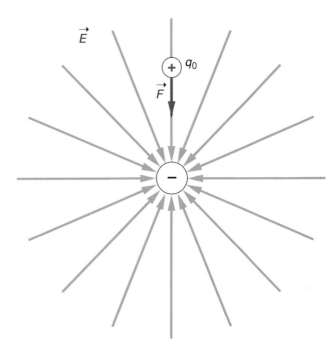

Figure 11.16 **The electric field lines associated with a negative charge are directed inward, as is indicated by the force on a positive test charge, q_0.**

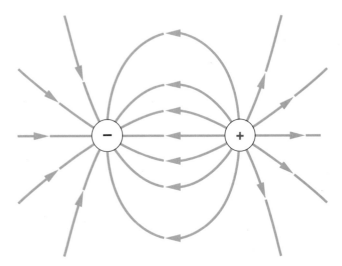

Figure 11.17 **The electric field lines associated with two equal but opposite charges (an electric dipole).**

11.5 ELECTRIC POTENTIAL

We have already noted the similarities and differences between the electrostatic force described by Coulomb's law and the gravitational force described by Newton's law of gravitation. In chapter 8, we defined the potential energy associated with the gravitational force, as well as the poten-

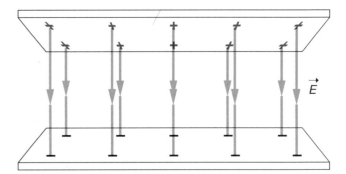

Figure 11.18 **Two parallel metal plates containing equal but opposite quantities of charge produce a uniform electric field in the region between them.**

tial energy of a spring. What about the electrostatic force? Can we talk about the potential energy of a charged particle being acted upon by an electrostatic force?

It turns out that the electrostatic force is a conservative force, which implies that we can define a potential energy for it. The potential energy associated with the electrostatic force is closely related to but not the same as the electric potential, which is often referred to simply as *voltage*. This is an extremely useful concept for discussing electric circuits, the subject of the next chapter, as well as many other phenomena.

What, then, is voltage and how is it related to electrostatic potential energy? We will approach these questions by first examining how the potential energy of a charged particle varies in a uniform electric field, using the same approach developed for computing potential energies in chapter 8.

Potential Energy in a Uniform Field

How do we go about producing a uniform electric field? The easiest way to obtain a uniform field is to arrange two parallel metal plates as shown in figure 11.18. If one plate is positively charged and the other is charged with an equal negative charge, then the field lines are straight lines originating at the positive charges and terminating at the negative charges, as shown in the diagram. You can test this conclusion yourself by thinking about the force on a positive test charge placed between the two plates, as we did in the previous section.

When the field lines are parallel to one another and evenly spaced, we say that we have a *uniform electric field*. The field does not vary in either direction or strength as we move from point to point within the region between the plates. It is constant within this region provided that we do not get too near to the ends of the plates.

Two conductors such as those in figure 11.18, separated by a region containing some insulating material such as air provide a useful means of storing charge. Such an arrangement is called a *capacitor,* which is essentially a device used to store charge. If the two conductors contain charges of opposite sign, the charges are attracted to one another and held in place by this attractive electrostatic force. The charges will not flow readily from the plates, therefore, unless a conducting path connects the two plates. Capacitors have numerous applications, particularly in electric circuits.

Suppose now that we place a positive charge in the uniform-field region of a parallel-plate capacitor. This charge will experience an electrostatic force in the direction of the electric field. The charge will be attracted toward the negative charges on the bottom plate and repelled by the positive charges on the top plate. If we released the charge, it would be accelerated towards the bottom plate. This acceleration has nothing to do with gravity; we assume that the gravitational force on the charged particle is very small compared to the electrostatic force.

If we apply an external force to move the charged particle in the opposite direction to the electric field, we can compute the work done by this external force. Since we have defined electric field as the electric force per unit positive charge, the electrostatic force acting on the charge is equal in magnitude to the charge times the field strength (qE). If we move the charge without accelerating it, then the external force must be equal in magnitude but opposite in direction to

the electrostatic force, so that the net force acting on the charge is zero. The external force, F, must therefore also be equal to qE (fig. 11.19).

It is easy, then, to calculate the work done by the external force. By our definition of work (see chapter 8), it is simply the force times the distance moved, $W = Fd$. When work is done to move an object (without accelerating it) against some conservative force, the potential energy of the object increases; it is like stretching a spring or lifting a mass. As with the other types of potential energy defined in chapter 8, the increase in potential energy is just equal to the work done by the external force. In this case then, the increase in potential energy is

$$\Delta PE = W = Fd = qEd.$$

Stated in words, the increase in potential energy is equal to the charge, q, times the field strength, E, times the distance moved, d.

The situation described here is exactly analogous to that involved in using an external force to lift an object against the force of gravity, thereby increasing its gravitational potential energy. The analogy is illustrated in figure 11.20. The gravitational potential energy always increases when we lift an object, of course, whereas the direction of increase for the electrostatic potential energy depends upon which plate is positively charged and the resulting direction of the electric field. If we had placed the positively charged plate on the bottom, so that the field direction was upwards, then the potential energy of a positive charge would increase if we moved the positive charge downwards.

Whenever work is done to move a charged particle against the direction in which it would naturally tend to move under the influence of the electrostatic force, its potential energy increases. If, for example, we move a negatively charged particle away from a positively charged particle, we increase the potential energy of the negative particle. As in the case of

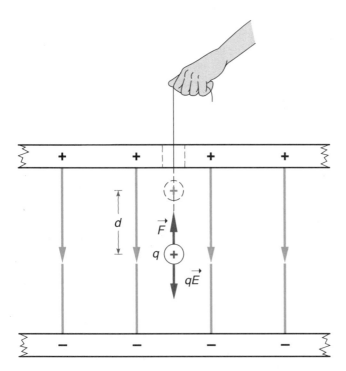

Figure 11.19 An external force, *F,* equal in magnitude to the electrostatic force *qE,* is used to move the charge *q* a distance *d* against the field.

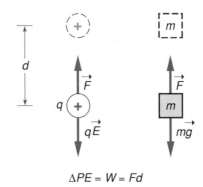

$$\Delta PE = W = Fd$$

Figure 11.20 The increase in potential energy (Δ*PE*) obtained when a charge *q* is moved against the electrostatic force is analogous to that which occurs when a mass *m* is lifted against the gravitational force.

a stretched spring, that increase in potential energy could easily be converted to kinetic energy if we released the charged particle and allowed it to accelerate towards the positive charge under the influence of the attractive electrostatic force. Other examples are provided in the questions at the end of this chapter.

Definition of Electric Potential

How is the concept of electric potential related to these ideas? Electric potential is related to electrostatic potential energy in much the same manner as the concept of electric field is related to the electrostatic force. We can regard the positive charge used in our discussion of electrostatic potential energy as a test charge used to determine how the potential energy varies with position. We define the change in electric potential, then, as follows:

The change in electric potential is equal to the change in electrostatic potential energy experienced by a positive charge divided by the magnitude of the charge:

$$\Delta V = \frac{\Delta PE}{q+}.$$

We have used the standard symbol V here to represent the electric potential. It is the change in potential energy per unit positive charge.

As you can see from the definition, the units of electric potential are those of energy per unit charge. In the metric system, this unit is called a volt (V), which is defined as one joule per coulomb (1 J/C = 1 volt). It is the unit and the corresponding term, *voltage,* then, that suggests the symbol V that is commonly used for electric potential.

As with the concept of electric field, no test charge or other charge need be present at a point in space in order to talk about the electric potential at that point. The change in electric potential from one point to another is equal to the change in potential energy *per unit positive charge* that would occur if a positive charge were moved between these two points. In other words, to find the change in potential energy for such a charge, we would multiply the change in electric potential by the magnitude of the charge:

$$\Delta PE = q\Delta V.$$

Electric potential and potential energy are closely related, but they are not the same thing. If the charge q happens to be negative, its potential energy will decrease when it moves in the direction of increasing electric potential.

If we want to give a specific value for electric potential rather than just stating a change in electric potential, we need to define a reference level or position where the potential is zero. The same problem arises in talking about either gravitational or electrostatic potential energy; the work done is equal to the *change* in potential energy rather than to a spe-

cific value of potential energy. If we define a position at which the potential is zero, however, then the change in potential in moving to any other position gives us the value of the potential at that other position. The position at which the potential is defined to be zero is called the *reference level;* other values of potential can then be defined in relation to that position.

A numerical example can help to clarify some of these ideas. Suppose that the constant electric field between two metal plates is equal to 10 000 N/C, a value that is easily achieved. Suppose also that we start a positively charged particle with a charge of $+ 5 \times 10^{-4}$ C at the bottom (negatively charged) plate and move it towards the top plate. (You could imagine that a string is tied to the charge and is pulling it upward, as in figure 11.21.) If the distance between the plates is 3 cm, then the work done in moving the charge to the top plate and the increase in potential energy is

$$W = \Delta PE = qEd = (5 \times 10^{-4} \text{ C})(10^4 \text{ N/C})(.03 \text{ m})$$

$$= 0.15 \text{ joules.}$$

The change in electric potential associated with this motion is equal to the change in electrostatic potential energy per unit charge:

$$\Delta V = \frac{\Delta PE}{q+} = \frac{0.15 \text{ J}}{5 \times 10^{-4} \text{ C}} = 300 \text{ volts.}$$

The electric potential has increased by 300 volts in moving the distance of 3 cm from the bottom to the top plate. The natural tendency of the charge if released, of course, would be to move in the opposite direction.

Finally, suppose that we choose the value of the electric potential to be zero at the bottom plate. This sets the reference level. The electric potential at the top plate is then +300 V. Halfway between the two plates the electric potential would be 150 V and so forth. There is a continuous increase in potential from 0 to 300 V along the distance from the bottom plate to the top plate.

Relationship Between Electric Potential and Electric Field

In the case of the uniform field, there is a simple relationship between the magnitude of the electric field and the change in electric potential. Since $\Delta PE = qEd$ and $\Delta V = \Delta PE / q$, this relationship has the following form:

$$\Delta V = \frac{qEd}{q} = Ed.$$

As our example has shown, however, the increase in electric potential, or voltage, occurs in the direction opposite to the direction of the field; because the potential energy of a positive charge increases when we move it against the field.

This relationship between E and ΔV can be used *only* in the case of a uniform field. If the field strength varies with

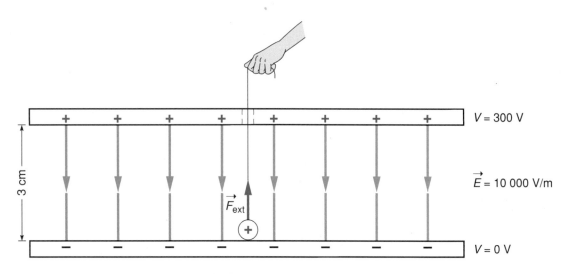

Figure 11.21   A positive charge is moved from the bottom plate to the top plate by an external force.

position, then the work computation is more involved, and more complicated relationships will be found. The field is approximately uniform, however, in many practical situations, and it is often convenient to recall and use this simple relationship. In fact, since $E = \Delta V / d$ in this case, we often express the value of the field strength in units of volts per meter (V/m), which turns out to be equal to newtons per coulomb (N/C).

Even when the field is not uniform, the electric potential always shows a maximum rate of increase in the direction directly opposite to the field direction. In the field associated with a positive charge, for example, the electric potential increases with motion towards the charge, whereas the field lines radiate outwards from the charge. Figure 11.22 shows the field for a positive charge with some values of electric potential shown at different distances from the charge. The reference level in this case is set by letting the potential be zero at an infinite distance from the charge.

In any situation, we can always determine how the potential varies by thinking about what would happen to a positive test charge being moved within the electric field in question. The electric potential always increases as we move towards positive charges and away from negative charges because the potential energy of a positive charge increases under those conditions. This provides a simple test that can strengthen your intuitive feel for what otherwise can seem to be a very abstract concept.

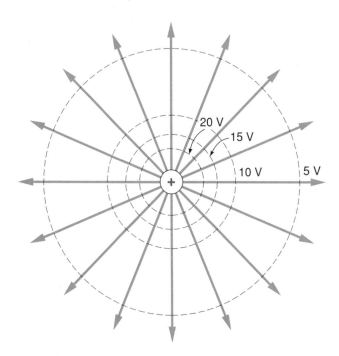

Figure 11.22   The electric potential (represented by the dashed equal-potential lines) increases as we move closer to the positive charge.

Box 11.2

Everyday Phenomenon:
Lightning

The Situation. We have all observed the awe-inspiring beauty and power of a good electrical storm. The flashes of lightning, followed at varying time intervals by claps of thunder, can be both fascinating and frightening. You have probably also heard of Ben Franklin's famous experiment with the kite and key, which demonstrated that lightning was an electrical phenomenon that could produce sparks and shocks identical to those experienced in electrostatic experiments. He was fortunate to survive that experiment.

What is lightning? How are thunderclouds capable of producing the impressive electrical discharges that we observe? Can we use the concepts developed in this chapter to provide a basic picture of what happens in an electrical storm? Such an understanding may help you avoid being struck by lightning.

The Analysis. Although the exact nature of the charging process is not well understood, extensive measurements have shown that most thunderclouds generate a separation of charge within themselves that produces a net positive charge near the top and a net negative charge near the bottom. It is thought that the highly turbulent convection processes taking place within the clouds are responsible for separating and transporting the charge. It is known that thunderclouds consist of rapidly rising and falling columns of air and water, with cells of rising air often being found adjacent to cells of falling air and water.

The charge separation that occurs within a thundercloud produces strong electric fields within the cloud as well as between the cloud and the earth. Since moist earth is a reasonably good conductor of electricity, a positive charge is induced on the surface of the earth below the cloud due to the negative charge on the bottom of the cloud. The electric field generated by this charge distribution (pictured in the drawing) can be several thousand volts per meter. Since the bottom of the cloud is usually several hundred meters above the surface of the earth, the potential difference that this represents ($\Delta V = Ed$) can easily be several million volts!

(Interestingly, there is an electric field of a few hundred volts per meter in the atmosphere near the surface of the earth even during fair weather. This field is weaker, however, and in the opposite direction to that which is usually generated between the cloud and the earth in an electrical storm.)

What happens, then, when a lightning strike occurs? Dry air is a good insulator, but moist air conducts electricity somewhat more readily. Any material will conduct, however, if the potential difference across the material is large enough. The very large potential difference between the cloud bottom and the earth (or between clouds) creates a situation in which the air will conduct charge between the cloud and the earth or other clouds. An initial flow of charge along some path that offers the best conducting properties over the shortest distance heats the air and ionizes (removes electrons from) some of the atoms along

Flashes of lightning illuminate the area. What is lightning and how is it produced?

The charge distribution within a thundercloud induces a positive charge on objects on the earth directly below the cloud.

that path. The charged particles represented by the ionized atoms now make the air much more conductive along that path, and a much greater flow of charge can then occur.

What actually takes place, then, is that the initial flow of charge (called the *leader*) makes the air a much better conductor than it would otherwise be along the path followed by the leader. The following strokes or discharges all take place along this same conducting path in very rapid succession, each one increasing the conductivity along that path. A very

large discharge or flow of charge can take place between the earth and the cloud in a very short period of time. Light is produced during the discharge by the ionized and excited atoms in a manner that is discussed in chapter 16. The sound wave that we know as thunder is produced at the same time, but takes longer to reach us because sound travels at a much lower speed than light.

Pointed conducting objects, such as a lightning rod, produce strong fields and potential differences in the vicinity of the point, and therefore can cause some local ionization of the air, which enhances the conductivity of air in that area. This phenomenon, together with the fact that a lightning rod is generally placed higher than surrounding objects, makes the rod part of a particularly favorable path for the flow of charge. Ben Franklin was intrigued by this property of pointed objects and was the inventor of the lightning rod. If the lightning rod is connected to the ground by a metal strap or cable, it can provide a path for the flow of charge which takes it safely around a house or building.

A large tree or a person standing on top of a tree-less hill can also play this role. People and trees are better conductors than the surrounding air, so they can become a part of the conducting path for a discharge. Standing under an isolated tree (such as on a golf course) during a lightning storm is therefore dangerous. A few golfers are killed every year in that manner. Your best bet is to be inside a building or car; if you are outdoors, be sure you are not standing under the highest conducting object in your immediate vicinity. Find a low (and preferably dry) spot and hunker down.

SUMMARY

Curiosity regarding the mildly shocking effects of simple electrostatic phenomena has led us to an understanding of the electrostatic force described by Coulomb's law. The difference between conductors and insulators and the concepts of electric field and electric potential are also involved in explaining the variety of electrostatic phenomena that are observed. The following ideas were introduced.

Electric Charge. Rubbing different materials together separates a quantity that we call *electric charge,* which is clearly capable of exerting a force upon other charges. There are two types of charge; following the single-fluid model introduced by Ben Franklin, they are called *positive* and *negative.* Like charges repel, and unlike charges attract.

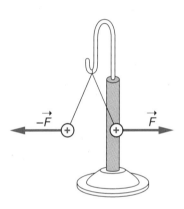

Conductors and Insulators. Different materials vary widely in their ability to permit the flow of charge. These differences between conductors and insulators help to explain why charging by induction works and why uncharged bits of paper are attracted to charged objects.

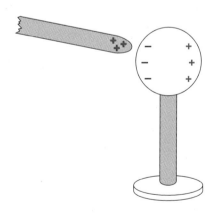

Coulomb's Law. By careful experiments, Coulomb was able to show that the electrostatic force between charged objects is proportional to the product of the two charges and inversely proportional to the square of the distance between the charges.

$$F = \frac{kq_1q_2}{r^2}$$

Electric Field. The electric field is defined as the electric force per unit positive test charge that would exist if there were a charge present at a point in space. Knowledge of the field at some point allows us to compute the force upon any charge placed at that point.

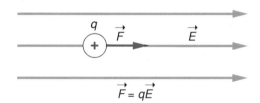

$$\vec{F} = q\vec{E}$$

Electric Potential. The electric potential is defined as the potential energy per unit positive charge that would exist at some point in space if there were a charge present at that point. The change in potential energy of a charge can be found, as before, by computing the work done to move the charge against the electrostatic force.

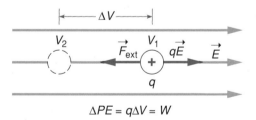

$$\Delta PE = q\Delta V = W$$

QUESTIONS

Q11.1 When two different materials are rubbed together, do the two materials acquire the same type of charge or different types of charge? Explain how you could justify your answer with a simple experiment.

Q11.2 Two pith balls are both charged by contact with a plastic rod that has been rubbed by cat's fur.

    a. What sign of charge will the pith balls have? Explain.

    b. Will the two pith balls attract one another or repel one another? Explain.

Q11.3 Two pith balls are charged by touching one to a glass rod that has been rubbed with a nylon cloth and the other to the cloth itself.

    a. What sign of charge will each pith ball have? Explain.

    b. Will the two pith balls attract one another or repel one another? Explain.

Q11.4 When a glass rod is rubbed by a nylon cloth, which of these two objects gains electrons? Explain.

Q11.5 When you comb your hair with a plastic comb, what sign of charge would you expect to be acquired by the comb? Explain.

Q11.6 If you touch the metal ball of a charged electroscope with an uncharged glass rod held in your hand, will the electroscope discharge completely? Explain.

Q11.7 When a metal ball is charged by induction using a negatively charged plastic rod, what sign of charge is acquired by the metal ball? Explain.

Q11.8 If, when charging by induction, you remove the charged rod from the vicinity of the metal ball before moving your finger from the ball, what will happen? Will the ball end up being charged? Explain.

Q11.9 Will bits of paper be attracted to a charged rod even if the bits of paper themselves have no net charge? Explain.

Q11.10 Can pith balls be initially attracted to a charged rod and then later repelled by the same rod, even though they have not touched any other charged object? Explain.

Q11.11 If the distance between two small charged objects is doubled, will the electrostatic force that one object exerts on the other be cut in half? Explain.

Q11.12 If two charges are both doubled in magnitude without changing the distance between them, will the force that one charge exerts upon the other also be doubled? Explain.

Q11.13 Can both the electrostatic force and the gravitational force be either attractive or repulsive? Explain.

Q11.14 Is it possible for an electric field to exist at some point in space at which there is no charge? Explain.

Q11.15 Two charges of equal magnitude but opposite sign lie along a line, as shown in the diagram. Using arrows, indicate the direction of the electric field at points *A*, *B*, *C*, and *D* on the diagram.

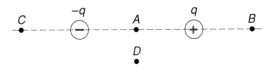

Q11.16 If we change the negative charge in the diagram shown for question 11.15 to a positive charge of the same magnitude, indicate the direction of the electric field at points *A*, *B*, *C*, and *D* that would result.

Q11.17 Three equal positive charges are located at the corners of a square as shown in the diagram. Using arrows, indicate the direction of the electric field at points *A* and *B* on the diagram.

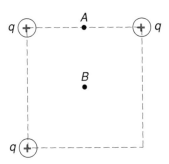

Q11.18 Is the electric field produced by a single positive charge a uniform field? Explain.

Q11.19 If a positive charge is moved toward a negative charge, does the potential energy of the positive charge increase or decrease? Explain.

Q11.20 If a negative charge is moved in the same direction as the electric field lines in some region of space, does the potential energy of the negative charge increase or decrease? Explain.

Q11.21 Does the electric potential increase or decrease with motion toward a negative charge? Explain.

Q11.22 Is electric potential the same thing as electric potential energy? Explain.

Q11.23 Will a negatively charged particle, initially at rest in an electric field, tend to move towards a region of lower electric potential if released? Explain.

EXERCISES

E11.1 An electron has a charge of -1.6×10^{-19} C. How many electrons would be needed to produce a net charge of -10^{-6} C?

E11.2 Two charged particles exert an electrostatic force of 5 N upon each other. What will the magnitude of the electrostatic force be if the distance between the two charges is reduced to $\frac{1}{3}$ of the original distance?

E11.3 Two charged particles exert an electrostatic force of 24×10^{-4} N upon each other. What will be the magnitude of the force if the distance between the two particles is increased to four times the original distance?

E11.4 Two positive charges, each of magnitude 3×10^{-6} C, are located a distance of 10 cm from each other.

- a. What is the magnitude of the force exerted upon each charge?
- b. Indicate the directions of the forces acting upon each charge in a diagram.

E11.5 A charge of $+5 \times 10^{-6}$ C is located 20 cm from a charge of -4×10^{-6} C.

- a. What is the magnitude of the force exerted upon each charge?
- b. Indicate the directions of the forces acting upon each charge in a diagram.

E11.6 An electron and a proton have equal but opposite charges of 1.6×10^{-19} C. If the electron and proton in a hydrogen atom are separated by a distance of 5×10^{-11} m, what are the magnitude and direction of the electrostatic forces exerted upon each particle?

E11.7 A uniform electric field is directed upwards and has a magnitude of 2.5 N/C. What are the magnitude and direction of the force on a -5.0-C charge placed in this field?

E11.8 A $+3 \times 10^{-6}$ C test charge experiences a downward electrostatic force of 9.0 N when placed at a certain point in space. What are the magnitude and direction of the electric field at this point?

E11.9 A -4×10^{-6} C charge is placed at a point in space where the electric field is directed towards the right and has a magnitude of 8.5×10^4 N/C. What are the magnitude and direction of the electrostatic force of this charge?

E11.10 A $+5$-C charge is moved from a position where the electric potential is 2 volts to a position where it is 10 volts. What is the change in potential energy of the charge associated with this change in position?

E11.11 The potential energy of a 2×10^{-6} C charge decreases from 0.04 J to 0.01 J when it is moved from point A to point B. What is the change in electric potential between these two points?

E11.12 The electric potential increases from 100 volts to 500 volts from the bottom plate to the top plate of a parallel-plate capacitor.

- a. What is the magnitude of the change in potential energy of a -3×10^{-4} C charge that is moved from the bottom plate to the top plate?
- b. Does the potential energy increase or decrease in this process?

CHALLENGE PROBLEMS

CP11.1 Three positive charges are located along a line as shown in the diagram. The 5-C charge at point A is 2 m to the left of the 2-C charge at point B, and the 3-C charge at point C is 1 m to the right of point B.

- a. What is the magnitude of the force exerted on the 2-C charge by the 5-C charge?
- b. What is the magnitude of the force exerted on the 2-C charge by the 3-C charge?
- c. What is the net force exerted upon the 2-C charge by the other two charges?
- d. If we regard the 2-C charge as a test charge used to probe the strength of the electric field produced by the other two charges, what are the magnitude and direction of the electric field at point B?

- e. If the 2-C charge at point B is replaced by a -7-C charge, what would be the magnitude and direction of the electrostatic force exerted on this new charge?

CP11.2 Suppose that two equal positive charges lie near one another as shown in the diagram on the next page.

- a. Using small arrows, indicate the direction of the electric field at the labeled points on the diagram. (Think in terms of the direction of the force that would be exerted upon a positive charge placed at each of these points.)

b. By drawing an equal number of field lines emerging from each charge, sketch the electric field lines for this distribution of charge. (See the diagrams in section 11.4.)

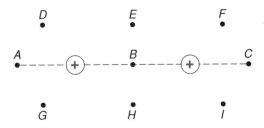

CP11.3 Suppose that one of the two charges in challenge problem 11.2 is twice as large as the other. Perform the procedures suggested in (*a*) and (*b*) of problem 11.2 for this new situation. (Your sketch of the field should now have twice as many field lines emerging from the larger charge as from the smaller charge.)

CP11.4 Suppose that the top plate of a parallel-plate capacitor has an electric potential of 0 V and the bottom plate has a potential of 600 V. The plates are 2 cm apart.

a. What is the change in potential energy of a 3×10^{-4} C charge that is moved from the bottom plate to the top plate?

b. What is the direction of the electrostatic force exerted upon this charge when it is between the plates?

c. What is the direction of the electric field between the plates?

d. What is the magnitude of the electric field between the plates?

CP11.5 Suppose that four equal positive charges are located at the corners of a square as shown in the diagram.

a. Using small arrows, indicate the direction of the electric field at each of the labeled points.

b. Would the magnitude of the electric field be equal to zero at any of the labeled points? Explain.

c. Would the electric potential at the center of the square be greater than at some point a long distance from the square? Explain.

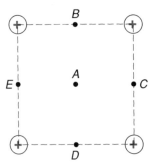

HOME EXPERIMENTS AND OBSERVATIONS

HE11.1 You can try the pith ball experiments described in section 11.1 with homemade equipment. Small, lightly wadded pieces of paper or pieces of styrofoam torn from a coffee cup can be used in place of the pith balls. These can be tied to pieces of thread and hung from any convenient support. Wooden pencils, plastic pens, or glass stirring rods can be used in place of the rods, and many different kinds of fabric are available in the form of clothing.

a. Using available materials on dry day, test to see which combinations of rod and fabric produce the best charge when rubbed.

b. Repeat the experiments described in section 11.1. Can you get both types of charge with the materials available?

HE11.2 In order to extend the observations of the previous experiment, you can construct a simple electroscope. Light aluminum foil can be used as the leaves, and these can be suspended from a paper clip by poking holes through the foil as shown. One end of the clip can be straightened, poked through a piece of cardboard, and then rebent. Place the

piece of cardboard on top of a drinking glass or a glass jar (see the drawing), and your electroscope is complete.

a. Test your electroscope with some of the materials suggested in the previous experiment.

b. Repeat the experiments on the conductors and insulators suggested in section 11.2.

c. Try charging a metal spoon by induction. The handle of the spoon should be wrapped in a napkin so that you will not discharge it when you handle it. Test the charge on the spoon with your electroscope.

12 Electric Circuits

Have you ever wondered how a flashlight works? The components are simple and familiar; a light bulb, a couple of batteries, and a cylindrical case with a switch. Its operation is also familiar; pushing the switch turns the light on or off, the batteries run down and need to be replaced or recharged, and occasionally the bulb burns out and must be replaced. But what is happening inside?

You turn electric switches on every day to produce light, heat, sound, or perhaps to run an electric motor. In fact, every time you start your car you are using an electric motor (the starting motor) powered by the battery. You know that electricity is involved in these situations, but exactly what takes place may be a mystery. It need not be.

Suppose that you have the essential components of a flashlight; the bulb, a battery, and a metallic conductor of some kind, such as a single piece of wire (fig. 12.1). Your task is to get the bulb to light. How would you do it? What principles would guide you in producing a working arrangement? If you have these items handy, see if you can do it.

This exercise presents a significant challenge for many people. Even those who can quickly get the bulb to light may not be able to clearly explain the principles involved. A thorough understanding of this very simple example is a good start to a basic understanding of electric circuits.

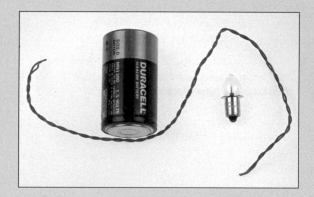

Figure 12.1   A battery, a wire, and a flashlight bulb. Could you get the bulb to light?

Chapter Objectives

Gaining an understanding of the concepts of electric circuit *and* electric current *will be our principal objectives in this chapter. Together with the concept of electric potential, or voltage, that was introduced in the previous chapter, these concepts are critical to our understanding of simple electrical devices. As the chapter outline indicates, we begin with these two concepts and then explore how electric current is related to voltage (Ohm's law) and to energy and power. These ideas are then applied to our use of electric power in household circuits.*

Chapter Outline

1 *Electric circuits and electric currents.* How is a complete electric circuit involved in getting a flashlight bulb to light? What is electric current, and in what ways is it analogous to the flow of water in a pipe?

2 *Ohm's law and resistance.* What is the relationship between current and voltage in a simple circuit, and how does this relationship (Ohm's law) define the concept of electrical resistance?

3 *Series and parallel circuits.* What is the difference between series and parallel connections of elements in a circuit, and what happens to the electric current in these different situations? What consequences do these ideas have for the way in which voltmeters and ammeters are used?

4 *Electric energy and power.* How can the concepts of energy and power introduced in chapter 8 be applied to electric circuits? What are the units used in discussing the use and cost of electrical energy in our homes?

5 *Alternating current and household circuits.* What is the nature of the alternating current commonly used in household and commercial applications? How are electric appliances connected to household circuits, and what safety considerations are involved?

12.1 ELECTRIC CIRCUITS AND ELECTRIC CURRENTS

A flashlight, an electric toaster, and the starting motor in your car all involve electric circuits and use electric current to accomplish their missions. The concepts of a circuit and current go hand in hand and are critical to understanding the operation of any electrical device. The battery-and-bulb exercise of the introduction provides a simple example for defining these ideas.

Getting the Bulb to Light

The battery-and-bulb exercise strips the flashlight down to its bare essentials; the bulb, the battery, and a single conductor. The rest of the flashlight is necessary only to hold it all together and provide a more convenient way of switching the light on and off. So how is it done? How do we get the bulb to light with just the wire?

Study Hint

If you have the materials handy, you should try this exercise before reading further. The delight of even college students in figuring out how to get the bulb to light is something not to be spoiled by reading on prematurely. Once you have achieved light (without, we hope, killing the battery), you may wish to experiment with other configurations and try to understand what distinguishes working arrangements from nonworking ones. This process will help to make the concept of a circuit more meaningful.

What many people fail to recognize in confronting this exercise is that the light bulb has two distinct connecting places that are electrically insulated from each other. It is necessary to make connection with both of them. It is also necessary to provide a complete path passing through the light bulb and connecting it to both ends of the battery. Such a *closed* or complete path is called a *circuit;* the word itself implies a closed loop.

Figure 12.2 shows three possible arrangements, one that works and two that do not. (Which is which?) In figure 12.2c, a wire runs from the bottom of the battery to the side of the light bulb, and the tip of the bulb rests on the other terminal of the battery. This is the arrangement that works; there is a complete circuit that passes through the bulb and the battery. Figure 12.2b shows a complete circuit, but it does not pass through the bulb. In this arrangement, the bulb will not light, but the wire will get warm. The battery will quickly die if the wire is left in place.

In the configuration in figure 12.2a, there is no complete path. Nothing at all will happen in this case; the bulb will not light, and the wire will not get warm. This case represents an incomplete, or *open,* circuit. To be complete, a circuit must have a closed path of conducting elements joining the two ends of the battery. Without a complete path, nothing happens.

In a flashlight, the circuit is essentially the same as the working arrangement in figure 12.2c. The bulb sits on top of the batteries in direct contact with the top battery. The side of the bulb is held within a metal sleeve, which is insulated from the rest of the flashlight canister. The switch connects this sleeve to the bottom of the battery, either through the canister itself or through a metal strip that runs to the bottom of the flashlight. Pushing the switch on closes the path. If you have a flashlight handy, take it apart and see if you can determine how this is accomplished.

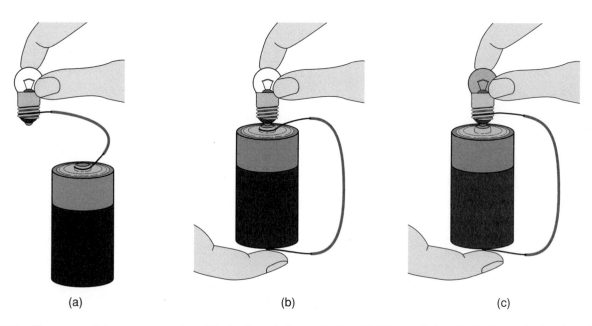

(a)   (b)   (c)

Figure 12.2   Three possible arrangements of the battery, bulb, and wire. Which one lights the bulb, and why does it work?

Electric Current

It is worthwhile to take a closer look at what is happening in the arrangement shown in figure 12.2c. As we will discuss in more detail in section 12.4, the battery is the energy source for this circuit. It uses energy from chemical reactions to produce a separation of positive and negative charges within the battery. This separation produces an increase in the electrostatic potential energy of the charges, which leads to a potential difference, as discussed in the previous chapter. Flashlight batteries typically generate a potential difference of 1.5 volts between the terminals.

Since there is an excess of positive charge at one end of the battery, and an excess of negative charge at the other, these charges have a tendency to rejoin by flowing from one terminal to the other if we provide a suitable conducting path. The positive terminal has an excess of positive charges that are attracted toward the negative charges at the negative terminal and vice versa. These charges can only flow via an *external* conducting path, however, because of opposing internal forces associated with the chemical reactions inside the battery. If we simply take a metal wire and connect it between the two terminals, charge will flow through the wire to the opposite terminal. In a metal wire, it is electrons, which are negatively charged, that flow. The charge carriers can be positive however, in other situations, such as positive ions in a chemical solution or *holes* in a semiconductor (see chapter 20).

Such a flow of electric charge is referred to as an *electric current*. Actually, to be more precise, current is the *rate* of flow:

Electric current is the rate of flow of electric charge. In symbols, $I = q / t$, where I is the commonly used symbol for electric current, q is the charge, and t is time. The direction of the current is defined as the direction of flow of positive charge.

The standard unit for electric current is the *ampere,* which is defined as one coulomb per second (1 A = 1 C/s); the coulomb is the unit of charge introduced in chapter 11. The ampere is often referred to as an *amp,* and the appropriate abbreviation is a capital *A*. From the definition, then, and the corresponding units, we see that the size of an electric current is related to how much charge flows in a given time.

Figure 12.3 shows two schematic views of charges flowing in a wire. If the charge carriers are positively charged, their direction of motion, by definition, would be the direction of the electric current. This direction is to the right, as pictured in figure 12.3a. In reality, however, the charge carriers in a metal wire are electrons, which are negatively charged. Negative charges flowing to the left have the same effect as positive charges flowing to the right, so the direction of the electric current is to the right in figure 12.3b also. Our definition of current therefore makes the direction of the current opposite to the direction of flow of the negatively charged electrons.

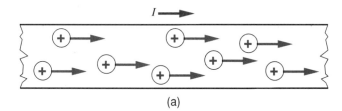

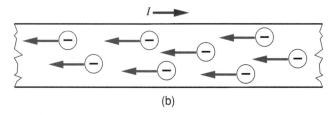

Figure 12.3 Positive charges moving to the right have the same effect as negative charges moving to the left. The electric current is, by definition, to the right in both cases.

If the battery in our simple circuit is fresh, the chemical reactions will continue to separate charge within the battery and the current will continue to flow. A metal wire is an excellent conductor, however, so if we directly connect the two terminals of the battery with a wire, a large current will flow. This will deplete the chemical fuel, and the battery will die after a short time. The wire will also become quite warm as a result of the large current. In order to avoid this problem, we need to place an element in our circuit (such as the light bulb) that will provide greater *resistance* to the flow of charge.

Looking closely at the light bulb, we see that it consists of a very thin wire filament enclosed inside the glass bulb. This filament is connected to two points inside the bulb; one end is connected to the metal cylinder that forms the lower sides of the bulb (fig. 12.4), and the other is connected to a metal post in the center of the bulb's base. These two points are electrically insulated from one another by a ceramic material surrounding the center post. The thin wire filament, although metal and a good conductor, restricts the flow of charge (the current) because of its very small cross-sectional area.

If we force the current to flow through the light bulb, as in figure 12.2c, a much smaller current will flow than when the wire is connected directly between the terminals. The thin wire filament restricts the amount of charge that can get through. The resistance of the filament is much greater than that of the thicker connecting wires. The wire filament, which is the bottleneck in the circuit, gets very hot as the battery forces charges through this constriction. Its high temperature causes it to glow, and thus we have light.

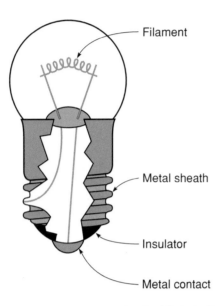

Figure 12.4 **A cutaway view of a flashlight bulb. The filament is connected to the metal post at the bottom as well as to the inside of the metal canister.**

The Analogy with Flowing Water

In discussing the battery-and-bulb circuit, we have described an electric current as the flow of electric charge. Picturing the flow of charge requires some imagination, however, since charge is invisible, and we cannot see inside the wire in any case. In situations such as this, analogies are often useful, provided that the analogy itself is easier to visualize or more familiar than the original situation. The analogy between electric current and water flowing in a pipe meets this criterion and is often helpful.

Figure 12.5a shows a pump that pumps water from a lower tank to a higher tank, thereby increasing the gravitational potential energy of the water (as discussed in chapter 8). The water might remain in the higher tank indefinitely unless we provide a path through which it can flow back to the lower tank. If this path consists of a large pipe, the water will flow back very quickly, and we might measure its rate of flow (the water current) in gallons per second or liters per second. The upper tank would be quickly depleted unless the pump runs continuously to replace the lost water. If, however, we place a narrow pipe at some point in the path of the returning water, this constriction limits the amount of water that will flow. The narrow pipe provides a greater resistance to the flow of water.

The analogy to the electrical case is fairly obvious and is represented in figure 12.5b. The pump corresponds to the battery; both produce an increase in potential energy—of water on one hand and of charge on the other. The thicker pipes are like the connecting wires in a circuit, and the narrow pipe is analogous to the filament in the light bulb. In fact, the water will become warmer as it flows through the constriction in the pipe in a manner similar to the warming of the wire filament. A valve placed at some point in the water system corresponds to a switch in an electric circuit. Table 12.1 summarizes the corresponding elements in our analogy.

Your intuitive understanding of an electric current may be greatly improved by recognizing its similarity to a water current. Both represent flows, and both can be defined as rates of flow—of water on one hand, and of charge on the other. In the case of electric current, we measure the rate at which charge flows through some point in a circuit. Likewise, in measuring a water current, we measure the rate at which water flows through a given point in the system.

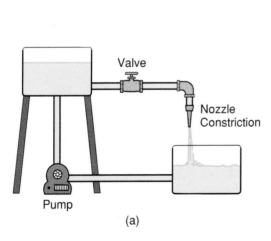

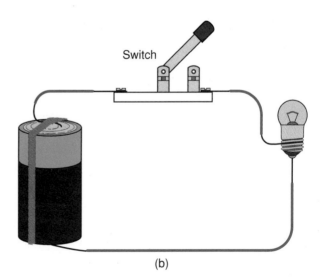

Figure 12.5 **The water-flow analogy to the battery-and-bulb circuit. Which elements correspond to which in the two systems?**

Table 12.1
Corresponding Components in the Analogy Between a Water-Flow System and an Electric Circuit

Water-Flow System	Electric Circuit
Water	Charge
Pump	Battery
Pipes	Wires
Narrow pipe	Wire filament
Valve	Switch

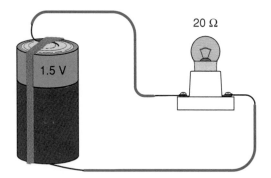

Figure 12.6 A simple circuit connecting a 1.5 V battery to a 20 Ω light bulb. What is the magnitude of the electric current?

An electric circuit must have a continuous path of conductors (the circuit) in order for charge to flow. If we break the circuit at any point, the current stops. The water system also requires a complete loop around which the water can flow unless we continually supply water to the upper tank from an external source. Like any analogy, the water-flow analogy has limits, but it is often helpful in picturing the concept of current.

12.2 OHM'S LAW AND RESISTANCE

What is it that actually determines the size of an electric current? In the previous section, we noted that the greater resistance of the wire filament restricts the flow of current. Is this the only effect, or does the voltage of the battery also play a role? Is it possible to predict how much current will flow for a given circuit?

Ohm's Law

It turns out that for most conducting elements in a circuit, there is a simple relationship between the current, the voltage, and the resistance of the element that we can use to predict the magnitude of the current. This relationship was discovered experimentally by Georg Ohm (1787–1854) during the 1820s and is known as Ohm's law:

The electric current through a given portion of a circuit is equal to the potential difference (voltage difference) across that portion divided by the resistance. In symbols,

$$I = \frac{\Delta V}{R},$$

where *R* is the resistance, and *I* and *V* stand for current and voltage, as before.

In other words, the current is directly proportional to the voltage difference and inversely proportional to the resistance.

Ohm's law actually represents a definition of the concept of electrical resistance as well as the experimental fact that

this quantity is approximately constant for different values of current and potential difference. In the previous section we used the term *resistance* in a qualitative sense as a property that restricts the flow of current. The quantitative definition of resistance can be obtained by using algebra to rearrange Ohm's law. If we multiply both sides of the Ohm's-law equation by *R*, we get the following expression:

$$I R = \Delta V.$$

Dividing both sides by *I* yields

$$R = \frac{\Delta V}{I},$$

which shows that resistance *R* is the ratio of the voltage difference to the current for a given portion of a circuit. The unit of resistance is therefore

$$R = \frac{\text{volts}}{\text{amperes}} = \frac{\text{V}}{\text{A}} = 1 \text{ ohm} = 1 \text{ }\Omega.$$

The name given to this unit is the *ohm*, often abbreviated with the symbol Ω, the Greek letter capital omega.

Thus if we know the resistance of a given portion of a circuit and the applied difference in voltage, it is a simple matter to compute the expected current. Suppose, for example, that a 1.5-volt battery is connected to a light bulb with a resistance of 20 Ω, as shown in figure 12.6. If the resistance of the battery itself is negligible, the expected current is

$$I = \frac{\Delta V}{R} = \frac{1.5 \text{ V}}{20 \text{ }\Omega} = 0.075 \text{ A},$$

or 75 milliamperes (mA).

Dead Batteries and Internal Resistance

The preceding calculation ignores the resistance of the battery itself. (It also ignores the resistances of the connecting wires, which are assumed to be very small.) If the battery is fresh, its resistance is normally very small and can often be neglected. As the battery is used, however, and its chemical fuel depleted, its internal resistance gets larger. In finding the current then, we really need to consider the total resistance of the circuit, including that of the battery.

Suppose, for example, that the battery has an internal resistance of 5 Ω. In a simple, single-loop circuit like that shown in figure 12.6, this resistance adds to the 20 Ω resistance of the lightbulb to give a total resistance of 25 Ω for the circuit. The current would then be

$$I = \frac{1.5 \text{ V}}{25 \text{ }\Omega} = 0.06 \text{ A} = 60 \text{ mA},$$

which is, of course, smaller than the 75 mA current computed earlier.

Does Ohm's law still hold for the light bulb itself in this case? It does indeed, but the voltage difference across the bulb is no longer 1.5 volts. Instead, from Ohm's law, we find that

$$\Delta V = I\,R = (0.06 \text{ A})\,(20 \text{ }\Omega) = 1.2 \text{ V}.$$

The voltage difference provided by the battery is reduced by the effect of its own internal resistance, since the current flowing through the light bulb must also pass through the battery.

In fact, if we calculate the decrease in voltage associated with the internal resistance of the battery, we find that

$$\Delta V = I\,R = (0.06 \text{ A})(5 \text{ }\Omega) = 0.3 \text{ V}.$$

The battery has the net effect of increasing the electric potential of charges flowing through it by 1.5 V – 0.3 V = 1.2 V, and 1.2 volts is the voltage difference that would actually be measured by a voltmeter placed across the terminals of the battery while the current is flowing (fig. 12.7). This is also the voltage difference across the filament of the light bulb, as we calculated in the previous paragraph.

The value of 1.5 volts is the voltage difference that would be measured across the terminals of the battery when no current is flowing through it. This quantity is sometimes referred to as the *electromotive force** of the battery and is usually represented by a script capital ε. The electromotive force represents the increase in potential energy per unit positive charge that is provided by the chemical reactions in the battery, and it is measured in volts (joules/coulomb). When a current is flowing through the battery, the voltage across its terminals is reduced by the amount of the voltage drop across the internal resistance, as we have just calculated.

For a complete current loop, the increase in electric potential produced by batteries or other sources of electrical energy must equal the sum of the decreases in voltage (voltage drops) across the resistances in the circuit. If this were not true, we would not get back to the original value of voltage if we went completely around the circuit. In symbols, this statement takes the following form:

$$\Sigma \varepsilon = \Sigma IR,$$

where the symbol Σ stands for *summation,* as before.

---

* This concept is discussed in more detail in section 12.4. The term is misleading, since it is not a force, but a potential difference.

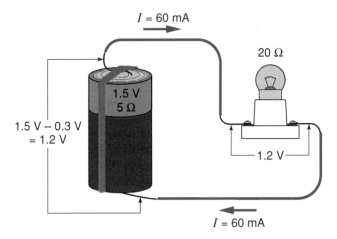

Figure 12.7 Voltage values for the battery-and-bulb circuit, assuming that the battery has an internal resistance of 5 Ω. The current is now 60 mA.

To find the current flowing in the loop, then, we divide the total electromotive force of the batteries by the total resistance of the loop, as we did above, to find the value of 60 mA for the battery-and-bulb circuit.

As a battery gets older, the internal resistance gets larger and larger. This increases the total resistance of the circuit, and the current that flows around the circuit becomes smaller and smaller, as predicted by the loop equation. As the current gets smaller, the bulb becomes dimmer, until finally it glows no more. In a dead battery the internal resistance has become so large that the battery can no longer produce a measurable current.

It is interesting to note that if we place a good voltmeter across the terminals of a dead battery, it still registers 1.5 volts as long as there is nothing else connected to the battery. (The 1.5-V potential difference is a characteristic of the electrodes and chemicals used in the battery.) As soon as we connect the battery to an external circuit (such as our light bulb), however, the voltage drop across the internal resistance of the battery reduces the voltage across its terminals to a very low and ineffective value. To be effective, a battery must have a low internal resistance so that it can produce a usable current.

12.3 SERIES AND PARALLEL CIRCUITS

An electric current, like a meandering stream, can split into different streams that rejoin at a later point. There are often advantages to arranging a circuit so that this occurs. How do we describe and analyze these different ways of connecting circuits? This question brings us to a discussion of series and parallel connections.

Series Circuits

The simple light-bulb circuit that we have been discussing is a single-loop, or *series,* circuit. This implies that there are no points in the circuit at which the current can branch into side streams or secondary loops. All of the elements lie in-line on a single loop; the current that passes through one element must also pass through the others. This is easy enough to see in figure 12.6, but it is often even easier to see in the schematic diagrams that represent circuits.

In figure 12.8, the battery-and-bulb circuit is shown again with the schematic diagram alongside. The symbols used in the schematic diagram to represent the various elements in the circuit are standard; they would be recognized anywhere in the world where people study or use electric circuits. The light bulb, for example is represented as a resistance, for which the standard symbol is a zigzag line, ⋀⋀⋀. It appears in this case inside a circle that represents the glass bulb.

The lines in the schematic diagram represent connecting wires and are usually drawn as straight lines, even though in an actual circuit they are not likely to be at all straight. We usually assume that the resistances of the connecting wires are small enough in comparison to the other resistances in the circuit that they can be ignored. The symbol for the switch bears an obvious schematic relationship to the actual device. The overall schematic diagram represents the components of the circuit and the connections between them in a manner that is both clear and easy to draw.

What happens if we put two more light bulbs in the circuit as shown in figure 12.9? When we put them in line like this, so that the current that flows through one must also flow through the others, we say that the bulbs are connected *in series.* Figure 12.9 shows both the circuit and the corresponding schematic diagram.

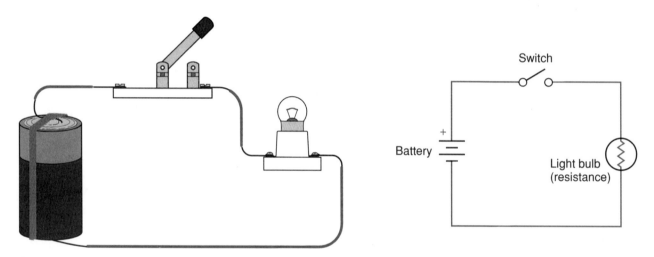

Figure 12.8   The battery-and-bulb circuit with its corresponding schematic circuit diagram.

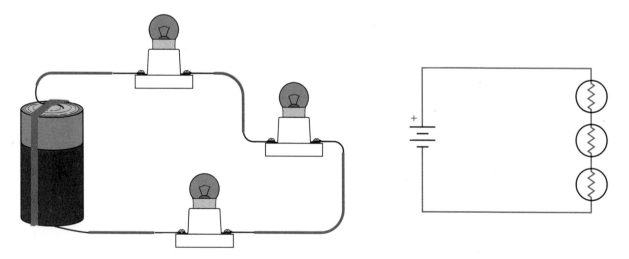

Figure 12.9   A series combination of three light bulbs and the corresponding circuit diagram.

In a series combination of resistances such as this, each resistance contributes to restricting the flow of current around the loop. Since the same current must flow through all three, the total series resistance of the combination, R_s, turns out to be the sum of the individual resistances:

$$R_s = R_1 + R_2 + R_3.$$

This is true no matter how many resistances are combined in series; we just add more terms to the sum if there are more resistances.

The reason that the resistances add in this manner is that the voltage decreases as the current passes through each element, according to Ohm's law. The total drop in voltage must then equal the sum of the individual voltage drops:

$$\Delta V = I R_s = I R_1 + I R_2 + I R_3.$$

Since the same current flows through each resistance and through the combination, dividing by I yields the simple summation formula shown in the previous paragraph. The effective series resistance, R_s, can then be used to calculate the voltage drop for the entire combination. In effect, we can replace the three resistances by a single value that has the same impact on the circuit. The sample exercise in box 12.1 illustrates this process.

Notice that the current calculated in box 12.1 is less than that which would result with just one bulb in the circuit. With just one bulb, the total resistance of the circuit would be 10 Ω, and the current would be 0.30 A, twice that calculated

---

Box 12.1

Sample Exercise

Suppose that two flashlight bulbs are placed in series with two flashlight batteries, also connected in series, as shown in the illustration. Each battery has a voltage (electromotive force) of 1.5 volts and each bulb has a resistance of 10 Ω. The internal resistances of the batteries are small enough to be ignored. What is the current flowing in the circuit?

$\varepsilon_1 = \varepsilon_2 = 1.5$ V.

$R_1 = R_2 = 10$ Ω.

$I = ?$

$R_s = R_1 + R_2$

$R_s = 10\ \Omega + 10\ \Omega = 20\ \Omega.$

Since each battery boosts the potential by 1.5 volts, the total potential difference of the two batteries in series is 3 volts. Thus

$$\sum \varepsilon = \sum I\,R = I\,R_s;$$

$$I = \frac{\sum \varepsilon}{R_s} = \frac{3\ \text{V}}{20\ \Omega} = 0.15\ \text{A}.$$

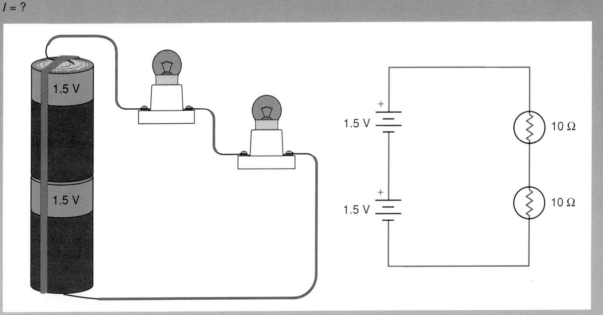

Two light bulbs connected in series with two batteries, also in series. What is the current?

for two bulbs. Doubling the resistance reduces the current to half of its former value. Because the current is smaller with two bulbs in series than with just one, the bulbs will burn less brightly in this arrangement. If we want brighter light, we must add more batteries in series. This is what is done, of course, in many flashlights, though they usually have just one bulb.

In addition to the fact that the lights are dimmer when connected in series, there is another serious disadvantage to connecting bulbs in this way. Suppose that one of them burns out? When a light bulb burns out, the filament breaks, and this produces a break in the circuit. Thus when bulbs are connected in series, if one burns out, the others go out also because no current can flow around the loop. This is usually highly undesirable, particularly in something like an automobile lighting system. It can also make it very difficult to find the burned-out bulb, as it was in the old-fashioned strings of Christmas tree lights that were connected in series.

Parallel Circuits

Another way to connect the bulbs, which avoids the problems just mentioned, is shown in figure 12.10a. Here the bulbs are not all in line around the loop; they are connected in *parallel* with one another. The key to recognizing this is to note that

there is now more than one loop in the circuit; the current can *branch*, or split up into different paths, like the meandering stream mentioned at the beginning of this section.

Does the total effective resistance of the circuit increase or decrease if we add bulbs in parallel with one another? The water-flow analogy may help your thinking here. Suppose that we add pipes in parallel in a water-flow system as shown in figure 12.10b. Does this increase or decrease the amount of water (the current) flowing in the system? What does this say about the resistance to the flow of water?

A little reflection should convince you that the flow of water is increased when we add pipes in parallel. In essence, we are increasing the total cross-sectional area through which the water can flow, and therefore decreasing the effective resistance to flow. The same is true in an electric circuit involving resistances connected in parallel. In fact, the relationship between the effective parallel resistance, R_p, and the individual resistances turns out to be

$$\frac{1}{R_p} = \frac{1}{R_1} + \frac{1}{R_2} + \frac{1}{R_3}.$$

We have to add the reciprocals of the resistances and then take the reciprocal of the sum in order to find R_p. This yields a value that is less than any of the individual resistances, as the example in box 12.2 illustrates.

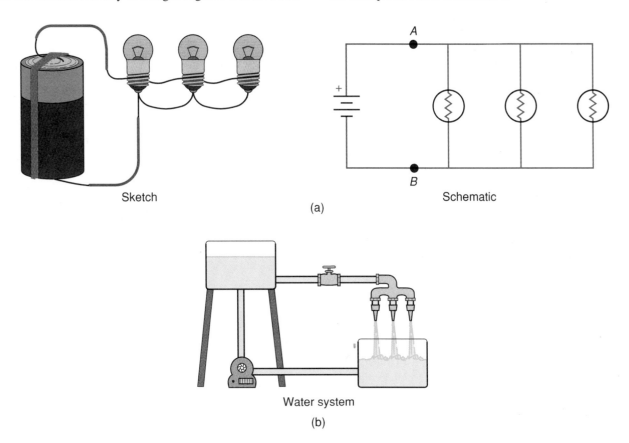

Sketch

Schematic

(a)

Water system

(b)

Figure 12.10 Bulbs connected in parallel, with the schematic diagram and the water-flow analogy also shown. The current now splits into three different paths.

Box 12.2

Sample Exercise

Suppose that the same two 10 Ω light bulbs used in box 12.1 are connected in parallel, as shown in the illustration. This parallel combination is then connected to the two flashlight batteries in series, as before. What is the total current flowing around the loop, and how much current passes through each bulb?

$R_1 = R_2 = 10 \ \Omega$

$\varepsilon = 1.5 \ \text{V} + 1.5 \ \text{V} = 3 \ \text{V}.$

$$\frac{1}{R_p} = \frac{1}{10 \ \Omega} + \frac{1}{10 \ \Omega}$$

$$= \frac{2}{10 \ \Omega} = \frac{1}{5 \ \Omega}$$

$$R_p = 5 \ \Omega.$$

$$\Sigma \varepsilon = R \, I;$$

$$I = \frac{\Sigma \varepsilon}{R_p} = \frac{3 \ \text{V}}{5 \ \Omega} = 0.6 \ \text{A}.$$

This is the total current flowing through the batteries and around the loop. The current through each bulb is found from Ohm's law by dividing the voltage drop across the parallel combination by the resistance of each bulb:

$$I_1 = \frac{\Delta V}{R_1} = \frac{3 \ \text{V}}{10 \ \Omega} = 0.3 \ \text{A} = I_2.$$

A current of 0.3 A flows through each bulb, for a total of 0.6 A through the parallel combination.

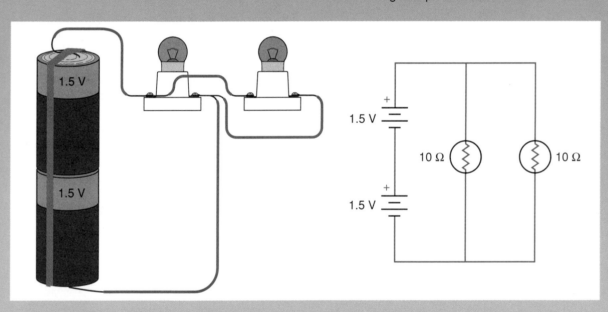

Two 10 Ω light bulbs connected in parallel with two 1.5 V batteries, which are connected in series. What are the currents?

This result can also be obtained from Ohm's law, using an argument similar to that used for the series combination. In this case, however, it is the voltage drop that must be the same for each resistance, since they are all connected between the same two points. The currents, on the other hand, must add to give the total current through the combination. This is because they split up and then recombine; a portion of the total current flows through each branch. From Ohm's law, $I = \Delta V / R$, so

$$I = \frac{\Delta V}{R_p} = I_1 + I_2 + I_3 = \frac{\Delta V}{R_1} + \frac{\Delta V}{R_2} + \frac{\Delta V}{R_3}.$$

Dividing by the voltage difference, ΔV, in this case yields the formula given in the previous paragraph.

In the example in box 12.2, the two bulbs had equal resistances; therefore, equal currents flowed through each bulb. If one bulb had a larger resistance than the other, the current would not divide equally between them. Instead, a larger portion of the current would flow through the bulb with the smaller resistance. The bulb with the larger resistance draws a smaller portion of the total current, as common sense would suggest. The calculation of the equivalent resistance of the combination R_p, however, and of the total current and the current through each bulb, would proceed just as shown in box 12.2.

As the example shows, the resistance of the parallel combination is indeed less than that of either bulb by itself. Parallel combinations decrease the resistance and increase the amount of current that will flow. This causes the bulbs to burn more brightly than in a series combination but also depletes the batteries more rapidly. There is no free lunch; we can choose dim light and a long lifetime for the batteries or bright light and a short lifetime, depending upon how we connect the bulbs to the batteries. The energy available from the batteries is the same in either case. There may be some parallels to life itself here!

A Word on Meters

How do we go about measuring currents or voltage differences? You may have used meters such as a voltmeter and ammeter in checking the electrical systems of your car or other trouble-shooting activities. An understanding of how these meters are used can reinforce your understanding of the nature of current and voltage.

The voltmeter is the easier of the two to use and also more commonly used in auto repair and other activities. Suppose, for example, that we wanted to measure the voltage difference across a light bulb in a circuit. The voltmeter has two leads, often coded red for positive and black for negative. It also has some form of display or readout, such as a needle and scale or a digital readout. The meters are often *multimeters,* which can measure voltage, current, and resistance; the appropriate function and range is selected with a switch. Figure 12.11 pictures both an analog multimeter that uses a needle and scale, and a more modern digital multimeter that uses a digital readout.

To measure voltage, the leads of the voltmeter are placed in *parallel* with the bulb, as shown in figure 12.12a. The reason for this is the nature of the voltage-difference concept, with emphasis on the word difference. A voltage difference is the difference in potential energy of electric charges between two points in a circuit. The voltmeter must therefore be connected between these two points, regardless of what other paths for current flow may be present.

When inserted in the circuit, the voltmeter should affect the circuit only slightly, to minimize change in the quantities we are trying to measure. Should the voltmeter have a large or small resistance then? A little thought should con-

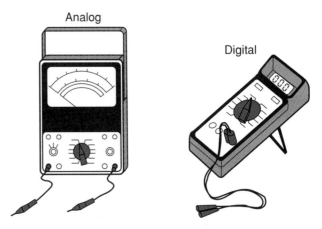

Figure 12.11 An analog multimeter (with needle and separate scales) and a digital multimeter are commonly used measuring instruments.

vince you that a large resistance is desirable since this guarantees that only a very small portion of the total current will flow through the voltmeter in this parallel combination. Voltmeters are therefore designed with large resistances; we can safely place a voltmeter directly across a battery to read its voltage without drawing much current from the battery.

Would this same type of connection work for measuring current with an *ammeter?* If we think of the water-flow analogy, we should recognize that to measure current, which is a rate of flow, we need to insert our gauge directly in the flow. To measure water flow, we need to cut the pipe and insert a flow gauge between the cut sections, so that the current being measured flows directly through the gauge. Likewise, in measuring electric current, we need to break the circuit and insert the meter where we want to measure the current. The ammeter is therefore connected *in series,* as shown in figure 12.13a.

Since the ammeter is inserted in series in the circuit and must have some resistance, it inevitably increases the total resistance of the circuit and decreases the current. It is desirable, then, for an ammeter to have a small resistance, so that its effect on the current is minimal. More care must be exercised in using an ammeter than a voltmeter because of this small resistance. If, for example, we placed an ammeter directly across the terminals of a battery, a large current would flow. This could damage the meter as well as rapidly deplete the battery.

With both ammeters and voltmeters, it is important to insert the meter in the proper direction, with the positive terminal, or lead, of the meter connected towards the positive terminal of the battery or power supply. (See figures 12.12 and 12.13.) If we get it backwards, the needle of an analog meter will deflect in the wrong direction, and this may also damage the meter. The positive and negative terminals of a meter are usually clearly marked.

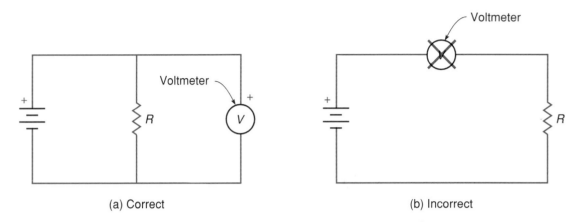

Figure 12.12   Correct and incorrect ways of inserting a voltmeter into a circuit. What happens when the voltmeter is inserted incorrectly?

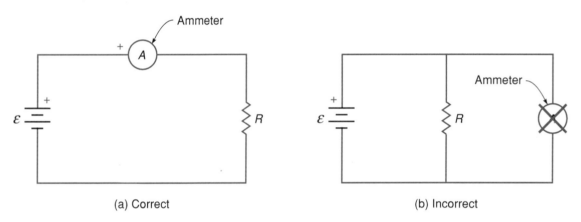

Figure 12.13   Correct and incorrect ways of inserting an ammeter into a circuit. What happens when the ammeter is inserted incorrectly?

The ways in which voltmeters and ammeters are inserted in a circuit, as we have seen, follow directly from the nature of the quantities being measured. Voltmeters measure differences and are therefore connected in parallel with the portion of the circuit whose potential difference is being measured. Ammeters, on the other hand, measure a flow and must be inserted directly in series with that flow.

12.4   ELECTRIC ENERGY AND POWER

We have talked about batteries as the sources of electrical energy in a circuit. We also use the term *power* in discussing our everyday use of electricity. Energy or power perspectives can be used to gain more insight into the behavior of electric circuits, just as we used these ideas in chapter 8 to examine mechanical situations. What happens, for example, to the energy that is supplied by a battery? What energy transformations take place?

Energy Transformations in a Circuit

Our familiar water-flow analogy can be used to good advantage to explore energy transformations. The energy transformations in the water-flow system are easy to describe and have already been alluded to in the first section of this chapter. Energy is supplied to the system via the pump, which in turn receives its energy from some external source. (Electricity, gasoline, or wind are energy sources commonly used for pumping water.) The pump increases the gravitational potential energy of the water by lifting it to a higher tank (fig. 12.14). As the water flows downhill, through the pipes to a lower tank or reservoir, the gravitational potential energy is transformed into the kinetic energy of the moving water.

Once the water has come to rest in the lower tank, however, it no longer has kinetic energy. The kinetic energy is dissipated eventually via frictional or *viscous* forces within the water or between the water and the inside surfaces of the pipes. Frictional forces generate heat, so that the energy added to the system by the pump ultimately increases the

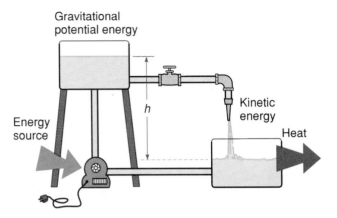

Gravitational
potential energy

h

Energy
source

Kinetic
energy

Heat

Figure 12.14 Energy transformations in the water-flow system. What happens ultimately to the energy introduced via the pump?

internal energy of the water and the surrounding pipes and air. This increase in internal energy shows up as an increase in the temperature of the water and its surroundings.

A similar picture can be painted for the electric circuit. In this case the energy is supplied by the battery, which in turn draws its energy from the potential energy stored in the chemical reactants it contains. (Electric energy can also be generated from mechanical energy, as we will see in the next chapter.) The battery, much like the pump, increases the potential energy of electric charges as it moves positive charges towards the positive terminal and negative charges towards the negative terminal. When we complete the circuit by providing an external conducting path from one terminal of the battery to the other, charge flows from points of higher potential energy to points of lower potential energy through the connecting wires and resistances.

As potential energy is lost, kinetic energy is gained in the form of increased velocity of the moving charges that make up the electric current. Ultimately, though, this kinetic energy becomes randomized through collisions with other electrons or the atoms in the wires and other resistances in the circuit. This increases the internal energy of the wires or resistances, which again shows up as an increase in temperature. We say that the kinetic energy has been converted to heat through these collisions.

Thus in both cases, the water-flow system and the electric circuit, we have energy transformations of the following form:

Source $\Rightarrow$ potential energy $\Rightarrow$ kinetic energy $\Rightarrow$ heat.

Pipes and resistances get warm as the current flows. In the case of electric circuits, of course, we often use the heat for some specific purpose: to light a light, to toast bread, or to heat our homes.

Electric Power

As indicated earlier, the potential difference produced by a battery when no current is being drawn from it is often called the *electromotive force* and is represented by the symbol ε. It is simply the potential energy per unit charge supplied by the battery (or other source of energy), and its units are therefore volts. The name *electromotive force* is not particularly apt, since we are talking about a potential difference and not a force, but traditions die hard, and this name is still generally used for this quantity.

A voltage difference represents a difference in potential energy per unit charge; thus if we multiply electromotive force by charge, we would get potential energy. If, on the other hand, we multiply by the current, which is the rate of flow of charge, we get the rate of use of energy. This may be more obvious if we consider the units involved in this operation:

$$(\varepsilon)(I) = \text{(volts)(amperes)}$$
$$= \text{(joules / coulomb) (coulomb / second)}.$$

The coulombs cancel in this product, so we have

$$(\varepsilon)(I) = \frac{\text{joules}}{\text{second}} = \text{watts}.$$

This is, by the definition given in chapter 8, the power, since power is the rate at which work is done or energy is utilized. The power supplied by any source of electrical energy is therefore equal to this product; that is, the power delivered is equal to the electromotive force times the current:

$$P = \varepsilon I.$$

Since a voltage difference times a current yields a power, the power dissipated in the resistances of a circuit can also be found if the current and voltage difference are known. In this case we express the voltage difference simply as ΔV, and the power expression takes the following form:

$$P = \Delta V \, I.$$

Using Ohm's law, we can also express the voltage difference as $\Delta V = IR$:

$$P = (\Delta V)(I) = (IR)\,(I) = I^2 R.$$

The power dissipated in the resistance R is therefore proportional to the square of the current I.

What is happening, then, in a simple circuit in terms of the power? Simply stated, the power delivered by the battery must equal the power dissipated in the resistances in a steady-state situation. In symbols,

$$P = \varepsilon I = I^2 R.$$

As long as the current remains constant, energy is supplied by the battery at a steady rate and is dissipated at the same rate in the resistances. There is no buildup of energy at any point in the circuit; the energy or power input equals the energy or power output (fig. 12.15).

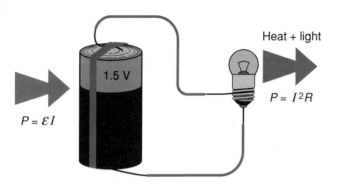

$P = \varepsilon I$

Heat + light

$P = I^2 R$

Figure 12.15 **The power supplied by the battery in an electric circuit is equal to the power dissipated in the resistances as heat.**

Box 12.3

Sample Exercise

What is the power dissipated in a 20 Ω light bulb that is powered by two 1.5 V batteries in series?

$$\varepsilon_1 = \varepsilon_2 = 1.5 \text{ V}.$$

$$R = 20 \ \Omega.$$

Since the voltages of the batteries in series add,

$$\varepsilon = 3 \text{ V} = RI.$$

$$I = \frac{\varepsilon}{R} = \frac{3 \text{ V}}{20 \ \Omega}$$

$$= 0.15 \text{ A}.$$

$$P = I^2 R = (0.15 \text{ A})^2 (20 \ \Omega) = 0.45 \text{ watts.}$$

This can be checked by calculating the power delivered by the batteries:

$$P = \varepsilon I = (3 \text{ V})(0.15 \text{ A}) = 0.45 \text{ watts.}$$

Distribution and Use of Electric Power

Whenever you turn on an electric light or use any electric appliance, you are using electric power. Usually, however, that power does not come from a battery, but from power lines (wires) that deliver power from a distant generating source (fig. 12.16). Thomas Edison's invention and development of the electric light bulb provided the original incentive for creating power-distribution systems utilizing a central power source. The ease with which electric power can be transmitted over considerable distances is one of its primary advantages over other forms of energy.

What are the sources of the energy that we use when we purchase electric power? The source might be the gravitational potential energy of water stored behind a dam. It could also be chemical potential energy stored in fuels such as coal, oil, or natural gas, or the nuclear potential energy stored in uranium. Like the chemical fuel in a battery, these latter sources can all be depleted since they involve fuels that are mined from the earth. (What is the energy source involved in lifting the water that is stored behind a dam? Can it be depleted?)

The potential energy stored in chemical or nuclear fuels is first converted to heat, which is then used to run a heat engine, as discussed in chapter 10. The heat engine usually takes the form of a steam turbine. In the case of hydroelectric power, the water stored behind the dam is run through water turbines to convert potential energy to kinetic energy. Whatever the source of energy, power plants all use electric generators that convert mechanical kinetic energy produced by the turbines to electric energy. (The principle of operation of electric generators is discussed in chapter 13 when we consider Faraday's law of electromagnetic induction.) The electric generator is the source of the electromotive force for the power-distribution system.

When you pay your electric bill, you are paying for the amount of energy you have used during the previous month. The unit of energy that is used for this purpose is the kilowatt-hour, which is obtained by multiplying a unit of power (the kilowatt) by a unit of time (an hour). This yields energy:

$$E = Pt = (1000 \text{ watts})(3600 \text{ s}) = 3.6 \times 10^6 \text{ J} = 1 \text{ kW·hr.}$$

We see that the kilowatt-hour is a much larger unit of energy than the joule, since it equals 3.6 million joules. It is a convenient size, however, for the amounts of electrical energy that we typically use in a home.

How much does it cost to light a 100-watt light bulb for one day? Electric rates vary across the country, from 2 to 3 cents per kilowatt hour to as much as 15 cents per kilowatt-hour, but an average rate might be about 8 cents per kilowatt-hour. If the light burns for 24 hours, the energy used would be

$$E = Pt = (100 \text{ watts})(24 \text{ hr}) = 2400 \text{ watt·hr} = 2.4 \text{ kW·hr.}$$

At 8 cents per kilowatt-hour, the cost would be

$$\text{Cost} = (8 \text{ cents/kW·hr}) (2.4 \text{ kW·hr}) = 19 \text{ cents.}$$

This may seem like a bargain, but as anyone who pays electric bills knows, it can add up quickly as the number of appliances multiply. Many appliances require larger amounts of power than a light bulb.

The first power-distribution systems were developed late in the nineteenth century, so the growth of our use of electric power has been primarily a twentieth-century phenomenon. The convenience of electricity for powering a large variety of appliances and the absence of the exhaust

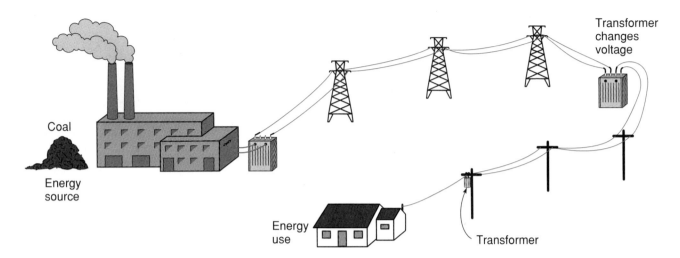

Figure 12.16 Electric power is generated at a central power plant and distributed to users via power lines.

gases produced by the direct use of chemical fuels are the major reasons for its widespread use. There are still parts of the world where electric power is not available, but it would be difficult for most of us to imagine life without it.

12.5 ALTERNATING CURRENT AND HOUSEHOLD CIRCUITS

What is the nature of the circuits involved in our everyday use of electric power? Are they similar in some ways to the simple circuits that we have described using batteries and light bulbs? You probably know that the current drawn from a wall outlet is *alternating current* (ac) rather than *direct current* (dc). What is the distinction between these forms of current? Can we use the same ideas employed in discussing simple battery-and-bulb circuits to describe household circuits that use alternating current?

Alternating Current

The term *direct current* implies that the current flows in a single direction, from the positive terminal of the battery or power supply to the negative terminal in a simple circuit. This is the natural result, of course, when a battery is used. Alternating current, on the other hand, continually reverses its direction: it flows first in one direction, then in the other, and then back again. The alternating current that we use in North America goes through 60 such back-and-forth cycles each second, so its *frequency* is 60 cycles/s or 60 hertz.

A galvanometer is an ammeter that is designed so that the needle deflects in either direction, depending upon the direction of the current. If we were to place a galvanometer in an alternating current circuit, the needle would fly back and forth rapidly if the meter could actually respond to the rapid changes in direction of the current. Since simple galvanometers generally cannot respond to such rapid changes, the usual result is that the needle vibrates but remains centered on zero. The average value of an alternating current is zero if the current magnitude varies in the same way on either side of zero as it goes through its cycle.

If we use an oscilloscope, an electronic instrument that can give us a plot of electrical voltages or currents as they vary with time, the graph of an alternating current would be like that shown in figure 12.17. On this graph, positive values of current represent one direction of flow, and negative values represent the other direction. This type of curve is called a *sinusoidal curve* because it can be represented mathematically by the trigonometric function, the sine. We encountered this type of curve earlier in our description of simple harmonic motion in chapter 8, and we will see it again in chapter 14 when we consider wave motion.

The fact that an alternating current is continually changing direction makes little difference for many electrical applications. If, for example, we consider the operation of a light bulb, the heating effect of the charges moving through the filament does not depend at all on which direction the charges are moving. Many electrical applications involve this heating effect (light bulbs, stoves, hair dryers, toasters, electric heaters, etc.). Other applications, such as those involving electric motors, do depend upon the nature of the current. We can design motors that operate on direct current, and we can also design (more easily) motors that operate on alternating current, but a dc motor will not operate on alternating current, and an ac motor will not operate on direct current.

Since the average value of an alternating current is zero, how do we usually characterize the quantity of alternating current? Because the power dissipated in a resistance is proportional to the square of the current, it makes sense to use an average of the squared current as a measure. (Squaring a negative number yields a positive value.) What we normally do, then, is to average over time the square of the current and then take the square root to obtain what is called the *effective current.* For a sinusoidal variation with time, the effective current is 0.707 times the peak value of the current, as indicated in figure 12.17. Because we have averaged the square of the current in obtaining this value, the effective current is the appropriate value to use in computing the average power by the formula $P = I^2R$.

If the variation of the voltage across an electrical outlet is plotted against time, we would get another sinusoidal curve, as shown in figure 12.18. The effective value of this voltage, obtained in the same manner as that for the current, is typically between 110 and 120 volts in North America. The standard household power supplied in this country is 115 volts, 60 hertz ac. Other voltages and frequencies are used in other parts of the world, so electric appliances such as electric shavers that employ electric motors may require adapters to function properly in Europe or other places.

Household Circuits

Suppose that you plug a light or other appliance into an electrical outlet. Does this create a series circuit or a parallel circuit with other appliances? Does the voltage available to the light bulb depend upon what else is connected to the same circuit? These are practical questions with implications for things that we do every day without giving them much thought.

As figure 12.19 indicates, household circuits are always wired in parallel. This assures that different appliances can be added or removed to the circuit without affecting the voltage available to each. Usually several outlets are wired together to make a single circuit with several possible loops. Each time you plug in an additional appliance, you create an additional loop in the parallel circuit.

As more appliances are added, however, the total current drawn from the circuit increases. This is because the total

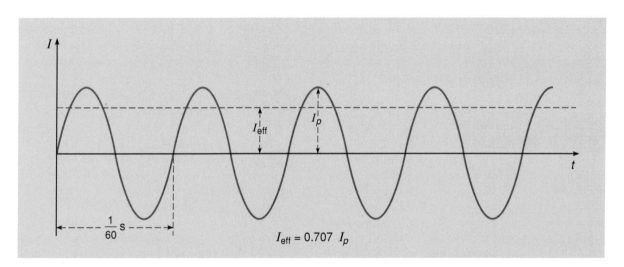

Figure 12.17 **The electric current plotted against time for an alternating current. The effective current I_{eff} is 0.707 times the peak current, I_p.**

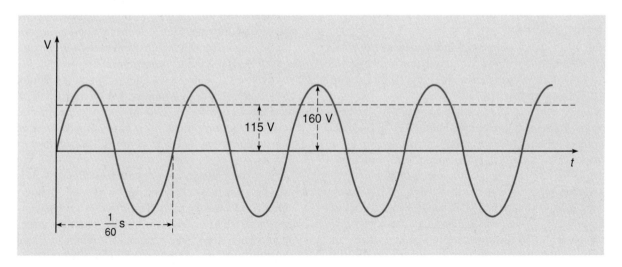

Figure 12.18 **Household (alternating) voltage plotted against time. The peak voltage is about 160 V if the effective voltage is 115 V.**

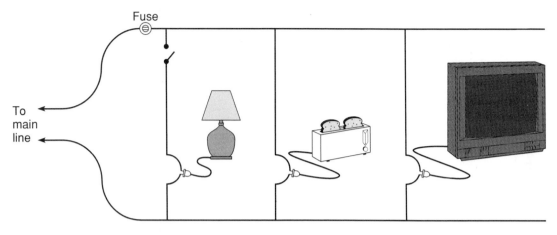

Figure 12.19 A typical household circuit may have several appliances connected in parallel with one another. A fuse or circuit breaker is in series with one leg of the circuit.

Box 12.4

Sample Exercise

A 60-watt light bulb is designed to operate on 120 V ac. What is the effective current drawn by the bulb, and what is the resistance of the bulb's filament?

$P = 60$ watts.

$\Delta V_{eff} = 120$ V.

$I_{eff} = ?$

$R = ?$

$P = I\Delta V.$

$I = \dfrac{P}{\Delta V} = \dfrac{60 \text{ watts}}{120 \text{ V}}.$

$I_{eff} = 0.5$ A.

From Ohm's law,

$\Delta V = IR$

$R = \dfrac{\Delta V_{eff}}{I_{eff}} = \dfrac{120 \text{ V}}{0.5 \text{ A}} = 240 \ \Omega.$

Table 12.2		
Power and Current Ratings of Common Appliances		
Appliance	**Power (watts)**	**Current (A)**
Stove	6000 (220 V)	27
Clothes dryer	5400 (220 V)	25
Water heater	4500 (220 V)	20
Clothes washer	1200	10
Dishwasher	1200	10
Electric iron	1100	9
Coffee maker	1000	8
Toaster	850	7
Hair dryer	650	5
Food processor	500	4
Large fan	240	2
Color TV	100	0.8
Small fan	50	0.4
Clock radio	12	0.1

effective resistance of the circuit decreases when resistances are added in parallel. The danger here is that too large a current could cause the wires to overheat. To protect against this, a fuse or circuit breaker is added in series with one arm of the circuit. If the current becomes too large, the fuse will blow, or the circuit breaker will trip. When this happens, the entire circuit is disrupted, and no current will flow to any of the attached appliances.

The current or power rating of an appliance indicates the maximum current that appliance will draw in normal operation. Our 60-watt light bulb, for example, will draw just half an ampere, as we showed in the sample exercise in box 12.4. Since a typical household circuit is fused at 15 to 20 A, several 60-watt bulbs could be included without danger of blowing the fuse.

A toaster or a hair dryer, on the other hand, generally requires much more current than the typical light bulb. A toaster often draws as much as 5–10 amperes by itself, and therefore can cause problems if connected to the same circuit as another appliance that involves heating elements. You will usually find current or power ratings printed somewhere on the appliance; table 12.2 gives some typical values. An effective line voltage of 120 V is used in computing most of the current ratings from the power ratings in this table. Appliances with larger power requirements, such as a stove, clothes dryer, or water heater, are usually connected to a 220 V line.

As you can see from the values in table 12.2, appliances with heating elements generally require more power than

Box 12.5

Everyday Phenomenon:
The Hidden Switch in Your Toaster

The Situation. Have you ever thought about why the toast suddenly pops up when you are toasting bread in an electric toaster, or why the coffeepot turns on and off to keep the coffee warm when you are using an electric coffeemaker? Where is the switch in the circuit?

Many appliances that involve heating elements also use a thermostat of some kind. A thermostat is a temperature-sensitive switch that breaks a circuit when the temperature reaches a certain point. How does a simple thermostat work, and where might we find one in a toaster? If you were to pull the plug and take a toaster apart (not a difficult operation usually), you might discover the answer to these questions.

The Analysis. If you take apart a toaster or coffeemaker (*making sure that it is unplugged first!*) and trace the wires within the appliance, you will usually find somewhere a two-part strip of metal as part of the circuit. This is most often located somewhere near the entering wires and at a point where it can sense the heat generated by the heating coils of the appliance. Although simple in appearance, this special strip of metal and its connections constitute the thermostat.

The two-part strip of metal is called a *bimetallic strip*. It is made of two different types of metal bonded together along their long dimension. Its role is almost as simple as its appearance. Because different types of metal expand at different rates as they are heated, the heat from the coil in the appliance causes one

Toaster, coffee makers, and electric heaters are among the appliances that contain thermostats. How do these thermostats work?

side of the bimetallic strip to grow longer than the other side. Since the two sides are bonded together, this differential rate of expansion causes the strip to bend, as shown in the drawing. The metal with the greater rate of expansion lies on the outside of the curve, and that with the smaller rate of expansion on the inside, thus compensating for the greater length of the metal with the greater rate of expansion.

those that use the power primarily to run an electric motor, such as a fan or food processor. Electronic devices, such as the TV or radio, require even less power. It is a good idea to check the power or current rating of appliances and to be aware of what other appliances are already present when you plug a toaster or similar appliance into a new location. The values given in table 12.2 vary depending upon the size and design of a particular appliance.

You do not have to be an electrician to appreciate the basic features of household circuits presented here. The fuse or breaker box in your house should indicate where each circuit is located in the house and the value of the current limit established by the fuse or circuit breaker. Replacing a fuse or resetting a breaker is a routine operation that all of us may need to do. Understanding the ideas presented in this chapter provide a basis for making judgments on the use of everyday appliances.

It is easy to see how such a device functions as a switch. If the strip itself is a part of the electric circuit, it may make contact with another metal tab when it is in its unbent state, but pull away and break the circuit as it warms up and bends. This form of simple bimetallic switch is shown in the drawing. In a toaster, the strip is often also part of a mechanical release system. When the toaster handle is pushed down, a ratchet arrangement holds it in place against the tension of a spring. The bending of the bimetallic strip releases the ratchet, causing the toast to pop up at the same time that the electric circuit is broken.

In devices such as a room thermostat, which require greater sensitivity to small temperature changes, the bimetallic strip is usually bent into a coil. This allows a much longer strip to be used in a small space. It also provides a convenient method for adjusting the set point of the thermostat. A knob or dial allows us to loosen or tighten the coil, thus changing the temperature at which the tightening or loosening caused by heating will break the circuit.

This very simple physical effect involving the different rates of thermal expansion of different metals finds extensive use in many electric appliances. There are probably some applications that have not yet been invented; perhaps you can acquire a patent and make your fortune!

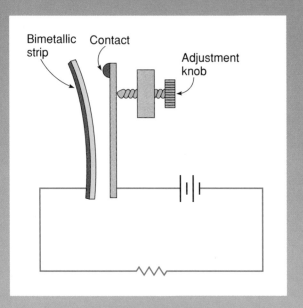

A bimetallic strip bends when it is heated because the two metals have different rates of thermal expansion. The bending of the strip can make or break a circuit.

SUMMARY

Starting with the simple example of lighting a flashlight bulb, this chapter introduced the concepts of electric circuit, electric current, and electric resistance. Ohm's law and the relationship between power and current were also used in analyzing series and parallel connections and some basic features of household circuits.

Electric Circuits and Current. An electric circuit is a closed conducting path around which charge can flow. The rate of flow of charge is called the *electric current,* which has units of charge per unit time, or amperes.

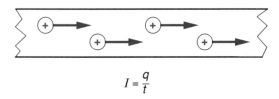

$$I = \frac{q}{t}$$

Ohm's Law and Resistance. Resistance is that property of a circuit element that opposes the flow of current. Ohm's law states that the current that flows through an element is proportional to the voltage difference across that element and inversely proportional to the resistance.

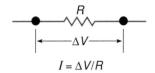

$$I = \Delta V/R$$

Series and Parallel Connections. When elements are connected in line so that the current that flows through one must also flow through the others, the elements are said to be connected *in series*. When the current can branch into different paths, the elements are said to be connected *in parallel*.

$$R_s = R_1 + R_2 + R_3$$

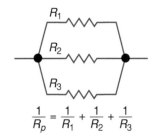

$$\frac{1}{R_p} = \frac{1}{R_1} + \frac{1}{R_2} + \frac{1}{R_3}$$

Energy and Power. Since voltage represents potential energy per unit charge, multiplying a voltage difference by a charge yields energy. Since power is the rate of use of energy, and current is the rate of flow of charge, multiplying a voltage difference by current yields power. The power supplied by a source must equal the power dissipated in the resistances.

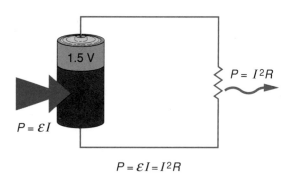

$$P = \mathcal{E}I = I^2R$$

Household Circuits. Household circuits use alternating current, which is continually reversing its direction, as opposed to direct current, which flows in one direction only. When we plug in appliances, we are connecting them in parallel with other appliances on the same circuit. The fuse or circuit breaker used to limit the current is in series with one arm of the circuit.

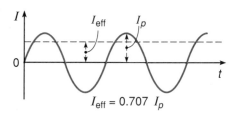

$$I_{eff} = 0.707\, I_p$$

QUESTIONS

Q12.1 Two arrangements of a battery, bulb, and wire are pictured here. Which of the two arrangements, if either, will light the bulb? Explain.

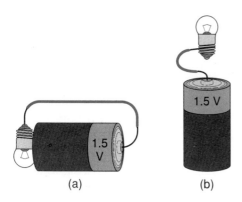

(a)   (b)

Q12.2 Suppose that you have two wires, a battery and a bulb. One of the wires is already in place in each of the arrangements pictured here. Indicate with a drawing where you would place the second wire in order to get the bulb to light. Explain your decision in each case.

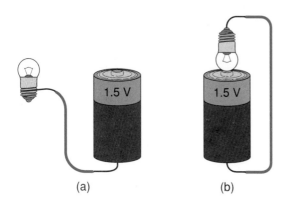

(a)   (b)

Q12.3 Two circuit diagrams are shown. Which one, if either, will cause the light bulb to light? Explain your analysis of each case.

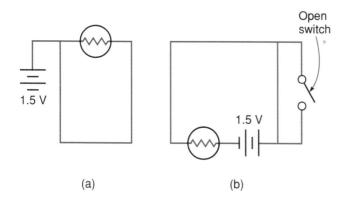

(a)   (b)

Q12.4 Consider the circuit shown here, where the wires are connected to either side of a wooden block as well as to the light bulb. Will the bulb light in this arrangement? Explain.

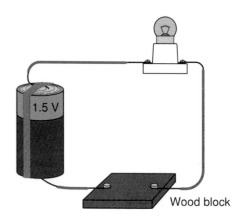

Wood block

Q12.5 Consider the circuit shown here. Could we improve the performance of this circuit by connecting a wire between points A and B? Explain.

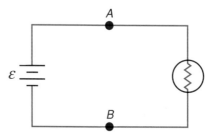

Q12.6 Suppose that we use a metal clamp to hold the wires in place in the battery-and-bulb circuit shown here. Will this be effective in keeping the bulb burning brightly? Explain.

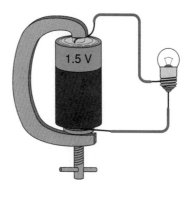

Q12.7 Consider the two signs shown here, located in different physics labs. Which of the two would be reason for the greater caution? Explain.

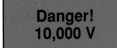

Danger! 100,000 Ω   **Danger! 10,000 V**

Q12.8 If we increase the potential difference across a resistance in a circuit, will the current flowing through that resistance increase, remain the same, or decrease? Explain.

Q12.9 A dead battery still indicates a voltage when a good voltmeter is connected across the terminals. Can the battery still be used, then, to light a light bulb? Explain.

Q12.10 When a battery is being used in a circuit, will the voltage difference across its terminals be less than when there is no current being drawn from the battery? Explain.

Q12.11 Two resistors are connected in series with a battery as shown in the diagram. R_1 is greater than R_2.

   a. Which of the two resistors, if either, has the greater current flowing through it? Explain.

   b. Which of the two resistors, if either, has the greatest voltage difference across it? Explain.

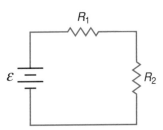

Q12.12 In the circuit shown here, R_1, R_2, and R_3 are three resistors of different values, R_3 is greater than R_2 and R_2, in turn, is greater than R_1. ε is the electromotive force of the battery, which has negligible internal resistance.

   a. Which of the three resistors has the greatest current flowing through it? Explain.

   b. Which of the two resistors, R_1 and R_2, if either, has the larger voltage difference across it? Explain.

   c. If we disconnect R_2 from the rest of the circuit, will the current through R_3 increase, decrease, or remain the same? Explain.

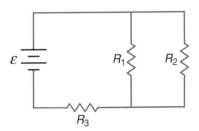

Q12.13 In the circuit shown here, the circle with a V in it represents a voltmeter. Which of the following statements is correct? (Comment upon each.)

   a. The voltmeter is in the correct position for measuring the voltage difference across R.

   b. No current will flow through the meter, so it will have no effect.

   c. The meter will draw a large current.

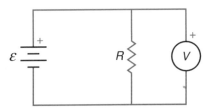

Q12.14 In the circuit shown here, the circle with an A in it represents an ammeter. Which of the following statements is correct? (Comment upon each.)

   a. The meter is in the correct position for measuring the current through R.

   b. No current will flow through the meter, so it will have no effect.

   c. The meter will draw a significant current from the battery.

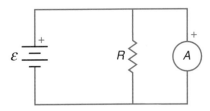

Q12.15 Which will normally have the larger resistance, a voltmeter or an ammeter? Explain.

Q12.16 Suppose that you have two appliances, an electric fan and an electric toaster, both designed to operate on 115 V, 60 hz alternating current. Which of the two, if either, could be operated successfully with an appropriate combination of storage batteries? Explain.

Q12.17 In using a dc voltmeter, it is important to connect the positive terminal of the meter in the correct direction in the circuit relative to the positive terminal of the battery. Is this likely to be true for the use of an ac voltmeter? Explain.

Q12.18 Which of the following appliances is most likely to cause an overload problem when connected to a circuit from which other appliances are already drawing current: an electric shaver, a coffee maker, or a television set? Explain.

Q12.19 Would it make sense to connect a fuse or circuit breaker in parallel with other elements in a circuit? Explain.

EXERCISES

E12.1 A 10-Ω resistor in a circuit has a voltage difference of 6 V across its leads. What is the current through this resistor?

E12.2 A current of 1.5 A is flowing through a resistance of 20 Ω. What is the voltage difference across this resistance?

E12.3 A current of 2.0 A flows through a resistor with a voltage difference of 36 V across it. What is the resistance of this resistor?

E12.4 In the circuit shown below, the 1-Ω resistance is the internal resistance of the battery and is in series, as shown, with the battery and the 11-Ω load.

    a. What is the current flowing through the 11-Ω resistor?

    b. What is the voltage difference across the 11-Ω resistor?

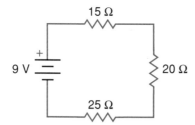

E12.5 Three resistances are connected to a 9 V battery as shown. The internal resistance of the battery is negligible.

    a. What is the current through the 15 Ω resistance?

    b. Does this same current flow through the 25 Ω resistance?

    c. What is the voltage difference across the 20 Ω resistance?

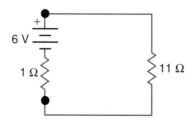

E12.6 Three resistances of 3 Ω, 6 Ω, and 2 Ω are connected in parallel with one another. What is the equivalent resistance of this combination?

E12.7 Three identical resistances, each 10 Ω, are connected in parallel with one another as shown. The combination is connected to a 30-V battery whose internal resistance is negligible.

    a. What is the equivalent resistance of this parallel combination?

    b. What is the total current through the combination?

    c. How much current flows through each resistor in the combination?

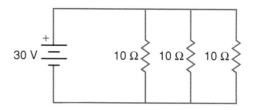

E12.8 Two resistances, one of 3 Ω and the other of 6 Ω, are connected in parallel with one another.

    a. What is the equivalent resistance of the combination?

    b. If there is a voltage difference of 6 V across the combination, what is the total current flowing through the combination?

    c. What are the currents through each resistor?

E12.9 A 12-V battery in a simple circuit is producing a current of 3 A through the circuit. How much power is being delivered by the battery?

E12.10 A 6-Ω resistor has a voltage difference of 10 V across its leads.

    a. What is the current through the resistor?

    b. What is the power being dissipated in this resistor?

E12.11 A 100-watt light bulb is designed to operate on an effective ac voltage of 110 V.

    a. What is the effective current through the light bulb?

    b. From Ohm's law, what is the resistance of the light bulb?

E12.12 A toaster draws a current of 5 A when it is connected to a 110-V ac line.

    a. What is the power consumption of this toaster?

    b. What is the resistance of the heating element in the toaster?

CHALLENGE PROBLEMS

CP12.1 In the circuit shown in the diagram, the internal resistance of the battery can be considered to be negligible.

    a. What is the equivalent resistance of the two-resistor parallel combination?

    b. What is the total current flowing around the circuit?

    c. What is the current through the 2-Ω resistor?

    d. What is the power dissipated in the 6-Ω resistor?

    e. Is the current through the 4-Ω resistor greater or less than that through the 2-Ω resistor?

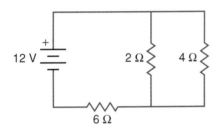

CP12.2 In the circuit shown here, the 3-V battery is opposing the 9 V battery as they are positioned. The total voltage of the two batteries will therefore be found by subtracting.

    a. What is the current flowing around the circuit?

    b. What is the voltage difference across the 5-Ω resistor?

    c. What is the power dissipated in the 7-Ω resistor?

    d. What is the power delivered by the 9-V battery?

    e. Is the 3-V battery discharging or charging in this arrangement?

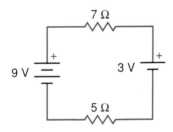

CP12.3 In the combination of 2-Ω resistors shown here, two different parallel combinations are in series with the middle resistor.

    a. What is the resistance of each of the two parallel combinations?

    b. What is the total equivalent resistance between points A and B?

    c. If there is a voltage difference of 6 V between points A and B, what is the current through the combination?

    d. What is the current through each of the resistors in the three-resistor parallel combination?

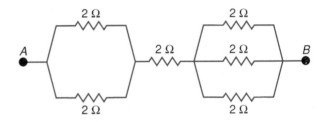

CP12.4 An 800-watt toaster, an 1100-watt iron, and a 500-watt food processor are all connected to the same 115-V household circuit, fused at 15 A.

    a. What is the current drawn by each of these appliances?

    b. If these appliances are all turned on at the same time, will there be a problem? Explain.

    c. What is the resistance of the heating element in the iron?

HOME EXPERIMENTS AND OBSERVATIONS

HE12.1 If you have a flashlight handy, you have a working example of a simple battery-and-bulb circuit.

   a. Open the flashlight and remove the batteries and bulb. Can you see how the switch operates? What makes contact when the switch is closed. Sketch the switch to make this clear.

   b. How are the connections made to the bulb in the flashlight?

   c. Leaving the bulb in its socket, can you get it to light by using external wires to make connections to the batteries? Sketch the arrangement that works.

HE12.2 If you have two flashlight bulbs available, a flashlight battery, and some wires, you can try your hand at building your own series and parallel circuits. The wires can be looped around the sides of the bulbs to hold them, and the contact with the base may be made to any metal object to which you can touch another wire. The bulb sockets available from a flashlight may be useful here.

   a. Construct a circuit with the two bulbs in series with one another and with the battery. Note the brightness of the bulbs when they are lit. (They may barely be glowing if you are using just one battery.)

   b. Now construct a circuit with the two bulbs in parallel with one another. Note the brightness of the bulbs in this case.

   c. Construct a circuit with just one bulb and compare the brightness of the bulb in this situation to that noted in (*a*) and (*b*).

HE12.3 If you have a toaster or a coffee maker available, see if you can find the bimetallic strip that turns the heating element off when a certain temperature is reached. (**Be sure to unplug the appliance before taking anything apart!**)

   a. How does this strip make or break the circuit? Sketch the arrangement of the strip within the appliance.

   b. How is the device turned on in the first place? Can you sketch the other features of the circuit in your toaster or coffee maker?

HE12.4 Locate all of the electric appliances that you use in your dormitory room or in a single room in your house.

   a. Can you find current or power ratings on these appliances? What are they?

   b. Even if power ratings are not present on each device, can you guess with the help of table 12.2 approximately how much current each appliance is likely to draw? List the appliances in order according to how much current each is likely to draw (from most to least).

   c. Is there likely to be a problem if all of the appliances are turned on at the same time on a single circuit fused at 20 A?

13 Magnets and Electromagnetism

Do you remember playing with refrigerator magnets when you were a child? Plastic letters and figures backed with small magnets can be arranged at your pleasure on the refrigerator door. They stick to the steel door because of the magnets, a fact that you have accepted since you were four years old or so. They are a wonderful teaching tool as well as a plaything, since the letters can be arranged to spell your name or simple words (fig. 13.1).

Because of such toys, we are generally much more familiar with magnetic forces than with electrostatic forces or electric currents. We have played with small horseshoe magnets and compasses and maybe even simple electromagnets that can be made from a steel nail, some wire, and a battery. We know that there is a force that attracts some types of metal, but not others, to magnets. Like the gravitational force and the electrostatic force, this force is effective even when the objects are not in direct contact with one another.

Although the magnetic force is familiar, it is also somewhat mysterious. Beyond the simple facts that we have just stated, your understanding of the nature of the magnetic force is likely to be limited. Why, for example, is it effective only with a few types of metal? In what ways is it similar to the electrostatic force, and how does it differ? On a more fundamental level, is there a relationship between electrical effects and magnetism? Our mention of electromagnets would suggest that there may be some connection.

There is indeed a connection between electric currents and magnetic forces. These connections were discovered during the nineteenth century and culminated in a theory of electromagnetism developed by the Scottish physicist, James Clerk Maxwell (1831–1879), during the 1860s. In the early part of the twentieth century, it became clear that the electrostatic force and the magnetic force are really just different aspects of one fundamental electromagnetic force.

An understanding of the relationships between electricity and magnetism has led to the invention of the electric motor and the electric generator, which play enormous roles in modern technology. You may have built a simple electric motor at some point in your educational career. What is its principle of operation? It is possible to gain a basic understanding of electric motors and generators from the ideas that we will explore in this chapter.

Figure 13.1   A child playing with magnet-backed letters that stick to the refrigerator door. What is the nature of the force that holds the letters to the door?

Chapter Objectives

Starting with a simple description of magnets that builds upon similarities with the electrostatic force, we will discuss the relationship between electric currents and magnetic forces. This will provide us with a more fundamental view of the nature of the magnetic force. Finally, we discuss Faraday's law, which describes how electric currents can be generated by the interaction of wire loops with magnetic fields (electromagnetic induction). Our principle objective is to gain an understanding of the nature of magnetic forces and fields and of Faraday's law.

Chapter Outline

1. *Magnets.* What are the poles of a magnet, and how is the magnetic force similar to the electrostatic force? Why does a compass work and what is the nature of the earth's magnetic field?

2. *Magnetic effects of electric currents.* What is the relationship between electric currents and magnetic forces and fields? How can we describe the magnetic force in terms of effects involving moving charges?

3. *Magnetic effects of current loops.* What are the magnetic characteristics of current loops? In what way is a loop of electric current similar to a bar magnet?

4. *Faraday's law: electromagnetic induction.* How can electric currents be produced by the interaction of loops of wire with magnetic fields? What is Faraday's law of electromagnetic induction?

5. *Generators and transformers.* How can Faraday's law be used to explain the operation of electric generators and transformers? What role do these devices play in the production and transmission of electric power?

13.1 MAGNETS

If you gather up a few magnets from around the home or office, you can establish some basic facts about their behavior with some simple experiments (fig. 13.2). You probably know already that magnets will attract paper clips, nails, or any metallic item that is made of iron or steel, but they will not attract items made of silver, copper, aluminum, or most nonmetallic materials.* Magnets will attract each other also, but only if the ends are properly aligned; if they are not, the force between them is repulsive rather than attractive.

The three common magnetic elements are the metals iron, cobalt, and nickel. The small magnets that are available for sticking things to refrigerator doors or metal cabinets are made of alloys of these three metals and others added to obtain desired properties. The first known magnets, however, were simply a form of iron ore called *magnetite,* which occurs naturally and is often weakly magnetized. The existence of magnetite was known in antiquity, and its properties had long been a source of curiosity and amusement.

Magnetic Poles

If you look closely at some of the magnets in your collection, you will find that they may have their ends labeled with the letters *N* and *S*. The letters stand for *north* and *south,* although if we explored the origin of these labels, we would find that they originally stood for *north-seeking* and *south-seeking*. If you suspend a simple bar magnet by a thread tied around its middle and allow it to turn freely around the point of suspension, it will eventually come to rest with one end pointing approximately northward. This is the end, or *pole,* that we label *N;* the other is then labeled *S*.

If you play with the magnets that have been labeled in this manner, you quickly discover that the opposite poles of two magnets attract one another; the north pole of one attracts the south pole of the other. If you hold the magnets firmly in each hand and bring the two north poles near one another, you can feel a repulsive force pushing them apart. The same is true of the two south poles. In fact, if one of the magnets is lying on a table and you try to bring two like poles together, the magnet on the table will suddenly flip around so that the opposite pole comes into contact with the pole of the approaching magnet.

These simple observations can be summarized in a rule that you probably first learned as part of a science unit in grade school:

Unlike poles attract one another, and like poles repel one another.

There is an obvious similarity here to the rule for the electrostatic forces between like and unlike charges (fig. 13.3).

---

* Ceramic materials that can be magnetized have been developed in recent years.

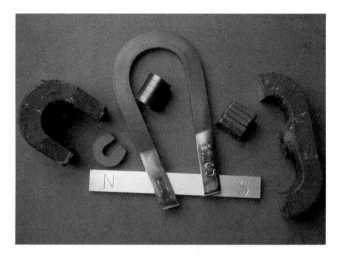

Figure 13.2   A collection of magnets. Where are the poles likely to be found on each magnet?

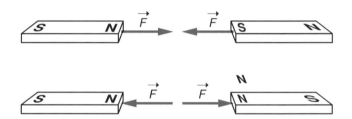

Figure 13.3   Unlike poles attract one another, and like poles repel one another. Magnetic poles obey the same rules as electric charges in this sense.

The rule for magnets was known well before the electrostatic rule, however.

It is fun to play with magnets, chasing them around with each other. Small disk or ring magnets usually have their poles on opposite sides of the disk. If you bring a bar magnet close to such a magnet with like poles facing each other, the disk may hop away or do a quick flip and attach itself to the bar magnet. Sometimes the flip is so quick that it is difficult to see. If the magnets are in the form of small rings, you can levitate them easily by stringing them on a thin rod, as shown in figure 13.4.

Some magnets seem to have more than two poles; there may be a south and north pole somewhere in the middle of a magnet. Iron filings sprinkled on top of a piece of paper lying on top of a magnet can be used to show where its poles are located. The filings will group most densely in the vicinity of the poles.

Coulomb's Law

The similarity between magnetic poles and electrostatic charges goes well beyond the rules for attraction and repulsion. If we attempt to measure how the force that two poles exert upon one another varies with distance or the strength of

Figure 13.4   Ring magnets levitating on a vertical rod. The rod keeps the magnets from flying off to the side.

the poles, we find that they behave according to a law of the same form as Coulomb's law for the electrostatic force. In fact, Coulomb himself was the one who made the initial measurements and first stated the force law.

Coulomb made his measurements with magnets by substituting long, thin bar magnets for the metal balls in his torsion balance (see chapter 11). The magnets must be long enough so that the poles we are not interested in are far enough from the point of measurement that we can ignore their effect on the force. The problem is that we can never isolate just a north pole or a south pole. If you break a thin bar magnet in half, you end up with two bar magnets, each one with a north and a south pole.

Coulomb's experiments showed that the force between two poles decreases with the square of the distance between the two poles, as does the electrostatic force. The force between poles is also proportional to a quantity called the *pole strength* of each pole. Some magnets are stronger than others, and the magnitude of the force exerted by one magnet on another depends on the pole strength of *both* magnets. Coulomb was able to determine relative pole strengths by comparing the force exerted by different magnets (at the same distance) on a magnet used as reference.

In symbols, we can state Coulomb's law for magnets as follows:

$$F \sim \frac{p_1 p_2}{r^2},$$

where p_1 and p_2 represent the pole strengths of the two magnets, and (as before) ~ stands for "proportional to." The rela-

tionship can be written as an equality if we define units for pole strength and insert an appropriate constant based upon our measurements. The important point, however, is that magnetic poles obey a force law that is identical in form to Coulomb's law for electrostatic charges.

Dipoles and Fields

The fact that we cannot completely isolate a single magnetic pole means that a magnet is always at least a *dipole*. Although there can be more than two poles, apparently there can never be fewer than two. Physicists have spent considerable effort looking for magnetic monopoles (particles consisting of a single isolated magnetic pole), but have found no conclusive evidence that such entities exist. Breaking a magnetic dipole in half always produces two dipoles.

This is one respect, then, in which magnetic poles are *not* similar to electric charges: positive or negative charges can be isolated. Electrons, for example, are negatively charged particles, and protons have a positive charge. Electric dipoles can be produced, however, which consist of equal but opposite charges separated by a small distance. The electric field lines produced by an electric dipole originate on the positive charge and terminate on the negative charge, as shown in figure 13.5.

Can we also associate field lines with a magnetic dipole? It turns out that we can indeed define a magnetic field in which the field lines show the same pattern for a magnetic dipole as the electric field lines do for the electric dipole. (The definition of magnetic field is considered in more detail in the next section.) For regions outside of the magnet, the magnetic field lines originate on the north pole and terminate on the south pole, as shown in figure 13.5. We can make this pattern visible by sprinkling iron filings on a piece of paper covering a bar magnet. The filings line up in the direction of the field.

If we place an electric dipole in an electric field that is produced by other, external charges, the dipole lines up with the external field. This is because the forces acting on each charge in the dipole combine to produce a torque that turns the dipole toward the direction of the external field (fig. 13.6). Similarly, a magnetic dipole lines up with an externally produced magnetic field. In fact, this is why the iron filings line up with the field lines around a magnet: the filings become polarized in the presence of the field, so that each one is a small dipole.

The Earth as a Magnet

A compass needle is also a magnetic dipole. The first compasses were made by balancing a thin crystal of magnetite on a support so that it could turn freely. As you are aware, the north, or north-seeking, pole of a magnet will point in a northerly direction, although not necessarily exactly due north. This simple discovery became a tremendous boon to

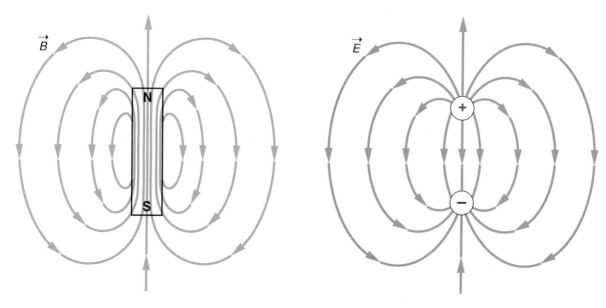

Figure 13.5   The magnetic field lines produced by a magnetic dipole form a pattern similar to the electric field lines produced by an electric dipole.

Magnetic field $\vec{B}$ Electric field $\vec{E}$

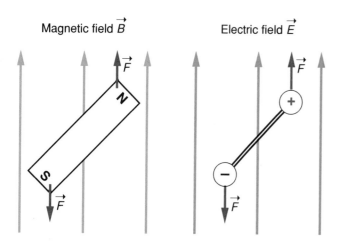

Figure 13.6   A magnetic dipole will line up with an externally produced magnetic field just as an electric dipole lines up with an electric field.

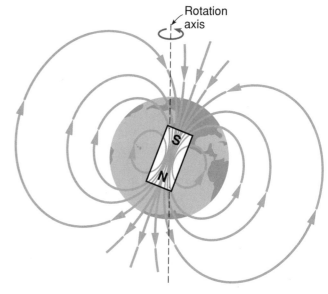

Figure 13.7   The magnetic field of the earth can be pictured by imagining that there is a bar magnet inside the earth (there is not, of course), oriented as shown here.

ocean navigation in the early Renaissance. Prior to that time, sailors were unable to navigate on cloudy days or nights.

Is the earth a magnet? What is the nature of the magnetic field to which the compass needle responds? Although the compass was invented by the Chinese, it was the English physician William Gilbert (1540–1603) who first thoroughly studied these phenomena in the West. Gilbert suggested that the earth does indeed behave like a large magnet. You can picture the magnetic field produced by the earth by imagining that there is a large bar magnet inside the earth, oriented as shown in figure 13.7. (Do not take this picture too literally; it is merely a device for describing the field.)

Since unlike poles attract, it must be the south pole of the earth's magnet that points in a northerly direction. The north-

seeking pole of the compass aligns itself northward along the field lines produced by the earth. The axis of the earth's magnetic field is not aligned exactly with the axis about which the earth rotates, however. Since the rotational axis is what defines true north, the compass needle does not point exactly northward at most locations. On the east coast of the United States, magnetic north is a few degrees to the west of true north, whereas on the west coast it is several degrees to the east of true north. Along some line in between, magnetic north and true north are identical. (The precise location of this line varies slowly with time.)

No one knows exactly how the earth's magnetic field is produced, although there has been considerable speculation on this question. Most of the models assume that electric currents associated with the motion of fluids in the core of the earth are responsible. But what do electric currents have to do with magnetic fields? The next two sections of this chapter explore that connection.

13.2 MAGNETIC EFFECTS OF ELECTRIC CURRENTS

The invention of the battery by Volta in the year 1800 made it possible to produce steady electric currents for the first time. Previously, currents usually took the form of very rapid discharges of charge that had been accumulated by electrostatic techniques. Connecting a long, thin wire with relatively high resistance across the terminals of Volta's battery permitted a much steadier flow of charge over a much longer period of time. This allowed scientists to study electric currents in ways that had not been possible before.

The similarities between magnetic effects and electrostatic effects discussed in the previous section led scientists to think that there might be a direct connection between electricity and magnetism. The same people were often involved in the study of both areas; William Gilbert, for example, was also an early explorer of electrostatic effects, and Charles Coulomb measured the force law for magnetic poles as well as electric charges. It was twenty years after the discovery of the battery, however, before the Danish scientist Hans Christian Oersted (1777–1851) made a striking discovery of a simple effect.

Oersted's Discovery

Oersted's initial discovery in 1820 of the magnetic effect produced by an electric current was made in the course of a lecture demonstration. Demonstrations often fail to go exactly as we have planned, but in this case the failure was fortuitous. Oersted was demonstrating some effects of electric currents and happened also to have a compass lying nearby. He noticed that the compass needle deflected when he completed a circuit consisting of a long wire and a battery.

Oersted had earlier used a compass in the vicinity of a current-carrying wire and had not noticed any such effect. Other scientists had also attempted to find such an effect without success. The deflection of the compass needle during the lecture was therefore a surprise. Not wishing to make a fool of himself in front of his students, Oersted decided to explore the situation more carefully after the lecture was over. He found that indeed there was a reproducible effect on the compass needle, provided that the current was sufficiently strong and the compass and wire had appropriate orientations.

The strange directional aspects of this newly discovered effect may explain why it had not been observed earlier by Oersted or other scientists, despite the suspected connection between electricity and magnetism. In order to get the maximum effect with the wire lying on a table, the wire must be oriented along a north-south line—the line along which the compass needle would point in the absence of any current. When the current is turned on, the needle deflects away from this line (fig. 13.8). The magnetic field produced by the current in the wire is perpendicular to the direction of the current!

Further study by Oersted and others showed that the magnetic field lines produced by a straight, current-carrying wire form circles centered on the wire. (Oersted did not talk about field lines, but about the direction in which the compass needle pointed at various locations.) When he placed the compass below the wire, the needle deflected in a direction opposite to its deflection when it was above the wire. A simple right-hand rule can be used to determine the direction of the field. If you imagine that you are holding the wire in your right hand with your thumb pointing in the direction of the electric current, then your fingers curl around the wire in the direction of the magnetic field lines produced by the current (fig. 13.9).

Not surprisingly, the effect gets weaker as the compass is moved away from the wire. A current of a few amperes is necessary to get a good deflection of the compass needle, even when the compass is just a few centimeters from the wire. Early batteries were not capable of sustaining very large currents. This fact, together with the unexpected direction of the effect, probably hindered earlier attempts to detect the magnetic effects of an electric current.

The Force on a Current-Carrying Wire

If we can produce a magnetic field with an electric current, should not a wire carrying an electric current also experience a magnetic force due to other magnetic fields? To put this question another way, if an electric current acts in some ways like a magnet, does it also experience a magnetic force in the presence of a magnet or another current-carrying wire?

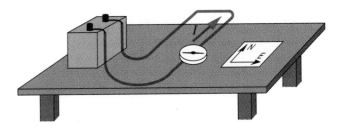

Figure 13.8 With the wire oriented along a north-south line, the compass needle deflects away from this line when there is a current flowing in the wire.

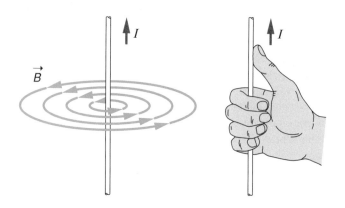

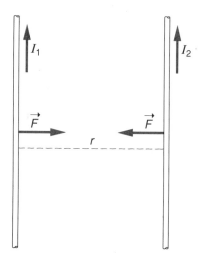

Figure 13.9 **The right-hand rule indicates the direction of the magnetic field lines that encircle a current-carrying wire. The thumb must point in the direction of the current, and the fingers represent the direction of the field lines.**

Figure 13.10 **Two parallel, current-carrying wires exert an attractive magnetic force upon each other when the two currents are in the same direction.**

This question was explored by André-Marie Ampère (1775–1836) in France, and by many other scientists who were excited by Oersted's discovery.

Ampère discovered that there was indeed a force exerted on one current-carrying wire by another. He also showed that this force was related to the magnetic effect discovered by Oersted, and could not be explained as an electrostatic effect. (A current-carrying wire is usually electrically neutral, with no net positive or negative charge.) Ampère measured the strength of the force between two parallel current-carrying wires and studied how it varied with the distance between the wires and the amount of current flowing in each.

Ampère's experiments showed that the force between two parallel wires was proportional to the two currents and inversely proportional to the distance between the two wires. In symbols, this relationship can be expressed as follows:

$$F \sim \frac{I_1 I_2}{r},$$

where electric current, as before, is represented by the symbol I, and r is the distance between the wires. The force exerted by one wire upon the other is attractive when the currents are flowing in the same direction and repulsive when they are flowing in opposite directions (fig. 13.10).

The relationship discovered by Ampère can be stated as an equality if we specify the units for electric current and define a constant of proportionality. We often express this equation as follows:

$$\frac{F}{l} = \frac{2k' I_1 I_2}{r},$$

where the constant k' is equal to 1.0×10^{-7} N/A^2, and F / l represents the force per unit length of the wire. (The longer the wire, the greater the force.) The strangely simple value of the constant k' is no accident; we define the unit of current,

the ampere (A) by using this relationship and measuring the force between two wires, as did Ampère.

One ampere (A) is the amount of current, flowing in each of two parallel wires separated by a distance of 1 meter, that produces a force per unit length on each wire of 2×10^{-7} newtons per meter.

The ampere is the basic unit of electromagnetism, then, and is standardized by making measurements of the magnetic force between current-carrying wires. The coulomb, the unit of charge, is defined from the ampere as follows:

$$1 \text{ coulomb} = 1 \text{ ampere-second},$$

since charge is the product of current and time, $q = It$. (This follows from the definition of current, $I = q / t$.)

Force on a Moving Charge

What is the basic nature, then, of the magnetic force? Magnetic forces are exerted by magnets on other magnets, by magnets on current-carrying wires, and by current-carrying wires on each other. Since electric current represents the flow of electric charge, we apparently obtain magnetic forces when charges are moving, but not when they are at rest. Is the motion of charge somehow a fundamental criterion for the existence of magnetic forces? This question was also addressed by Ampère during the 1820s.

If we think about the direction of the force exerted by one current-carrying wire upon another, we see that this force is perpendicular to the direction of the current and also perpendicular to the direction of the magnetic field. Since we define the direction of electric current as the direction of flow of positive charge, we can imagine that there is a magnetic force exerted upon the moving charges in the wire. This force must be perpendicular to both the direction of the motion of the charges and to the direction of the magnetic field to be consistent with the observed force on the wire (fig. 13.11).

The magnetic force on a current-carrying wire has its maximum value when the wire is also perpendicular to the direction of the field. All of these facts suggest that a magnetic force is exerted upon a moving charge that is proportional to the quantity of the charge, the velocity of the moving charge, and the strength of the magnetic field. In symbols this relationship takes the form

$$F = qvB,$$

where q is the charge, v is the magnitude of the velocity of the charge, and B represents the strength of the magnetic field. For this relationship to be valid, however, the velocity must be perpendicular to the field.

This expression for the magnetic force actually defines the magnitude of the magnetic field. Dividing by q and v, we can put it in the following form:

$$B = \frac{F}{qv_\perp},$$

where $v_\perp$ stands for the component of the velocity that is perpendicular to the field. The units of magnetic field are thus N/(C)(m/s), which are equivalent to N / A·m. This unit is now called a tesla (T). Just as we defined the electric field as the electrostatic force per unit charge, the magnetic field is the magnetic force per unit charge and unit velocity. If the velocity of the charge is zero, there is no magnetic force!

The perpendicularities in our description of the magnetic force are a natural and unavoidable feature of this force. Another right-hand rule, illustrated in figure 13.12, is often used to describe the direction of the magnetic force. If you point your right index finger in the direction of the velocity of a positive charge and your middle finger in the direction of the magnetic field, your thumb points in the direction of the magnetic force on the moving charge. This force is always perpendicular to both the velocity and the field directions. The force on a negative charge is in the opposite direction to that on a positive charge moving in the same direction.

Because the magnetic force on a moving charge is perpendicular to the velocity of the charge, this force does no work on the charge and does not increase its kinetic energy. The acceleration of the charge that results is a centripetal acceleration that will change the direction of the charge's velocity. If the charge is moving in a direction perpendicular to a uniform magnetic field, the magnetic force bends the path of the charged particle into a circle. The radius of the circle is determined by the mass and speed of the particle and can be found by applying Newton's second law. (See challenge problem 13.2.)

Since the direction of electric current is that of the flow of positive charge, the right-hand rule used to find the direction of the force on a moving charge can also describe the direction of the force on a current-carrying wire. In this case, the index finger points in the direction of the current, and the middle finger and the thumb have the same interpretation as before. Here again, the force is perpendicular to both the direction of the current and to the field.

In fact, the force on a segment of wire can also be described in terms of the magnetic field by a simple equation that follows from $F = qvB$. The product of the charge and the velocity (qv) is replaced by the product of the current and the length of the segment of wire (Il) to yield

$$F = IlB,$$

where I is the current, l is the length of the segment of wire, and B is the strength of the magnetic field (see box 13.1). The direction of the current must be perpendicular to the field for this expression to be valid.

At the fundamental level, then, the magnetic force is a force that *moving* charges exert on one another. Since electric currents involve moving charges, we can also say that the magnetic force is exerted by electric currents on one another, and we can always replace the product qv with the product Il in any

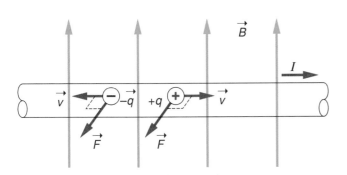

Figure 13.11 The magnetic force exerted upon the moving charges of an electric current is perpendicular to both the velocity of the charges and to the magnetic field.

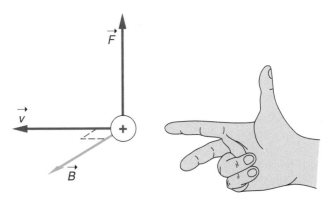

Figure 13.12 If the index finger of the right hand points in the direction of the velocity of the charge, and the middle finger in the direction of the magnetic field, the thumb indicates the direction of the magnetic force acting upon the charge.

Box 13.1

Sample Exercise

A straight wire with a length of 15 cm carries a current of 4.0 A. The wire is oriented perpendicularly to a magnetic field with a magnitude of 0.50 T. What is the magnitude of the force exerted on the wire by the magnetic field?

l = 15 cm = .15 m. $F = IlB$

I = 4.0 A.   = (4.0 A)(0.15 m)(0.50 T)

B = 0.50 T.   = 0.30 N.

(The direction of this force is perpendicular to both the current in the wire and to the magnetic field and is described by the right-hand rule.)

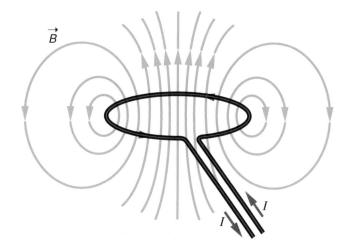

Figure 13.13 **When a current-carrying wire is bent into a circular loop, the magnetic fields produced by different segments of the wire act to produce a strong magnetic field near the center of the loop.**

expression involving magnetic fields or forces. What about magnets, however? Do they also involve moving charges? The next section provides some insight to that question.

13.3 MAGNETIC EFFECTS OF CURRENT LOOPS

Up to this point, the current-carrying wires that we have considered have been straight, although they must bend somewhere in order to complete an electrical circuit. Many applications of electromagnetism, however, involve loops of wire. Wire coils are used in electromagnets, electric motors, electric generators, transformers, and a host of other applications.

What happens when we bend a current-carrying wire into a coil? What does the magnetic field look like, and how is the coil affected by an external magnetic field? The flurry of experimental activity that occurred during the 1820s following Oersted's discovery included exploration of these questions. André Ampère was active in this work as well as many other investigators.

The Magnetic Field of a Current Loop

The magnetic field lines around a straight wire form circles centered on the wire. Imagine the process of bending such a wire into a circular loop. What happens to the field lines? Very near the wire, the field lines are still circles, presumably. Toward the center of the loop, however, the field contributions produced by different segments of the wire begin to add as vectors. Since they are all pointed approximately in the same direction near the center of the loop, they combine to give a field that is quite strong in the vicinity of the center.

The resulting field is pictured in figure 13.13. The field lines are close together near the center of the loop, indicating a strong field in that region. The field is also strong near the current-carrying wire itself. Further away from the loop, the field weakens, as would be expected. As was the case for the straight wire, the magnetic field lines of the current loop curve around until they meet themselves, forming continuous loops.

You may notice that the field pictured in figure 13.13 is very similar to the field for a bar magnet, pictured in figure 13.5. In fact, if the bar magnet is short enough, as shown in figure 13.14, the field patterns are essentially identical. We might conclude, then, that a current loop is a magnetic dipole because its field is the same as that of the more familiar dipole, the bar magnet.

Torque on a Current Loop

The similarity between a current loop and a bar magnet goes beyond the similarity of the fields produced. If we place a current loop in an external field, it experiences a torque, just as a bar magnet would if it were not aligned with the field initially. It would be possible, then, to use a loop or several loops of current-carrying wire as a compass needle, because its axis (perpendicular to the plane of the loop) would line up with an external field, just as a normal compass needle does.

We can see the origin of the torque on a current loop most easily by considering a loop in the form of a rectangle, as shown in figure 13.15. Each of the segments of the rectangular loop is a straight wire, and the force on each segment is given by the expression $F = IlB$. The directions of the forces are given by the right-hand rule stated at the end of section 13.2. Limber up your right hand and verify that the force directions are those shown in the diagram.

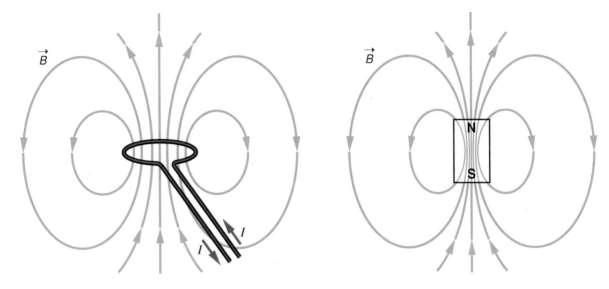

Figure 13.14 The magnetic field produced by a current loop is essentially identical to that produced by a short bar magnet (a magnetic dipole).

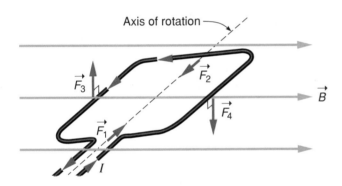

Figure 13.15 **The forces on each segment of a current-carrying rectangular loop of wire combine to produce a torque that tends to rotate the loop so that its plane is perpendicular to the external magnetic field.**

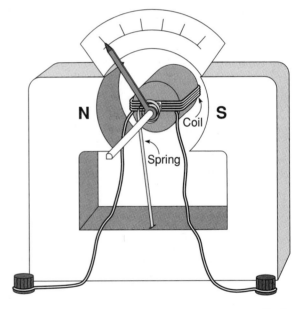

Figure 13.16 **An electric meter consists of a coil of wire, a permanent magnet, and a restoring spring to return the needle to zero when there is no current flowing through the coil.**

Each of the four sides of the loop shown in figure 13.15 experiences a magnetic force. The forces on the two ends of the loop ($\vec{F}_1$ and $\vec{F}_2$) produce no torque about the center of the loop, however, because their lines of action pass through the center of the loop. Their lever arms about any axis passing through the center of the loop are zero; thus they produce no torque. The forces on the other two sides ($\vec{F}_3$ and $\vec{F}_4$) have lines of action that do not pass through the center of the loop, provided that the plane of the loop is not perpendicular to the external magnetic field. From the diagram we can see that these two forces combine to produce a net torque that tends to rotate the loop about the axis shown, so that its plane ends up perpendicular to the magnetic field. (See chapter 7 for a discussion of torque and rotational motion.)

Since the magnetic forces on the loop segments are proportional to the electric current flowing around the loop ($F = IlB$),

the magnitude of the torque on the loop is also proportional to the current. This fact makes the torque on a current-carrying loop or coil useful for measuring electric current. Most simple electric meters have a coil of wire with a needle attached to it. A permanent magnet provides the magnetic field, and a spring returns the needle to the zero position when there is no current (fig. 13.16).

This torque is also the basis of operation of electric motors. In order to keep the coil turning, however, the current in the coil must reverse directions every half turn.

Otherwise the coil would come to rest with the plane of its loop perpendicular to the external magnetic field. An alternating current is well suited to operating electric motors, therefore, but we can also design motors that operate on direct current. Box 13.3 describes the design of a simple dc motor. Electric motors are found everywhere, from the starting motor in your car to the motors in elevators, kitchen appliances, vacuum cleaners, washers, dryers, and electric shavers.

Coils and Electromagnets

We have been talking about loops of current, but also coils, which consist of several loops all wound in the same direction. What happens when we add several loops to form a coil? Is the magnetic field stronger than that for a single loop? What effect do we get by winding a coil of wire on an iron or steel core?

It is not difficult to see that the magnetic field produced by a coil of wire will be stronger than that produced by a single loop carrying the same current. The magnetic fields produced by each loop are all in the same direction near the axis of the coil, and they therefore combine. The resulting field strength will be proportional to the number of turns, N, that are wound on the coil (fig. 13.17). The torque on the coil, when it is in an external field, will also be proportional to both the current and the number of turns in the coil.

The effectiveness of a coil of wire as a magnet was discovered by the French scientist Dominique Arago (1786–1853), very soon after Oersted's discovery. Arago first noticed that iron filings were attracted to a current-carrying copper wire much as they would be to a magnet. Following a suggestion by Ampère, he also found that the effect was enhanced if the wire was wound into the shape of a helix or coil. An even larger effect was obtained, however, if the helix was wound around a steel needle. The ability to attract iron filings in this case was better than that of most natural magnets.

Arago's simple electromagnet was shown to have a north pole and a south pole like those of a bar magnet. In fact, the magnetic field of a coil of wire is identical outside of the coil to that produced by a bar magnet of the same length and strength. This similarity led Ampère to suggest that the magnetism of a naturally magnetic material such as magnetite might be caused by current loops in the atoms that make up the material. If, for some reason, these atomic current loops could all line up with one another and lock into those positions, a permanent magnet would result.

We should point out that in Ampère's time nothing at all was known about the structure of atoms. This knowledge was not obtained until the beginning of the current century, almost a hundred years later, yet Ampère's theory is remarkably close to the modern view of what happens in ferromagnetic materials such as iron, nickel, and cobalt, and their alloys. We now know that the current loops are associated with the spins of the electrons in the atoms. We have only recently been able to understand why these spins tend to line

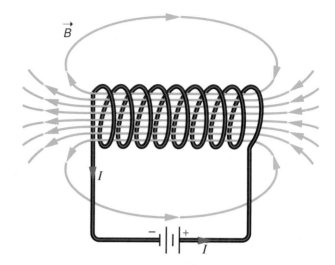

Figure 13.17 **A current-carrying coil of wire produces a magnetic field that is greater than that of a single loop and is proportional in strength to the number of loops in the coil.**

up in ferromagnetic materials, but not in other materials.

Within ten years of Oersted's initial discovery of the magnetic effect of an electric current in 1820, many of the phenomena of electromagnetism had been thoroughly explored and described. André Ampère was the leader in much of this work, and he was also responsible for developing the mathematical theory describing the magnetic forces on current-carrying wires and coils. His theory describing the magnetism in natural magnets as being due to atomic current loops completed the link between electricity and magnetism. Magnetic effects could all be regarded as due to the action of electric currents, or moving charges.

13.4 FARADAY'S LAW: ELECTROMAGNETIC INDUCTION

The work of Oersted, Ampère, and others had firmly established that magnetic forces are associated with electric currents. In the modern terminology involving fields, we say that a magnetic field is *produced* by an electric current. What about the reverse process, though? Is it possible to produce an electric current using a magnetic field?

One person who asked this question was Michael Faraday (1791–1867), who got his start in science as an assistant to the English chemist Sir Humphrey Davy. Davy's primary interest, and much of Faraday's early work, concerned the chemical action of electric currents, or *electrolysis*. Faraday was not trained in mathematics and therefore was not inclined to develop or use mathematical descriptions of electrical phenomena. He was, however, a brilliant experimentalist. He set out methodically to explore the possibility of producing an electric current from magnetic effects.

Faraday's Experiments

Faraday was aware that magnetic effects were enhanced by coils of wire. He started his extensive series of experiments by winding two unconnected coils of wire upon the same wooden cylinder. One coil was connected to a battery, and the other to a galvanometer. (A galvanometer is an electrical meter designed to measure both the direction and the amount of current flow; its zero point is in the middle of the scale, so that it can deflect in either direction.) His purpose was to see whether he could detect a current in the coil connected to the galvanometer when a current was flowing in the coil connected to the battery.

The results of these first experiments were completely negative; no current was detected in the second coil. Undeterred, Faraday persisted by winding longer and longer coils. Finally, when his coils reached a length of approximately 200 feet of copper wire, he did notice an effect. There was still no evidence of a steady current being produced in the second coil, but there was a very brief and feeble deflection of the galvanometer needle when he connected the first coil to the battery. There was another momentary deflection in the opposite direction when he broke contact with the battery.

This was not the effect that Faraday had expected, but experimentalists cannot be choosy about what nature provides. Through further experiments, Faraday found that he could get a considerably stronger deflection (although still momentary) if he wound both coils upon an iron core rather than a wooden cylinder as shown in figure 13.18. Faraday wound two coils on either side of a welded iron ring. One coil, the primary, was connected to a battery; the other, the secondary, to his galvanometer. Again, the galvanometer needle deflected in one direction when the primary circuit was closed by making contact with the battery and in the opposite direction when the primary circuit was broken.

Once again, no current was detected in the secondary coil when there was a steady current in the primary coil. It was only if this primary current was *changing* as the circuit was completed or broken that a deflection occurred. The strength of the deflection was proportional to the number of turns of wire wound on the secondary coil as well as to the strength of the battery used with the primary coil. Faraday began to formulate the idea that an electric current could be induced by magnetic fields that were changing with time, rather than by the steady-state presence of a magnet or electric current.

Faraday pursued this idea with many other experiments. One of these involved moving a permanent bar magnet in and out of a hollow helical coil of wire attached to a galvanometer. When the magnet was moved in, the galvanometer needle deflected in one direction; when it moved out, the needle deflected in the opposite direction (fig. 13.19). When the magnet was not moving, no deflection resulted. This is an easy effect to demonstrate with equipment available in most physics labs.

Faraday's Law

The results of all of Faraday's experiments indicate that an electric current is induced in a coil or circuit when the magnetic field passing through the circuit is changing. Because the amount of current flowing in the secondary circuit depends upon the resistance of that circuit, we generally choose to express this fact in terms of the induced voltage, however, instead of the current. In order to construct a quantitative statement to describe this effect, we also need to think about how we should define the amount of magnetic field passing through the circuit.

This latter problem is solved by defining the concept of magnetic flux. For a simple loop of wire lying in a plane perpendicular to the magnetic field, *magnetic flux* is the product of the magnetic field B times the area A bounded by the circuit or loop of wire (fig. 13.20). In symbols, the flux definition takes the following form:

$$\Phi = BA,$$

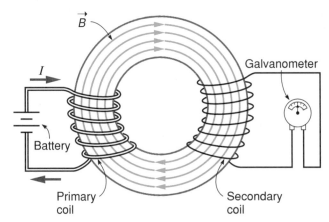

Figure 13.18 **The magnetic field produced by the primary coil was channeled through the secondary coil by the iron ring used in one of Faraday's experiments.**

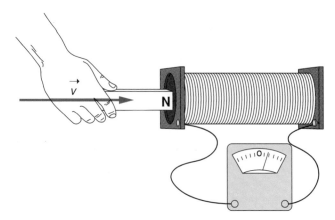

Figure 13.19 **A bar magnet moved in or out of a helical coil of wire produces an electric current in the coil.**

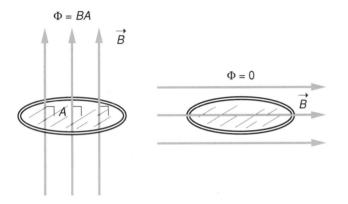

$\Phi = BA$

$\vec{B}$

$\Phi = 0$

A

$\vec{B}$

Figure 13.20 **The magnetic flux through the loop of wire has its maximum value when the field lines are perpendicular to the plane of the loop; it is zero when the field lines are parallel to the plane of the loop and therefore do not cross the plane.**

where the Greek letter Φ (phi) is the standard symbol used to represent flux.

We must place an important qualification on this definition, however. The maximum flux is obtained when the field lines pass through the circuit in a direction perpendicular to the plane of the circuit. If the field lines are parallel to this plane, there is no flux because no field lines actually pass through the circuit, as shown in figure 13.20. To take this fact into account, we use only the component of $\vec{B}$ that is perpendicular to the plane of the loop in calculating the flux.

We can now make a quantitative statement of Faraday's law that summarizes the results of his experiments:

A voltage (electromotive force) is induced in a circuit when there is a changing magnetic flux passing through the circuit. The induced voltage is equal to the *rate of change* of the magnetic flux. In symbols,

$$\varepsilon = \frac{\Delta \Phi}{\Delta t}.$$

The more rapidly the magnetic flux passing through the circuit changes, the larger the induced voltage. This aspect of the effect can be readily observed in the experiment involving the moving magnet. If we move the magnet in and out of the coil quickly, we get larger deflections of the galvanometer needle than if we move it more slowly. The magnetic flux passing through a coil of wire passes through each loop in the coil, so the total flux through a coil is actually equal to the number of turns of wire, N, in the coil times the flux through each turn or $\Phi = NBA$. (Think about professional basketball to remember this expression!) The more turns of wire in the coil, therefore, the larger the induced voltage.

The example in box 13.2 provides a numerical illustration of these ideas. A coil of wire can be used to assess the strength of a magnetic field in a manner similar to the situation pictured

---

Box 13.2

Sample Exercise

A coil of wire with 50 turns has a uniform magnetic field of 0.40 T passing through the coil perpendicular to its plane. The coil encloses an area of 0.03 m². If the flux through the coil is reduced to zero by removing it from the field in a time of 0.25 seconds, what is the induced voltage in the coil?

$N = 50$ turns.

$B = 0.40$ T.

$A = 0.03$ m².

$\Delta t = 0.25$ s.

The initial flux through the coil is

$\Phi = NBA = (50)(0.40 \text{ T})(0.03 \text{ m}^2)$

$= 0.60$ T·m².

The induced voltage is equal to the rate of change of flux:

$$\varepsilon = \frac{\Delta \Phi}{\Delta t} = \frac{(0.60 \text{ T·m}^2 - 0)}{0.25 \text{ s}} = \textbf{2.4 volts.}$$

---

in this example. The magnetic flux can be quickly reduced to zero by either removing the coil from the field or by turning it through a quarter turn so that the field lines are parallel to the plane of the coil. If we know the time required to make this change, then the strength of the magnetic field can be found from a measurement of the induced voltage.

Lenz's Law

Can we predict the direction of the induced current in the coil? The rule for doing so goes hand in hand with Faraday's law and is credited to Heinrich Lenz (1804–1865):

The direction of the induced current generated by a changing magnetic flux produces a magnetic field that opposes the change in the original magnetic flux.

If the magnetic flux is decreasing with time, the magnetic field produced by the induced current will be in the same direction as the original external field, thus opposing the decrease. On the other hand, if the magnetic flux is increasing with time, the magnetic field produced by the induced current will be in the opposite direction to the original external field, thus opposing the increase.

Lenz's law is illustrated in figure 13.21. When the magnet is moving into the coil and the flux is increasing with time, the induced current is in a counterclockwise direction in the coil. The magnetic field produced by this induced current is therefore upwards through the area bounded by the

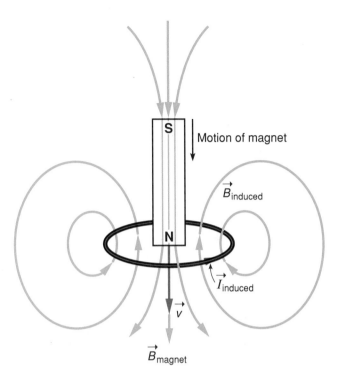

Figure 13.21 The induced current in the loop of wire produces an upward magnetic field inside the loop that opposes the increase in the downward field associated with the moving magnet.

coil. This opposes the increase in the downward field associated with the moving magnet. Lenz's law also explains why the galvanometer needle deflected in opposite directions when Faraday closed or opened the primary circuit in his experiments.

Self-Induction

Faraday first reported the results of his discoveries on electromagnetic induction in the year 1831. One year later, the discovery of a related effect was reported by Joseph Henry (1797–1878), who was working at Princeton University in the United States. Henry was experimenting with electromagnets when he noticed that the spark or shock obtained when an electromagnet was connected to a battery was larger than that obtained by touching just a short wire across the terminals of the battery. The best sparks were obtained when the circuit was broken.

Henry proceeded to explore this phenomenon in more detail. He found that a long wire yielded a larger spark than a short wire and that coiling the wire into a helix yielded an even larger spark. If the wire was wound upon a steel nail to make an electromagnet, the effect was larger still. He roughly calibrated the strength of the shock obtained by indicating how far up his arms it could be felt. If it could be felt just at the fingers, it was not very strong; one that could be felt at the elbows was obviously stronger! You can get a much stronger shock when connecting the wires of an electromagnet to a battery than by touching the terminals of the battery directly with your finger.

What Henry had discovered was *self-induction*. The changing magnetic flux through a coil of wire that is produced when the coil is connected to or disconnected from the battery produces an induced voltage in the same coil. The induced current in the coil and its associated magnetic field oppose the changing magnetic flux. The magnitude of the induced voltage is described by Faraday's law and is therefore proportional to the number of turns in the coil and to the rate of change of current in the coil. The voltage induced when the circuit is broken can be several times larger than the voltage of the battery itself.

This self-inductive effect has many applications. Coils of wire can be used in circuits to smooth out changes in current that may be occurring. The back voltage induced when the current is increasing causes the change in current to take place more slowly than it would without the coil. Thus an *inductor,* or coil, effectively adds to a system an electrical inertia that reduces rapid changes in current. Induction coils are also used to generate large voltage pulses like those needed for the spark plugs in an automobile.

13.5 GENERATORS AND TRANSFORMERS

Our everyday use of electric power is so commonplace that we do not usually stop to think about where the energy comes from or how it is generated. We know that it comes to us via overhead or underground wires and that it powers our appliances, lights our homes and offices, and sometimes heats or cools our homes. We may be vaguely aware of the transformers that are visible on utility poles and have some concept of their function.

Electric generators and transformers both play important roles in the production and use of electric power. The basic principle of operation of both of these devices happens to be Faraday's law of electromagnetic induction. A closer look at how they work will give you a better understanding of Faraday's law, as well as of the nature of our power-distribution system.

The Electric Generator

We have already mentioned the use of electric generators in our discussion of energy resources in chapter 10. The basic function of a generator is to convert mechanical energy, such as might be obtained from a water or steam turbine in a power plant, to electric energy. How is this done? How is Faraday's law involved in describing how a generator works?

The simplest form that an electric generator can assume is that pictured in figure 13.22. Mechanical energy is supplied from some external source (it could be your hand) to turn a

Figure 13.22   A simple generator consists of a coil of wire that generates an electric current when turned between the pole faces of permanent magnets.

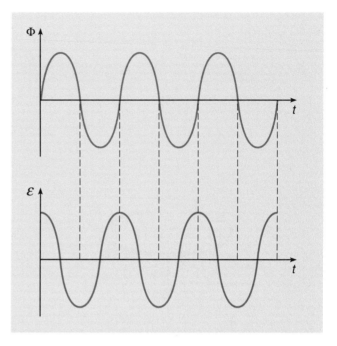

Figure 13.23   Graphs of the magnetic flux, Φ, and the induced voltage ε for an electric generator, plotted on the same time scale.

coil of wire that is located between the pole faces of a permanent magnet. The magnetic flux through the plane of the coil has its maximum value when the plane of the coil is perpendicular to the magnetic field lines. As the coil is turned to a position where the plane of the coil is parallel to the field lines, the flux becomes zero. If we continue to turn the coil past this point, the field lines pass through the coil in the direction opposite to the initial direction (relative to the coil).

The coil's motion, then, causes the magnetic flux passing through the coil to change continuously from a maximum in one direction, to zero, to a maximum in the opposite direction, to zero, etc. According to Faraday's law, a voltage is therefore induced in the coil. The magnitude of this induced voltage depends on the rate at which the magnetic flux is changing as well as on the factors associated with the magnet and coil that determine the size of the maximum flux ($\Phi = NBA$). The faster the coil is turned, therefore, the larger the maximum value of the induced voltage, since increased speed causes the magnetic flux to change more rapidly.

Figure 13.23 shows a graph of the continuously varying magnetic flux plotted against time. If the axle of the coil is turned at a steady rate, we get a smooth variation of the flux, as shown in the graph. Below the flux graph is a graph of the induced voltage, plotted on the same time scale. The induced voltage is equal to the *rate of change* of magnetic flux, so it has its maximum values where the slope of the flux curve is the greatest, and its zero values where the flux is momentarily not changing.

The ideas here are similar to those introduced in chapter 2, when we defined acceleration as the rate of change of velocity. The slope of the velocity-versus-time graph was then equal to the instantaneous acceleration. Likewise, here the slope of the flux-versus-time curve is the instantaneous rate of change of flux, and is equal to the induced voltage by Faraday's law. Where the slope is steep, the flux is chang-

ing rapidly, producing a large induced voltage; where the slope is zero, the rate of change of flux is momentarily zero. Comparison of the two curves in figure 13.23 should help to clarify this relationship.

The electric generators used in power plants are similar in design to the simple one we have described here. They have more than one coil, usually, and the magnets are generally electromagnets rather than permanent magnets, but the principle of operation is the same. Electric generators are also found in your car; the power generated is used to keep the battery charged, and operate the lights and other electrical systems, such as an air conditioner.

As you can see from the graphs, the voltage that is naturally produced by a generator is an *alternating* voltage. This is one reason that we have chosen to use alternating current in our power-distribution system. For power production, the rotational velocity of the rotating coils in our generators must be maintained at a specific value in order to produce the 60 hz (60 cycles/s) alternating current that we use routinely in this country.

Transformers

Another advantage of alternating current for purposes of power distribution is that we can easily use transformers to modify the voltage. Transformers are familiar sights; you have seen them on utility poles, at electric substations, and as voltage adapters for low-voltage devices like electric calculators

Figure 13.24 Transformers on utility poles and at electrical substations are familiar sights. Smaller ones are common components of electrical appliances.

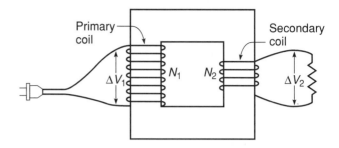
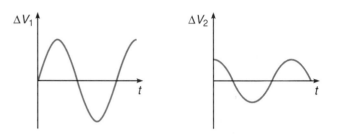

Figure 13.25 A source of alternating voltage applied to the primary coil of a transformer produces a changing magnetic flux, which induces a voltage in the secondary coil.

and model trains (fig. 13.24). Their function is to step the voltage up or down to suit the needs of a particular application.

Some of Faraday's early experiments involved prototypes of the modern transformer. The iron-ring version, with primary and secondary coils wound on either side of an iron ring, can serve as our model. By Faraday's law, a voltage is induced in the secondary coil if the current (and its associated magnetic field) of the primary coil is changing. That condition is naturally present if we power the primary coil with an alternating-current source (fig. 13.25).

What is the relation of the voltage change obtained with a transformer to the number of turns in the primary and secondary coils? A very simple relationship follows from Faraday's law: The voltage induced on the secondary coil is proportional to the number of turns on the secondary coil, since this determines the total magnetic flux through the coil. It is also proportional to the voltage on the primary coil, since this determines the size of the current and its associated magnetic field. It is inversely proportional, however, to the number of turns on the primary coil.

The reason that the voltage induced on the secondary coil is inversely proportional to the number of turns on the primary coil can be regarded as a self-inductive effect. The more turns there are on the primary coil, the harder it is to produce a rapid change in current in the primary coil because of the back voltage that is produced by self-induction. This effect limits the current and the size of the magnetic field that will be produced by the primary coil, which in turn determines the magnetic flux passing through the secondary coil.

The relationship, then, for the voltage induced in the secondary coil can be stated in symbols as follows:

$$\Delta V_2 = \Delta V_1 \left(\frac{N_2}{N_1} \right),$$

where N_1 and N_2 are the number of turns of wire on the primary and secondary coils, and ΔV_1 and ΔV_2 are the associated voltages. This relationship is often written as a proportion: $\Delta V_2 / \Delta V_1 = N_2 / N_1$. The ratio of the number of turns on the two coils determines the ratio of the voltages.

If, for example, we want to run an electric train at 12 volts, but the electric power provided at the wall socket is at 120 volts, we would use a transformer that had ten times as many turns on the primary coil as on the secondary coil. This would yield a voltage in the secondary coil that would be one-tenth of that in the primary, or 12 volts. A television set requires much more than 120 volts to power the picture tube, so it incorporates a transformer with many more turns on the secondary than the primary.

If we can get higher voltages than the input voltage by using a transformer, are we somehow getting more power out than what we put in? The answer, of course, is no; there is still no free lunch. The power produced in the secondary circuit is always less than, or at best equal to, the power provided to the primary coil. Since electric power can be expressed as the product of the voltage times the current, conservation of energy provides a second relationship that is useful for analyzing transformers:

$$P_2 = I_2 \Delta V_2 \leq I_1 \Delta V_1 = P_1$$

The reason that the power in the secondary coil is often slightly less than that provided to the primary is that some energy losses are associated with heating of the iron core in the transformer and with resistive losses in the wires. There are actually electric currents (eddy currents) induced in the iron core itself that generate heat. The price, then, of a larger voltage in the secondary coil is a smaller current in the secondary than that which must be provided to the primary coil.

On the other hand, if we are using a smaller voltage in the secondary circuit, then the secondary current can be larger than that provided to the primary coil.

High voltages are chosen for long-distance transmission of electric power. The higher the voltage, the lower the current for a given amount of power being transmitted. Since the joule heat losses in a wire are related to the current ($P = I^2R$), smaller currents mean that less energy is lost to heating effects in the transmission wires. It is not unusual to transmit power at very high voltages such as 230 kV (230 kilovolts, or 230 000 volts).

Such high voltages are not convenient, however, for distribution within a city or for use in homes or buildings.

Transformers located at electric substations step down the voltage to 7200 volts for distribution within a city or town. Transformers on the utility poles (or partially underground) step this voltage down once again to 220–240 volts for entry into buildings. This can be split within the building to yield both the 110 volts commonly used for household circuits and the 220 volts required for stoves, dryers, and furnaces.

The original electrical distribution system in this country was a 110-volt, direct-current system designed by Thomas Edison for a portion of New York City in 1882. There was a controversy for several years thereafter as to whether dc or ac systems were the most appropriate for the development of

Box 13.3

Everyday Phenomenon:
Direct-Current Motors

The Situation. As a child, you may have built a simple dc motor that would run on flashlight batteries. Cheap kits are available for this purpose, and this project is often done as a science activity in the middle grades of public schools. The thrill of getting it to run and understanding why it works is available to anyone beyond 8 years of age or so.

How does a dc motor work? How is it that the rotor keeps moving in the same direction without using an ac power source? How can we vary the speed of the motor? Your present understanding of electromagnetic effects should be sufficient to answer these questions.

The Analysis. There are many ways to build a dc motor, but the simplest ones usually take the form illustrated to the right. A coil of wire is mounted so that it can rotate between the pole faces of a permanent horseshoe magnet. The coil is connected to a battery via sliding contacts that come up on either side of a split ring that is mounted on the axle of the rotor. In the simplest versions, the rotor is sometimes just a steel nail wound with a coil of insulated wire.

A simple way to understand how the motor works is to view the rotor as an electromagnet. The south pole of the rotor is attracted to the north pole of the horseshoe magnet causing the rotor to turn. When the rotor reaches the vertical position where the two poles are the closest, however, the contacts from the battery cross the gap between the two halves of the split ring that is attached to the coil. The direction of the current in the coil is then reversed because opposite ends of the coil are now connected to the positive and negative terminals of the battery. This causes the polarity

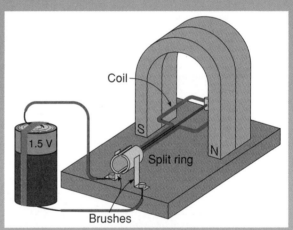

A simple dc motor consists of a wire-wound rotor mounted on an axle between the pole faces of a permanent magnet. The split ring causes the current to reverse directions every half turn.

of the rotor to reverse; what had been the south pole becomes the north pole and vice versa.

As its momentum carries the rotor past the vertical position, its newly created north pole is now attracted to the south pole of the horseshoe magnet, causing the rotor to continue turning in the same direction. The torque on the rotor, then, can be viewed as due to the attraction of opposite poles of the two magnets, the rotor and the horseshoe magnet. Every half turn, the current in the rotor coil reverses direction due to the effect of the split-ring contacts.

electric power. Ultimately the proponents of ac systems prevailed. The ability to transmit electric power at high voltages and to use transformers to step down these voltages at the point of use was a major reason for this choice.

You should understand that a transformer is fundamentally an *alternating* current device; Faraday's law requires that there be a *changing* magnetic flux if there is to be an induced voltage. If we attempted to use a transformer with a battery or other dc source, we would have to continually make and break the primary circuit somehow in order to induce a voltage in the secondary. There are ways to do this, but they add to the complexity and cost of a system and are less efficient than using alternating current.

Direct current does have at least one advantage over alternating current for transmitting power. As we will see in the final section of the next chapter, an alternating current flowing in a wire acts like an antenna that radiates electromagnetic waves. (This is the source of the static that you pick up on your car radio when you drive near a transmission line.) These waves carry away energy and therefore add to the energy losses that result from joule heating. This is why high-voltage direct current is sometimes used in transmitting power over large distances; one of the intertie lines between the Pacific Northwest and California is a dc line.

This split-ring arrangement is a crucial feature of a direct-current motor and is called a *split-ring commutator*. It consists of two half-circle metal bands, each connected to one end of the coil and mounted upon an insulating cylinder. The metal half-rings make sliding contact with fixed-position metal strips, or *brushes* on either side of the cylinder. An alternating-current motor does not need a split-ring commutator because the current reverses direction every half cycle due to the nature of alternating current.

Although we have described the rotor as an electromagnet, we could also view the torque on the rotor as resulting from the magnetic forces on current elements on either side of the coil, as shown in figure 13.15 of section 3.3. When the current in the coil reverses direction, these forces also reverse direction, so that the torque can be maintained in the appropriate direction. Viewing the origin of the torque in this manner is really equivalent to describing it as an electromagnet, because a current loop is a magnetic dipole, and magnetic dipoles are inevitably, at some level, made up of current loops.

It turns out that the speed of a dc motor is related to the applied voltage, which is convenient when a variable-speed motor is needed. At first glance, this might not seem surprising, but the explanation for this dependence actually follows from Faraday's law rather than Ohm's law. If you compare the description of a motor to that of a generator (section 13.5), you may be struck by how similar these two devices are. In fact, we could use a motor as a generator by supplying mechanical energy from an external source to turn the rotor.

Because a motor behaves like a generator, a back voltage is induced in the coil of the rotor due to Faraday's law and the changing magnetic flux through the coil. The magnitude of this induced voltage increases with the rotational velocity of the coil, just as in a generator. A larger voltage from the power source is thus required to overcome the larger induced voltage. Higher speeds, therefore, require higher applied voltage.

When the rotor reaches a constant rotational velocity, the net torque on the rotor must be zero, since there is no rotational acceleration. The average torque generated by the magnetic forces must then equal the frictional torques plus the other load torques involved with the work the motor is doing. The more work the motor does, the more electric energy must be provided. Since electric power is equal to the voltage times the current, and the voltage is determined primarily by the rotational velocity of the rotor, larger work loads for the motor require greater currents.

As you can see, an analysis of why and how a direct-current motor works is not difficult to provide. The voltage determines the rotational velocity, and the product of the voltage and current determines the power output. The variable-speed feature of dc motors provides an advantage in flexibility; the rotational velocity of an ac motor is determined by the frequency of the applied voltage and cannot be readily varied.

SUMMARY

The recognition that magnetic forces are associated with electric currents brought a new unity to the study of electricity and magnetism in the early part of the nineteenth century. The primary emphasis of this chapter is on the basic nature of magnetic forces and fields, and the relationship between electricity and magnetism. This relationship includes the phenomenon of electromagnetic induction, which was discovered by Faraday and is necessary to explain the operation of electric generators and transformers.

Magnets. In a manner similar to electrostatic forces, the magnetic forces between the poles of magnets obey Coulomb's law and the rule that like poles repel and unlike poles attract. The earth itself behaves like a large magnet with its south pole pointing northward.

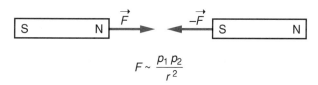

$$F \sim \frac{p_1 \, p_2}{r^2}$$

Magnetic Forces and Electric Currents. Oersted's discovery that magnetic forces are associated with electric currents led to the description of the magnetic field of a current-carrying wire. A magnetic force is exerted upon either a current-carrying wire or a moving charge when they are located in a magnetic field.

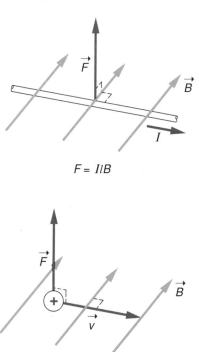

$$F = IlB$$

$$F = qvB$$

Current Loops. A current-carrying loop of wire produces a magnetic field that is identical to that of a short bar magnet. Current loops and coils can be regarded as magnetic dipoles, and the magnetism of natural magnets can be regarded as due to atomic current loops. There is a magnetic torque exerted on a current loop in an external field that is similar to that exerted upon a bar magnet.

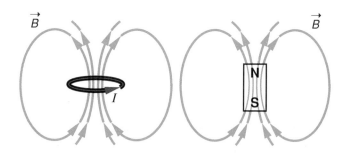

Electromagnetic Induction. Faraday discovered that a voltage is induced in a circuit that is equal to the rate of change of magnetic flux passing through the circuit (Faraday's law). Magnetic flux is equal to the product of the component of the magnetic field that is perpendicular to the plane of the circuit and the area bounded by the circuit.

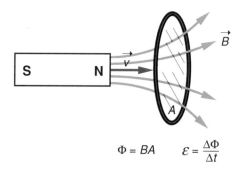

$$\Phi = BA \qquad \varepsilon = \frac{\Delta \Phi}{\Delta t}$$

Generators and Transformers. A generator is used to convert mechanical energy to electrical energy and naturally produces an alternating current. Transformers are used to step alternating voltages up and down. Faraday's law is the basic principle of operation for both of these devices. For transformers, $\Delta V_2 / \Delta V_1 = N_2 / N_1$.

QUESTIONS

Q13.1 The north pole of a hand-held bar magnet is brought near to the north pole of a second bar magnet that is lying on a table. How will the second magnet tend to move? Explain.

Q13.2 If the distance between the south poles of two long bar magnets is reduced by half, will the repulsive force acting upon these poles be doubled? Explain.

Q13.3 Is it possible for a bar magnet to have just one pole? Explain.

Q13.4 Does a compass needle always point directly northward in the absence of other nearby magnets or currents? Explain.

Q13.5 If we regard the earth as a magnet, does its north pole coincide with the geographical north pole? Explain.

Q13.6 If a horizontal wire is oriented along a magnetic north-south line and a compass is placed above the wire, will the needle of the compass deflect when a current flows through the wire? Explain.

Q13.7 If a horizontal wire is oriented along an east-west line (perpendicular to magnetic north) and a compass is placed above the wire, will the needle of the compass deflect when a current flows through the wire? Explain.

Q13.8 Is the force exerted by one current-carrying wire upon another due to electrostatic effects or to magnetic effects? Explain.

Q13.9 If a uniform magnetic field is directed horizontally towards the north, and a positive charge is moving westward through this field, is there a magnetic force on this charge? If so, in what direction? Explain.

Q13.10 If a uniform magnetic field is directed horizontally towards the east, and a negative charge is moving eastward through this field, is there a magnetic force on this charge? If so, in what direction? Explain.

Q13.11 Imagine that you are looking downward at the top of a circular loop of wire whose plane is horizontal. If the wire carries a current in the clockwise direction, what is the direction of the magnetic field at the center of the circle? Explain.

Q13.12 If you were to represent the current loop of question 13.11 as a bar magnet or magnetic dipole, in what direction would the north pole be pointing? Explain.

Q13.13 A current-carrying rectangular loop of wire is placed in an external magnetic field with the directions of the current and field as shown in the diagram. In what direction will this loop tend to rotate as a result of the magnetic torque that is exerted on it? Explain.

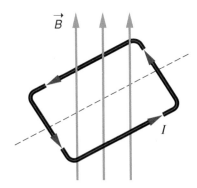

Q13.14 Since the magnetic fields of a coil of wire and a bar magnet are essentially identical, are there loops of current inside of natural magnetic materials such as iron or cobalt? Explain.

Q13.15 If Faraday wound enough turns of wire on the secondary coil of his iron ring, did he find that a large, steady-state, direct current in the primary coil would induce a current in the secondary coil? Explain.

Q13.16 Is a magnetic flux the same thing as a magnetic field? Explain.

Q13.17 A horizontal loop of wire has a magnetic field passing upwards through the plane of the loop. If this magnetic field is increasing with time, is the direction of the induced current clockwise or counterclockwise (viewed from the top) as predicted by Lenz's law? Explain.

Q13.18 Two coils of wire are identical except that coil *A* has twice as many turns as coil *B*. If a magnetic field is increasing with time at the same rate through both coils, which coil (if either) has the larger induced voltage? Explain.

Q13.19 Suppose that the magnetic flux through a coil of wire varies with time as shown in the graph. Using the same time scale, sketch a graph showing how the induced voltage varies with time. Where does the induced voltage have its largest magnitude? Explain.

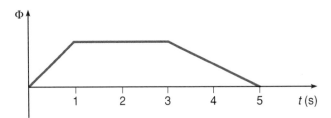

Q13.20 If the magnetic field produced by the magnets in a generator is constant, is there a changing magnetic flux through the generator coil when it is turning? Explain.

Q13.21 Does a simple generator produce a steady direct current? Explain.

Q13.22 A simple generator and a simple electric motor have very similar designs. Do they have essentially the same function? Explain.

Q13.23 Can a transformer be used as shown in the diagram to step up the voltage of a battery? Explain.

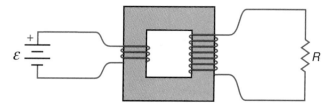

Q13.24 If we step up the voltage of an alternating-current source by using a transformer, can we increase the amount of electric energy that can be obtained from the source? Explain.

EXERCISES

E13.1 Two long bar magnets lying on a table with their south poles facing one another exert a force of 6 N upon each other. If the distance between these two poles is tripled, what is the new value of this force?

E13.2 Two long parallel wires, each carrying a current of 10 A, lie a distance of 10 cm from each other. What is the magnetic force per unit length exerted by one wire upon the other?

E13.3 If the distance between the two wires in exercise 13.2 is doubled, how does the force per unit length change?

E13.4 A particle with a charge of 0.03 C is moving at right angles to a uniform magnetic field with a strength of 0.4 T. The velocity of the charge is 600 m/s. What is the magnitude of the magnetic force exerted on the particle?

E13.5 An electron ($q = -1.6 \times 10^{-19}$ C) moving with a velocity of 4×10^6 m/s at right angles to a magnetic field experiences a magnetic force of 2.4×10^{-12} N. What is the magnitude of the magnetic field?

E13.6 A straight segment of wire has a length of 20 cm and carries a current of 5 A. It is oriented at right angles to a magnetic field of 0.6 T. What is the magnitude of the magnetic force on this segment of wire?

E13.7 The magnetic force on a 15-cm straight segment of wire carrying a current of 3 A is 1.0 N . What is the magnitude of the component of the magnetic field that is perpendicular to the wire?

E13.8 A coil of wire with 50 turns has a cross-sectional area of 0 .04 m². A magnetic field of 0.6 T passes through the coil perpendicular to the plane of the coil. What is the total magnetic flux passing through the coil?

E13.9 A loop of wire enclosing an area of 0.20 m² has a magnetic field passing through its plane at an angle to the plane. The component of the field that is perpendicular to the plane is 0.4 T, and the component that is parallel to the plane is 0.6 T. What is the magnetic flux through this coil?

E13.10 The magnetic flux through a coil of wire changes from 4.0 T·m² to zero in a time of 0.5 seconds. What is the magnitude of the average voltage induced in the coil during this change?

E13.11 A coil of wire with 25 turns and a cross-sectional area of 0.02 m² lies with its plane perpendicular to a magnetic field of magnitude 1.2 T. The coil is rapidly turned in a time of 0.25 s so that its plane is now parallel to the magnetic field.

    a. What is the magnitude of the initial total magnetic flux through the coil?

    b. What is the magnitude of the flux after the coil is turned?

    c. What is the average magnitude of the induced voltage in the coil while it is being turned?

E13.12 A transformer has 20 turns of wire in its primary coil and 80 turns in its secondary coil.

    a. Is this a step-up or step-down transformer?

    b. If an alternating voltage with an effective voltage of 110 volts is applied to the primary, what is the effective voltage induced in the secondary?

E13.13 A step-down transformer is to be used to convert a line ac voltage of 120 volts to 6 volts in order to power an electric train. If there are 200 turns in the primary coil, how many turns should there be in the secondary coil?

CHALLENGE PROBLEMS

CP13.1 Two long parallel wires carry currents of 10 A and 20 A in opposite directions as shown in the diagram. The distance between the wires is 5 cm.

    a. What is the magnitude of the force per unit length exerted by one wire on the other?

    b. What are the directions of the forces on each wire?

    c. What is the total force exerted on a 30-cm length of the 20-A wire?

    d. From this force, compute the strength of the magnetic field produced by the 10-A wire at the position of the 20-A wire ($F = IlB$).

    e. What is the direction of the magnetic field produced by the 10-A wire at the position of the 20-A wire?

    f. Knowing how the force between two current-carrying wires varies with distance, find the strength of the magnetic field produced by the 10-A wire at a distance of 10 cm from the wire.

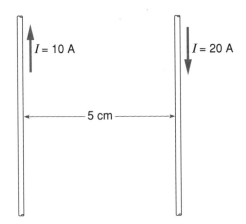

$I = 10$ A $I = 20$ A

        5 cm

CP13.2 A small, charged metal sphere with a charge of +0.06 C and a mass of 25 g (0.025 kg) enters a region in which there is a magnetic field of 0.5 T. The sphere is traveling with a velocity of 150 m/s in a direction perpendicular to the magnetic field, as shown in the diagram.

    a. What is the magnitude of the magnetic force exerted upon the sphere?

    b. What is the direction of the magnetic force exerted on the sphere when it is at the position shown?

    c. Will this force cause the magnitude of the velocity of the sphere to increase? Explain.

    d. What is the magnitude of the acceleration of the charged sphere?

    e. Since centripetal acceleration is equal to v^2 / r, what is the radius of the curve through which the particle will move under the influence of the magnetic force?

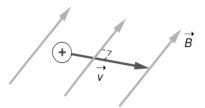

CP13.3 A rectangular coil of wire has dimensions of 6 cm by 10 cm and is wound with 25 turns of wire. It is turned between the pole faces of a horseshoe magnet that produces an approximately uniform field of strength 0.4 T, such that sometimes the plane of the coil is perpendicular to the field and sometimes it is parallel to the field.

    a. What is the area bounded by the rectangular coil?

    b. What is the maximum value of the total magnetic flux that passes through the coil as it turns?

    c. What is the minimum value of the total flux through the coil as it turns?

    d. If the coil makes one complete turn each second and turns at a uniform rate, what is the time involved in changing the flux from the maximum value to the minimum value?

    e. What is the average value of the voltage generated in the coil as it passes from the maximum to the minimum value of flux?

CP13.4 A transformer is designed to step up line voltage of 110 volts to 990 volts. The primary coil has 50 turns of wire.

    a. How many turns of wire should there be in the secondary coil?

    b. If the current in the primary coil is 15 A, what is the maximum possible current in the secondary coil?

    c. Are we likely to actually achieve the current calculated in (*b*)? Explain.

HOME EXPERIMENTS AND OBSERVATIONS

HE13.1 Look around your home or dormitory to see how many magnets you can find. Small packages of magnets are often sold in variety stores, and they are used around the house to stick things on refrigerators, hang tools, or just as toys.

a. If one of your magnets is a bar magnet, locate its north-seeking pole by suspending it from a string and noting which end points north.

b. Use this magnet, or any other whose north and south poles can be identified, to find the poles of your other magnets, and label them north or south with a crayon.

c. Verify that any two like poles will repel and unlike poles attract for any combination of your magnets.

d. Try to determine which of your magnets is the strongest by seeing through what distance it will attract a paper clip or other small steel item. (The largest magnet may not be the strongest!)

HE13.2 If you have a small compass available, use it to explore the magnetic field of magnets and currents. The compass needle will always point in the direction of the net magnetic field, wherever it is located.

a. Using a compass with your strongest magnet, find out how far you must move the compass from the magnet before the field of the magnet is no longer noticeable and the compass needle simply points north.

b. Bringing the compass much closer to the magnet, use the compass to determine the direction of the magnetic field at various points around the magnet. Make a sketch showing the direction of the compass needle at various points around the magnet. (If the compass is close enough to the magnet, the effect of the earth's field should be small compared to that of the magnet.)

HE13.3 Find a flashlight battery and a long wire. Stretch a portion of the wire in a straight horizontal line oriented along the north-south direction. (It may help to tape the wire to a wooden table to hold it in place.) Place a compass on top of the wire near the middle of this straight section.

a. Touching the free ends of your wire to the ends of the battery, notice whether the compass needle deflects. How far does it deflect when the compass is sitting on the wire? (*Do not* leave the battery connected to the wire more than a few seconds. If you do, the battery will quickly run down, and the wire may get quite warm.)

b. How far from the wire must the compass be moved in order for the deflection of the compass needle to be 45° (from the original northward direction)? At this point, the magnetic field of the wire is equal to that of the horizontal component of the earth's field. (If the deflection of the compass needle is not greater than 45° when the compass is sitting directly on the wire, you will not be able to make this observation.)

HE13.4 Using a long insulated wire, wrap several turns (as many as 50 to 100) around a large steel nail. If the free ends of the wire are connected to a flashlight battery, you have an electromagnet. (Again, avoid leaving the wire connected to the battery for very long, in order to keep from depleting the battery.)

a. Compare the strength of your electromagnet to that of some of the permanent magnets you may have available. (At what distance can each pick up a paper clip?)

b. Note whether you can see a spark and feel a light shock when you break connection with the battery. Are the spark and shock greater with your electromagnet than with an uncoiled piece of wire?

Part 4

WAVE MOTION

and Optics

14 Making Waves

If you have ever visited a beach at an ocean or a large lake, you have probably delighted in the idle exercise of watching the waves come in. A cliff or some other high spot is a good vantage point for this amusement. From there you can track a single wave crest as it moves in and finally breaks upon the shore (fig. 14.1). Because of their regularity, watching the waves can be very relaxing, even hypnotic.

As you watched the waves, you may have been struck by a curious aspect of their behavior. Although water appears to move towards the shore, no water actually accumulates on the beach. What is happening then—what is actually moving? Is this apparent motion of the water somehow deceiving us?

If you stand in the surf and let the waves break over you, you certainly get the sense that the waves are carrying energy. A good wave can knock you over or carry you along. A person on a surfboard can gain a large velocity by riding with the crest of a wave. The kinetic energy gained in this manner must come from the wave.

Although the behavior of waves near the shore can be complex, water waves do exhibit many of the general features that are associated with wave motion of all types. Light, sound, radio waves, and waves on guitar strings are all examples of wave motion that have much in common with the waves that we observe at the beach. Our understanding of the nature and properties of wave motion has implications for all areas of physics, including atomic and nuclear physics. An enormous range of phenomena can be explained in terms of waves.

Figure 14.1 The waves move in and break upon the shore. Why does the water not accumulate there?

Chapter Objectives

The primary objective of this chapter is to provide an understanding of the basic nature of waves and their properties, including the properties of velocity, wavelength, period, and frequency, as well as the phenomena of reflection, interference, and energy transmission. We will also examine some specific types of wave motion in more detail, including waves on a string, sound waves, and electromagnetic waves.

Chapter Outline

1 *Wave pulses and waves.* What is a wave pulse, and what are its general properties? How are wave pulses related to longer periodic waves? What distinguishes a longitudinal wave from a transverse wave?

2 *Waves on a string.* What are the general features of a simple harmonic wave traveling on a rope or string? How are its properties of frequency, period, wavelength, and velocity related to one another? What determines the wave velocity?

3 *Interference and standing waves.* What happens when two or more waves combine, and what do we mean by interference? What is a standing wave, and how is it produced?

4 *Sound waves.* What is the nature of sound waves, and how are they produced? How are the musical properties of pitch and harmonics related to the wave properties of sound?

5 *Electromagnetic waves.* What is the nature of electromagnetic waves, and how are they produced? In what ways are radio waves and light waves similar, and how do they differ?

14.1 WAVE PULSES AND WAVES

Although waves on a beach show many characteristics that are common to any wave motion, the details of their motion are actually quite complex, particularly near a beach. That complexity adds to their beauty and interest, but for a beginning discussion of the nature of waves, a simpler example is better. A Slinky, that toy spring that walks down stairs, is an ideal medium for simple waves.

The original Slinkies were metal springs, but cheaper plastic ones are available now. They are standard equipment in most physics storerooms, but you may have one left over from your childhood or be able to borrow one from a younger brother or sister. Having your own Slinky available can help you develop a feeling for wave phenomena, even if you have already seen one used in a lecture demonstration.

Wave Pulses in a Slinky

If a Slinky is laid out on a smooth table with one end held motionless, it is easy to produce a single traveling pulse. Holding the other end of the Slinky in one hand with the Slinky slightly stretched, move the near end back and forth quickly, along the line of the Slinky. As you do this, you can see a disturbance move from the near end to the fixed end of the Slinky (fig. 14.2).

The motion of the pulse created in this manner is easily visible. It travels down the Slinky and may be reflected at the fixed end and come back towards its starting point before dying out. But what is actually moving? The Slinky itself is not going anywhere. The pulse moves through the Slinky, and portions of the Slinky are moving as the pulse passes through them. After the pulse dies out, however, the Slinky is exactly as it was before the pulse began.

If we take a closer look at what is happening within the Slinky, the nature of the pulse becomes clearer. Moving the end of the Slinky back and forth creates a local compression: the rings of the spring are closer together in this region than in the rest of the Slinky. It is this region of compression that moves along the Slinky and constitutes the pulse that we

see. Individual loops of the spring move back and forth as this region of compression goes by.

Some general features of wave motion are present in this simple example of a pulse on a Slinky. The wave or pulse moves through the medium, in this case the Slinky, but the medium itself goes nowhere. Water waves move towards the beach, but no water accumulates. What moves is a disturbance within the medium, which may be a local compression, a sideways displacement, or some other form of local change in the state of the medium. The disturbance moves with a definite velocity, which is determined by properties of the medium.

In the case of the Slinky, the velocity is partly determined by the tension in the Slinky. You can easily confirm this if you stretch the Slinky by different amounts and note the effect on the velocity of a pulse. The pulse travels more rapidly when the Slinky is highly stretched than when it is only slightly stretched. Give it a try! The other factor that determines the pulse velocity in the Slinky is its mass per unit length. A pulse travels more slowly on a steel Slinky than on a plastic Slinky because the mass for a given length of the spring is greater for steel than for plastic.

Another general feature of wave motion is the transmission of energy through the medium. The work done at one end of the Slinky increases both the potential energy of the spring and the kinetic energy of individual loops. This region of higher energy moves along the Slinky and could be used to ring a bell attached to the far end or to perform other types of work (fig. 14.3). Energy carried by water waves does substantial work over time in eroding and shaping a shoreline. Energy transmission is an extremely important aspect of any wave motion.

Longitudinal and Transverse Waves

The wave pulse that we have described in the Slinky is called a *longitudinal* wave. In a longitudinal wave the displacement or disturbance in the medium is *parallel* to the direction of travel of the wave or pulse. In the Slinky, the loops of the

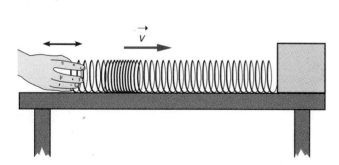

Figure 14.2 With one end of the Slinky fixed, a simple back-and-forth motion of the other end produces a traveling pulse.

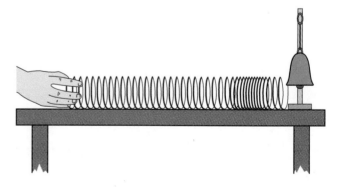

Figure 14.3 Energy transmitted along the Slinky by the wave pulse can be used to ring a bell or to perform other kinds of work.

Longitudinal

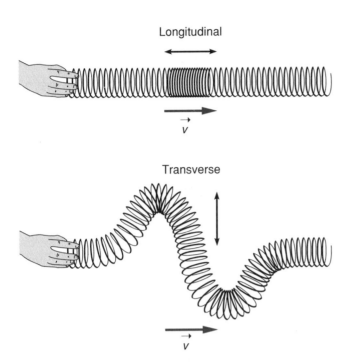

Transverse

Figure 14.4 **In a longitudinal pulse, the disturbance is parallel to the direction of travel; in a transverse pulse, the disturbance is perpendicular to the direction of travel.**

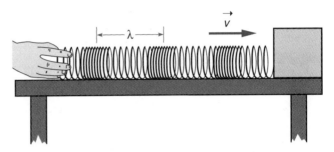

Figure 14.5 **A series of pulses produced at regular time intervals generate a periodic wave on the Slinky. The wavelength, λ, is the distance between the pulse centers.**

spring move back and forth along the axis of the Slinky, and the pulse itself also travels along this axis.

You can also produce a *transverse* pulse on a Slinky or spring. In a transverse wave the disturbance or displacement takes place in a direction *perpendicular* to the direction of travel of the wave (fig. 14.4). If you move your hand up and down in a direction perpendicular to the axis of the spring, you create a transverse pulse. (This works best with a Slinky when it is highly stretched; a long, thin spring is actually more effective than a Slinky for producing a transverse pulse or wave.) Like the longitudinal pulse, the transverse pulse moves down the Slinky and also transmits energy along the Slinky.

Waves on a string, which we discuss in the next section, are generally transverse in nature, as are electromagnetic waves, which we discuss in the last section of this chapter. As you will see, polarization effects are associated with transverse waves, but not longitudinal waves. Sound waves (section 14.4) are longitudinal in nature and are quite similar in some ways to longitudinal waves on a Slinky. Water waves have both longitudinal and transverse properties.

Periodic Waves

Up to this point, we have discussed single wave pulses on a Slinky. If, instead of moving your hand back and forth just once, you continue to produce pulses, you send a series of longitudinal pulses down the Slinky. If equal time intervals

separate the pulses, you then produce a *periodic wave* on the Slinky (fig. 14.5).

The time between pulses is referred to as the *period* of the wave and is often represented by the symbol *T*. The *frequency* is the number of pulses or cycles per unit of time and is equal to the reciprocal of the period, 1 / *T*. If the time between pulses is 0.5 seconds, for example, then the frequency is

$$f = \frac{1}{T} = \frac{1}{0.5 \text{ s}} = 2 \text{ cycles/s} = 2 \text{ hz},$$

where *f* represents frequency. These are the same symbols and meanings that were given to these quantities in chapter 8 when we discussed simple harmonic motion. The hertz (hz) is the commonly used unit for frequency and is equal to one cycle per second.

As the pulses move down the Slinky, they are spaced at regular distance intervals if they have been created at regular time intervals. The distance between the same point in successive pulses is called the *wavelength*. This distance is shown in figure 14.5 and is labeled with the Greek letter λ (lambda), which is the commonly used symbol for wavelength.

A pulse in a periodic wave travels a distance of one wavelength in a time equal to one period before the next pulse is created. The velocity of the wave can therefore be expressed in terms of these quantities. The velocity is equal to one wavelength (the distance traveled between pulses) divided by one period (the time between pulses):

$$v = \frac{\lambda}{T} = f\lambda.$$

The last portion of this equality is true because the frequency, *f*, is equal to 1 / *T*, the reciprocal of the period.

The relationship just stated is true for any periodic wave and is very useful for finding the frequency or the wavelength when the other quantity is known. The wave velocity depends on properties of the medium and is often known from other considerations. The velocity of electromagnetic waves in free space, for example, has a fixed value that is independent of the wavelength or frequency. The use of the relationship between velocity, wavelength, and frequency is illustrated in the next section, when we discuss waves on a string, and also in later sections.

14.2 WAVES ON A STRING

Imagine a heavy rope tied to a wall or post. If you hold the other end of the rope in your hand and move it up and down, you could create either a wave pulse or a periodic wave that would travel down the rope towards its fixed end (fig. 14.6). These waves would be transverse because the displacement of the rope as the wave travels through a given point is perpendicular to the original line of the rope and to the direction in which the wave travels. The situation is essentially the same as that of producing a transverse wave on a Slinky or spring; we can think of the rope as a very stiff spring.

The transverse waves that travel on a stretched string or rope are examples of one of the simplest types of wave motion. The disturbance in this case is a displacement of the rope from its straight-line position. This is very easy to visualize and graph, which is why the study of this type of wave is so useful for understanding wave phenomena.

Waves and Graphs

Suppose that you give the rope, which is tied to the wall at the opposite end, a simple up-and-down motion. The pulse that is produced on the rope might look like the one in figure 14.6. The leading edge of the pulse corresponds to the beginning of the up-and-down motion; the trailing edge represents the end of that motion. Like the pulse on the Slinky, this disturbance travels down the rope. In this case, however, the picture of the rope can be thought of as a graph, with the vertical axis representing the vertical displacement, *y*, of the rope, and the horizontal axis the horizontal position, *x,* of the pulse on the rope.

A single picture of the pulse on the rope does not tell the whole story, of course; it is like a snapshot showing the displacement of the rope at just one instant in time. The pulse is moving, and at some later instant it will be farther down the rope at a different horizontal position. We would have to draw a series of graphs at different times to represent this fact. Although the pulse may gradually decrease in size due to frictional effects, its shape remains essentially the same as it moves along the rope.

If you give the rope more than just one up-and-down motion, repeating the motion at regular time intervals, you will produce a periodic wave like the one in figure 14.7. The wavelength, λ (shown on the graph), is the distance covered by one complete cycle of the wave. As with the single pulse, this wave pattern moves to the right along the rope, retaining its shape. When the leading edge of this wave reaches the

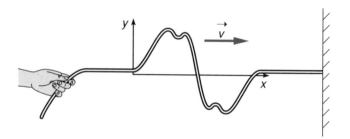

Figure 14.6 **At any given instant as a transverse pulse moves along a stretched rope, it can be thought of as a graph of vertical displacement plotted against horizontal position.**

wall, it will be reflected and start to move back towards your hand. As this happens, the reflected wave interferes with the waves still traveling toward the right, and the picture becomes more complex.

The shape of the wave pulses in figure 14.7 depends on the motion of the hand or other oscillator that generates the wave. It could be much more complex than the simple shape in this drawing. There is, however, a special shape that plays a particularly important role in the analysis of wave motion. If you move your hand up and down so that it is undergoing simple harmonic motion, the displacement of the end you hold then varies sinusoidally* with time, as discussed in chapter 8. The resulting periodic wave therefore has a sinusoidal form, and we call it a *harmonic* wave (fig. 14.8).

Like other waves, the sinusoidal wave travels along the rope until it is reflected at the fixed end. Although moving your hand with perfect simple harmonic motion is not easy, it is not too difficult to produce a reasonable approximation and to get a wave that looks like the one in figure 14.8. One reason for this is that individual segments of a rope tend naturally to move with simple harmonic motion, because the restoring force pulling the rope back towards the center line is approximately proportional to the vertical displacement from the center line. (This was the condition for simple harmonic motion discussed in chapter 8.)

There is another reason, however, that harmonic waves play an important role in the discussion of wave motion. It turns out that any periodic wave can be represented as a sum of harmonic waves of different wavelengths and frequencies. We call this process *Fourier,* or *harmonic analysis,* which is essentially the process of breaking a complex wave down into its simple harmonic components. Harmonic waves can be thought of, then, as building blocks for more complex waves.

---

* As discussed in chapter 8, the term *sinusoidal* refers to the shape of the curve, which is that of the trigonometric sine function. The shape of the curve is all that matters here; no knowledge of trigonometry is needed.

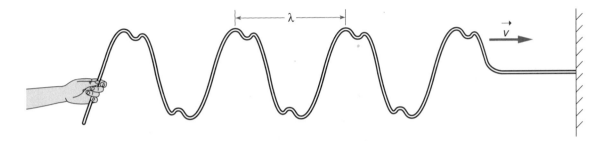

Figure 14.7 **A periodic wave moving along a stretched rope. The peak-to-peak distance between adjacent pulses is the wavelength, λ.**

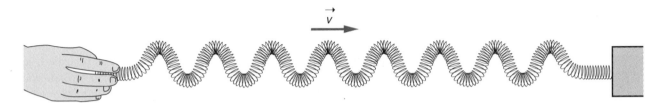

Figure 14.8 **A harmonic wave results when the end of a rope or long spring is moved back and forth in simple harmonic motion.**

Wave Velocity

Like the waves on a Slinky, the waves on a rope move along the rope with a velocity that is independent of the shape or frequency of the pulses. What determines this velocity? In order to answer this question, we need to think about the process that causes the disturbance to propagate along the rope. Why do the pulses move?

If we picture just a single pulse moving along the rope, it is obvious that portions of the rope lying in front of this pulse are at rest before the pulse gets there. Something must cause these portions to accelerate as the pulse approaches. In fact, the reason the pulse moves is that lifting the rope causes the tension in the rope to acquire an upward component. This upward component acts upon the segment of the rope to the right of the raised portion, as shown in figure 14.9. The resulting upward force causes this next segment to accelerate upwards, and so on down the rope.

The velocity of the pulse depends on the rate of acceleration of succeeding segments of the rope; the more rapidly they can be started moving, the more rapidly the pulse will move down the rope. By Newton's second law, this acceleration is proportional to the magnitude of the force, and inversely proportional to the mass of the segment. The force will be related to the tension in the rope: greater tension produces a larger accelerating force. The acceleration is also related to the mass of the segment: a greater mass produces a smaller acceleration. These considerations dictate that the velocity of a pulse on the rope will increase with the tension in the rope and decrease with the mass per unit length of the rope or string.

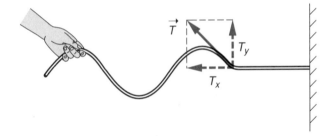

Figure 14.9 **As the raised portion of a pulse approaches a given point on the rope, the tension in the rope acquires an upward component. This causes the next segment of the rope to accelerate upwards.**

The exact expression for the wave velocity turns out to involve the square root of the tension divided by the mass per unit length. In symbols, this takes the following form:

$$v = \sqrt{\frac{F}{\mu}},$$

where F is the tension in the rope (a force), and μ is the Greek letter *mu* which is often used to represent mass per unit length. This quantity would be found by dividing the total mass of the rope by its length, or $\mu = m/L$.

If you stretch the rope more tightly, you should expect the wave velocity to increase, just as it did for the Slinky. A thicker rope, on the other hand, with a larger mass per unit length, should produce a slower wave velocity. This is the reason that a heavy rope is more effective for demonstrating wave motion than a light string: the waves move too rapidly on a light string for us to follow them visually.

Velocity, Frequency, and Wavelength

The expression developed in section 14.1 relating the wave velocity to the frequency and wavelength ($v = f\lambda$) is useful for predicting the wavelength that will result for a wave on a rope or string. The velocity, as we have just seen, depends on the tension and the mass per unit length, but is independent of the frequency or wavelength. A certain frequency, then, will determine the wavelength or vice versa.

Suppose that we consider a rope with an overall length of 10 meters and total mass of 2 kg. That represents a reasonably heavy rope, which should, depending on the tension, produce waves with a velocity that is slow enough to observe visually. The mass per unit length ($\mu = m / L$) is 2 kg/10 m or 0.20 kg/m.

A tension of at least 50 N would be required to keep the rope from sagging too much if it is fixed at one end and held by a hand at the opposite end. (The weight of the rope itself [mg] is almost 20 N.) With these values, the wave velocity for a wave on this rope would be

$$v = \sqrt{\frac{F}{\mu}} = \sqrt{\frac{50 \text{ N}}{0.20 \text{ kg/m}}} = 15.8 \text{ m/s}.$$

It will take less than a second, then, for a pulse to travel the 10-meter length of the rope, but the motion should be observable.

If you move the end of the rope up and down with simple harmonic motion, you will produce a harmonic wave on the rope with the same frequency as the motion of your hand. If you move your hand with a frequency of 4 cycles per second (4 hz), this will also determine the wavelength of the harmonic wave. Since $v = f\lambda$,

$$\lambda = \frac{v}{f} = \frac{15.8 \text{ m/s}}{4 \text{ hz}} = 3.95 \text{ m}.$$

We should expect to see waves, then, with a wavelength of approximately 4 meters. Two and a half complete cycles of the wave would fit along the 10-meter length of the rope. Lower frequencies would result in longer wavelengths, and higher frequencies in shorter wavelengths.

If we want to observe these waves, we must look quickly, because within a second the wave reaches the fixed end of the rope and is reflected. The reflected wave interferes with the simple pattern of the wave traveling in the original direction. We would need either a longer rope or some means of damping out the reflected wave if we wanted a more leisurely view of the wave.

Although waves on a rope can be demonstrated, their real advantage is the ease with which we can picture them graphically and get a physical sense of how they are produced. In lecture or laboratory demonstrations, we often use a long, but not very stiff spring in place of the rope. This gives us a larger mass per unit length and a slower wave velocity. Things happen far too quickly on a string or light rope to permit us to follow them visually.

14.3 INTERFERENCE AND STANDING WAVES

When a wave reaches the fixed end of a rope, it is reflected, as mentioned in the previous section, and travels back down the rope in the opposite direction to the incoming wave. If only a single pulse is involved, we can observe the returning pulse quite clearly. If the wave is periodic, then the returning wave *interferes* with the incoming wave, and the resulting pattern becomes more complex and difficult to analyze.

The same thing happens when water waves come into a beach. Waves reflected at the beach interfere with those coming in and create a more complex pattern than that created by incoming waves at some distance from the beach. This process, in which two or more waves combine, is called, naturally enough, *interference*. What happens when waves interfere? Can we predict what the resulting wave pattern will be?

Two Waves on a Rope

Waves on a rope provide examples of interference that are easy to visualize and therefore useful in introducing some of the basic concepts of wave interference. Suppose that we have a rope like that shown in figure 14.10, which consists of two identical segments that are smoothly joined to form a single rope of the same mass per unit length as the two initial segments. If you hold one segment in the left hand and the other in the right, you can generate waves on each segment that will combine when they reach the junction.

If you move both hands up and down in time with one another through the same vertical height, the waves generated on each segment should be identical. What happens, then, when they reach the junction? We might assume that since each wave by itself would generate a wave of its own height on the portion of the rope to the right of the junction, the combined effect of the two waves will be a wave with the same frequency and wavelength but twice the height of the initial waves. This is, in fact, what happens in this situation.

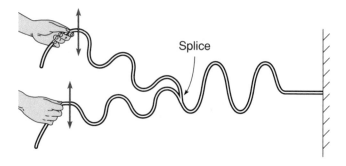

Figure 14.10 Identical waves, traveling on two identical ropes that are spliced together, combine to produce a larger wave.

This additive effect of two or more waves traveling in the same medium is called the *principle of superposition*. Simply stated:

When two or more waves combine, the resulting disturbance or displacement is equal to the sum of the individual disturbances.

The principle of superposition holds for most types of waves. In some situations the resulting disturbance might be so large that the medium in which the wave is traveling could not respond to that extent. The net disturbance in these situations would be less than that predicted by the principle of superposition. For most situations, however, the principle of superposition is valid, and it forms the basis for our analysis of interference phenomena.

When two waves are produced in time with one another, as pictured in figure 14.10, they are said to be *in phase*, or to have the same phase. When they reach the junction, both waves are going up or down at the same time. It is possible, however, to produce waves that are not in phase. If, for example, you move the two segments so that one is going up while the other is going down, and these *opposite* motions are in time with one another, the two resulting waves are said to be *completely out of phase* (see fig. 14.11).

What happens in this situation? If the two waves have the same height, then the principle of superposition predicts that the net disturbance will be zero. When the displacements of the two waves are added at the junction, the displacement of one is positive (up) while the other is negative (down), and they cancel one another. The sum is always zero, and no wave is propagated beyond the junction.

Thus, the result of adding two or more waves certainly depends on the phases of the waves as well as the amplitude, or height. The two situations pictured for the rope are the extremes; in one case the waves are completely in phase, in the other they are completely out of phase. In the first case we get complete addition, or *constructive interference;* in the other we get complete cancellation, or *destructive interference*.

It is possible, of course, for the two waves to be neither completely in or out of phase. In these situations the resulting wave has an amplitude (height) somewhere between zero and the sum of the two amplitudes. For example, if one of two equal-amplitude waves is at half of its maximum height at the same time that the other is at its full height, the combined wave will have an amplitude equal to 1.5 ($\frac{3}{2}$) times the amplitude of the two initial waves.

Standing Waves

Although it is somewhat difficult to achieve the conditions necessary for two waves traveling in the same direction on separate ropes to combine as we have described, interference of two or more waves traveling in the same direction and combining at some point in space is quite common with water waves, sound waves, or light waves. The difference in phase between the waves determines whether the interference will be constructive, destructive, or somewhere in between, just as we have described for the waves on the rope.

What does happen more frequently with waves on strings or ropes is the interference of two waves traveling in opposite directions, which occurs when a wave is reflected at a fixed support. What happens in this situation? Can the principle of superposition be used to predict the form of the combined wave?

Figure 14.12 shows two waves of the same amplitude and wavelength traveling in opposite directions on a string. If we pick different points on the string and consider how these two waves will add at different times, we can gain some insight regarding the resulting wave. At point *A*, for example, we can see that the two waves will cancel each other at all times. One wave is always positive while the other is negative by an equal amount as the two waves approach this point from opposite sides. At this particular point, then, the string will not move at all!

If we move a quarter of a wavelength in either direction from point *A*, we see a very different result. At point *B*, for example, we can see that both waves will be in phase at that

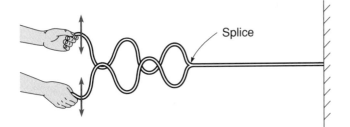

Figure 14.11 Two waves, exactly out of phase in their up and down motions, combine to produce no net disturbance on the rope beyond the splice.

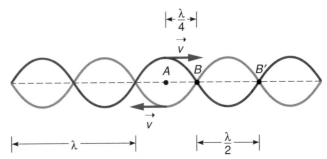

Figure 14.12 Two waves of the same amplitude and wavelength are shown traveling in opposite directions on a string. A node results at point *A* and an antinode at point *B*.

particular point for all times as they approach this point from opposite sides. When one is positive, so is the other and when one is negative, the other is also. The two waves always add, then, at this point producing a displacement that is twice as large as the one that would be produced by just one of the waves.

The interesting thing about these two points is that they remain fixed in space along the string. Point *A,* at which there is no motion, is called a *node.* There are nodes at regular intervals along a string, separated by half the wavelength of the two traveling waves. This can be confirmed by moving half a wavelength in either direction from point *A* and seeing that the two waves cancel there also. The nodes do not move; they are at fixed locations on the string.

The same is true for points such as *B,* where the waves add to yield a large amplitude. These points are called *antinodes,* and they are also found at fixed locations along the string separated by half a wavelength. The picture that emerges appears in figure 14.13. The two waves traveling in opposite directions produce a fixed pattern with regularly spaced nodes and antinodes. At the antinodes, the string is oscillating with a large amplitude; at the nodes, it is not moving at all. At points between the nodes and antinodes, the amplitude has intermediate values.

This pattern of oscillation is called a *standing wave* because the pattern does not move. The two waves that produce this pattern are moving, of course, in opposite directions. They interfere, however, in such a way as to produce a standing or fixed pattern. Standing waves can be observed for all types of wave motion and always involve the interference of waves traveling in *opposite* directions. The interference of a reflected wave with an incoming wave is the usual cause of this condition.

A Guitar String

The author plays a guitar and thus has found much enjoyment over the years in generating standing waves on a string.

A guitar, a piano, or any other stringed instrument consists of strings or wires of different weights that are fixed at both ends with tuning pegs to adjust the tension. If you generate a wave by plucking a string, the wave is reflected back and forth on the string, producing a standing wave.

The frequency of the string's oscillation is equal to the frequency of the sound wave that is produced by the string; as we will see in the next section, frequency is related to the musical *pitch* that we hear. A higher frequency represents a higher-pitched note. What conditions determine the frequency of the guitar string? How are standing waves involved?

The standing wave on a plucked guitar string obviously has nodes at both ends; it is fixed at both ends and cannot oscillate at these points. The simplest standing wave, and the one that usually results when the string is plucked, is one with nodes at either end and an antinode in the middle (fig. 14.14a). Since the distance between nodes is half a wavelength of the traveling waves that interfere to form the standing wave pattern, their wavelength must be twice the length of the string (2*L*).

This simplest standing wave is referred to as the *fundamental* wave, or the *first harmonic* (fig. 14.14a). As we have

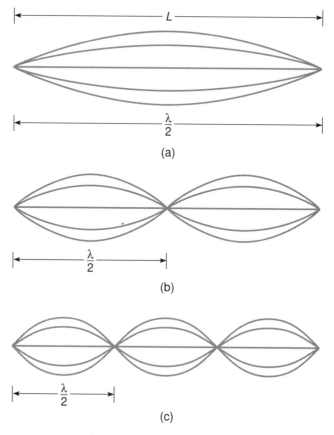

(a)

(b)

(c)

Figure 14.14 **The first three harmonics represent the three simplest standing-wave patterns that can be generated on a string fixed at both ends.**

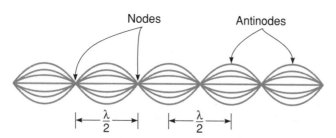

Nodes

Antinodes

Figure 14.13 **The pattern of oscillation produced by two waves traveling in opposite directions is called a *standing wave.* The nodes and antinodes do not move along the string. The distance between adjacent nodes or antinodes is equal to half the wavelength of either of the original waves.**

just seen, the wavelength of the interfering traveling waves is dictated by the length of the string. What about the frequency, though? It turns out that the frequency can be found from the relationship between velocity, frequency, and wavelength, $v = f\lambda$. The velocity is determined by the tension in the string, F, and the mass per unit length of the string, μ, through the relationship for waves on a string, $v = \sqrt{F/\mu}$.

We have before us, then, all of the factors that determine the frequency of the waves. The frequency is given by

$$f = \frac{v}{\lambda} = \frac{v}{2L}.$$

A longer string, therefore, results in a lower frequency. This is why the bass strings on a piano are much longer than the treble strings. On a guitar, you can change the effective length of the strings by *fretting* them; that is, by placing your finger firmly on the string along the neck of the guitar. Shortening the effective length of the string produces a higher frequency and a higher-pitched tone.

The other factors that affect the frequency are the tension in the string and the weight of the string, which together determine the wave velocity. A higher tension leads to a higher wave velocity and therefore a higher frequency, as we can see from the relationship in the previous paragraph. You can easily confirm this by tightening a tuning peg on a guitar. A heavier string, on the other hand, produces a lower wave velocity and therefore a lower frequency. The bass strings on a steel-string guitar or piano are always made by wrapping wire around a core wire in order to produce a larger mass per unit length. Box 14.1 shows a computation of the fundamental frequency for a guitar string.

Although the fundamental frequency for a guitar string dominates if you just pluck the string, it is possible to produce the other two patterns shown in figure 14.14, as well as patterns with even more nodes. To get the second harmonic, for example, you touch the string lightly at the midpoint at the same time that you pluck it.

This creates the second pattern, shown in figure 14.14b, with nodes at the center and at either end. The wavelength of the traveling waves for this pattern is equal to the length of the string, L. Since this wavelength is just half that of the fundamental, the frequency is twice that of the fundamental. Musically, the tone produced by doubling the frequency is one octave above the fundamental.

If you touch the string lightly at a position that is one-third the length of the string from one end, you get the pattern shown in figure 14.14c, which has four nodes (counting those at either end) and three antinodes. The wavelength of the traveling waves is two-thirds the length of the string, or $2L/3$. This is one-third the wavelength associated with the fundamental, which was $2L$. The frequency is thus three times that of the fundamental and $\frac{3}{2}$ times that of the second harmonic. Musically this is not a complete octave above the second harmonic, but rather an interval called a *fifth* above that tone.

During the sixties, guitars were found in almost every dormitory room. They are a little less common now, but there are still quite a few around. If you have one handy, try generating some of these harmonics and noting the pitch that results. It requires no particular skill as a musician and will help to clarify these ideas. If you look closely at the string, you can see the standing wave patterns (fig. 14.15).

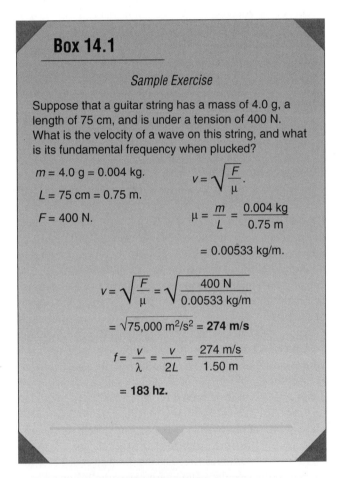

Box 14.1

Sample Exercise

Suppose that a guitar string has a mass of 4.0 g, a length of 75 cm, and is under a tension of 400 N. What is the velocity of a wave on this string, and what is its fundamental frequency when plucked?

$m = 4.0 \text{ g} = 0.004 \text{ kg}.$

$L = 75 \text{ cm} = 0.75 \text{ m}.$

$F = 400 \text{ N}.$

$v = \sqrt{\dfrac{F}{\mu}}.$

$\mu = \dfrac{m}{L} = \dfrac{0.004 \text{ kg}}{0.75 \text{ m}}$

$= 0.00533 \text{ kg/m}.$

$v = \sqrt{\dfrac{F}{\mu}} = \sqrt{\dfrac{400 \text{ N}}{0.00533 \text{ kg/m}}}$

$= \sqrt{75{,}000 \text{ m}^2/\text{s}^2} = \textbf{274 m/s}$

$f = \dfrac{v}{\lambda} = \dfrac{v}{2L} = \dfrac{274 \text{ m/s}}{1.50 \text{ m}}$

$= \textbf{183 hz.}$

Figure 14.15 The strings on a steel-string guitar have different weights. The base string has been plucked, producing the blur near the middle where the amplitude of the motion is the greatest.

14.4 SOUND WAVES

We have discussed sound waves that are generated by oscillating strings on a guitar or piano. You certainly can think of many other ways of generating sound waves, such as firing a pistol, using your voice, or banging a stick on a metal pot. Small children are experts at finding ways to generate sound waves—the louder the better!

Obviously, sound waves must be able to travel through air, since they do reach our ears. What is the nature of sound waves, though, and how do they manage to propagate through air? How fast do they travel? Do they interfere like waves on a string to form standing waves? Exploring these and other questions will provide a basic understanding of the nature of sound.

Producing a Sound Wave

If you look closely at the speaker in your stereo system or in a loudspeaker, you would see a mechanism like that pictured in figure 14.16. There is a flexible, cardboardlike material (the diaphragm) mounted in front of a permanent magnet that is fixed to the housing of the speaker. A coil of wire is attached to the base of the diaphragm so that it is centered on the end of the permanent magnet. An oscillating current applied to the wire coil causes it to behave as a electromagnet that is alternately attracted to and repelled by the permanent magnet. This causes the diaphragm to oscillate with the same frequency as that of the applied electric current.

What effect does the oscillating diaphragm have upon the adjacent air? As the diaphragm moves forward, it compresses the air in front of it; as it moves backwards, it produces a region of lower pressure. The compressed region, in turn, pushes against air that is further removed from the diaphragm, increasing the pressure there. Thus this region of increased pressure propagates through the air, as do the regions of reduced pressure. The disturbance could take the form of a single pulse of increased pressure followed by a region of reduced pressure, but if the diaphragm is moving back and forth repeatedly, it produces a continuous periodic wave involving pressure variations.

This wave of pressure variations is a sound wave. In the regions of elevated pressure, the molecules in the air are closer together, on the average, than they are in the regions of reduced pressure. The picture that emerges is something like that shown in figure 14.17, although the variations in air density are exaggerated. Below the picture is a graph of pressure versus position; for a simple harmonic wave this pressure graph takes a simple sinusoidal form. To complete the picture, you must imagine that the whole pattern is moving away from the source, where new regions of elevated and reduced pressure are being constantly generated.

The molecules making up air are in constant motion in all directions, as is true in any gas. In addition to this random motion, however, there must be a back-and-forth motion of

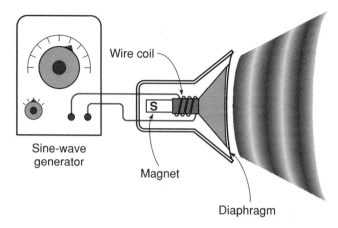

Figure 14.16 **An oscillating current applied to the coil of wire attached to the diaphragm of a loud speaker causes the diaphragm to oscillate as it is attracted to or repulsed by the permanent magnet, thus generating a sound wave.**

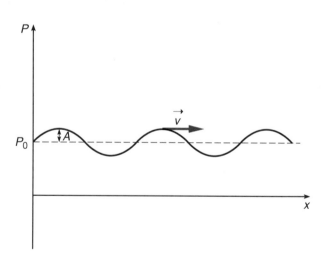

Figure 14.17 **Variations in air pressure (and density) move through the air in a sound wave. The graph shows pressure plotted against position.**

the molecules along the line of the wave's path in order to create the regions of higher and lower density. A sound wave is therefore a *longitudinal* wave; the displacement of the molecules is parallel to the direction of propagation of the wave. It is much like a wave on a Slinky when the end of the Slinky is moved back and forth rather than up and down. The coils of the Slinky move back and forth along the line of propagation as the molecules in air do, and there are moving regions of increased density (compression) in the Slinky just as there are in the air.

The Velocity of Sound

How fast do sound waves travel, and what factors determine the velocity of sound? The first half of this question turns out to be easier to answer than the second. In room-temperature air, sound waves travel with a speed of approximately 340 m/s, or roughly 750 MPH (1100 ft/s). If you have ever watched someone in the distance pounding on metal or chopping wood, you have probably noticed that the sound reaches your ear a split second after you see the collision of hammer on metal or axe on wood. If we stand at the finish line of the 100-meter dash, we see the flash of the starter's pistol before we hear the shot. Sound travels quickly, but not nearly as quickly as light.

This same phenomenon is at work when we hear a clap of thunder a few seconds after we see the flash of lightning. Since light is extremely fast, the light flash reaches us almost instantaneously. The sound wave, on the other hand, takes about 3 seconds to cover 1 kilometer (or 5 seconds to cover 1 mile), given the preceding values for the velocity of sound. Counting the seconds between the flash and the thunder can therefore tell us how far away the lightning strike was (fig. 14.18). If the flash and the thunderclap occur almost simultaneously, you may be in trouble!

The factors that determine the velocity of sound are related to how rapidly one molecule can transmit changes in velocity to nearby molecules, so as to propagate the wave. In air, then, temperature is a major factor, since air molecules have higher average velocities at higher temperatures and collide more frequently. An increase in temperature of 10 C° increases the velocity of sound by approximately 6 m/s.

If we propagate sound waves through gases other than air, the masses of the molecules or atoms make a difference in the velocity of propagation. Hydrogen molecules, for example, have a small inertial mass and are much easier to accelerate than the nitrogen and oxygen molecules that are the primary components of air. The velocity of sound in hydrogen is almost four times greater than it is in air (for similar pressures and temperatures).

Sound waves can also travel through media other than gases, often with considerably higher velocities. The velocity of sound in water, for example, is four to five times faster than it is in air. Molecules of water are much closer together than molecules in a gas, so the propagation of a wave does not depend on random collisions. Sound travels even more rapidly through a steel bar or other metal in which the atoms are tightly bound within a crystal lattice. Sound waves in rock or metal generally have velocities four to five times faster than in water and fifteen to twenty times faster than in air.

If someone strikes a long steel rail with a hammer at one end and you listen for the sound at the other end, you might

Figure 14.18 **Timing the interval between a lightning flash and the associated clap of thunder provides an accurate estimate of your distance from the lightning strike.**

hear two bangs. The first one comes through the steel rail itself and reaches us a moment before the one that comes through the air. The actual difference in time depends on how far you are from the blow.

Organ Pipes and Soft-Drink Bottles

Can we observe interference phenomena, such as standing waves, involving sound waves? We can indeed; the operation of many musical instruments depends upon creating standing waves within a tube or pipe. An organ pipe, a trombone slide, and the several meters of metal tubing in a sousaphone are all there for that purpose. A soft-drink bottle provides the handiest example, however, for observing this phenomenon.

If you place your lips near the edge of a soft-drink bottle, as shown in figure 14.19, and blow softly across the opening, you can generate a standing sound wave in the bottle. Since the bottle is closed at one end, there should be a displacement node at the bottom of the bottle. (A displacement node is one in which there is no longitudinal motion of the medium. Displacement nodes occur at the positions of pressure antinodes.) The rigid glass bottom will not permit longitudinal motion of the air at this point. On the other hand, we would expect there to be a displacement antinode somewhere near the opening of the bottle, since that is where we are exciting the oscillation.

Figure 14.19 Placing your lower lip alongside of the rim of an empty soft-drink bottle and blowing softly across the opening produces a standing wave in the bottle.

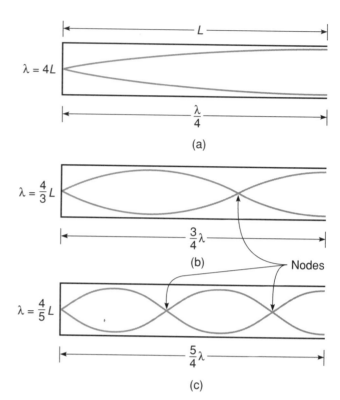

Figure 14.20 The standing wave patterns for the first three harmonics are shown for a tube or pipe open at one end and closed at the other. The curves represent the amplitude of back-and-forth molecular motion at each point in the tube.

The simplest standing wave that could exist in a pipe that is open at one end would then show a variation in displacement amplitude like that plotted in figure 14.20a. The curved line is a graph of the size of the variation in displacement plotted against position. There is a node at the closed end and an antinode near the open end. There is just a quarter-wavelength distance between a node and an antinode, so the wavelength of the traveling sound waves that are interfering to form this standing wave must be approximately four times the length of the tube. Since we know that the velocity of sound in air is about 340 m/s, we can use the wavelength and the velocity to determine the frequency.

For a tube 25 cm in length (more or less the length of a 16 oz soft-drink bottle), the wavelength of the interfering sound waves would be approximately 1 meter. This should produce a frequency of

$$f = \frac{v}{\lambda} = \frac{340 \text{ m/s}}{1 \text{ m}} = 340 \text{ hz.}$$

Since middle C on the musical scale is 264 hz, the tone generated should be a step or two above middle C. (E above middle C has a frequency of 330 hz.)

With some practice, you can generate harmonics above the simplest, or fundamental, frequency. The diagram of figure 14.20b shows the standing wave pattern for the next highest harmonic. This one has an antinode near the opening and another one inside the tube, and two nodes within the tube. Since three-fourths of the wavelength is contained within the tube, the wavelength of the traveling wave should be approx-

imately four-thirds times the length of the tube ($4L / 3$). The frequency of the sound waves generated for the second harmonic in a 25-cm tube should be

$$f = \frac{v}{\lambda} = \frac{340 \text{ m/s}}{0.333 \text{ m}} = 1020 \text{ hz.}$$

This is three times the fundamental frequency and corresponds to a tone in the next octave (near high B).

Similar reasoning can be used to predict the frequencies of higher harmonics. We could also analyze the standing wave patterns produced in a tube closed at both ends or in one open at both ends. The actual tone produced by a bottle or a musical instrument is usually a mix of the various possible harmonics. The nature of the mix determines the quality or richness of the resulting tones.

As noted earlier, the pitch of a musical tone is directly related to the dominant frequency of the associated sound wave. The wave impinging on our ears causes our ear drums to vibrate. Through an intricate process within the ear, these vibrations are converted to electrical signals in the nerves going to the brain. Individuals vary, however, in the types of discrimination that the brain can make among these signals. A combination of genetics and experience seems to be at work in determining who will have a "good ear" and who will be "tone deaf."

Standing waves represent one type of interference that is readily observable with sound waves. Sound waves traveling in the same direction can also produce either constructive or destructive interference, depending upon the phase relation-ship between the interfering waves. "Dead spots" which are sometimes a problem in auditoriums, are produced by destructive interference. The acoustic design of concert halls is a complex art that must take these interference problems into account.

Box 14.2

Everyday Phenomena:
A Moving Car Horn and the Doppler Effect

The Situation. We have all had the experience of standing near a busy street and hearing a car horn as the car goes by. If you try to reproduce that sound vocally, you hum at one pitch to represent the car horn as it approaches and then at a lower pitch to rep-resent the receding car. In other words, you hear a lower-frequency sound wave after the car has passed than the one you hear as it approaches.

What is happening here? Does the frequency of the sound wave produced by the car horn actually change in this process? This hardly seems likely. There must be something about the motion of the car and the horn that affects the frequency of the pitch that we hear. How can we explain this change in frequency?

The Analysis. The top part of the second drawing shows the wave crests, or wavefronts, associated with the car horn when the car is not moving. Each curve represents a surface along which the air pres-sure is at its maximum value as the pressure changes due to the sound wave. The distance between these curves is the wavelength of the sound wave; it is determined by the frequency of the horn and the velocity of sound in air. The wave velocity dictates how far a crest will move in a given time, while the fre-quency of the horn determines when the next crest will appear.

The frequency that we hear is equal to the rate at which the wave crests reach our ear. A little thought should convince you that this rate is determined by the distance between the wave crests (the wave-length) and by the wave velocity. You can think of these wave crests as impinging upon your ear in much the same way that water waves wash up on the shore. The greater the velocity, the higher the rate at which the wave crests reach your ear. The longer the wavelength, however, the lower the rate (frequency) at which they reach your ear. The quantitative rela-

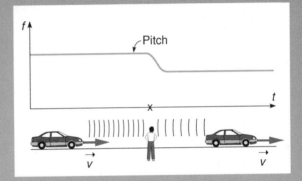

The pitch of the car horn seems to change from a higher to a lower pitch as the car passes.

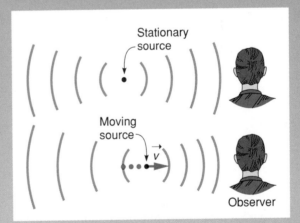

Wavefronts for a stationary car horn (top) and for one that is moving towards the observer (bottom). Motion of the source changes the wavelength on both sides of the source.

tionship is found from the familiar formula $v = f\lambda$, which yields $f = v / \lambda$ for the frequency.

Continued

Box 14.2 Continued

What happens, then, when the horn is moving? The lower portion of the drawing shows the case in which the horn is moving towards you, the observer. Between the time that one crest and the next are emitted by the horn, the horn has moved a short distance. As the diagram shows, this has the effect of shortening the distance between successive wave crests. Even though the horn is still emitting the same frequency as before, the wavelength of the sound wave traveling through the air is now shorter.

Because the wavelength of the waves produced by the moving horn are shorter than those produced by the stationary horn, the wave crests reach your ear at a faster rate; there is less distance between wave crests, but the waves are still moving at the same velocity as before. The faster rate at which the wave crests reach your ear is detected as a higher frequency by your ear. The frequency of the sound that you hear when the horn is moving towards you is therefore higher than the one you hear when the horn is stationary. Again, $f = v / \lambda$, so a shorter wavelength produces a higher frequency.

This change in the detected frequency of a wave resulting from the motion of either the source or the observer is called the *Doppler effect*. Using similar reasoning, you can see that the wavelength in air will get longer when the horn is moving away from you. A longer wavelength produces a lower frequency as detected by the observer. The frequency that you hear as a car approaches you is higher than the natural frequency of the horn, and the frequency you hear as the car moves away from you is lower than the natural frequency. The frequency that you hear if the car is at rest is between these two frequencies.

There is also a Doppler effect if the observer is moving relative to the air in which the wave is traveling. An observer who is moving towards the wave source, will intersect wave crests more rapidly than a stationary observer, and will therefore detect a higher frequency. A receding observer detects a lower frequency. The Doppler effect occurs for light and other types of wave motion as well, but it is most familiar in the common experience of listening to sounds produced by moving vehicles.

14.5 ELECTROMAGNETIC WAVES

What do light, radio waves, microwaves, and X rays all have in common? They are all forms of electromagnetic waves. Together they represent an enormous range of phenomena that have become extremely important in our modern technological world. Some understanding of their basic nature is therefore crucial to any claim to scientific literacy.

The prediction of the existence of electromagnetic waves and a description of their nature was first published by the Scottish physicist James Clerk Maxwell (1831–1879) in the year 1865. Maxwell was an enormously talented theoretical physicist who made contributions in many areas of physics, including electromagnetism, thermodynamics, the kinetic theory of gases, color vision, and astronomy. He is best known, however, for his treatise on electric and magnetic fields, in which he formally introduced the concept of fields and presented a complete theory of electromagnetism in terms of such fields. The description of electromagnetic waves, which included a prediction of their velocity, was an important bonus of this theoretical masterpiece.

Fields and Waves

To understand the nature of electromagnetic waves, you must review your understanding of the concepts of electric field and magnetic field. Both fields can be produced by charged particles, but *motion* of the charge is necessary to generate a magnetic field. The electric field associated with a charged particle is present regardless of whether the charge is moving. These fields represent a property of the space in the vicinity of the charges and are useful for predicting the forces on other charges that may be present (and perhaps moving) at some point in space, as we have seen in chapters 11 and 13.

Suppose, for example, that there is charge flowing up and down in two lengths of wire connected to an alternating-current source, as shown in figure 14.21. If the current reverses direction rapidly enough, an alternating current will flow in this arrangement even though it appears to be an open circuit. Charge will begin to accumulate in the wires, but before the accumulated charge gets too large, the current reverses and the charge flows back and begins to build up in the opposite sign. We have, then, both a changing accumulation of charge and a changing electric current.

The magnetic field generated by this arrangement will be in the form of circular field lines centered upon the wires, as shown. This field is constantly changing in both magnitude and direction, however, as the current changes. From Faraday's law, Maxwell knew that a changing magnetic field generates a voltage in a circuit whose plane is perpendicular to the magnetic field lines. A voltage implies an electric field, and even in the absence of a circuit, a changing magnetic field will generate an electric field at any point in space where the magnetic field is changing.

We should expect, then, that a changing electric field will be generated by the changing magnetic field, according to Faraday's law. Maxwell recognized a symmetry in the behavior of electric and magnetic fields, and assumed that a changing electric field would also generate a magnetic field. This phenomenon—Faraday's law in reverse—was a new effect introduced by Maxwell in his equations describing the behavior of electric and magnetic fields. Experimental measurements confirmed the existence of this effect.

Maxwell was then in a position to recognize that a wave involving these fields could propagate through space. A changing magnetic field produces a changing electric field, which in turn produces a changing magnetic field, and so on. In a vacuum, the process can go on indefinitely and affect charged particles at much greater distances from the source than would be possible with fields generated by nonchanging currents or charges. This is essentially how an electromagnetic wave is produced; the wires in figure 14.21 serve as a transmitting antenna for the waves. A second antenna, placed at a considerable distance from the source, can be used to detect the waves.

Although Maxwell predicted the existence of such waves in 1865, the first successful experiment to demonstrate that they could be produced by electrical circuits was performed by Heinrich Hertz (1857–1894) in 1888. Hertz's original antennas were in the form of circular loops of wires instead of straight wires, but he also used straight wires in later experiments. He demonstrated that a wave produced by the source circuit could be detected with another circuit at some distance from the source. This represented the experimental discovery of radio waves.

Figure 14.22 presents a closer look at the nature of simple electromagnetic waves. If we picture the magnetic field as lying in the horizontal plane, which is consistent with figure 14.21, then the electric field generated by the changing magnetic field would be in the vertical plane. These two fields are everywhere perpendicular to each other, and are also perpendicular to the direction of travel, which is along an axis away from the source antenna. Electromagnetic

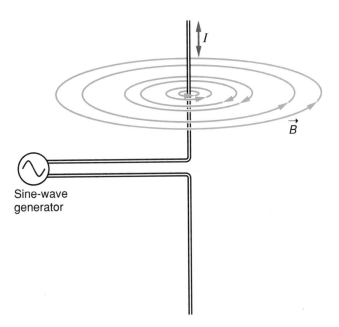

Figure 14.21 A rapidly alternating current in the vertical wires generates changing magnetic fields.

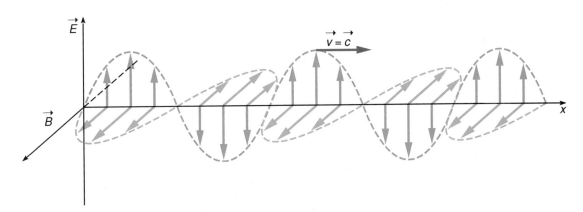

Figure 14.22 The time-varying electric and magnetic fields in a simple electromagnetic wave are in directions perpendicular to each other as well as to the wave velocity.

waves are therefore transverse waves. The magnitudes of the electric and magnetic fields are pictured here as varying sinusoidally and in phase with one another. Just as with waves on a string or sound waves, more complex waves can be regarded as the sum of sinusoidal waves of different frequencies.

Like the other types of waves that we have studied, the sinusoidal wave pattern moves. This means that the picture in figure 14.22 shows the field magnitudes and directions at just one instant in time and along just one line in space. The same kind of variation occurs in all directions perpendicular to the antenna. As the sinusoidal pattern moves, some field values increase while others decrease. As the fields go through zero, they change direction and then begin to increase in the opposite direction. It is these coordinated changes in space and time of the electric and magnetic fields that constitute the electromagnetic wave.

Wave Velocity

In predicting the existence of electromagnetic waves from his theory of electric and magnetic fields, Maxwell was also able to predict their velocity—the velocity of the sinusoidal pattern depicted in figure 14.22. The velocity of these waves in a vacuum can be computed from just two constants: the constant k appearing in Coulomb's law, and the magnetic force constant k' appearing in Ampère's expression for the force between two current-carrying wires. Maxwell's theory predicted that the wave velocity should be equal to the square root of the ratio of these two numbers:

$$v = \sqrt{\frac{k}{k'}} = \sqrt{\frac{9 \times 10^9 \ \text{N·m}^2/\text{C}^2}{10^{-7} \ \text{N/A·m}}}$$

$$= \sqrt{9 \times 10^{16} \ \text{m}^2/\text{s}^2} = 3 \times 10^8 \ \text{m/s}.$$

The striking fact about this value, other than its incredible size of 300 million meters per second, is that it corresponded, within the limits of experimental uncertainty, with

the known value of the velocity of light. This velocity had been accurately measured by different workers not too many years before Maxwell's work. This coincidence led Maxwell to suggest that light itself was a form of electromagnetic wave—the first direct connection between the fields of optics and electromagnetism.

Measuring the velocity of light was no easy task in Maxwell's day. Galileo had been one of the first to attempt to measure it a couple of centuries earlier. He sent an assistant with a shuttered lantern to a distant hill with instructions to open his lantern when he first saw the light from a similar lantern operated by Galileo, who would measure the time required for the light to travel to his assistant and back. This attempt was doomed to failure; the reaction times involved in opening the lanterns were much greater than the traveling time of the light.

Although estimates of the velocity of light had been obtained from astronomy, the first successful land-based measurement of this velocity was accomplished by Fizeau (1819–1896) in 1849. He used a toothed-wheel apparatus like the one shown in figure 14.23. A light beam passes through the gap between the teeth of the rotating wheel and is then reflected from a distant mirror. The beam will be blocked upon its return if the wheel has rotated just far enough so that a tooth is in the place where the gap had been. By measuring the rotational velocity of the wheel necessary to accomplish this, and knowing the distance that the light beam travels to get back to the wheel, Fizeau could calculate the velocity of light.

The speed of Fizeau's wheel allowed a tooth to move into the former position of a gap in less than 1/10 000 of a second. Even at that rate, he had to place his reflecting mirror more than 8 kilometers (about 5 miles) from the wheel. Fizeau used telescopes to recollimate the beam at the mirror and to observe the returning beam. Recognizing that the light beam traveled over 10 miles in a time less than 1/10 000 of a second in Fizeau's experiment may give you some appreciation for the magnitude of the velocity of light!

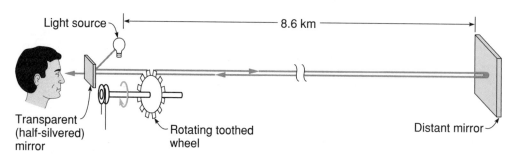

Figure 14.23 **A schematic diagram of Fizeau's toothed-wheel apparatus for measuring the velocity of light. As the rapidly spinning wheel turns through a small fraction of a revolution, the returning light beam is blocked by a tooth on the wheel.**

The velocity of light is an important constant of nature that deserves its own symbol: the symbol c is usually used to represent the velocity of light in a vacuum. Its value, $c = 2.997\ 924\ 58 \times 10^8$ m/s, is known with great precision; it is very close to 3.0×10^8 m/s, which is the value we usually quote and remember. Light and other forms of electromagnetic waves travel somewhat more slowly in media such as glass or water. The velocity of such waves in air is very close to that in a vacuum, however. Although light travels extremely rapidly on the scale of earthbound events, it still takes several years for the light from some of the nearer stars to reach us.

The Electromagnetic Wave Spectrum

We have noted that both radio waves and light waves are types of electromagnetic waves. Are they essentially the same thing, or do they differ in some significant respect? The primary difference between radio waves and light waves is in their wavelengths and frequencies. Radio waves have relatively long wavelengths, several meters or more, whereas light waves have very short wavelengths, less than a micron (one-millionth of a meter).

Since different types of electromagnetic waves all travel with the same velocity in a vacuum or in air, their frequencies are related to their wavelengths by the following relationship:

$$v = c = f\lambda.$$

Thus radio waves with a wavelength of 10 meters would have a frequency of

$$f = \frac{c}{\lambda} = \frac{3 \times 10^8 \text{ m/s}}{10 \text{ m}} = 3 \times 10^7 \text{ hz},$$

or 30 megahertz. An AM radio station broadcasting at a frequency of 600 kilohertz would produce wavelengths of

$$\lambda = \frac{c}{f} = \frac{3 \times 10^8 \text{ m/s}}{600 \times 10^3 \text{ hz}} = 500 \text{ m}.$$

Light waves have much higher frequencies than radio waves because of their shorter wavelength. Orange light with a wavelength of 6×10^{-7} m, for example, has a frequency of

$$f = \frac{c}{\lambda} = \frac{3 \times 10^8 \text{ m/s}}{6 \times 10^{-7} \text{ m}} = 5 \times 10^{14} \text{ hz}.$$

Light wavelengths range from 4×10^{-7} m at the violet end of the spectrum to 7×10^{-7} m at the far red end of the visible spectrum. Electromagnetic waves with somewhat longer wavelengths are called *infrared* waves; those with somewhat shorter wavelengths than 4×10^{-7} are called *ultraviolet* light. X rays and gamma rays have even shorter wavelengths than ultraviolet waves, but they are also electromagnetic waves.

Figure 14.24 shows the wavelengths and frequency bands for various portions of the electromagnetic spectrum. The waves associated with different parts of this spectrum differ not only in wavelength and frequency, but also in the way they are generated and what materials they will travel through. X rays, for example, will pass through materials that are opaque to visible light. Radio waves will also pass through walls that light cannot penetrate.

Interference phenomena can be observed for waves in any portion of the spectrum. In fact, an interference experiment performed by Thomas Young (1773–1829) in the year 1803 established that light could be regarded as a wave phenomenon. In his famous double-slit experiment, Young split a beam of light into two beams by passing it through two

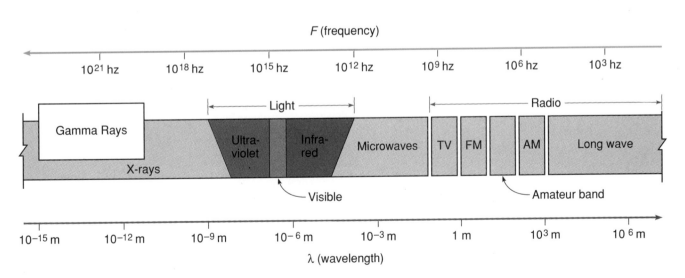

Figure 14.24 The electromagnetic wave spectrum. Both wavelengths and frequencies are shown for different portions of the spectrum.

closely spaced parallel slits. When the two beams were recombined, they interfered to produce a pattern of bright and dark lines representing constructive and destructive interference.

Radio waves can also interfere with one another. This explains why you can sometimes receive a radio broadcast several hundred miles from the station but not at some points in between. Radio waves reflected from charged-particle belts in the atmosphere can interfere with those reaching you directly. Depending on the distances they have traveled, the two waves may be either in phase or out of phase.

If the two radio waves travel essentially the same distance, they would reach you in phase and you would receive a strong signal. If the reflected wave travels a greater distance to reach you than the direct wave, however, it will not be at the same point in its cycle as the direct wave when the two combine. If they are exactly out of phase, you get destructive interference, and cannot receive the station. At some greater distance, the waves may come back in phase and produce a strong signal.

The wavelengths of ordinary radio waves may be several hundred meters, so large differences in the distance traveled are necessary to make a significant difference in phase. Microwaves, however, which can be thought of as high-frequency and short-wavelength radio waves, can have wavelengths of just a few centimeters. This makes it easy to do interference experiments using microwaves on a tabletop or lab bench in student laboratories. A metal reflector can serve to simulate the role of the atmosphere in reflecting radio waves, and moving the detector away from the source will make the signal fade in and out just as a radio signal does.

Although the means of producing different types of electromagnetic waves vary enormously for the different portions of the spectrum, at some level they all involve an oscillating current or an accelerated charged particle, either in an electric circuit, as for radio waves, or within an atom, as for light, X rays, and gamma rays. You are radiating electromagnetic waves yourself in the infrared portion of the spectrum, as does any warm body. In this case, oscillating atoms within the molecules of your skin serve as the antennas. You have been burned by electromagnetic waves (ultraviolet light) coming from the sun. These waves play an enormous role in our everyday experience.

SUMMARY

A mechanical wave can be thought of as a moving disturbance that propagates energy through some medium. The medium is distorted or displaced in some manner as the wave passes through, but in most cases the medium itself goes nowhere. Water waves, waves on a Slinky, waves on a string or rope, sound waves, and electromagnetic waves all share features such as reflection or interference that are common to any wave motion. These waves differ in the type of medium involved, the nature of the disturbance that is propagating, and in the wave velocity.

Wave Pulses and Waves. A Slinky can be used to illustrate some basic features of wave motion, including the generation of single pulses as well as continuous waves in either a longitudinal or transverse mode. For longitudinal waves, the disturbance is along the line of travel; for transverse waves it is perpendicular to the direction of travel. The wave velocity is equal to the frequency times the wavelength of the wave.

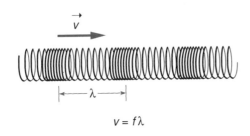

$$v = f\lambda$$

Waves on a String. Transverse waves can be generated on a rope or string. If the end of the rope is moved in simple harmonic motion, the wave has a simple sinusoidal shape. The wave velocity depends upon the tension in the rope and the mass per unit length ($\mu = m/L$) of the rope.

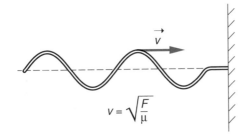

$$v = \sqrt{\frac{F}{\mu}}$$

Interference and Standing Waves. When two or more waves combine, the disturbances add to form a new wave. The interference can be constructive, producing a larger amplitude if the waves are in phase, or it can be destructive, producing a smaller or zero amplitude if the waves are out of phase. Two waves traveling in opposite directions produce a standing wave.

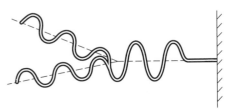

Sound Waves. Sound waves are longitudinal waves involving the propagation of pressure variations through air or other media. The velocity of sound is about 340 m/s in room-temperature air. Standing waves can be formed in organ pipes or soft-drink bottles. The length of the pipe determines the wavelength and thus the frequency of the various harmonics.

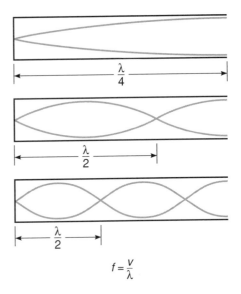

$$f = \frac{v}{\lambda}$$

Electromagnetic Waves. James Maxwell predicted the existence of electromagnetic waves as a result of his theory of electric and magnetic fields. These waves are produced by oscillating charges and involve the propagation of variations in the electric and magnetic fields. They have a velocity of 3×10^8 m/s, the velocity of light. Radio waves, microwaves, light (infrared, visible, and ultraviolet), and X rays are all forms of electromagnetic waves.

QUESTIONS

Q14.1 A wave pulse is transmitted down a Slinky, but the Slinky itself does not change position. Is there a transfer of energy that takes place in this process? Explain.

Q14.2 A slowly moving engine bumps into a string of coupled railroad cars standing on a siding. A wave pulse is transmitted down the string of cars as each one bumps into the next one. Is this wave transverse or longitudinal? Explain.

Q14.3 A wave can be propagated on a blanket by holding adjacent corners in your hands and moving the end of the blanket up and down. Is this wave transverse or longitudinal? Explain.

Q14.4 If you increase the frequency with which you are moving the end of a Slinky back and forth, does the wavelength of the wave on the Slinky increase or decrease? Explain.

Q14.5 If you increase the wave velocity of a wave on a Slinky by increasing the tension, but keep the same frequency of back-and-forth motion, does the wavelength of the wave increase or decrease? Explain.

Q14.6 Is it possible to produce a transverse wave on a Slinky? Explain.

Q14.7 Is it possible to produce a longitudinal wave on a rope? Explain.

Q14.8 Suppose that you double the mass per unit length of a rope by braiding two ropes together.

  a. What effect does this have upon the wave velocity of a wave on this rope? Explain.

  b. For a given frequency of oscillation, what effect does doubling the mass per unit length have on the wavelength of a wave on the rope? Explain.

Q14.9 Suppose that you increase the tension in a rope, keeping the frequency of oscillation of the end of the rope the same. What effect does this have on the wavelength of the wave that is produced? Explain.

Q14.10 Two ropes are smoothly joined to form a single rope that is attached to a wall. If the ends of the two ropes are moved up and down so that one is going up while the other is going down, will the interference of these two waves at the point where the two ropes join be constructive or destructive? Explain.

Q14.11 It is possible to form standing waves on a rope attached to a wall by moving the opposite end of the rope up and down at an appropriate frequency. Where does the second wave come from that interferes with the initial wave to form the standing wave? Explain.

Q14.12 A standing wave is produced on a string fixed at both ends so that there is a node in the middle as well as at either end. Will the frequency of this wave be greater than or less than that of the fundamental frequency, which has nodes only at the ends? Explain.

Q14.13 Is the distance between the antinodes of a standing wave equal to the wavelength of the two waves that interfere to form the standing wave? Explain.

Q14.14 Is it possible, without varying the length, mass, or tension of the string, to form standing waves of different frequencies on a given string fixed at both ends? Explain.

Q14.15 Is it possible to produce a transverse sound wave in air? Explain.

Q14.16 Suppose that we increase the temperature of the air through which a sound wave is traveling.

  a. What effect does this have on the velocity of the sound wave? Explain.

  b. For a given frequency, what effect does increasing the temperature have upon the wavelength of the sound wave? Explain.

Q14.17 If the temperature in an organ pipe increases above room temperature, thereby increasing the velocity of sound waves in the pipe but not affecting the length of pipe significantly, what effect does this have on the frequency of the standing waves produced by this pipe? Explain.

Q14.18 Is it possible for a sound wave to travel through a vacuum? Explain.

Q14.19 What characteristic of the electromagnetic waves predicted by Maxwell's theory led him to suggest that light might be an electromagnetic wave? Explain.

Q14.20 Is it possible for an electromagnetic wave to travel through a vacuum? Explain.

Q14.21 For which of the characteristics of waves—velocity, wavelength, and frequency—is light similar to radio waves, and for which does it differ? Explain.

Q14.22 A cannon is fired at you from a mile or so away. Which will you perceive first, the sound of the cannon or the flash associated with its firing? Explain.

Q14.23 A band playing on a flatbed truck is approaching you in a parade. Will you hear the same pitch for the various instruments as someone down the street who has already been passed by the truck? Explain.

EXERCISES

E14.1 Suppose that water waves coming toward a dock have a velocity of 3 m/s and a wavelength of 1.5 m. With what frequency do these waves meet the dock?

E14.2 A wave on a rope is pictured in the diagram.

a. What is the wavelength of this wave?

b. If the frequency of the wave is 5 hz, what is the wave velocity?

|← —————— 6 m —————— →|

E14.3 A wave on a string has a velocity of 30 m/s and a period of 0.10 s.

a. What is the frequency of the wave?

b. What is the wavelength of the wave?

E14.4 Suppose that a guitar string has a length of 80 cm, a mass of 0.160 kg, and a tension of 50 N.

a. What is the mass per unit length of this string?

b. What is the velocity of a wave on this string?

E14.5 A string with a length of 1 m is fixed at both ends.

a. What is the longest possible wavelength for the waves that interfere to form a standing wave on this string?

b. If waves travel with a velocity of 150 m/s on this string, what is the frequency associated with this longest wavelength?

E14.6 Suppose that the string of exercise 14.5 is plucked so that there are two nodes in addition to those at either end and therefore three antinodes along the length of the string.

a. What is the wavelength of the interfering waves for this mode?

b. Given the same wave velocity as in exercise 14.5, what is the frequency associated with this mode?

E14.7 Sound waves have a velocity of 340 m/s in room-temperature air. What is the wavelength of the sound waves for the musical tone middle C, which has a frequency of 264 hz?

E14.8 What is the frequency of a sound wave with a wavelength of 0.50 m traveling in room-temperature air with a velocity of 340 m/s?

E14.9 An organ pipe, closed at one end and open at the other, has a length of 1 m.

a. What is the longest possible wavelength for the interfering sound waves that form a standing wave in this pipe?

b. What is the frequency associated with this standing wave if the velocity of sound is 340 m/s?

E14.10 Microwaves used in laboratory experiments often have a wavelength of about 1 cm. What is the frequency of these waves?

E14.11 What is the wavelength of the radio waves for a station broadcasting at 1200 kilohertz?

E14.12 What is the frequency of violet light waves with a wavelength of 4.0×10^{-7} m?

E14.13 X rays often have a wavelength of about 10^{-10} m. What is the frequency associated with such waves?

CHALLENGE PROBLEMS

CP14.1 A certain rope has a length of 20 m and a mass of 1.5 kg. It is fixed at one end and held taut at the other with a tension of 50 N. The end of the rope is moved up and down with a frequency of 5 hz.

  a. What is the mass per unit length of the rope?

  b. What is the velocity of waves on this rope?

  c. What is the wavelength of waves having a frequency of 5 hz?

  d. How many complete cycles of these waves will fit on the rope?

  e. How long does it take for the leading edge of the waves to reach the other end of the rope and start coming back?

CP14.2 A guitar string has an overall length of 1.20 m and a total mass of 20 g (0.020 kg) before it is strung on the guitar. Once on the guitar, however, there is a distance of 70 cm between its fixed end points. It is tightened to a tension of 800 N.

  a. What is the mass per unit length of this string?

  b. What is the wave velocity of waves on the tightened string?

  c. What is the wavelength of the traveling waves associated with the fundamental standing-wave mode (nodes just at either end) for this string?

  d. What is the frequency of the fundamental mode?

  e. What are the wavelength and frequency of the next harmonic, which has a node in the middle of the string?

CP14.3 A pipe that is open at both ends will form standing waves, if properly excited, with displacement antinodes near both ends of the pipe. Suppose that we have an open pipe 60 cm in length.

  a. Sketch the standing wave pattern for the fundamental standing-wave mode for this pipe. (There will be a displacement node in the middle and antinodes at either end.)

  b. What is the wavelength of the sound waves that interfere to form this mode?

  c. If the velocity of sound in air is 340 m/s, what is the frequency of this sound wave?

  d. If the air temperature increases so that the velocity of sound is now 350 m/s, by how much does the frequency change?

  e. Sketch the standing wave pattern and find the wavelength and frequency for the next harmonic in this pipe. (Let $v = 340$ m/s.)

CP14.4 It is possible to form standing waves by reflecting microwaves from a metallic reflector placed perpendicularly to the direction of the incoming microwave beam. Suppose that we use a source of microwaves with a wavelength of 3 cm. We can assume that there will be an antinode at the position of the source and a node at the position of the reflector.

  a. As we move the source away from the reflector, at what distance of the source from the reflector should the first standing waves appear?

  b. As we continue to move the source away from the reflector, at what distances will the next few standing waves appear?

  c. If we had a tiny microwave detector that we could move along the line between the source and the reflector, describe the variation you would observe in the strength of the signal received by this detector for the second standing wave that is encountered. (As we move the detector away from the reflector, where is the signal the strongest, and where is it the weakest?)

HOME EXPERIMENTS AND OBSERVATIONS

HE14.1 If you have access to a Slinky, either through a younger brother or sister or through your local physics lab, try producing some of the effects described in section 14.1.

    a. Can you estimate the velocity of single longitudinal pulses produced on the Slinky?

    b. How does this velocity change as the Slinky is stretched?

    c. Does a transverse pulse travel with the same velocity as longitudinal pulse?

    d. Can you produce a continuous longitudinal wave on the Slinky?

HE14.2 Water waves can be easily created by moving your hand in a bathtub or other small pool of water. The wave crests can be directly observed.

    a. Try moving your hand at different frequencies. How does the wavelength vary with frequency?

    b. Can you estimate the velocity of a wave pulse?

    c. Using both hands, you can create two waves that may come together and interfere. If you move your hands in unison, you should observe constructive interference along the center line between the two waves. What effect does this have on the wave that you observe along the center line?

    d. Can you produce destructive interference by moving your hands so that one is going up while the other is going down?

HE14.3 Using rubber bands of different weights, and a cigar box, shoe box, or similar container, create a crude banjo or stringed instrument. The rubber bands are stretched around the open box and are plucked above the opening.

    a. How does the pitch of your various rubber bands vary with their weight or thickness? What is the effect of increasing the tension?

    b. You should be able to observe a standing-wave pattern on a rubber band. Can you produce the second harmonic by touching the band lightly in the middle at the same time that you pluck it? How does the pitch of this mode compare to that of the fundamental mode?

    c. Can you create higher harmonics by touching the band at different points?

HE14.4 Empty a soft-drink bottle and practice blowing over the opening as described in section 14.4 until you are able to produce a consistent tone.

    a. How does the pitch of this tone vary if you put water in the bottle? What is the relationship of the pitch when the bottle is half filled to that when it is empty?

    b. Try producing higher harmonics by blowing harder and reducing the opening in your lips. (This is easy for a flute player, but takes some practice for other mortals.) How is the pitch of a higher harmonic related to that of the fundamental?

    c. By filling bottles to different levels, you can produce all the notes of a one-octave scale. Gather a few friends and try playing "Three Blind Mice" or some other simple tune.

15

Light and Image Formation

Have you ever looked in a mirror and wondered how you are able to stare yourself in the face? Or, if it is early in the morning, how that awful mussed-up face pretending to be you can appear behind that glass plate? You know that you are observing an image that, depending on the quality of the mirror, may or may not be an accurate reflection of reality (fig. 15.1). You are free to believe what you wish.

You have probably also been impressed by the beauty of a rainbow on a showery afternoon, as discussed in chapter 1. This is also an image of sorts; the phenomenon is similar to the separation of colors that occurs when using a prism. The image formed by a mirror involves the reflection of light, and that formed by a prism involves the refraction or bending of light. Both of these phenomena are at work in the formation of a rainbow.

The last section of chapter 14 introduced the idea that light waves are a form of electromagnetic waves and described general features of these waves. How can such waves be involved in the formation of images in a mirror, by a slide projector, or in your eye itself? What do eyeglasses and raindrops have in common? Can we predict where images will be formed and how they will appear?

Such questions lie in the realm of what is usually called *geometric optics.* This discipline treats light waves using rays that are perpendicular to the wavefronts, as described in more detail in the following section. The laws of reflection and refraction are the basic principles that allow us to predict how and where images will be formed.

Figure 15.1   An early-morning look in the mirror. How does that discouraging image get there?

Chapter Objectives

The primary objective of this chapter is to provide an understanding of the laws of reflection and refraction and their role in image formation. To do this, we will need to discuss the relationship between waves and rays, and to show how rays can be traced to define the nature and location of images. In the process we will examine the behavior of mirrors and lenses as well as the operation of simple optical instruments such as cameras, magnifiers, microscopes, and telescopes.

Chapter Outline

❶ *Reflection and images.* **What is the relationship between rays and waves? What does the law of reflection say about the behavior of light rays when they strike a mirror, and how does this explain the formation of images by a mirror?**

❷ *Refraction of light.* **How does the law of refraction describe the bending of light rays when they pass from air into glass, or vice versa? How can this law explain the misleading appearance of underwater objects or the separation of colors by a prism?**

❸ *Focusing light rays with mirrors.* **How can the law of reflection and ray-tracing techniques be used to predict the location and nature of images formed by curved mirrors. What is the relationship between the positions of the object and image?**

❹ *Lenses and eyeglasses.* **How can the law of refraction explain image formation by simple lenses? How is the eye similar to a camera, and how do eyeglasses work to improve our vision?**

❺ *Microscopes and telescopes.* **How do simple magnifying glasses work, and how can lenses and mirrors be combined to produce microscopes and telescopes? What are the basic functions of these instruments?**

15.1 REFLECTION AND IMAGES

How is your image in the bathroom mirror produced? We know that it involves the behavior of light in some way. This is easily ascertained by simply turning off the bathroom light: the image disappears, and instantly reappears when you turn the light back on. Light waves from the bathroom light must be bouncing off of your face to the mirror and then reflecting back to your eyes.

Wavefronts and Rays

If we consider just one point on your face and trace what happens to the waves that are reflected from that point, we may get a clearer idea of what is happening. Since the skin on your face is somewhat rough (at least on a microscopic scale), light is reflected, or *scattered* in all directions from any given point. The tip of your nose, for example, behaves as though it were a source of light waves that spread out uniformly from that point. You can think of these waves as being similar to the ripples that spread on a pond when you drop a rock into the water.

The light waves scattered from your face are electromagnetic waves, of course, not water waves, but they have crests and valleys (where the electric and magnetic fields are stronger or weaker) that move outward from the source point just as water waves do. If we connect points on the wave that are at the same point in their oscillation, we define a wavefront. We often choose the crest of the wave for this purpose, since it is clearly visible in water waves. The next wavefront behind the leading one is the next point at which the waves are at this same point in their cycle; it is therefore separated from the previous wavefront by a distance of one wavelength, as shown in figure 15.2. These wavefronts of light waves move away from the source point at the speed of light.

We could describe almost everything that happens to these waves by tracing what happens to these wavefronts. It turns out to be easier, however to treat their behavior in terms of rays that are perpendicular to the wavefronts. If the waves are all traveling in the same medium (air, for example), the wavefronts move forward uniformly, and the resulting rays are straight lines, as shown in figure 15.2. These rays are easier to draw than the curved wavefronts and thus make the tracing process easier.

When light rays strike your face, they scatter in all directions from each point on your face. These points thus act as sources of diverging light rays, as shown in figure 15.3. The light rays go from your face to the mirror, where they are reflected in a more regular manner because of the smoothness of the mirror. Your eyes are receiving light that has come from the light bulb, but has been reflected by both your face and the mirror before getting back to your eyes.

Law of Reflection

What happens, then, when light rays and wavefronts strike a smooth reflecting surface such as a plane mirror? The waves are reflected, of course, and after reflection they are traveling away from the mirror with the same velocity that they had before reflection. Let us consider plane wavefronts

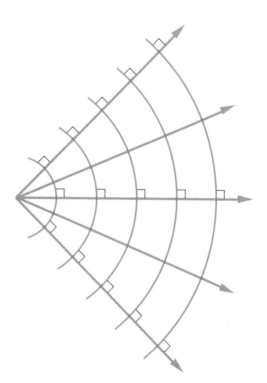

Figure 15.2   Light rays are drawn so that they are everywhere perpendicular to the wavefronts. If the waves travel with uniform velocity, the rays are straight lines.

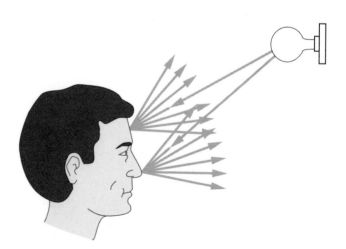

Figure 15.3   Any point on your face acts as a secondary source of light rays that are reflected in all directions from that point.

with no curvature, as shown in figure 15.4, where the wavefronts and rays are approaching the plane mirror at an angle rather than head-on.

Since the wavefronts are approaching the mirror at an angle, some parts of a wavefront are reflected sooner than others, as you can see in figure 15.4. As they are reflected, the wavefronts travel away from the mirror with the same spacing and velocity, but in a new direction. The angle between the wavefront and the mirror is the same, however, for the emerging wave as for the incoming wave. This is true because the outgoing wavefronts travel the same distance in a given time as the incoming ones, since their velocities are equal. For the wavefront in figure 15.4 that is bent in the middle upon reflection, the distance that the right half travels away from the mirror is equal to the distance that the left half travels toward the mirror in one cycle, which produces equal angles on either side of the center point.

The picture is easier to visualize in terms of rays. The angle that a ray makes to a line drawn perpendicular to the surface of the mirror is the same as the angle that the wavefront makes to the surface of the mirror. As with the wavefronts, the angle that the reflected ray makes to this perpendicular line is equal to the angle that the incoming, or *incident,* ray makes to this line. Using the word *normal* to mean "perpendicular" (as we did in chapter 4 when we discussed normal forces), we often refer to this perpendicular line as the *surface normal.*

What we have just described is the essence of the law of reflection. A more formal statement follows:

When light is reflected from a plane reflecting surface, the angle the reflected ray makes with the surface normal is equal to the angle that the incident ray makes with the surface normal.

In other words the angle of reflection is equal to the angle of incidence. The reflected ray also lies within the plane defined by the incident ray and the surface normal; it does not deviate in or out of the plane of the page in our diagram.

Image Formation

How can the law of reflection help us to explain how an image is formed in a plane mirror? What happens to the light rays that are scattered from your nose? Tracing these rays from their origin and following them through reflection from the mirror provides an idea of how the image is formed. This process is illustrated in figure 15.5.

If we extend the reflected rays backward, we see that they intersect at a point behind the mirror. As you will see in more detail later, your eye has the ability to focus diverging light rays. As it collects and focuses these reflected rays, it perceives an image that appears to lie at this point of intersection. In other words, as far as your eye can tell, these light rays come from that point. You therefore see the tip of your nose as lying behind the mirror. The same argument holds for

any other point on your face; they all seem to lie behind the mirror.

Using simple geometry, you can easily see that the distance of this point of intersection from the mirror surface is equal to the distance of the original object from the mirror surface. This fact depends upon the law of reflection, as illustrated in figure 15.6, in which two rays coming from the top of the woman's head are traced as before. The ray that comes into the mirror parallel to the floor and perpendicular to the mirror is reflected back along the same line. The angle of incidence for this ray is zero, and so is the angle of reflection.

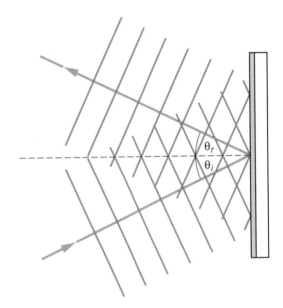

Figure 15.4 Plane light waves approaching a mirror at an angle travel with the same velocity both before and after hitting the mirror. The angle of reflection equals the angle of incidence.

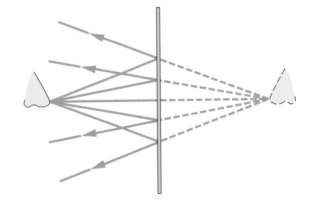

Figure 15.5 A few rays are traced from their origin at the nose to show their reflection from the mirror. They diverge after reflection as though they were coming from a point behind the mirror.

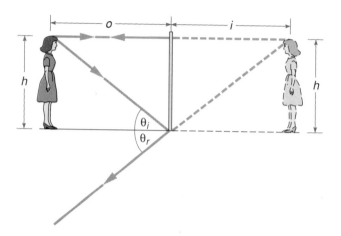

Figure 15.6   Two light rays coming from the top of the woman's head appear to diverge, after reflection, from a point that is as far behind the mirror as the woman is in front of the mirror.

The other ray is shown as being reflected from a point on the mirror that is even with the woman's feet, and it is also reflected at an angle equal to the angle of incidence. When these two rays are extended, their intersection locates the image position. They also form two identical triangles on either side of the mirror, as shown in figure 15.6. Since the angles are equal and the height of the woman and her image are the same, the long sides of these identical triangles must also be equal. The image must therefore be located behind the mirror at an image distance, *i*, equal to the object distance, *o*, of the woman from the front of the mirror.

Since we all have access to plane mirrors, you can verify some of these ideas by observing your own image and those of other objects as you move relative to the mirror. The mirror does not have to be as tall as you are, for example, in order for you to see your entire height. Do you see more of yourself as you move towards the mirror or away from the mirror? What about other objects? Where must you be positioned to see various other objects in the room? Can you explain these observations in terms of the law of reflection and which rays reach your eyes?

The image formed by a plane mirror is referred to as a *virtual* image because the light never actually passes through the point at which the image is located. In fact, the light never gets behind the mirror at all; it just appears to come from points behind the mirror as it is reflected. The image can also be characterized as being upright (right-side-up) and as having the same size as the object. There is a reversal, however, of right and left; what appears to be your right hand in the mirror is the image of your left hand, and vice versa. Take another look; you look at such images every day, but you probably have never given much thought to their nature.

15.2 REFRACTION OF LIGHT

A plane mirror is the simplest kind of image-forming device. We know, however, that images can be formed in other ways: with prisms, lenses, or maybe just a tank of water. These examples all involve substances that are transparent to visible light. What happens to light rays when they encounter the surface of a transparent object? Why do we get a misleading impression about the location of underwater objects? The law of refraction provides the answers to these questions.

The Law of Refraction

Suppose that we consider light waves traveling initially in air and then encountering the plane surface of a piece of glass. What happens to these waves as they pass into the glass and then continue traveling through the glass? Experimental measurements have shown that the velocity of light in glass, water, or similar transparent substances is less than its velocity in a vacuum or in air. (The velocity of light in air is very close to that in a vacuum.) This fact indicates that the distance between wavefronts (the wavelength) will be shorter in glass or water than it is in air, as pictured in figure 15.7, because the waves travel a shorter distance in one cycle, given their smaller velocity.

The difference in the velocity of light in different substances is usually described by a quantity called the *index of refraction*, represented by the symbol *n*. The relationship is simple: the velocity of light, *v*, in the substance is related to the velocity of light, *c*, in a vacuum as follows:

$$v = \frac{c}{n}.$$

In other words, to find the velocity of light in a substance, we divide the velocity of light in a vacuum ($c = 3.0 \times 10^8$ m/s) by the index of refraction of that substance. For glass, the index of refraction is typically in the vicinity of 1.5, so the velocity of light in glass is approximately

$$v = \frac{c}{1.5} = 0.667 \, c,$$

or about two-thirds the velocity of light in a vacuum.

What effect does this reduction in velocity and wavelength have upon the direction of light rays as they pass into glass? If we consider wavefronts and their corresponding rays approaching the surface at an angle, as in figure 15.8, we can see clearly that the rays will be bent as they pass from air to glass. The bending occurs because the wavefronts do not travel as far in one cycle in the glass as they do in air. As the diagram shows, the wavefront that is halfway into the glass travels a smaller distance in glass than it does in air, causing it to bend in the middle. Thus the ray, drawn perpendicular to the wavefront, is also bent.

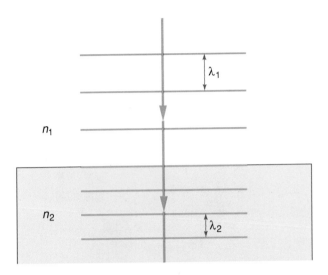

Figure 15.7 The wavelength of light waves traveling from air into glass is shorter in the glass than it is in air because of the smaller velocity of light waves in glass.

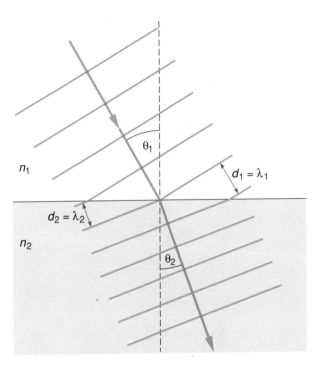

Figure 15.8 Wavefronts approaching a glass surface at an angle to the surface are bent as they pass into the glass. The angle of refraction, θ_2, is less than the angle of incidence, θ_1.

The amount of bending is dictated by the change in the velocity of light as the waves pass into the glass, which in turn depends upon the indices of refraction of air and glass. This can be seen qualitatively by looking at the picture in figure 15.8. The greater the difference in velocity, the greater the difference in how far the wavefronts travel in the two substances. A larger difference in the indices of refraction of the two substances therefore produces a larger bend in the wavefront and ray.

The *law of refraction* describing this bending can be stated in qualitative* terms as follows:

When light passes from one transparent medium to another, the rays are bent towards the surface normal (the axis drawn perpendicular to the surface) if the velocity of light is lower in the second medium than in the first. The rays are bent away from this axis if the velocity of light is greater in the second medium than in the first.

If the angles are small, the quantitative relationship takes a simpler form than the one shown in the footnote. For small angles, the sine function is proportional to the value of the angle itself, so we can write

$$n_1\theta_1 \approx n_2\theta_2.$$

The product of the index of refraction of the first medium and the angle of incidence is *approximately* equal to the product of the index of refraction of the second medium times the angle of refraction. As the index of refraction of the second medium increases, the angle of refraction must decrease, which means that the ray is bent closer to the axis (the surface normal) for larger indices of refraction.

For light waves traveling from glass to air, the bending is in the opposite direction; the rays are bent away from the surface normal, according to the law of refraction. This can be understood by simply reversing the directions of the rays and wavefronts in figure 15.8. The increase in velocity as the wave travels from glass to air causes the ray to bend away from the axis.

Underwater Images

The bending of light rays that occurs at the interface of two transparent substances is responsible for some deceptive appearances. Suppose, for example, that you are standing on a bridge over a stream looking down at a fish. Water has an index of refraction of about 1.33, whereas air has an index of refraction of approximately 1.0. Light traveling from the fish to your eyes is therefore bent away from the surface normal, as shown in figure 15.9.

Tracing light rays from a point on a fish as they pass from water to air shows that they diverge more strongly in air than in water. If we extend these light rays backward, we see that they now appear to come from a point that is closer to the

* The quantitative relationship for this bending is usually described using the trigonometric function, the sine, which will only be familiar if you have some background in trigonometry. It is usually written in symbolic form as $n_1 \sin(\theta_1) = n_2 \sin(\theta_2)$, where n_1 and n_2 represent the indices of refraction of the two media, and θ_1 and θ_2 represent the angle of incidence and the angle of refraction, which are shown on the diagram in figure 15.8.

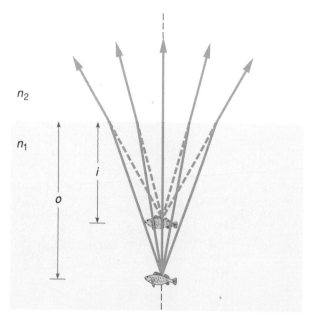

Figure 15.9 Light rays coming from the fish are bent as they pass from water to air, so that the rays appear to diverge from a point closer to the surface.

Figure 15.10 A straw appears to be bent when part of it is above the water surface and the rest below. Is the straw really bent?

surface of the water than their actual point of origin. Since this is true for any point on the fish, the fish appears to be closer to the surface than it actually is. If you were attempting to shoot the fish with a rifle (illegal in most states), you would be likely to miss unless you were shooting straight down at the fish.

The apparent distance of the fish beneath the surface can be predicted from the law of refraction if we know the actual distance. The argument involves triangle relationships and the assumption that the angles of incidence and refraction are small, so that the rays lie close to the axis. The result is that the apparent distance of the image, i, as seen from the air is related to the actual object distance, o, and the indices of refraction as follows:

$$i = o\left(\frac{n_a}{n_w}\right)$$

Since the index of refraction of air ($n_a = 1.0$) is less than that of water ($n_w = 1.33$), the apparent distance is also less than the actual distance, as figure 15.9 clearly illustrates. If the fish is 1.0 m below the surface, its apparent distance below the surface will be

$$i = (1.0 \text{ m}) \left(\frac{1.0}{1.33}\right) = 0.75 \text{ m}.$$

This apparent location of the fish is the position of the image of the fish. Light rays scattered from the fish appear to come from the point of the image rather than from the actual position of the fish. Thus we see a virtual image, like the image seen in a mirror, since the light rays do not actually

pass through the image position; they only appear to come from that point. The bending of light rays at a surface between two different transparent substances thus provides another example of image formation.

The misleading position of objects viewed underwater is something we observe daily but often fail to notice simply because the experience is so common. A straight stick or rod appears to bend if part of it is above water and the rest below. Each point underwater appears to be closer to the surface than it actually is. A straw or spoon in a glass of water or other beverage likewise appears to bend, as shown in figure 15.10. We are used to the deception and seldom give it a second thought.

Total Internal Reflection

Another interesting phenomenon occurs when light rays travel from water to air or from glass to air. As we have already indicated, the light rays bend away from the axis as they pass to the medium (air in these examples) with the lower index of refraction. What happens, though, if the rays are bent so much that the angle of refraction is 90°? The angle of refraction cannot be any larger than that and still result in rays passing into the second medium.

This situation is pictured in figure 15.11. As the angle of incidence for rays traveling in the glass gets larger, so does the angle of refraction. Ultimately, the angle of refraction is 90°. As this point is approached, the refracted ray just skims the surface as it emerges from the glass. If any angle of incidence inside the glass is larger than this, the ray does not escape the glass at all; instead it is reflected. The angle of incidence for which the predicted angle of refraction is 90° is called the *critical angle*. Rays incident at angles at or beyond the critical angle are reflected back inside the glass.

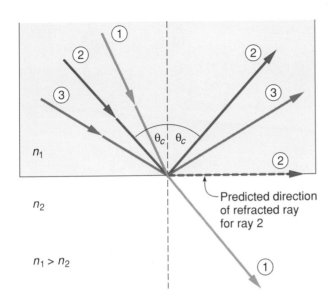

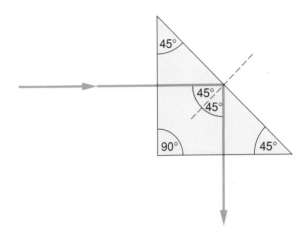

Figure 15.12 **A prism cut with two 45° angles can be used as a mirror, since the light is totally reflected as shown.**

Figure 15.11 **When light travels from glass to air, the angle of incidence that would produce an angle of refraction of 90° (ray 2), is called the *critical angle*, θ_c. Rays incident at equal or greater angles than θ_c are totally reflected.**

This phenomenon is called *total internal reflection.* One hundred percent of the light is reflected inside the substance having the larger index of refraction in this case, making the surface between this substance and air an excellent mirror. For glass, which has an index of refraction of 1.5, the critical angle is approximately 42°. A prism cut with two 45° angles, as shown in figure 15.12, can therefore be used as a reflector. Incident light that is perpendicular to the first surface strikes the long surface at an angle of incidence of 45°, which is greater than the critical angle. It is totally reflected at this surface, so the surface acts as a plane mirror.

Prisms and Dispersion

A final phenomenon associated with the law of refraction is also familiar. You know that light can be separated into different colors by a prism, producing a rainbow effect. What is happening here? How do we get the colors of the rainbow when we started with ordinary white light?

The answer is simple. It turns out that the index of refraction for different materials varies with the wavelength of light; different wavelengths are thus bent by different amounts. The wavelength of light, in turn, is associated with the color that we perceive. Red light, at one end of the visible spectrum, has longer wavelengths and lower frequencies than the violet light at the opposite end of the spectrum. The index of refraction for violet or blue light is greater than that for red light for most kinds of glass, as well as other transparent substances, such as water. Blue light is therefore bent more than red light, and the intermediate wavelengths

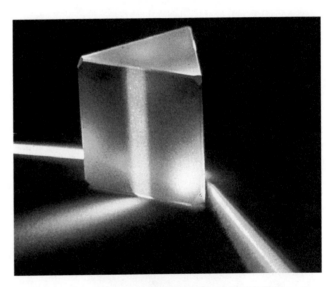

Figure 15.13 **Light rays passing through a prism are bent at both surfaces, with blue light being bent more strongly than red.**

associated with the colors green, yellow, and orange are bent by intermediate amounts.

When light passes through a prism at an angle, the light is bent as it enters the prism and also as it leaves. The maximum deflection of the light rays occurs when they pass through the prism symmetrically, as shown in figure 15.13. The rays are bent towards the surface normal as they enter the prism, and then away from the normal of the second surface as they leave the prism, consistent with the law of refraction. Since different wavelengths have different indices of refraction, the violet and blue rays are bent the most and lie at the bottom of the resulting spectrum of colors, as shown.

The variation of the index of refraction with wavelength is called *dispersion,* and it exists for all transparent materials, including water, glass, and clear plastics. It is responsible for the colors that you see when light passes through a fish tank

Box 15.1

Everyday Phenomenon:
Rainbows

The Situation. In chapter 1, we used the phenomenon of a rainbow to contrast the scientific
approach to explanation with other possible
approaches. We have all seen rainbows and been
awed by their beauty. We know that they occur when
the sun is shining at the same time that it is raining
nearby. If conditions are right, we can see an entire
semicircular arc of color with red on the outside and
violet on the inside. Sometimes we can also see a
fainter bow of color forming a secondary arc outside
of the primary one, with the colors in reverse order, as
shown in the illustration.

How is a rainbow formed? What are the necessary
conditions for observing a rainbow, and where should
we look to find one? Can the laws of reflection and
refraction be used to explain the phenomenon? We
are now in a position to answer these questions.
(See also figure 1.1).

The Analysis. The secret to understanding the rainbow lies in considering what happens to light rays
when they enter a raindrop, as shown in the second
drawing. When a light ray strikes the first surface of the
raindrop, some of it is refracted into the drop. Since the
amount of bending depends upon the index of refraction, which in turn depends upon the wavelength of the
light, violet light is bent more than red light at this first
surface. This effect is essentially the same as the dispersion that occurs when light passes through a prism.

After being refracted at the first surface, the light
rays travel through the drop and encounter its back
surface. At this point the rays are partially internally
reflected; some of the light passes out of the drop,
and some is reflected back towards the front surface
as shown. At the front surface, the rays are refracted
again, with more dispersion taking place as the rays
leave the drop.

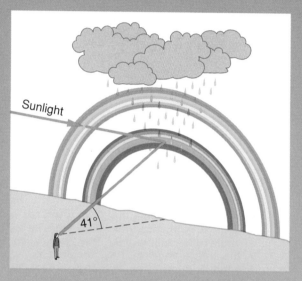

**The primary rainbow has red on the outside and
violet on the inside. The secondary rainbow,
sometimes visible, has the colors reversed.**

It may seem paradoxical that red light, which is bent
the least in the refractions, is actually diverted through
a larger angle than violet light in its overall path through
and back out of the raindrop. The reason is easy to
understand, however, from the diagram. The smaller
bending at the first surface causes the red rays to strike
the back surface at a greater angle of incidence than
the violet rays do. The red rays are therefore reflected
through a greater angle also, according to the law of
reflection. This larger angle of reflection dominates in
determining the overall deflection of the ray.

When we view a rainbow, the sun must be behind
us, because we are observing reflected rays. We see

or around the edges of a lens, such as that in an overhead projector. It is also responsible for the beautiful displays of color
seen in rainbows, as explained in box 15.1.

15.3 FOCUSING LIGHT RAYS WITH MIRRORS

Most of you have had the experience of using a shaving or
makeup mirror that magnifies features on your face. What is

going on here? How is the magnification accomplished, producing an image that appears larger than the actual object?

Shaving or makeup mirrors that produce magnification
involve curved surfaces rather than plane surfaces. The curvature is usually spherical; that is, the surface of the mirror
makes up a portion of a spherical surface. Such a reflecting
surface focuses light rays in a way that produces images that
are different from those obtained with a plane mirror. Simple
ray-tracing techniques can provide an understanding of the
origin and nature of these images.

different colors at different points in the sky because, for a given color, the raindrops must be at the appropriate angle for the light rays to reach our eyes. Since red rays are deviated by the largest angle overall, we see red light reflected from raindrops at the top of the rainbow or at the greatest distance from the center line. Violet light comes to us from raindrops nearest to the center line as is also illustrated in the diagram. The other colors, of course, lie in between, producing the colorful arc that we are used to seeing.

The secondary rainbow, which is usually much fainter than the primary rainbow, is produced by a double reflection inside the raindrops, as shown in the final diagram. Light rays entering near the bottom of the first surface of the drop at an appropriate angle strike the back surface at a large enough angle of incidence that they are reflected twice before being refracted back out through the front surface. The light rays cross over in this case and are deviated through larger overall angles than any of the colors in the primary bow. For this reason, the secondary bow is seen outside of the arc of the primary bow. The secondary bow is also weaker in intensity than the primary bow, because some light is lost in each reflection. A third (tertiary) bow is theoretically possible, but normally is much too weak to be observed.

The sun must be reasonably low in the sky for either the primary or secondary rainbow to be observed. This is why, during the summer, a late afternoon shower often provides the best opportunity for viewing a rainbow. Rainbows can be seen at almost any time during the day from an airplane, however, and sometimes make a complete circle rather than just an arc. If you can understand why this is so, you have mastered the explanations provided here.

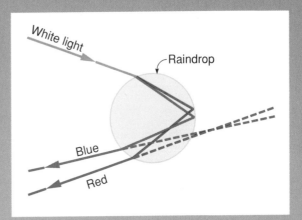

Light rays entering a raindrop are refracted by different amounts at the first surface, reflected at the back surface, and refracted again as they leave the raindrop.

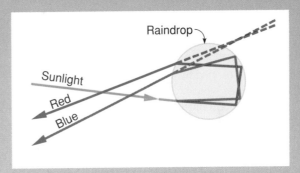

Light rays entering near the bottom of the first surface of a raindrop may be reflected twice before reemerging. These rays produce the secondary rainbow.

Ray Tracing with a Concave Mirror

Mirrors that produce magnification are concave; we are facing the inside of a spherical reflecting surface when we use them. Their focusing properties can be recognized by examining what happens to rays that approach the reflecting surface traveling parallel to the axis, as shown in figure 15.14. The center of curvature of the spherical surface lies on this axis, as shown, and the law of reflection dictates where each ray will go.

Each ray traced in figure 15.14 obeys the law of reflection; the angle of reflection is equal to the angle of incidence. The surface normal for each ray is found by drawing a line from the center of curvature of the sphere to the surface. Since a radius of a sphere is always perpendicular to its surface, this provides the necessary perpendicular line.

As you can see from the diagram, each ray is reflected so that it crosses the axis at approximately the same point as all the other rays. This point of intersection is called the *focal point* and is labeled with the letter *F*. It is the point at which rays coming in parallel to the axis are focused. Since any of the

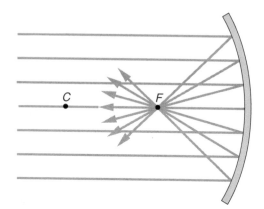

Figure 15.14 Light rays that approach a spherical concave mirror traveling parallel to the axis are reflected so that they all pass approximately through the same focal point, *F*.

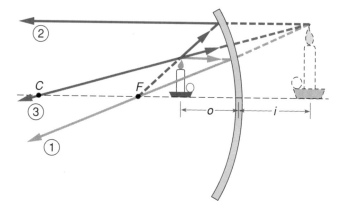

Figure 15.15 Three light rays are traced from the top of a simple object. The point of intersection of the extended rays defines the location of the top of the image.

rays in the drawing could be reversed in direction and still obey the law of reflection, we can also see that rays that pass through the focal point on their way to the mirror will be reflected so that they come out parallel to the axis of the mirror.

We can also see from figure 15.14 that the distance of the focal point, *F*, from the point at which the axis meets the mirror is approximately half the distance of the center of curvature, *C*, from this point. These two points, *F* and *C*, can be used to good advantage in tracing rays to locate and describe images formed by the mirror. When you use such a mirror to scrutinize your face, you generally place your face inside the focal point of the mirror. This produces the magnified image that you normally observe.

Figure 15.15 shows how rays can be traced to locate the image of a simple object that is located inside the focal point of the mirror. We can easily trace three rays if we know the positions of the center of curvature and the focal point. Any two rays would be sufficient to locate the image; the third confirms our result. The rays are the following:

1. A ray coming from the top of the object parallel to the axis and being reflected through the focal point.
2. A ray coming in along a line passing through the focal point and being reflected parallel to the axis.
3. A ray coming in along a line passing through the center of curvature and being reflected back along itself.

The third ray comes in and out along the same line because it strikes the mirror perpendicular to its surface; the angles of incidence and reflection are both zero in this case.

When these three rays are extended backward, we see that they all appear to come from a point that lies behind the mirror. This point defines the position of the top of the image. The bottom of the image lies on the axis, since a light ray coming from that point and traveling along the axis would be reflected straight back along the axis. Light rays from intermediate points of the object appear to come from intermedi-

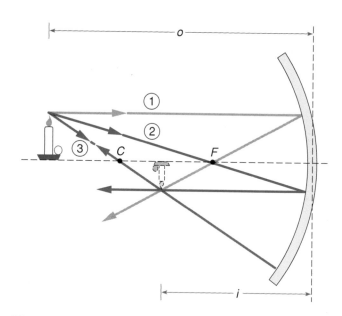

Figure 15.16 Light rays coming from an object that is located beyond the focal point of a concave mirror converge to intersect at an image point in front of the mirror, thus forming a real image.

ate points between the top and bottom of the image. We therefore see the image as lying behind the mirror and, as is obvious from the diagram, magnified.

The resulting image is upright, magnified, and virtual. It is a virtual image because, just as in the case of the plane mirror, the light rays do not actually pass through the image points. The rays, of course, never get behind the mirror; they remain on the side from which they came. The image, on the other hand, appears to lie behind the mirror.

It is also possible to form real images with a concave mirror. This occurs when the object is located somewhere beyond the focal point of the mirror, as illustrated in figure 15.16. In this case, when the three rays are traced, we see that they converge rather than diverge as they leave the mirror. The rays coming from the top of the object intersect at a point on the same side of the mirror as the object and then diverge again from this point. If our eyes collect these rays, we see an image that lies in front of the mirror.

Since we are used to looking at images that lie behind a mirror, it is a little harder for us to focus on one lying in front. This image can be observed, however, using a curved makeup or shaving mirror. As you move the mirror away from your face, the original magnified image becomes larger and finally disappears. It is replaced by an upside-down image that grows smaller as you continue to move away from the mirror. This image is called a *real image* because the light rays actually do pass through the image points and then diverge again from these points. As is also clear from the diagram, the image is inverted (upside-down) and reduced in size (smaller than the object).

Object and Image Distances

Can we predict where an image will be found for a given location of an object? The easiest way to do so is simply to carefully trace the rays as already illustrated. Using triangle relationships and the law of reflection, however, we can also develop a quantitative relationship between the object distance, o, the image distance, i, and the focal length, f, of a mirror. These distances are all measured from the vertex, the point where the axis meets the mirror. The focal length, f, is the distance of the focal point, F, from this vertex. The relationship can be stated as follows:

$$\frac{1}{o} + \frac{1}{i} = \frac{1}{f};$$

in words, the reciprocal of the object distance plus the reciprocal of the image distance is equal to the reciprocal of the focal length.

In figure 15.6 these distances are all positive quantities. In general, if a distance lies on the same side of the mirror as the light rays themselves, the distances are positive. If they lie on the opposite side, as the image distance does for a virtual image, the distance is negative. The sample exercise in box 15.2 demonstrates the use of this relationship between the object and image distances.

The triangles in figures 15.15 and 15.16 can also be used to find a relationship between the image height, I, and object height, O, that is, the magnification, m. This relationship is simpler than that between the object and image distances, but also involves these distances. Magnification is defined as the ratio of the image height to the object height and is equal to

$$m = \frac{I}{O} = \frac{-i}{o}.$$

Thus in the example of box 15.2, where the image distance is −10 cm and the object distance + 5 cm, the magnification is

---

Box 15.2

Sample Exercise

An object lies 5 cm to the left of a concave mirror with a focal length of 10 cm. Where is the image? Is it real or virtual?

$o = 5$ cm.
$f = 10$ cm.
$i = ?$

$$\frac{1}{o} + \frac{1}{i} = \frac{1}{f};$$

$$\frac{1}{i} = \frac{1}{f} - \frac{1}{o}$$

$$= \frac{1}{10 \text{ cm}} - \frac{1}{5 \text{ cm}}$$

$$= \frac{1}{10 \text{ cm}} - \frac{2}{10 \text{ cm}}$$

$$= -\frac{1}{10 \text{ cm}};$$

$$i = -\textbf{10 cm}.$$

Since the image distance is negative, the image lies 10 cm behind the mirror and is virtual. The situation is much like that pictured in figure 15.15.

$$m = \frac{-i}{o} = -\frac{(-10 \text{ cm})}{(5 \text{ cm})} = +2.0.$$

In other words, the image height is twice that of the object. The positive sign of the result indicates that the image in this case is upright. Negative magnifications represent inverted images, and magnifications less than 1.0 represent reductions in size.

Convex Mirrors

Up to this point, we have been considering concave mirrors, which curve inward, toward the viewer. What happens if the mirror curves in the opposite direction? Such convex mirrors are also commonly used as in stores to give the clerks a view of the store aisles. These mirrors produce a diminished image, but a large field of view. It is this wide-angle view that makes them useful.

Figure 15.17 shows rays approaching a convex mirror traveling parallel to the axis of the mirror. The center of curvature, C, lies behind the mirror in this case, and lines drawn from the center of curvature are perpendicular to the surface of the mirror, as before. The law of reflection dictates that these parallel rays will be reflected away from the axis as

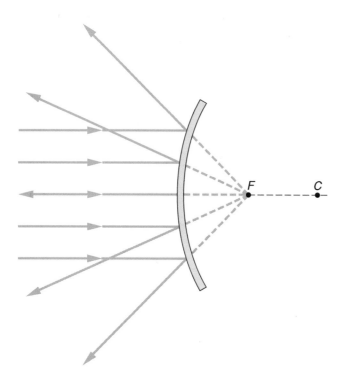

Figure 15.17   Light rays traveling parallel to the axis of a convex spherical mirror are reflected so that they appear to come from a focal point, *F*, that lies behind the mirror.

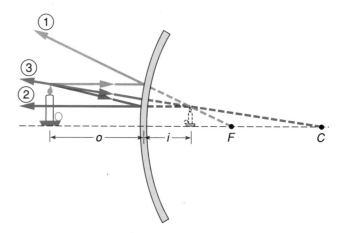

Figure 15.18   Three rays can be traced to locate the virtual image formed by a convex mirror. The rays diverge as though coming from an image point lying behind the mirror.

shown. When these reflected rays are extended backwards, they all appear to have come from approximately the same point, *F*, the focal point.

A convex mirror is therefore a diverging mirror; parallel light rays leave the mirror in a diverging pattern rather than converging as they do with a concave mirror. We can use the same ray-tracing techniques to locate an image, however, as we did with the concave mirror. Figure 15.18 illustrates this process. The ray coming from the top of the object traveling parallel to the axis (1) is reflected as though it came from the focal point. A ray coming in towards the focal point (2) is reflected parallel to the axis, and the ray coming in towards the center of curvature (3) is reflected back along itself, since it is perpendicular to the surface.

When extended backward, these three rays all appear to come from a point behind the mirror, thus locating the top of the image. The image is virtual since it lies behind the mirror, and it is obviously upright and reduced in size. Check it out the next time you are in a store that has one of these mirrors mounted above an aisle. The images are small but cover a broad area.

The object/image distance formula introduced for concave mirrors can also be used to locate images for a convex mirror. The one difference is that the focal length of convex mirror must be treated as a negative quantity, since the focal point lies behind the mirror. The image formed by a convex mirror always lies behind the mirror, and therefore the image distance will always be negative for any object distance that you choose. The only exceptions to this rule occur when rays coming from the object interact with other imaging devices (lenses or concave mirrors) before hitting the mirror.

15.4 LENSES AND EYEGLASSES

We encounter the images formed by mirrors daily and seldom give them a second thought. We are probably even less aware of the images formed by lenses, even though many of us have them hanging on our noses or sitting on our eyeballs in the form of corrective eyeglasses or contact lenses. We also encounter them in cameras, overhead projectors, opera glasses, and even simple magnifying glasses.

How do lenses form images? Lenses are usually made of glass or plastic; thus it is the law of refraction rather than reflection that governs their behavior. The bending of light rays as they pass through a lens is responsible for their image-forming effects. The descriptions of these effects, however, are very similar to those used in our discussion of curved mirrors.

Positive Lenses

A simple convex lens (fig. 15.19), is the easiest case to consider. Light rays coming into such a lens are bent, according to the law of refraction, at each surface. If both surfaces are convex, both bend the rays toward the axis. The easiest way to see this is to imagine that each section of the lens behaves as though it were a prism. The prism angle (the angle between the two sides) becomes larger toward the top of the lens, so that light coming through the lens near the top is bent more than that passing through near the middle.

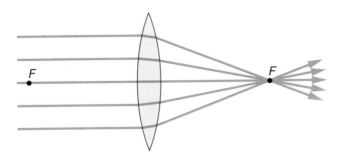

Figure 15.19 Parallel light rays passing through a simple convex lens are bent towards the axis so that they all pass approximately through the focal point, *F*. A lens has two focal points, one on either side.

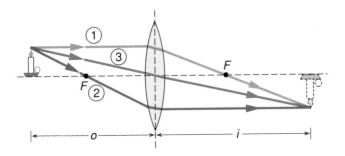

Figure 15.20 Three rays can be traced from the top of an object through a positive lens to locate the image. If the object is beyond the focal point, an inverted real image is formed on the opposite side of the lens.

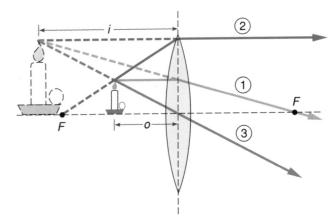

Figure 15.21 A magnified virtual image is formed when the object lies inside the focal point of a positive lens. The emerging light rays appear to diverge from a point behind the object.

Light rays coming in parallel to the axis are bent by different amounts as they pass through the lens so that they all pass approximately through a single point, *F*, on the opposite side of the lens. We call this point the *focal point,* just as we did for the mirrors. The distance from the center of the lens to the focal point is called the focal length, *f,* as before. Because of the symmetry of the lens, however, there is also a focal point on the other side of the lens associated with light rays coming from the opposite direction. Rays that come from either focal point and pass through the lens come out parallel to the axis.

To see how images are formed by such a lens, we can use ray-tracing techniques that are very similar to those used for mirrors. The process is illustrated in figure 15.20 for an object lying outside of the focal point of the lens. Three rays are traced:

1. A ray coming from the top of the object traveling parallel to the axis is bent so that it passes through the focal point on the far side of the lens.
2. A ray coming through the focal point on the near side comes out parallel to the axis.
3. A ray coming in through the center of the lens is undeviated; it passes through the lens without being bent.

The sides of the lens near the center are essentially parallel to one another; the prism angle here is therefore zero, and the lens behaves like a plane sheet of glass in this region. This is why the ray passing through the center can be drawn as a straight line.

The image lies on the opposite side of the lens from the object in figure 15.20 and is therefore a real image, since the light rays do pass through the image point. If we placed a screen at that point, we would see the inverted image on the screen. This is essentially what happens when we use a slide projector; the slides must be inserted upside-down in order to produce normal images on the screen. The magnification of the image depends upon the object and image distances; the image can be either increased or decreased in size. You may wish to try tracing rays for different object distances to see which ones produce magnified images and which ones produce images with magnifications less than 1.0.

The object/image distance formula that we introduced for mirrors can also be used to predict the image position. The focal length for a converging lens is a positive quantity, just as it was for a converging (concave) mirror. A positive image distance, *i,* for a lens lies on the opposite side of the lens from the object, however, since this is where the light actually travels. The object distance, *o,* image distance, *i,* and the focal length, *f,* are all positive quantities in the situation pictured in figure 15.20.

If the object lies inside the focal point of a positive lens, we can get a virtual image, as is illustrated in figure 15.21. In this situation, the image distance is a negative quantity since it does not lie on the same side of the lens as the emerging light. The sample exercise in box 15.3 treats this situation.

When we view the image of an object placed inside the focal point of a positive lens, we see a magnified image, as indicated by the result of the sample exercise. We are then using the lens as a magnifying glass. The image is magni-

Box 15.3

Sample Exercise

An object 2 cm in height lies 10 cm to the left of a positive lens with a focal length of 20 cm. Where is the image and what is its height?

$f = +20$ cm.

$o = +10$ cm.

$O = 2$ cm.

$i = ?$

$I = ?$

$$\frac{1}{o} + \frac{1}{i} = \frac{1}{f};$$

$$\frac{1}{i} = \frac{1}{f} - \frac{1}{o}$$

$$= \frac{1}{20 \text{ cm}} - \frac{1}{10 \text{ cm}}$$

$$= \frac{1}{20 \text{ cm}} - \frac{2}{20 \text{ cm}}$$

$$= -\frac{1}{20 \text{ cm}};$$

$$i = -20 \text{ cm}.$$

$$m = -\frac{i}{o} = -\frac{(-20 \text{ cm})}{10 \text{ cm}} = +2.$$

$$I = mO = (2)(2 \text{ cm}) = 4 \text{ cm}.$$

The image lies 20 cm to the left of the lens and is magnified to a height of 4 cm (twice the height of the object). It is upright because the magnification is positive.

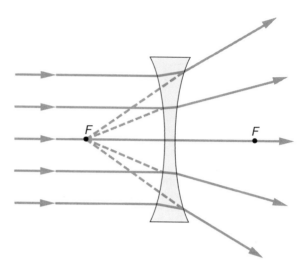

Figure 15.22 Light rays traveling parallel to the axis are bent away from the axis by a negative lens so that they appear to come from a common focal point, *F.*

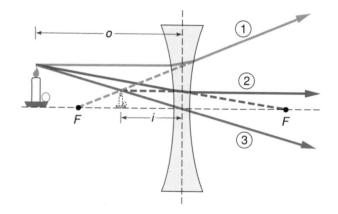

Figure 15.23 Three light rays are traced from the top of an object to locate the image formed by a negative lens. The virtual image lies on the same side of the lens as the object.

fied but it also lies behind the object, farther away from our eyes; the greater distance makes it easier for us to focus on the image than on the object itself. This is an advantage, particularly for older people who have lost the ability to accommodate their focus to view objects that are close to their eyes. When you hear someone over fifty years of age kidding that their arms are too short to read without glasses, this loss of accommodation ability is their problem. A magnifying glass can help.

Negative Lenses

As we have seen, a positive lens is a *converging* lens; it bends light rays towards the axis. What happens if we change the curvature of the lens surfaces so that they are concave rather than convex? If we consider the lens to be made up of different prism sections, as before, these sections are upside-down compared to those of a positive lens. As you can see in figure 15.22, these prism sections bend light rays away from the axis rather than towards it. The lens is therefore a *diverging,* or negative, lens.

Light rays coming in parallel to the axis are bent away from the axis by the lens in figure 15.22 so that they all appear to be diverging from a common point. This point is one of two focal points of the negative lens; the other lies on the opposite side at the same distance from the center of the lens. Light coming in towards the focal point on the far side of the lens is bent so that it comes out parallel to the axis. As you might have guessed, the focal length, *f,* of a negative lens is defined as a negative quantity.

We can trace the same three rays that we traced for the positive lens to locate an image for the negative lens, as is shown in figure 15.23. The ray coming from the top of the object parallel to the axis (ray 1) is bent away from the axis in this case, so that it appears to come from the focal point on the near side of the lens. A ray coming in towards the focal point on the far side (ray 2) is bent so that it comes out parallel to the axis. The third ray passes through the center of the lens undeviated, as before.

The resulting image lies on the same side of the lens as the object and is upright and reduced in size, as figure 15.23 shows. This can also be verified by using the object/image distance formula, treating the focal length as a negative quantity. The image is virtual because the rays appear to come from the image point but do not (except for the undeviated ray) actually pass through that point. This is true regardless of the object distance; a negative lens used by itself always forms a virtual, diminished image.

If you look through a negative lens held near a printed page, the letters appear smaller than their actual size. It is easy to distinguish a negative lens from a positive lens, therefore; one makes the print smaller and the other magnifies it. The latter is usually more useful, but negative lenses have important applications also, particularly for people who are nearsighted.

The Eye and Eyeglasses

Most of us will wear eyeglasses at some time in our lives, and many of us have worn them since adolescence or even earlier. What is it that goes wrong with our vision that requires corrective lenses? To answer that question, we need to explore the optics of the eye itself.

Our eyes contain positive lenses that focus light rays on the back surface of the eyeball when they are working properly. As shown in figure 15.24, the eye actually contains two positive lenses: the cornea, which is the curved surface forming the front of the eye, and the accommodating lens, which is attached to muscles inside the eye. Most of the bending of light rays actually takes place as the light passes through the cornea; the accommodating lens is more of a fine-tuning device.

There is a good analogy between the eye and a camera. A camera uses a compound positive lens system to focus light rays on the film at the back of the camera. The lens system in a camera can be moved back and forth in order to focus on objects at different distances from the camera. In the eye, the distance between the lens system and the back surface of the eye is fixed, so that we need a variable focal-length or "zoom" lens, the accommodating lens, to focus on objects at different distances.

This system of two positive lenses forms an inverted image on the retina, which is the layer of receptor cells on the back inside surface of the eye. The retina plays the role of the film in a camera; it is the sensor that detects the image formed by the lenses. The light reaching these cells in the retina initiates nerve impulses that are carried to the brain. The brain processes the nerve impulses received from both eyes and interprets the image according to its experience. Most of the time this interpretation is straightforward, and what we see is what we get; at times, however, the brain can produce misleading impressions of what is out there.

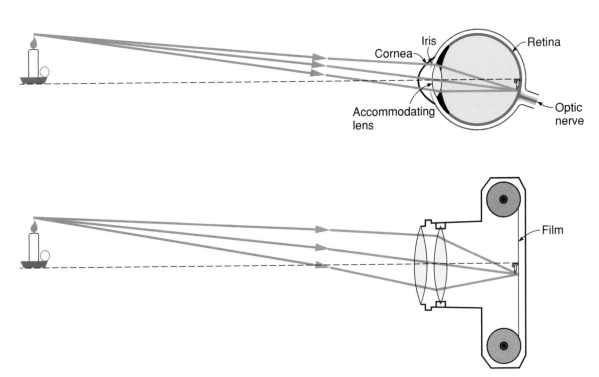

Figure 15.24 Light rays entering the eye from a distant object are focused on the back surface of the eye (the retina) by the cornea and the accommodating lens, much as light rays are focused on the film by a camera lens.

Even though the image on the retina is upside-down, the brain interprets it as being right-side-up. Interestingly, if we fit people with inverting lenses that turn the image on the retina right-side-up, they initially see things upside-down. After some time, however, the brain makes an adjustment and they begin to see things right-side-up again. Everything is then fine until you take the inverting lenses off—then everything appears to be upside-down again until the brain readjusts! A great deal of processing is required on the signal received by our eyes to yield what we actually perceive.

The most common visual problem in our society is nearsightedness, or *myopia*. The eyes of a nearsighted person bend the light rays from a distant object too strongly, causing them to focus in front of the retina, as shown in figure 15.25a. By the time the rays reach the retina, they are diverging again and no longer form a sharp focus. Things then appear fuzzy, although we sometimes fail to notice because we grow accustomed to this indistinct view of things. A myopic person sees near objects distinctly because the light rays are diverging more strongly from a near object than from a distant object before they enter the eye.

A negative lens corrects for the tendency of the eye to converge the light rays too strongly (fig. 15.25b). Since a negative lens diverges light rays, it compensates for the overconvergence of the lenses in the eye so that the image of a distant object is focused on the retina. The images are therefore sharper, and the world becomes clearer. For a nearsighted person who has not worn glasses before, the difference can be a real eye-opener. A farsighted person has the opposite problem; the eye does not converge light rays

strongly enough, and images of near objects are formed behind the retina. Positive lenses are required to correct this problem.

As we age, the accommodating lenses lose their flexibility. We gradually lose the ability to change the converging power of our eyes and cannot focus on near objects, since light rays diverge more strongly from near objects than from distant objects. At this point we need bifocals, in which the top half of the lens has one focal length and the bottom half another. We look through the bottom half when we are doing close work and through the top half for viewing distant objects. Bifocals are also prescribed for younger people who need to do a lot of close work, such as reading. There is some evidence that this may slow the development of myopia.

15.5 MICROSCOPES AND TELESCOPES

Lensmaking was an art that developed during the Renaissance; before then it was not possible to correct visual problems such as nearsightedness or farsightedness or to magnify objects with a magnifying glass. Once lenses became commonplace, however, it did not take long for people to discover that they could be combined to make optical instruments. Microscopes and telescopes are the most commonly used of these instruments; all of us have probably used at least one of these devices at some time.

Microscopes

How are lenses combined to form a microscope, and what does this combination do for us? A microscope consists of two positive lenses usually held at either end of a connecting tube, as shown in figure 15.26. If you have ever used a microscope, you are aware that the object being viewed is placed quite near to the first lens, which is called the *objective* lens.

The objective lens forms a real image, provided that the object lies beyond the focal point. If the object lies just outside of the focal point, then the real image formed by this lens has a relatively large image distance, and the image is magnified. This can be verified by tracing rays or by using the object/image distance formula. As noted in the previous section, this real image is inverted.

Since the light rays actually pass through a real image and then diverge again from that point, this real image becomes the object for the second lens in the microscope. It is viewed through the eyepiece lens, or *ocular,* which is also a positive lens. The eyepiece is essentially a magnifying glass for observing the real image formed by the objective lens. The real image is focused just inside the focal point of the eyepiece, which produces a magnified virtual image. Since the image distance for this virtual image is large and negative, the eyepiece has the effect of moving the image away from your eye, so that you can focus upon it more readily.

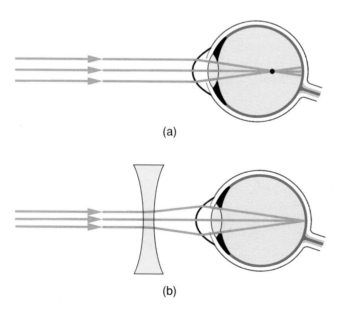

(a)

(b)

Figure 15.25 **In a nearsighted person, parallel light rays from a distant object are focused in front of the retina. A negative lens placed in front of the eye can correct this problem.**

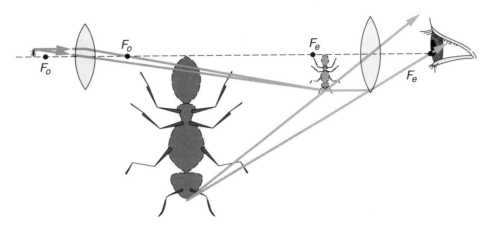

Figure 15.26 A simple microscope consists of two positive lenses separated by a connecting tube. The real image formed by the first lens is then viewed through the second lens; thus it is magnified by both lenses.

Thus, the lenses in a microscope cooperate to produce the magnification. The objective lens forms a magnified real image, and this image is magnified again by the eyepiece. The overall magnification of the microscope is found by multiplying these two magnifications together. This multiplying effect makes it easy to achieve magnifications of over one hundred times the original object size. The magnifying power of a microscope is determined primarily by the power of the objective lens. A high-power objective lens has a very short focal length; thus the object must be placed very close to the lens to achieve the high magnification. Microscopes often have two or three different objective lenses of different powers mounted on a turret, as shown in figure 15.27.

The inverted image formed by an ordinary microscope can be a nuisance. When you are trying to observe different portions of a slide, you have to move the slide in the opposite direction to that in which the image appears to move. This takes some practice—the natural tendency is to move the slide in a direction that corresponds to what you are seeing. If you use a microscope frequently, however, you adjust to this problem, and the reversed movement seems natural.

Some microscopes employ additional lenses or prisms to reinvert the image being viewed. This is important in dissecting microscopes or others that are designed to be used while the object is being manipulated in some manner. These microscopes usually have a relatively low magnifying power, however, since the object cannot be brought very close to the objective lens if it is to be worked upon.

The invention of the microscope opened up a whole new world for biologists as well as for other scientists. Microorganisms that were too small to be seen with the naked eye or with a simple magnifying glass became visible when viewed through a microscope. Seemingly clean pond water was revealed to be teeming with life. The structure of a fly's wing or various kinds of human tissue suddenly became apparent. This is a striking example of the way in which developments in one area of science can have a dramatic impact on other areas.

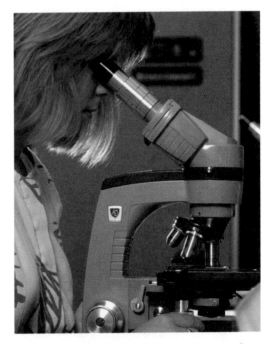

Figure 15.27 A laboratory microscope often has three or four different objective lenses mounted upon a rotatable turret. Light passes through the object slide from a source below the slide.

Telescopes

Whereas the development of the microscope opened up the world of the very small, the earlier invention of the telescope had an equally dramatic impact in opening up the world of distant objects. Astronomy was the primary beneficiary in this case. A simple astronomical telescope, like a microscope, can be constructed from two positive lenses. How then does a telescope differ from a microscope in its design and function?

Distant objects, such as stars, are very large but also so far away that they appear to be very small. One obvious difference, then, between the use of a microscope and a telescope is that objects viewed with a telescope are much farther from the objective lens. As shown in figure 15.28, the objective lens of a telescope, like that of a microscope, forms a real image of the object, which is then viewed through the eyepiece. Unlike the microscope, however, the real image formed by a telescope is reduced rather than magnified.

If the real image formed by the objective lens is reduced, how can there be an advantage to using a telescope? The answer is that the image is much closer to the eye than the original object. Even though the real image is smaller than the object itself, it forms a larger image on the retina of the eye when viewed through the eyepiece than the object does because the image is so much closer to the eye. This idea is illustrated in figure 15.29, which shows two objects of equal height at different distances from the eye. By assuming that the images of both objects are focused on the retina, and tracing just the central, undeviated ray from the top of each object, we see that the nearer object forms a larger image on the retina.

When you want to see fine detail on an object, you usually bring it closer to your eye in order to take advantage of the larger image formed on the retina. Since the size of the image on the retina is proportional to the angle that the object forms at the eye, we say that we have achieved an *angular magnification* by bringing the object nearer. We are limited, however, in how close we can bring the object by the focusing power of the eye. The eyepiece of either a telescope or microscope gets around this problem by forming a virtual image farther away from the eye that makes the same angle

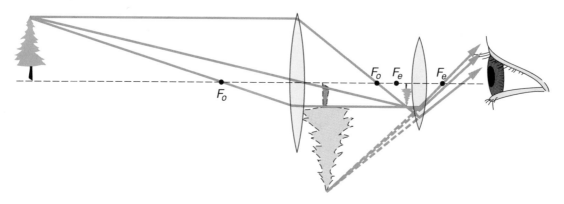

Figure 15.28 **The objective lens of a telescope forms a real, demagnified image of the object, which is then viewed through the eyepiece. The real image is much closer to the eye than the original object.**

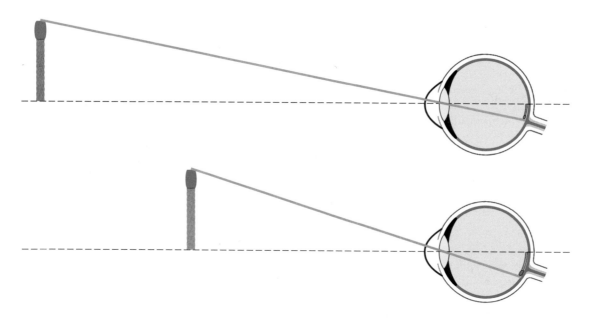

Figure 15.29 **Two objects of the same size, at different distances from the eye form different-sized images on the retina. We can see more detail on the nearer object.**

at the eye as the original real image did. A magnifying glass achieves essentially the same function.

The magnifying effect of a telescope, then, is essentially an angular magnification. The image formed by the telescope forms a larger angle at the eye than does the original object because it is closer to the eye. This produces a larger image on the retina and allows us to see more detail on the object, even though the image we are observing is much smaller than the actual object. Stars and galaxies are enormous, but they appear small to the naked eye because they are so far away.

The overall angular magnification produced by a telescope can be described by a very simple relationship involving the focal lengths of the two lenses:

$$M = \frac{f_o}{f_e},$$

where f_o is the focal length of the objective lens, f_e is the focal length of the eyepiece lens, and M is the angular magnification. From this relationship, we can see that it is desirable to have a large focal length for the objective lens of a telescope in order to produce a large angular magnification. The focal length of the objective lens for a microscope, on the other hand, is usually very short. This allows the object to be brought very close to the lens, producing a larger linear magnification, m, for the image formed. This is the fundamental difference in the design of telescopes and microscopes.

The large telescopes used in astronomy use concave mirrors instead of lenses for the objective. The objects of interest are usually very dim, and the telescope must collect as much light as possible. This requires a large objective lens or mirror with a large *aperture*, or opening, for the incoming light. Since it is easier to make and physically support large mirrors than large lenses, mirrors are usually used for this purpose.

Binoculars and Opera Glasses

The image formed by an astronomical telescope is inverted, like the image formed by a microscope. This is not a big problem if the objects we are viewing are stars or planets; since these are spherical objects, up and down do not make too much difference to us. If we are viewing objects on land, however, seeing the image upside down can be confusing. For this reason, telescopes designed for use on land usually employ some means of reinverting the image.

The most familiar form of land, or *terrestrial*, telescope is a pair of *prism binoculars*. As the term implies, prism

binoculars employ reflecting prisms to transform the light paths so that the images are right-side-up again. This is accomplished by a double reflection, and describing it involves a three-dimensional ray-tracing picture that we will not attempt to illustrate here. The presence of the prisms produces a displacement in the light path as the rays travel through the binoculars; thus the tubes do not have a straight-line appearance. The wider spacing of the objective lenses compared to that of the eyepieces generally indicates the presence of prisms.

Opera glasses are a simpler form of terrestrial telescope. The two tubes in this case are straight, and the reinversion of the image is accomplished by using negative lenses instead of positive lenses for the eyepieces (fig. 15.30). This has the additional advantage of making the tubes shorter, because the negative lenses must be placed in front of the real-image position. The disadvantage of opera glasses is that they produce a very narrow field of view and relatively weak magnification. They do fit into a purse or pocket more readily than prism binoculars, however.

The two tubes in either binoculars or opera glasses allow us to use both eyes when viewing distant objects. This preserves some of the three-dimensional aspects of what we see. In normal vision, your eyes form slightly different images of what you are viewing because each eye sees objects from a slightly different angle. The brain interprets these differences as being produced by three-dimensional features of the scene you are viewing. Try closing one eye when you are viewing near objects, and then reopen that eye. Can you see the difference? A person with just one functional eye sees a flatter world.

Figure 15.30 Prism binoculars and opera glasses employ different means of reinverting the image so that it is upright.

SUMMARY

The propagation of light can be treated, for many purposes, in terms of rays traveling perpendicular to the wavefronts. The laws of reflection and refraction are the basic principles that determine what happens to these rays. Using these principles, this chapter explains how images are formed by mirrors and lenses, and how these elements can be combined to make simple optical instruments.

Reflection.   The law of reflection states that the angle that a reflected ray makes to an axis drawn perpendicular to the surface equals the angle made by the incident ray. The image formed by a plane mirror lies the same distance behind the mirror as the object is from the front of the mirror.

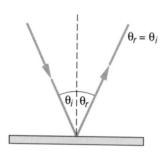

Refraction.   A light ray passing into glass or water from air is bent towards the axis by an amount that depends upon the index of refraction, n. Because of this bending, the image of an underwater object appears to lie closer to the surface than its actual distance. The index of refraction depends upon the wavelength of the light. This causes different amounts of bending for different colors, which is called *dispersion*.

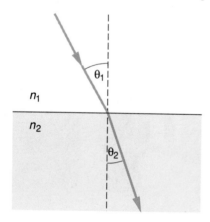

Focusing Mirrors.   A mirror with a curved surface can focus light rays so that parallel incoming rays pass through or appear to come from a single focal point. A concave mirror can produce magnified images, and a convex mirror pro-

duces reduced images. The image distance and object distance are related by a simple formula that also involves the focal length.

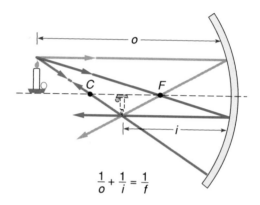

$$\frac{1}{o} + \frac{1}{i} = \frac{1}{f}$$

Lenses and Eyeglasses.   Lenses can also focus light rays to form images. A convex, or positive, lens converges light rays and can be used as a magnifying glass. A concave, or negative, lens diverges light rays and forms reduced images. Negative lenses are used to correct the image position for nearsighted people; positive lenses are used for farsighted people.

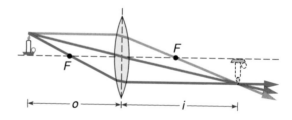

Microscopes and Telescopes.   Lenses and mirrors can be combined to form optical instruments. A microscope uses a positive lens to form a magnified image of a small object, which is then viewed through a second positive lens. An astronomical telescope also uses two positive lenses, but the first lens forms a reduced image that lies much closer to the eye than the original object. This produces a larger image on the retina of the eye when viewed through the eyepiece, since the image makes a larger angle at the eye than was made by the original object (angular magnification).

QUESTIONS

Q15.1 Does light actually pass through the position of the image formed by a plane mirror? Explain.

Q15.2 Does a plane mirror focus light rays? Explain.

Q15.3 If you want to view your full height in a plane mirror, must the mirror be as tall as you are? Explain using a ray diagram.

Q15.4 Objects *A, B,* and *C* lie in the next room hidden from direct view of the person shown in the diagram. A plane mirror is placed on the wall of the passageway between the two rooms, as shown. Which of the objects will the person be able to see in the mirror? Explain using a ray diagram.

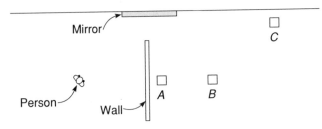

Q15.5 When two plane mirrors are joined at right angles to one another, three images of an object can be seen. The image of the object formed by each mirror can serve as an object for the other mirror. Where is the third image located? Explain using a ray diagram.

Q15.6 A light ray traveling in water (*n* = 1.33) passes from the water into a rectangular piece of glass (*n* = 1.5). Is the light ray bent towards the surface normal (the axis drawn perpendicular to the surface) of the glass, or away from that axis? Explain.

Q15.7 A fish swimming underwater looks up at an object lying a couple of feet above the surface of the water. Does this object appear to the fish to lie nearer to the surface or farther from the surface than its actual distance? Explain.

Q15.8 A light ray traveling in glass for which the critical angle is 42°, strikes a surface between the glass and air at an angle of 45° to the surface normal. Is this ray refracted into the air? Explain.

Q15.9 Do light waves of different colors all travel at the same velocity in glass? Explain.

Q15.10 Is reflection or refraction responsible for the separation of colors in a rainbow? Explain.

Q15.11 Do all light rays (traveling at any angle to the axis) that strike a concave (converging) mirror pass through the focal point? Explain.

Q15.12 An object is located at the center of curvature of a concave mirror. Trace three rays from the top of the object to locate the image formed by the mirror. Is the image real or virtual, upright or inverted?

Q15.13 If an object is placed inside the focal point of a concave mirror (closer to the mirror), is the image formed closer to or farther away from the eye of the viewer than the object? Explain.

Q15.14 Can an object be located in front of a convex (diverging) mirror at a distance that will cause a real image to be formed? Explain.

Q15.15 Is it possible to form a real image with a positive (converging) lens? Explain.

Q15.16 An object is located at a distance from a positive lens that is twice the focal length of the lens. Trace three rays from the top of the object to locate the image. Is the image real or virtual, upright or inverted?

Q15.17 An object is located at the left-side focal point of a negative lens. Trace three rays from the top of the object to locate the image. Is the image real or virtual, upright or inverted?

Q15.18 Light rays approach a negative lens so that they are converging towards the focal point on the far side. Are these rays diverging when they leave the lens? Explain.

Q15.19 Does a nearsighted person have trouble seeing near objects? Explain.

Q15.20 Would you use a positive lens or a negative lens to correct the vision of a farsighted person? Explain.

Q15.21 Does each of the two lenses used in a microscope produce a magnification of the object being viewed? Explain.

Q15.22 Does each of the two lenses used in a telescope produce a magnification of the object being viewed? Explain.

Q15.23 Is it possible to produce an angular magnification of an object by simply bringing it closer to your eye? Explain.

EXERCISES

E15.1 A fish lies 2 m below the surface of a clear pond. If the index of refraction of water is assumed to be 1.33 and that of air approximately 1.0, how far below the surface does the fish appear to be to a person looking down from above?

E15.2 A rock appears to lie just 20 cm below the surface of a smooth stream when viewed from above the surface of the stream. Using the indices of refraction given in exercise 15.1, what is the actual distance of the rock below the surface?

E15.3 An insect is embedded inside a glass block ($n = 1.5$) so that it is 2 cm below a plane surface of the block. How far from this surface does this insect appear to be to a person looking at the block?

E15.4 A concave mirror has a focal length of 12 cm. An object is located 6 cm from the surface of the mirror.

    a. How far from the mirror is the image?

    b. Is the image real or virtual, upright or inverted?

E15.5 A concave mirror has a focal length of 15 cm. An object is located 30 cm from the surface of the mirror.

    a. How far from the mirror is the image?

    b. Is the image real or virtual, upright or inverted?

    c. Trace three rays from the top of the object to confirm your numerical results.

E15.6 A convex mirror has a focal length of –10 cm. An object is located 20 cm from the surface of the mirror.

    a. How far from the mirror is the image?

    b. Is the image real or virtual, upright or inverted?

E15.7 A positive lens has a focal length of 6 cm. An object is located 3 cm from the lens.

    a. How far from the lens is the image?

    b. Is the image real or virtual, upright or inverted?

    c. Trace three rays from the top of the object to confirm your numerical results.

E15.8 A positive lens has a focal length of 20 cm. An object is located 30 cm from the lens.

    a. How far from the lens is the image?

    b. Is the image real or virtual, upright or inverted?

    c. Trace three rays from the top of the object to confirm your numerical results.

E15.9 A negative lens has a focal length of – 20 cm. An object is located 30 cm from the lens.

    a. How far from the lens is the image

    b. Is the image real or virtual, upright or inverted?

E15.10 A magnifying glass with a focal length of +4 cm is placed 2 cm above a page of print.

    a. At what distance from the lens is the image of the print?

    b. What is the magnification of this image?

E15.11 The objective lens of a microscope has a focal length of 1.0 cm. An object on the microscope slide is placed at a distance of 1.2 cm from the lens.

    a. At what distance from the lens is the image formed by the objective lens?

    b. What is the magnification of this image?

E15.12 The objective lens of a telescope has a focal length of 25 cm (0.25 m). An object is located 10 m from the lens.

    a. At what distance from the objective lens is the image formed by this lens?

    b. What is the magnification of this image?

    c. If the eyepiece has a focal length of 5 cm, what is the overall angular magnification produced by this telescope?

E15.13 A telescope that produces an overall angular magnification of 20X has an eyepiece lens with a focal length of 4 cm. What is the focal length of the objective lens?

CHALLENGE PROBLEMS

CP15.1 A fish is viewed through the glass wall of a fish tank. The index of refraction of the glass is 1.5, and that of the water in the tank is 1.33. The fish lies a distance of 20 cm behind the glass. Light rays coming from the fish are bent as they pass from the water to the glass and then again as they pass from the glass to air ($n = 1.0$). The glass is 0.5 cm thick.

    a. Considering just the first interface between the water and the glass, how far behind the glass is the image of the fish formed by the bending at this interface?

    b. Using this image as the object for the second interface between the glass and air, how far behind the front surface of the glass does this "object" lie?

    c. Considering the bending of light at this second interface between the glass and air, how far behind the front surface of the glass does the fish appear to be?

CP15.2 An object that is 2 cm tall is located 30 cm from a concave mirror with a focal length of 15 cm. Since the focal length is just half of the radius of curvature of the mirror, the object is located at the center of curvature.

    a. Using the object/image distance formula, find the image distance for this object.

    b. Calculate the magnification of the image.

    c. Is the image real or virtual, upright or inverted?

    d. Trace two rays to locate the image and confirm your conclusions from parts *a*, *b*, and *c*.

    e. Why does it not make sense in this case to trace the third ray that is often used to confirm the image position?

CP15.3 An object is located at the focal point of a positive lens with a focal length of 10 cm.

    a. What is the image distance predicted by the object/image distance formula?

    b. Trace two rays to confirm the conclusion of part *a*.

    c. Will the image be in focus in this situation? Explain.

CP15.4 An object with a height of 0.5 cm lies 5 cm to the left of a positive lens with a focal length of 4 cm.

    a. Using the object/image distance formula, calculate the image distance for this object.

    b. What is the magnification of this image?

    c. Trace three rays to confirm your conclusions from parts *a* and *b*.

    d. Suppose that this image serves as the object for a second lens that also has a focal length of +4 cm. The second lens is placed 2 cm to the right of the image serving as its object.

    e. Where is the image formed by this second lens, and what is its magnification?

    f. What is the overall magnification produced by this two-lens system?

HOME EXPERIMENTS AND OBSERVATIONS

HE15.1 Fill a clear glass almost to the top with water and put various objects in it.

   a. Do the objects appear to be shorter than their actual length when viewed from above the glass?

   b. Do the objects appear to be shorter than their actual length when viewed through the sides of the glass? What distortions do you notice when viewing the objects through the sides?

   c. If a 6-inch plastic ruler is inserted to the 3-inch mark in the water, can you predict how much shorter than its actual length the ruler will appear to be when viewed from above? Does your observation seem to confirm this prediction?

HE15.2 If you have a magnifying (concave) mirror handy (a shaving or makeup mirror), try moving the mirror slowly away from your face.

   a. What changes in the image of your face do you observe as you move the mirror away?

   b. The image should become very blurred and indistinct when your face is located near the focal point of the mirror. Can you estimate the focal length of the mirror by finding the distance at which the image disappears before reappearing in inverted form?

   c. As you move the mirror away so that the object distance is larger than the focal length, where does the image appear to lie?

HE15.3 Find two small plane mirrors (like those often carried in a purse). Place the two mirrors adjacent to each other so that they touch along one edge, making an angle of 90°. Place a small object, such as a paper clip, in front of the two mirrors.

   a. How many images do you see in the two mirrors when the angle between the mirrors is a right angle (90°).

   b. As you decrease the angle between the two mirrors, describe what happens to the number of images that you can see.

   c. Using the idea that each of the images formed can serve as an object for the other mirror, can you explain your observations? (Ray diagrams may be useful.)

HE15.4 If you wear glasses or have a friend willing to lend you a pair, try looking at a page of print through these glasses.

   a. Holding the glasses a few centimeters above the page of print, does the print appear to be magnified or reduced in size? Are the lenses positive or negative?

   b. Is the owner of the glasses nearsighted or farsighted? If they are your own glasses, do you need them more for viewing near objects or distant objects?

HE15.5 If there is an overhead projector in your classroom or other handy location, examine it carefully, and describe the optical system involved.

   a. What optical elements (lenses or mirrors) are present in this system?

   b. What is the function of each of these elements? (Holding a white card or stiff paper at various locations between the elements when the projector is in use may help your analysis.)

   c. Can you produce a ray diagram showing how the rays coming from the object (the transparency being viewed) converge or diverge on their way to the screen?

Part 5

THE ATOM

and its Nucleus

16

The Structure of the Atom

Have you ever seen an atom? You have certainly heard people talk about atoms and have probably seen pictures of atomic models like the one shown in figure 16.1, but do you have any real idea of why we think atoms exist? Perhaps the question that really should be posed is do you believe in atoms? If so, why?

Most of us believe in atoms on the basis of the pronouncements of textbooks or teachers dating back to our elementary-school days. You may be shocked to realize that many of those teachers may never have seriously questioned their belief in atoms or wondered where our evidence for the existence of atoms originated. Why, then, should you believe in atoms or in physicists' descriptions of their structure?

Although the idea that matter is made up of tiny submicroscopic particles called atoms has a long history (dating at least to the early Greeks, a few hundred years before the birth of Christ), virtually nothing was known about their structure until the early part of the twentieth century. In fact, just before the turn of the century, there was an ongoing debate among physicists as to whether atoms did indeed exist, or were merely a convenient fiction used primarily by chemists. The evidence for their existence must not have been overwhelming at that time.

During a period of years extending from about 1895 to 1930, a series of discoveries and theoretical developments occurred that revolutionized our view of the nature of the atom. We went from a state of knowing almost nothing about the structure of atoms (and even questioning their existence) to a firmly based theory of the structure of the atom that was capable of explaining an enormous range of physical and chemical phenomena. This revolution in our view of the atom is one of the greatest achievements of human intellectual endeavor, with wide-ranging implications to our economy and technology. Its story deserves to be understood by more than a small fraction of our population.

Although we cannot see atoms directly and therefore do not recognize them as part of our everyday experience, atomic phenomena do play a large role in our everyday world. The operation of a television set, chemical changes that occur in our bodies, the use of diagnostic X rays, and many other common phenomena can all be understood in terms of our modern knowledge of atomic behavior. One chapter cannot possibly do justice to even a small slice of this knowledge, but it can introduce some of the fundamental ideas. Most important, we will consider the basic question of why we believe in the existence of atoms and in our models of atomic structure. How have these ideas developed?

Figure 16.1   A stylization of an atom. Does an atom really look like this?

Chapter Objectives

The primary objective of this chapter is to introduce some of the evidence for the existence of atoms and to describe some of the discoveries that have led to an understanding of the structure of atoms. We will begin with evidence from chemistry and proceed to the discoveries of the electron, X rays, natural radioactivity, the nucleus of the atom, and atomic spectra. We will then examine the role of the Bohr model of the atom (1913) and its relationship to the modern view provided by the theory of quantum mechanics (1925–30).

Chapter Outline

❶ *The existence of atoms: evidence from chemistry.* What information does the study of chemical reactions provide about the existence and nature of atoms. How was the periodic table of the elements developed?

❷ *Cathode rays, electrons, and X rays.* How are cathode rays produced, and what are they? How did the study of cathode rays lead to the discovery of the electron and of X rays?

❸ *Radioactivity and the discovery of the nucleus.* How was natural radioactivity discovered, and what is its nature? What role did it play in the discovery of the nucleus of the atom?

❹ *Atomic spectra and the Bohr model of the atom.* What are atomic spectra, and what role did they play in the understanding of atomic structure? What are the basic features of Bohr's model of the atom, and what were its successes?

❺ *Particle waves and quantum mechanics.* What were the limitations of the Bohr model, and how does the theory of quantum mechanics address these problems? What do we mean when we say that particles have wave properties, and what role did this idea have in the development of quantum mechanics?

16.1 THE EXISTENCE OF ATOMS: EVIDENCE FROM CHEMISTRY

Why should we believe in the existence of things that we have never personally seen? Why did many scientists talk with confidence of the existence and properties of the atoms of different substances during the nineteenth century when they knew essentially nothing about their actual structure? Are there things in our everyday experience that should cause us to believe in atoms?

Many of the ideas used in modern science involve things that we cannot see directly; we infer their existence from observations that, taken together, provide convincing evidence regarding the behavior and nature of these entities. In the case of atoms, much of that early evidence came from the study of chemistry. Although we do not often give them much thought, chemical processes are very common in our everyday experience. If the concept of atoms were not already available, we might find it necessary to invent the idea just to explain some of these phenomena.

Early Chemistry

Chemistry involves the study of the differences in substances and how they can be combined to form still other substances. During early Greek civilization, people speculated on what might be the elementary substances from which all other materials are made. Fire, earth, water, and air were early candidates, but clearly that idea required some refinement. Earth, in particular, was capable of taking on a large variety of forms.

One of the most striking demonstrations in elementary chemistry is that of taking a small crystal or tablet of dye or food coloring and dropping it into a glass of water (fig. 16.2). Rather quickly, the color diffuses, until the originally clear water becomes a uniformly colored fluid. Clearly a change has taken place. How might we explain that change?

Even without any prior instruction in the language of chemistry, we might find ourselves assuming that tiny particles of the dye are migrating within the water, moving through spaces between particles of the water that are not apparent. Similar ideas might be used to explain the disappearance of sugar or salt that is placed in water or other fluids.

The idea that almost any solid substance can be crushed into a fine powder is also familiar. By subjecting that powder to heat (or fire), we can restore the solid form, although perhaps modified relative to the original substance. If that powder is combined with powder from some other substance, however, and then heated, the resulting product can be quite different from either of the original substances. Baking provides simple examples of that process, but early experiments in purifying and modifying metals were perhaps the most important efforts of early chemistry. Alchemists were enticed by the elusive prospect of making gold from more common metals.

Figure 16.2   A tablet of food coloring is dropped into a glass of water. How might we explain what happens?

Is it possible to powder something indefinitely into smaller and smaller particles? Although the evidence on this point was not convincing to early chemists, it was tempting to think that this was not the case. Since certain elementary substances always seemed to be retrievable from their experiments, it was natural to assume that there were irreducible particles of these substances that retained their form. The notion that each of these elementary substances, or *elements,* were made up of tiny particles, or *atoms,* that were all the same for a given element but differed for different elements, was an attractive model for explaining chemical phenomena. Such atoms could then combine with atoms of other elements to form different substances, but could always be retrieved from these substances by sufficient heating or other processes.

A systematic study of how various elements combined with each other revealed certain regularities and rules that formed the basis of early chemical knowledge. Clearly some elements were more similar in their properties and reactions than others; elements could thus be grouped or classified. This suggested that the atoms of these elements must be similar in structure, but the details of that structure, and even the size of the atoms, were completely unknown and seemingly inaccessible.

The Importance of Weighing

The birth of modern chemistry is usually dated by the work of the French scientist, Antoine Lavoisier (1743–94), who is often called the father of modern chemistry. Lavoisier's major contribution was the discovery of the conservation of mass in chemical reactions, and thus the establishment of the importance of weighing the reactants and products in the study of chemistry. That process has since become a ritual of almost any chemical experiment.

Although the idea that mass is conserved in chemical changes might seem self-evident now because of our familiarity with the idea, it was not at all clear in Lavoisier's time. The reason was simple; most chemical experiments were performed in open air, and oxygen and other gases in the air were involved in the reactions. Since the quantities of these reacting gases were not controlled or measured, the masses of the solid or liquid substances involved in the reactions usually did not seem to be conserved. Air itself is not a simple substance, and this fact was only beginning to be recognized at that time.

One of the most common of all chemical reactions, for example, is the burning, or *combustion,* of carbon compounds such as wood or coal (fig. 16.3). That reaction involves the combination of oxygen from the air with the carbon in the coal or wood to form carbon dioxide (a gas) and water vapor (also a gas). If we are not aware of the role of the gases, we can be easily misled to think that we have lost some mass in the course of the combustion.

Lavoisier performed a series of experiments in which he carefully controlled and weighed the quantities of gas that participated either as reactants or products. (Anyone who has ever worked with gases knows that this is not a simple process; leaks are notoriously difficult to detect or prevent.) The results of these experiments showed clearly that the total mass of the products was equal to the total initial mass of the reactants; no mass was lost or gained. In the process, Lavoisier was able to clearly distinguish oxygen (or "highly respirable air") from carbon dioxide and water vapor, and to

Figure 16.3 The burning of wood is a chemical reaction. What substances are reacting? What is produced?

provide the first accurate understanding of combustion reactions. The results of this work were published in 1789 as *Traite Elementaire de Chimie.*

Dalton and Atomic Weights

Although Lavoisier's brilliant career was tragically cut short (literally) when he lost his head to the guillotine in the French revolution, his leading work was followed shortly by another very important insight due to the English chemist, John Dalton (1766–1844). Dalton tried to make sense of some of the regularities that were observed in the ratios of the weights of chemical reactants and products. His ideas depended heavily upon the experimental work of other chemists, who were engaged in the new tradition, established by Lavoisier, of weighing carefully all of the reactants and products.

Dalton was intrigued by the fact that when chemical reactions took place, the reactants always seemed to combine in the same proportions by weight for a given reaction. For example, when carbon combined with oxygen to form the gas that we now call *carbon dioxide,* the ratio of the mass of oxygen to the mass of carbon that would react was always approximately 8 to 3. In other words, if 3 grams of carbon were involved, 8 grams of oxygen were required to complete the reaction, no more and no less. If more than 8 grams of oxygen were present, some would be left over; if less than 8 grams of oxygen were present, some carbon would be left over.

The same thing could be said for many other reactions, although the ratios were different than 8 to 3. In the case of hydrogen combining with oxygen to form water, the mass ratio is 8 to 1: 8 grams of oxygen are required to react completely with 1 gram of hydrogen. (This particular ratio was not accurately known by Dalton.) Each reaction had its own definite proportion of the mass of one substance that would react with another. This observation is often referred to as Dalton's *law of definite proportions.*

Although others had commented upon these facts, Dalton recognized that a model involving atoms could explain these observations. If we assume that each element is made up of tiny particles (or atoms) that are identical in mass and form, and that each element has a different atomic mass, a simple picture emerges. A chemical compound might then involve the combination of a few atoms of one element with a few atoms of another to form *molecules*—several atoms of different elements that are bound together somehow. The characteristic masses of the atoms would then be responsible for the regular proportions by mass that were observed in the reactions.

This idea is illustrated in figure 16.4. Suppose (as we now know) that the chemical compound of water is formed by two atoms of hydrogen combined with one atom of oxygen to form a water molecule (H_2O in modern notation). If the mass of an oxygen atom is 16 times that of a hydrogen atom, this would account for the observed proportion of oxygen to

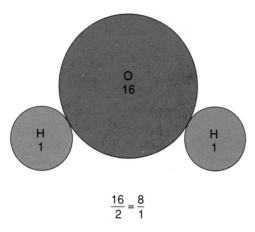

$$\frac{16}{2} = \frac{8}{1}$$

Figure 16.4 **Two atoms of hydrogen combine with one of oxygen to form water (H_2O). The mass ratio is 8 to 1 if the atomic mass of oxygen is 16 times that of hydrogen.**

Table 16.1		
The Relative Atomic Masses of Some Common Elements		
Element	**Chemical symbol**	**Relative atomic mass**
Hydrogen	H	1.01
Helium	He	4.00
Carbon	C	12.01
Nitrogen	N	14.01
Oxygen	O	16.00
Sodium	Na	22.99
Chlorine	Cl	35.45
Iron	Fe	55.85
Lead	Pb	207.2

hydrogen of 8 to 1 (16 to 2), since each molecule has two atoms of hydrogen for every one of oxygen. Similar pictures could be developed for any other reaction.

The 8 to 3 proportion for the reaction of oxygen and carbon can likewise be explained by assuming that the mass of a carbon atom is 12 times that of hydrogen, and that two atoms of oxygen (each one 16 times the mass of hydrogen) combine with one atom of carbon to form a carbon dioxide molecule (CO_2). The mass ratio would then be $2 \times 16 = 32$ parts of oxygen to 12 parts of carbon, or 32 to 12, which is equal to 8 to 3 (dividing each number by 4). One reaction by itself is not enough to establish the ratio of the atomic masses, but the study of several reactions can provide a self-consistent picture. This is what Dalton attempted to do in his treatise entitled *A New System of Chemical Philosophy* published in 1808.

It is important to recognize that Dalton's atomic hypothesis did not establish the actual weight or mass of individual atoms. We still did not know how large one atom was. What it did provide was a means of determining the *relative masses* of the atoms of one element to those of another. That task occupied chemists through much of the remainder of the nineteenth century. Like any good theory, Dalton's model was a very productive guide for chemical research. It also presented the concept of atoms in a new and more detailed light; some atoms were heavier than others, and *atomic mass* was a property of a given element.

Table 16.1 lists the relative masses of the atoms of several common elements, often referred to as *atomic weights*.* (A more complete list is found in the periodic table in the inside back cover of this text.) It is interesting to note that

---

* Given the distinction between the terms *weight* and *mass* introduced in chapter 4, *mass* is the more appropriate term. In chemistry, however, the term *atomic weight* has a long tradition.

many of the relative masses are approximately whole numbers, but that others are not. Therein lay another intriguing mystery and a clue, perhaps, to the structure of the atom.

The Periodic Table

As chemists gathered more and more information on the atomic weights and chemical properties of the various elements, some other interesting regularities began to emerge. Chemists had known for some time that families of elements exhibited similar chemical properties. Chlorine, fluorine, and bromine (called *halogens*), for example, formed similar compounds when combined with highly reactive metals such as sodium, potassium, or lithium (*alkali* metals). The atomic weights of the elements within one of these families were very different, however.

When all of the elements were listed in order of increasing atomic weight, however, the members of a given family seemed to pop up at more or less regular intervals in the list, particularly for the lighter elements. Although others had tried to make sense of these regularities, the person who succeeded in producing the most useful organization of this information was the Russian chemist Dimitry Mendeleyev (1834–1907). Mendeleyev's scheme is now called the periodic table of the elements; it was first published in 1869.

To understand Mendeleyev's table, we might first imagine that we have listed all of the known elements by increasing atomic weight on a long strip of paper. Then, to make the table, we cut the strip at various points and lay the strips out in rows. We begin by cutting at every place that we encounter an alkali metal in the list. We line these strips up so that the alkali metals are all in a column on the left side of the table (fig. 16.5).

In order to get the halogens to line up, we need to cut the remaining strips again, somewhere near their midpoints,

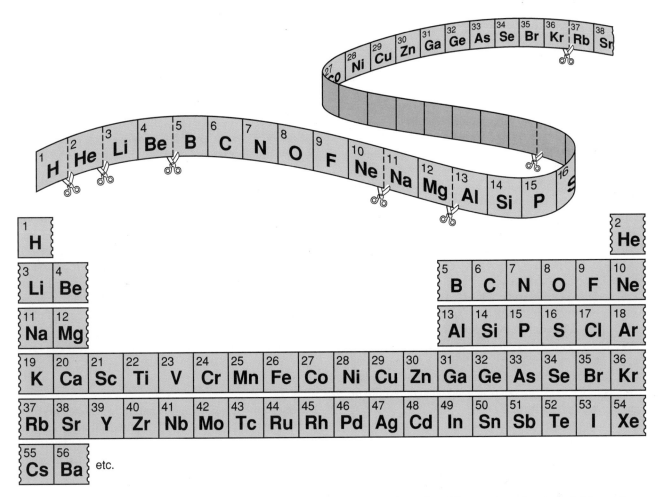

Figure 16.5 The periodic table can be formed by listing the elements in order of increasing atomic weight and then cutting the list at certain points. Elements with similar chemical properties are then aligned in columns.

because there are more elements in some rows than in others. The halogens are then arranged above one another in a column near the right side of the table. (In Mendeleyev's original table, the halogens formed the column on the far right, but that was because the noble-gas elements had not yet been discovered.) In the finished table, elements with common chemical properties line up in columns above one another, but the order of atomic weights is preserved in the rows and throughout the table. (See the complete periodic table on the inside back cover.)

Although the periodic table was an intriguing way of organizing knowledge of the chemical elements, it raised more questions than it answered. Atoms of the elements in a given row must in some way have similar properties, but chemists still knew virtually nothing about the structure of atoms. They were driven to drawing little hooks and rings on their atoms as they attempted to explain how they combined, but they knew that these pictures were probably not accurate. A body of knowledge was growing that begged explanation, but the explanation did not come until the early part of the twentieth century.

16.2 CATHODE RAYS, ELECTRONS, AND X RAYS

By the end of the nineteenth century, chemists were quite comfortable with the concept of atoms and knew a good deal about their relative masses and properties, if not their actual structure. Physicists, on the other hand, were less sure. Many physicists were not aware of the details of the chemical evidence, and some denied the need to assume the existence of atoms.

Physicists did use the concept of atoms, however, in explaining gas behavior. If a gas were pictured as a collection of randomly moving atoms or molecules, statistical techniques could be used to infer properties of the gas that were in good agreement with experimental evidence. For this purpose, atoms were pictured as small, hard spheres that could move about rapidly within a container. Their collisions with the walls of the container produced forces on the wall that could be used to calculate the pressure of the gas.

Near the end of the nineteenth century, however, a series of discoveries were made in physics that would ultimately prove crucial to the understanding of atomic structure. This part of the story begins with the study of *cathode rays,* which were the focus of much curiosity in the latter half of the century.

Cathode Rays

Cathode rays are actually quite common around us, although you may not be aware of it. The heart of a television set, the picture tube, is a cathode-ray tube (or *CRT,* as it is known in the electronics business). The discovery of cathode rays resulted from the coming together of two different technologies: the production of good vacuums with improved vacuum pumps, and improved understanding of electrical phenomena.

Johann Hittorf (1824–1914) was one of the first to describe cathode rays. In a paper published in 1869, he described in detail what happens when you place a high voltage across two electrodes sealed in a glass tube connected to a vacuum pump (fig. 16.6). As the air is pumped out of the tube, you first observe a glow discharge in the vicinity of the *cathode,* the negative electrode. As you continue to reduce the pressure of the gas in the tube, the glow spreads across the entire area between the two electrodes, producing a very colorful display. The colors of the glow discharge depend upon the nature of the gas that was originally in the tube.

As you continue to evacuate the tube to still lower pressures, the glow discharge disappears. What you actually observe is a dark region that starts to form in the vicinity of the cathode and then moves across the tube towards the anode as the pressure is reduced. When the dark region has moved completely across the tube, a new phenomenon appears. Instead of the gas glowing, there is now a faint glow on the glass wall of the tube on the end opposite the cathode. The color and nature of this glow depend somewhat upon the chemical composition of the glass.

Since the darkening began in the vicinity of the cathode and spread across the tube, it seemed natural to suspect that something was being emitted from the cathode that was responsible for the glow on the opposite wall of the tube. Thus, the invisible radiation was referred to as *cathode rays.* The nature of this invisible something was the subject of much curiosity and research during the remainder of the century.

One of the simpler experiments that can be performed with cathode rays is deflecting the beam with a magnet. If the cathode rays are focused into a narrow beam by appropriate shaping and positioning of the cathode and anode, the beam can be moved around with a magnet. If, for example, the north pole of a magnet is brought in from the top, as shown in figure 16.7, the spot of light created by the beam is deflected to the left on the face of the tube. This is consistent with the assumption that the cathode rays consist of negatively charged particles, as you can confirm by applying the right-hand rule introduced in chapter 13 for magnetic forces.

Discovery of the Electron

The questions surrounding the nature of cathode rays were largely resolved by the English physicist J. J. Thomson (1856–1940). Thomson performed a series of experiments designed to measure the masses of the presumed negatively charged particles making up the cathode-ray beam. One of his methods was to pass the beam through crossed electric and magnetic fields of known strength. The combined effect of the electric and magnetic fields on the beam allowed him to estimate the velocity of the particles, since the magnetic force depends upon velocity, but the electric force does not.

Knowing the velocity and the extent of deflection produced by the magnetic field, he could use Newton's second

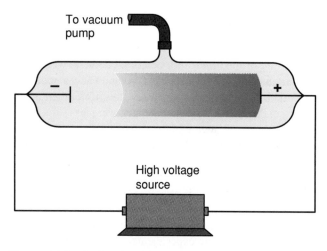

Figure 16.6 A simple cathode-ray tube consists of two electrodes sealed in a glass tube. A glow discharge appears between the electrodes as the tube is evacuated. With further evacuation, the discharge disappears, and a glow appears on the end of the tube opposite the cathode.

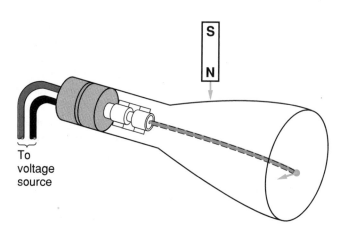

Figure 16.7 If the north pole of a magnet is brought down towards the top of a cathode-ray tube, the spot of light is deflected to the left across the face of the tube.

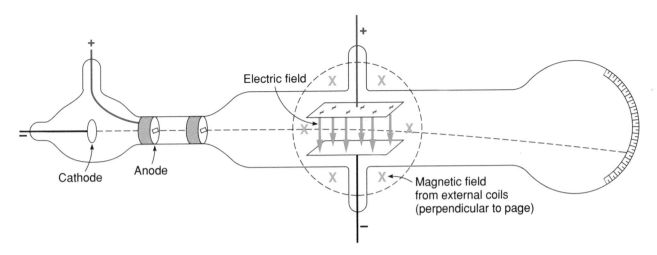

Figure 16.8   J. J. Thomson used both electric and magnetic fields to deflect the cathode-ray beam in a tube specially designed to measure the mass of the cathode-ray particles.

law of motion to estimate the mass of the particles (fig. 16.8): the acceleration of the particles by the magnetic force is inversely proportional to their mass. Since the magnetic force also depends upon the charge of the particles, which was not known, Thomson actually ended up measuring the ratio of the charge to the mass, q/m. He published the results of this work in 1897.

The striking features of his results were the apparently small mass of these particles and the fact that all of them seemed to have the same ratio of charge to mass, which suggested that the particles were all identical. The lightest element in the periodic table is hydrogen, which has an atomic weight of approximately 1.0. If Thomson assumed that the charges of hydrogen ions and cathode-ray particles had the same magnitude, then the mass of the cathode-ray particles was apparently only 1/1830 that of the hydrogen ion, or almost two thousand times smaller!

Not only were these particles identical for a given cathode, but the same charge-to-mass ratio was found even if the cathode were made of a different metal. The same particles seemed to be present, then, in all kinds of metal; Thomson checked this by repeating the experiment with cathodes made from various metals. This fact, along with their small mass, suggested that these particles must be common constituents of different types of atoms.

We now call the negatively charged particles of the cathode-ray beam *electrons,* and Thomson is credited with discovering the electron on the basis of these experiments. In other words, a cathode-ray beam is a beam of electrons, and each electron is now recognized to have a mass equal to 9.1×10^{-31} kg and a charge of -1.6×10^{-19} C. The significance of this discovery was that it defined the first known *subatomic* particle, a particle smaller than the smallest known atom. In the attempt to discover the structure of atoms, the electron became the first possible candidate for a building block of atoms.

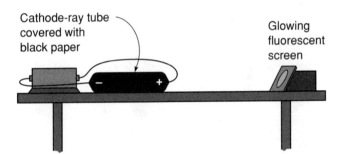

Figure 16.9   Roentgen noticed that a fluorescent material would glow when placed near his covered cathode-ray tube. The glow appeared only when the cathode-ray tube was turned on.

X Rays

The study of cathode rays produced other dividends in addition to the discovery of the electron. A German physicist, Wilhelm Roentgen (1845–1923), discovered another type of radiation associated with the cathode-ray tube. His discovery created a sensation in the popular press as well as within the scientific community

Roentgen's discovery was, as is often the case, partly an accident. For reasons that he never made completely clear, he was experimenting with a cathode-ray tube that he had covered with black paper. Nearby on his workbench he had a piece of paper coated with a fluorescent material, barium platinocyanide. He noticed that the paper glowed in a dark room when the cathode-ray tube was turned on, even though no light was escaping from the tube itself (fig. 16.9). The glow stopped when the tube was turned off.

It was already well known that cathode rays themselves could not travel very far in air; nor could they travel through the glass walls of the tube. The fluorescence appeared, however, even when the paper screen was located as far as two meters from the tube. The new radiation causing the fluorescence, then, could not be the cathode rays themselves. For lack of knowledge of their precise nature, he called them X rays, since the letter *X* is often used to represent an unknown quantity.

The most striking feature of these X rays was their penetrating power. They passed readily through the glass walls of the tube, apparently, but also through various other obstacles placed in their path. In some of his earliest experiments, Roentgen showed that he could produce a shadow of the bones in his hand by placing his hand between the end of the cathode-ray tube and his fluorescent screen (fig. 16.10). He also showed that the X rays were capable of exposing a covered photographic plate; he took pictures of the outlines of brass weights placed inside a wooden box. Roentgen published the results of his initial experiments with X rays in 1895.

This ability to "see" through objects that were opaque to visible light was what excited the imagination of the popular press. Everyone, scientists included, wanted to see an X-ray tube at work. Within a few years of the discovery, doctors were using X rays to take pictures of broken bones and other types of dense tissue. Unfortunately, they knew very little of the hazards of repeated exposure to X rays, so many doctors and dentists suffered from severe radiation effects in these early years of their use.

Roentgen performed an extensive series of experiments with his newly discovered radiation in an attempt to pin down its nature. As a result of his work, and that of other scientists, it was eventually decided that X rays were electromagnetic waves with very short wavelengths and very high frequencies. They were produced by collisions of the cathode rays (electrons) with the walls of the cathode-ray tube or with the anode of the tube. The strongest X-ray beams were produced by placing the metal anode at a 45° angle to the electron beam and using high voltages to excite the tube (fig. 16.11).

Although the discovery of X rays was highly significant because of the applications and nature of X rays themselves, its greatest significance in the story of how we learned about the structure of the atom was that it led directly to the discovery of yet another kind of radiation. This new type of radiation was called *natural radioactivity,* which actually consisted of three distinct forms of radiation. The next section describes how this discovery provided a powerful probe for getting inside of an atom.

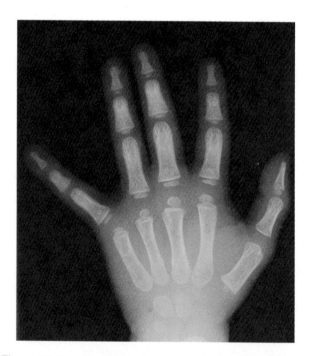

Figure 16.10 **Roentgen discovered that a shadowgram of the bones in the hand could be produced by passing X rays through the hand.**

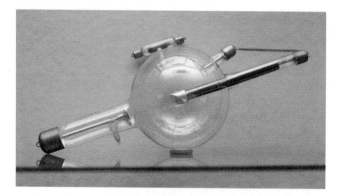

Figure 16.11 **An X-ray tube found in a diagnostic X-ray machine uses an angled anode to project X rays through the side of the tube.**

Box 16.1

Everyday Phenomena:
Electrons and Television

The Situation. Television plays a large role in modern life; it is the primary source of entertainment and news for a large portion of our population. For many of us, spending some time watching the dancing light of the "boob tube" is an everyday pursuit. Surely, then, we should have some idea of how this device works.

Could you explain the basic operation of a television set to a younger brother or sister? What is inside that picture tube? Although the details of its operation may be complex, the basic principles are quite simple. The discussion of cathode rays in section 16.2 provides the starting point.

The Analysis. The heart of most television sets is a cathode-ray tube, or *CRT*. (We now also have television sets that project images or use a liquid crystal display in place of the CRT.) The basic features of a cathode-ray tube were described in section 16.2; it consists of an evacuated tube containing electrodes across which a high voltage is placed. This produces a beam of electrons in the tube, which create flashes of light when they strike the glass wall of the tube.

A modern cathode-ray tube used in a television set, pictured in the drawing, is a more sophisticated version of the simple tube that we described earlier. The electrodes that produce and focus the electron beam are located near the tube socket on the left side of the diagram and are referred to as the *electron gun*. The gun contains a cathode (negatively charged), behind which lies a *filament*. The filament uses an electric current to heat the cathode, which causes a higher rate of emission of electrons.

Beyond the heated cathode in the electron gun lies an anode, which is positively charged and has a hole in its center. Electrons are accelerated from the cathode to the anode by the voltage difference placed across these two electrodes. Those that pass through the hole in the initial anode make up the electron beam. These electrons, in turn, are focused into a narrow beam by further electrodes (or *electron lenses*) located beyond the anode. (The screen itself also serves as an anode in most picture tubes, and is at a much higher voltage than the anodes in the electron gun.)

At the point where the narrow beam hits the screen, a bright spot of light is produced. This effect is enhanced by coating the inside of the front surface of

A family performing the common ritual of communing with a television set.

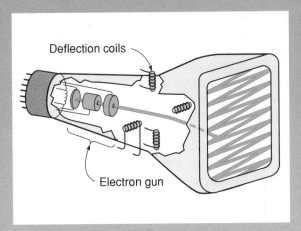

A cutaway view of a modern picture tube shows the electron gun and the magnetic coils used to deflect the beam to different points on the screen.

the tube with a special *phosphor,* a material that emits light when struck by fast-moving particles. Magnetic coils, usually arranged in a yoke that fits around the tube, are used to deflect the electron beam so that it strikes different points on the face of the tube, producing a pattern of dark and light spots.

The electron beam can be moved very rapidly from one point to another on the face of the tube, while the intensity of the beam is varied to produce varying degrees of brightness at different points. This is usu-

Continued

Box 16.1 Continued

ally done in a zigzag scan pattern that moves the beam back and forth across the face of the tube in a fraction of a second. In the system used in the United States, 525 horizontal scans are required to make one picture, and this process is repeated thirty times a second. The pattern of varying brightness at the different spots makes up the picture that we view.

The process just described would produce a black-and-white picture. In order to produce a colored picture, three different phosphors are used for three different colors. Each spot on the face of the tube is, in fact, three closely spaced spots or lines, and three different electron guns are used, one for each of the colors. Different combinations of these three colors produce the range of color that we perceive. If you look closely at the face of a color-television picture tube (with the set turned off!), you can see the pattern of vertical lines containing the three different phosphors.

The graininess of a television picture is determined largely by the size of the individual spots that are excited by the electron beam. Making these spots smaller and scanning the face of the tube more rapidly produces a higher-quality picture. High-resolution television is an emerging technology that may someday produce a revolution in the television industry. Conversion to a high-resolution system, however, would require television stations to broadcast signals containing larger amounts of information than present signals carry.

The information used to produce both the pictures and the sound that we perceive is carried to us by electromagnetic waves lying in the shorter-wavelength portion of the radio-wave spectrum. The signals can also be transmitted via cables or reflected from satellites (in microwave form) and picked up by dish antennas at remote locations. The availability of multiple stations and programming, and the range of technologies used to record and transmit the signals, would have seemed like an absurd dream to people just one hundred years earlier, when radio waves were discovered.

16.3 RADIOACTIVITY AND THE DISCOVERY OF THE NUCLEUS

Most of us have heard of the phenomenon of radioactivity, and perhaps have learned to fear it as a result of publicity surrounding nuclear power, nuclear weapons, and radon in our homes and buildings. For most of the time that humans have lived upon this earth, however, they were blissfully unaware of the presence of radioactivity. The story of its discovery, and how that led to the discovery of the nucleus of the atom, is a very important chapter in the history of science. The fields of atomic and nuclear physics, which did not exist prior to the beginning of the twentieth century, arose from these discoveries.

The Discovery of Radioactivity

Henri Becquerel (1852–1908), a French scientist, discovered natural radioactivity in 1896. His experiments were directly motivated by Roentgen's discovery of X rays the previous year. Becquerel and his father before him had been engaged for many years in the study of phosphorescent materials which glow in the dark after being exposed to visible or ultraviolet light.

Many of the phosphorescent materials that Becquerel was studying were compounds containing uranium, the heaviest element known at that time. Becquerel wondered whether penetrating radiation similar to Roentgen's X rays might be emitted by his phosphorescent compounds. Accordingly, he tried a simple experiment in which he exposed some of these compounds to sunlight for a while and then placed them on top of photographic plates that were wrapped in black paper so that no light could reach them. Sure enough, the photographic plates were exposed in the vicinity of the pieces of phosphorescent material (fig. 16.12). Some type of radiation was apparently passing from these materials through the black paper to expose the film.

Although this itself was an interesting discovery, there was more to come. Further experiments by Becquerel showed that not all phosphorescent materials could expose a photographic plate, only those that contained uranium or thorium, another heavy element, could do so. Furthermore, somewhat by accident, Becquerel discovered that it was not necessary to expose these materials to light in order to produce the effect.

Becquerel had prepared some samples that he intended to expose to sunlight. The sun was not shining in Paris on that particular day, however, so he put them away in a drawer

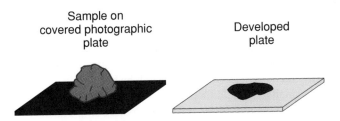

Figure 16.12 When Becquerel placed a sample of phosphorescent material on a covered photographic plate, the developed plate showed a silhouette of the sample, indicating that it had been exposed by rays passing through the black paper cover.

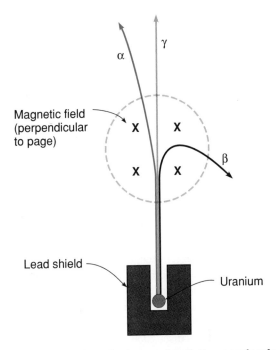

Figure 16.13 When the beam of radiation coming from a uranium sample passes through a magnetic field, it splits into three components, which were named α (alpha), β (beta), and γ (gamma).

for a few days together with the covered photographic plate. When he returned to this project several days later, he decided to develop that plate before proceeding, just to be safe, fully expecting that it would not be exposed. To his great surprise, he discovered that the plate was very heavily exposed in the vicinity of the uranium samples. Apparently prior exposure to sunlight was not necessary to produce the effect.

He was even more surprised to discover that the uranium samples retained the ability to expose film indefinitely, even if kept in a dark box or drawer for weeks. The longer the samples were left in the dark on top of a covered photographic plate, the heavier the exposure of the film. The phosphorescent effect, on the other hand, disappeared quite rapidly (in just a few minutes) after the samples were removed from the source of light. The penetrating radiation coming from the uranium samples did not seem to be connected with the phosphorescence at all.

Becquerel named this new radiation *natural radioactivity* because it seemed to be produced continuously by compounds containing uranium or thorium with no need for special preparation. It was a very puzzling phenomenon for many physicists at that time, because there was no apparent source of energy to produce the radiation. Where did these rays come from? How could they continue to be emitted when we were not adding energy to the samples in any obvious manner? Was this radiation somehow a property of the atoms themselves?

Three Types of Radiation

Along with the discovery of X rays, the discovery of natural radioactivity generated a lot of new experimental activity and theoretical speculation. Many scientists were involved, most notably the Curies, Marie and Pierre, and a young Australian-born physicist, Ernest Rutherford (1871–1937). The Curies, by painstaking chemical techniques, were able to isolate two more radioactive elements: radium and polonium. Both were contained in samples of uranium and thorium,

but they were much more radioactive than uranium or thorium themselves.

Rutherford became interested in the nature of the radiation itself. One of his earliest experiments with this new phenomenon established that at least three varieties of radiation came from uranium samples. By placing a uranium sample at the base of a hole drilled in a piece of lead, he could produce a beam of the radiation. When this beam was passed through a magnetic field produced by a strong magnet, it split into three components, as shown in figure 16.13.

Rutherford used the first three letters of the Greek alphabet α (alpha), β (beta), and γ (gamma) to name these three components. One of the components, alpha, was deviated slightly to the left, as shown in figure 16.13. This was the direction one would expect if the radiation consisted of positively charged particles. The location of the beam could be detected with photographic film or, more conveniently, with a zinc-sulfide screen, which produces flashes of light when struck by the beam.

Another component of the beam, the beta component, was bent more strongly in the opposite direction, as would be expected for negatively charged particles. Further study indicated that these beta rays were actually electrons, which had recently been discovered by J. J. Thomson. The final component, the gamma rays, were undeviated by the magnetic field. These turned out to be X rays, a very short-wavelength variety of electromagnetic wave.

The exact nature of the alpha rays remained a mystery, however, until it was clarified by an experiment performed by Rutherford and a student assistant, Royd. The fact that these rays or particles were deviated only slightly by the magnetic field indicated that they were much more massive than the electrons in the beta portion of the beam. They were also the primary component emitted by the radium obtained from uranium samples.

The experiment reported by Rutherford and Royd in 1908 established that these alpha rays were actually ionized atoms of helium, the second lightest element in the periodic table. They determined this by placing a small sample of radium inside a very thin-walled tube, which in turn was sealed inside of a somewhat larger tube. The alpha particles could pass through the thin-walled tube, but could not escape from the larger tube. The larger tube contained electrodes across which a high voltage could be introduced to produce a gas discharge in the alpha-particle gas that accumulated. The particular colors present in this discharge were characteristic of helium, which had not been present initially in the tube. (See section 16.4 for a discussion of atomic spectra.)

Discovery of the Nucleus

The clarification of the nature of the alpha radiation was important because Rutherford quickly realized that alpha particles would make effective probes for studying the structure of the atom itself. Because they were much more massive than electrons, and also highly energetic, it might be possible to get them inside of an atom. By firing a beam of alpha particles at a thin metal foil, for example, and noting what happens to the beam, he might be able to deduce features of atomic structure. Such an experiment is called a *scattering* experiment.

The basic scheme of Rutherford's scattering experiments

is illustrated in figure 16.14. An alpha-emitting substance such as radium or polonium is placed at the base of a hole in a lead shield in order to produce a beam of alpha particles. This beam is directed at a very thin foil of gold or some other metal. The scattering of the alpha particles is detected by a small hand-held scope with a zinc-sulfide screen at one end and a magnifying eyepiece at the other end. The experimenter counts the flashes of light (scintillations) produced by alpha particles striking the screen at various angles from the initial direction of the beam.

The initial results of these experiments were not very surprising nor informative. Most of the alpha particles went straight through the gold foil without deviating very much. A few were scattered through larger angles, but the number fell off rapidly as the angle from the initial direction of the beam was increased. These results seemed to be consistent with the prevailing view of the atom at that time, in which the mass and positive charge of the atom were pictured as being distributed uniformly throughout the volume of the atom. Electrons, which were known to be present in atoms, were thought to be located here and there within this volume, much like raisins in a plum pudding. Such an arrangement was not thought to be dense enough to have much effect upon a beam of energetic alpha particles.

Just to be sure, however, Rutherford suggested to one of his students, an undergraduate named Marsden, that he try looking for scattered alpha particles on the other side of the foil, the side from which the beam originated. After a few days in the dark lab, squinting at occasional flashes of light seen through the detecting scope, Marsden reported to Rutherford that there were indeed a few alpha particles scattered at these much larger angles. Rutherford was tempted not to believe him. Further checking by Marsden and a more senior research associate, Hans Geiger, verified their presence, however.

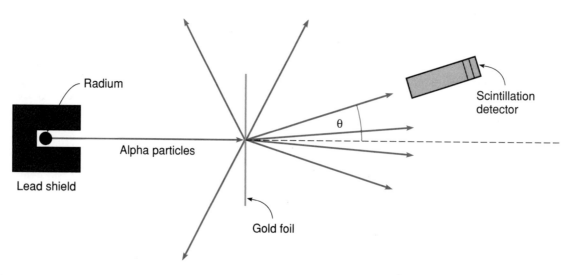

Figure 16.14   A beam of alpha particles is scattered from a thin gold foil in the scattering experiment performed by Rutherford's assistants.

Figure 16.15 The contents of a Christmas present could be probed by firing a rifle into it and noting how the bullets were scattered by the contents.

Much later, Rutherford said that it was as if someone had fired bullets into a piece of tissue paper and the bullets came bouncing back. It was a totally unexpected result. The analogy that is often used to explain the interpretation of this scattering experiment is illustrated in figure 16.15. In this case we are attempting to learn what is inside of a Christmas present that comes bearing the label "Do not open until Christmas." We may lift and shake the package to gain some sense of its weight and nature. Another (somewhat more destructive) test could be made by firing a .22-caliber rifle at the package and noting what happens to the bullets as they emerge (fig. 16.15). This is a scattering experiment, similar in concept to that performed by Rutherford and his assistants.

If we had already determined that the package was not very heavy, we would be very surprised to find some of the bullets, even just a few, coming back toward us. This would indicate the presence somewhere in the package of some small but dense objects with enough mass to reverse the momentum of a rapidly moving bullet. Since we know that the package is not very heavy and that many of the bullets go right on through, the objects responsible for the large-angle scattering of a few bullets must be quite small. Small steel balls held within a porous packing material might do the job.

The same type of reasoning applies to the atom. If most of the alpha particles go on through, but a few are scattered through large angles, there must be a few very dense but small centers somewhere within the atoms, massive enough to reverse the momentum of the rapidly moving alpha particles. In order to explain the quantitative results of the scattering experiment, Rutherford had to assume that these massive centers were very small indeed. By this time individual atoms were known to be approximately 10^{-10} m in diameter. These massive centers had to have a diameter

roughly ten thousand times smaller than that in order to explain the data!

The analysis of these scattering experiments led to the discovery of the nucleus of the atom. The nucleus was presumed to be a very dense, positively charged center of the atom, containing most of its mass and all of its positive charge. The rest of the atom consisted of the negatively charged electrons arranged somehow around this center. The electrons were responsible for most of the size of the atom, but for very little of its mass. To get a sense of the scale of things, imagine an atom enlarged to the size of a football field (roughly 100 m, counting the end zones). The nucleus would then be approximately the size of a pea placed on the 50-yard line.

Rutherford's analysis of the alpha particle scattering experiments performed by Geiger and Marsden was published in 1911. The idea that the atom has a tiny nucleus, containing most of its mass and all of its positive charge, presented a radical new view of the atom. The tools were now at hand to develop a model of the atom that might explain its chemical properties and other features. The first successful model, reported by Niels Bohr in 1913, is described in the next section.

16.4 ATOMIC SPECTRA AND THE BOHR MODEL OF THE ATOM

If the atom has a positively-charged nucleus, and electrons (with their negative charges) are arranged somehow around this nucleus, it is natural to think that the atom might be analogous to the solar system. In the solar system, the planets are held in orbit about the sun by the gravitational force, which is proportional to the inverse of the square of the distance between the planets and the sun, $1/r^2$. (see chapter 5.) In the case of the atom, the electrons would be attracted to the nucleus by the electrostatic force, which by Coulomb's law is also proportional to the inverse square of the distance. Perhaps the atom is just a miniature solar system.

Although this idea was intriguing, there were some problems with it, not the least of which was the fact that an orbiting or oscillating electron was expected to act like a transmitting antenna and radiate electromagnetic waves. Thus the atom would lose energy and quickly collapse, according to what was understood in the early 1900s of the behavior of charged particles. Physicists were aware, however, that under some circumstances atoms did emit electromagnetic waves in the form of light. The patterns of the light emitted by the smallest atom, hydrogen, were particularly interesting because of their simplicity.

Niels Bohr (1885–1962) was working with Rutherford at the time that the nucleus was discovered. It was his model of the atom that first suggested answers to these problems and provided an explanation of the wavelengths of light (the *spectrum*) emitted by hydrogen. The publication of the Bohr model of the atom in 1913 opened a tremendously exciting period of research that culminated in our current understanding of atomic structure.

Figure 16.16 A high voltage placed across the electrodes of a gas-discharge tube produces a colorful glow. The colors produced are characteristic of the type of gas in the tube.

The Hydrogen Spectrum

The study of the light emitted by different substances had begun over fifty years before Bohr's work. If different substances were heated in the flame of a Bunsen burner and the emitted light observed through a prism, they produced characteristic colors or wavelengths. These characteristic wavelengths constitute the *spectrum* of that particular substance. For gases, the most convenient way of producing this spectrum was in a gas-discharge tube.

We encountered the phenomenon of a gas discharge in section 16.2 when we discussed cathode rays. When a high voltage is placed across electrodes sealed inside of a tube containing a gas at low pressure, a colorful discharge is observed (fig. 16.16). This is what is happening in a fluorescent light, although in that case the tube is given a fluorescent coating in order to produce a more uniform distribution of wavelengths.

If we observe the light emitted by a gas discharge through a prism or diffraction grating, we see that the spectrum consists of a series of specific, discrete wavelengths. (A diffraction grating uses interference phenomena to separate wavelengths, and also provides an accurate means of measuring wavelengths.) If the source itself is long and thin, as pictured in figure 16.16, these wavelengths show up as colored lines. Each gas has its own wavelengths or lines, which can be used as a very reliable means of identifying the particular substance.

The spectrum of hydrogen, as mentioned earlier, is particularly simple. The visible portion consists of just four

lines; a red line, a blue-green line, and two violet lines, one of which is quite difficult to see (fig. 16.17). In 1884, a Swiss teacher, J. J. Balmer (1825–1898), discovered that the wavelengths of these four lines could all be computed from a simple formula:

$$\lambda = (364.56 \text{ nm}) \, \frac{m^2}{m^2 - 4}$$

where m was just an integer having the values 3, 4, 5, and 6 for each of the four observed lines. (No underlying theory suggested these numbers; Balmer simply found that they worked to generate the observed wavelengths.)

For example, if we let $m = 3$, we get

$$\lambda = (364.56 \text{ nm}) \times \frac{9}{(9 - 4)} = \left(\frac{9}{5}\right)(364.56 \text{ nm}) = 656.2 \text{ nm}.$$

This is the measured wavelength of the red line in the hydrogen spectrum. Letting $m = 4$ produces $\lambda = 486.1$ nm, which is the measured wavelength of the blue-green line, and so on. (A nanometer [nm] is equal to 10^{-9} m.) This simple formula generated the observed wavelengths of the hydrogen spectrum with surprising accuracy.

Balmer's formula was not based upon any underlying theory; it was merely a numerical means of computing the wavelengths of the observed lines. Shortly after he developed the formula, however, Balmer learned of the detection and measurement of additional lines in the hydrogen spectrum by other scientists. These lines were in the ultraviolet portion of the spectrum, so they were not visible, but they

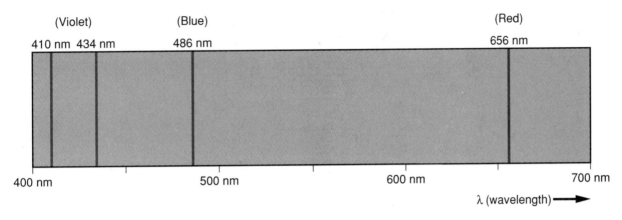

(Violet)   (Blue)   (Red)

410 nm   434 nm   486 nm   656 nm

400 nm   500 nm   600 nm   700 nm

λ (wavelength) ⟶

Figure 16.17   **The hydrogen gas-discharge spectrum shows four lines in the visible portion of the spectrum: a red line, a blue-green line, and two violet lines.**

could be recorded on photographic film. His formula accurately predicted the positions of these lines as well.

Balmer also noted that the number 4 appearing in the formula was the square of 2, which suggested that the formula might be written more generally by substituting n^2 for 4, where n is also an integer. However, in Balmer's day, no spectral lines were known for situations in which n was anything but 2. Much later, other spectral series were discovered for hydrogen that fit a modified form of Balmer's formula published in 1908 by Rydberg and Ritz. This formula is usually written as follows:

$$\frac{1}{\lambda} = R \left(\frac{1}{n^2} - \frac{1}{m^2} \right),$$

where n and m are both integers, and R is called the *Rydberg constant:* $R = 1.097 \times 10^7 \ \text{m}^{-1}$. For $n = 1$, we get a series in the more distant ultraviolet portion of the spectrum, and for $n = 3$ or 4, we get lines in the infrared. None of these are visible to the unaided eye, of course.

Balmer's formula pointed out a simple regularity in the spectrum of hydrogen that cried out for explanation. J. J. Thomson, in working with his so-called plum-pudding model of the atom, had attempted to explain this regularity without success. Bohr, working with the new ideas provided by Rutherford, took a fresh approach to the problem.

Quantization of Light Energy

Although the discovery of the nucleus and the regularities in the spectrum of hydrogen played crucial roles in Bohr's model, another new idea was at least as important. It was introduced tentatively by Max Planck (1858–1947) in the year 1900, and was later strengthened by Albert Einstein (1879–1955). This idea also originated in the study of spectra, in this case the spectrum produced by a heated *blackbody.*

A blackbody is nothing more than a hole or cavity carved into a metal or ceramic material that can be heated to high

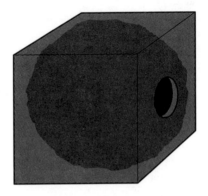

Figure 16.18   **A blackbody radiator consists of a hole carved in some material that can be heated to high temperatures. When heated, it emits a continuous spectrum of electromagnetic radiation.**

temperatures (fig. 16.18). Such a hole appears black, and the spectrum that it emits when heated turns out not to depend upon the material in which the cavity is carved, but merely upon the temperature. The spectrum emitted is continuous (no discrete lines), but the average wavelength emitted becomes shorter as the temperature is increased. At high enough temperatures, the emitted light becomes visible; it first appears "red hot," and at even higher temperatures it appears "white hot," meaning that the average wavelength is near the middle of the visible spectrum.

Planck and other theorists were engaged in an attempt to explain the distribution of wavelengths that emerged from a heated blackbody. Planck arrived at a formula that succeeded in predicting the proper distribution and its dependence upon temperature. However, in attempting to provide a rationale for his formula, he was forced to a radical conclusion. This was the idea that light could not be absorbed or emitted from the surface of a blackbody in continuously varying energies, but only in discrete chunks, or *quanta,* whose energy depended upon the frequency or wavelength.

To be more precise, at a given frequency (which is related to the wavelength by $f = c / \lambda$, as discussed in chapter 14), the only energies allowed come in packages having the size

$$E = hf,$$

where f is the frequency, and h is a constant called *Planck's constant*. The value of this constant is extremely small; in metric units it is

$$h = 6.626 \times 10^{-34} \text{ J·s.}$$

According to Planck's theory, for a particular frequency, f, light could be emitted with an energy of hf, or $2hf$, $3hf$, and so on if more than one quanta were emitted, but not at any energy between these values. Making this assumption allowed him to provide a theoretical justification for his formula, which very successfully predicted the distribution of blackbody radiation.

This idea was disturbing to Planck himself, as well as other physicists at that time. There had previously been no reason to believe that light waves could not be emitted in continuously varying energies, depending only upon how much energy was available. The idea that this process was restricted to discrete energy chunks was indeed radical. In 1905, however, Einstein showed that this concept could be used to explain a number of other phenomena. The idea of light quanta (or particles of light, which we now call *photons*) having energies $E = hf$, was thus available, if not fully accepted, at the time of Bohr's attempt to develop a new model of the atom.

The Bohr Model

Bohr's accomplishment was to take all of these ideas—the discovery of the nucleus, knowledge of the electron, the regularities in the hydrogen spectrum, and the new quantum ideas of Planck and Einstein—and to combine them in a new model of the atom. He started with the planetary model mentioned at the beginning of this section, in which the electron in the hydrogen atom is pictured as orbiting about the nucleus. The electrostatic force provides the necessary centripetal acceleration.

His first bold step was to assume that, for some reason, there were certain stable orbits that did not radiate electromagnetic waves continuously as expected from classical physics. Instead, he imagined that light was emitted from the atom in quantum jumps from one stable orbit to another (fig. 16.19). Since the energy of a quantum of light or photon, as given by Planck and Einstein, was $E = hf$, the energy of the emitted photon would be equal to the difference in energies of the two stable (or quasi-stable) orbits, and this energy could be used to compute the frequency or wavelength of the emitted photon. For a given radius of orbit, these energies could be calculated from ordinary Newtonian mechanics.

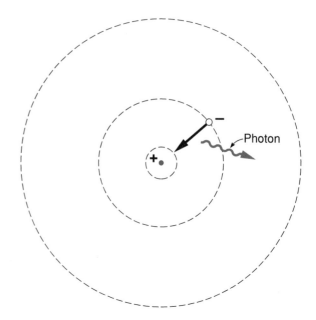

Figure 16.19 **Bohr pictured the electron as orbiting the nucleus in certain quasi-stable orbits. Light is emitted when the electron jumps from one orbit to another.**

Putting these ideas in symbolic form, we can represent the energy of the emitted photon as follows:

$$E = hf = \text{E}_{\text{initial}} - \text{E}_{\text{final}},$$

where the latter two energies are those of the initial and final stable orbits. This also allows us to compute the wavelength of the emitted photon, because $v = c = f\lambda$, or

$$\frac{c}{\lambda} = f,$$

where c is the velocity of light. Combining these two equations gives us an expression of the following form:

$$\frac{1}{\lambda} = \left(\frac{1}{hc} \right)(\text{E}_{\text{initial}} - \text{E}_{\text{final}}).$$

A comparison of this formula to the Rydberg-Ritz formula for the lines in the hydrogen spectrum suggested that the energies of the stable orbits must all be given by some constant divided by an integer squared, $E = E_1 / n^2$, where E_1 is the energy value for $n = 1$. Further comparisons to what classical physics would predict for the frequencies emitted by an orbiting electron led Bohr to the surprising conclusion that the condition for a stable orbit could be stated even more simply in terms of the angular momentum. The only orbits allowed were those for which the angular momentum, L, was given by the following expression:

$$L = n\left(\frac{h}{2\pi} \right)$$

where n is an integer and h is, once again, Planck's constant.

Although the mathematical details of computing the energies have been omitted, these are the essential features of the Bohr model:

1. Electrons were pictured as orbiting the nucleus in certain stable orbits, given by the condition $L = n (h / 2\pi)$.

2. Light was emitted when an electron jumped from one stable orbit to another.

3. The frequencies and wavelengths of the emitted light could be computed from the energy differences between the two orbits, yielding, for example, the observed wavelengths in the hydrogen spectrum (see the example exercise in box 16.2).* The energy values for the first few stable orbits are shown in figure 16.20.

One of the most striking successes of the Bohr theory was that it could predict the correct value of the Rydberg constant from quantities such as the mass of the electron, the charge of the electron, Planck's constant, and the speed of light. Bohr's model was an instant sensation in the physics commu-

nity; its introduction produced intense activity in both experimental and theoretical physics. Much of the experimental work was focused upon making more accurate measurements of the atomic spectra of different elements. The theoretical work was concerned with extending Bohr's model to atoms other than hydrogen and attempts to understand the periodic properties observed in the periodic table of the elements.

Despite its impressive successes, the Bohr model also left many unanswered questions, the most bothersome of which was just why these few stable orbits, described by the Bohr condition for angular momentum, should exist. Also, the attempts to extend the Bohr model to elements other than hydrogen met with only limited success. Physicists now recognize that the Bohr model was inaccurate in many of its details. Its real significance was that it opened the door to research that ultimately produced our modern theory of the atom. The development of this theory, *quantum mechanics*, is briefly described in the next section.

---

* The energy levels indicated in figure 16.20 and in box 16.2 are expressed in units of electron-volts rather than joules. An *electron-volt* (eV) is the amount of kinetic energy gained when an electron is accelerated through an electric potential difference of 1 volt. Since the kinetic energy gained is equal to the potential energy lost,

$$\Delta KE = \Delta PE = q\Delta V = (1.6 \times 10^{-19} \text{ C})(1\text{V}) = 1.6 \times 10^{-19} \text{ J}.$$

Thus 1 eV $= 1.6 \times 10^{-19}$ J; this unit is often used for energies involving atomic phenomena because of its convenient size. The energy levels for the hydrogen atom are all negative due to the negative potential energy associated with a negative charge in the vicinity of a positive charge.

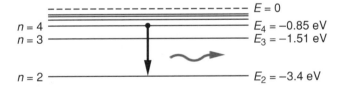

Figure 16.20   The energies for the different quasi-stable orbits are shown in an energy-level diagram. The blue-green Balmer line is produced by the jump indicated here.

Box 16.2

Sample Exercise

Using the energy values shown in figure 16.20, calculate the wavelength of the photon that is emitted in the transition from the $n = 4$ energy level to the $n = 2$ energy level in the Bohr model of the atom.

$E_2 = -3.4$ eV.

$E_4 = -0.85$ eV.

$\lambda = ?$

The energy difference between these two levels is

$$\Delta E = E_4 - E_2 = -0.85 \text{ eV} - (-3.4 \text{ eV})$$

$$= 3.4 \text{ eV} - 0.85 \text{ eV} = 2.55 \text{ eV}.$$

This is the energy of the emitted photon. Converted to joules, the photon energy is

$$E = \Delta E = (2.55 \text{ eV})(1.6 \times 10^{-19} \text{ J/eV}) = 4.08 \times 10^{-19} \text{ J}.$$

We find the wavelength, λ, from the following expression:

$$E = hf = \frac{hc}{\lambda}.$$

Multiplying both sides by λ and dividing by E yields

$$E\lambda = hc$$

$$\lambda = \frac{hc}{E}$$

$$= \frac{(6.626 \times 10^{-34} \text{ J·s}) (3 \times 10^8 \text{ m/s})}{(4.08 \times 10^{-19} \text{ J})}$$

$$= 4.87 \times 10^{-7} \text{ m} = \textbf{487 nm}.$$

16.5 PARTICLE WAVES AND QUANTUM MECHANICS

The intense activity and the unanswered questions generated by Bohr's model of the atom attracted many young physicists to the field of atomic physics. There was an obvious need for a more comprehensive model of the atom that might explain why only certain orbits were stable. That need was filled when quantum mechanics was developed in 1925. It was actually developed from two independent approaches, which were quickly shown to be fundamentally the same in their structure and predictions.

The approach that is usually described followed from the work of Louis de Broglie (1892–1987) and Erwin Schrödinger (1887–1961). De Broglie lit the spark by asking a simple but radical question: If light waves sometimes behave like particles (as shown by Planck and Einstein), could not particles sometimes behave like waves? The ideas that followed from that question produced a revolution in our thinking about basic physical principles.

De Broglie Waves

The question posed by de Broglie was inspired by the concept of the photon introduced by Planck and Einstein. The work of Maxwell in 1865 had shown that light could be described as an electromagnetic wave. On the other hand, light seemed to behave in some situations as though it were made up of discrete and localized particle-like bundles of energy, which we now call *photons*. Certain experiments involving the interaction of light with other matter were most simply explained by using this particle-like view of light.

Einstein was most instrumental in pointing out this aspect of light. His 1905 paper, mentioned earlier, discussed a number of phenomena that could be treated in this way, the simplest of which is the photoelectric effect—a phenomenon in which light shining on an electrode in an evacuated tube causes an electric current to flow across the tube. This effect has often been used in producing electric-eye devices for opening doors when a person interrupts the light beam, or for similar applications.

Einstein showed that this effect could be simply explained if we assumed that one quantum, or photon, of light, having energy $E = hf$, as suggested by Planck's work, could eject one electron upon hitting the electrode. This simple assumption could explain the observed frequency-dependence of the photoelectric effect, as well as other features. Similar experiments could also be treated in this manner by attributing an energy, $E = hf$, and a momentum, $p = h / \lambda$, to the photon.

Although this idea was simple, the physics community was slow to accept it because particles and waves were thought to be very different phenomena, and it was difficult to understand how light could behave as a particle in some respects and a wave in others. An ideal wave extends infinitely in space, whereas an ideal particle is completely

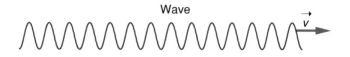

Wave

Particle

Figure 16.21   An ideal wave goes on indefinitely, whereas an ideal particle is just a point with no volume or extension in space.

localized, a simple point in space (fig. 16.21). Real waves, of course, have a finite length, and real particles have some extension in space, but the concepts are still very different.

De Broglie suggested that perhaps things that we had traditionally thought of as particles, such as the electron, could in some circumstances behave like waves. In particular, he suggested that we merely reverse the relationships just mentioned that describe the energy and momentum of a photon in order to find the frequency and wavelength associated with a particle. Inverting the energy relationship yields

$$E = hf \;\Rightarrow\; f = \frac{E}{h}.$$

Inverting the momentum relationship yields

$$p = \frac{h}{\lambda} \;\Rightarrow\; \lambda = \frac{h}{p},$$

where p is the momentum, and h is Planck's constant. If we knew the energy and momentum of an electron, for example, we could compute a frequency and wavelength from these relationships.

This suggestion by de Broglie might have passed unnoticed if it were not for a striking result that he obtained from the idea. If he treated the electron as a wavelike entity orbiting the nucleus of the hydrogen atom, he could explain the condition for stable orbits in Bohr's atomic model by picturing a standing wave wrapping around the circular orbit. This idea is illustrated in figure 16.22.

In order to form a circular standing wave, the wavelength would have to be restricted to values such that an integer number of wavelengths would fit onto the circumference of the circle, $2\pi r$, where r is the radius of the orbit. In symbols, we have

$$n\lambda = 2\,\pi r,$$

where n is an integer. However, by the de Broglie relationship, the wavelength, λ, should be equal to h / p, where p represents momentum. Substituting this expression for λ and using algebra to rearrange the equation yields:

$$L = pr = n \left(\frac{h}{2\pi} \right),$$

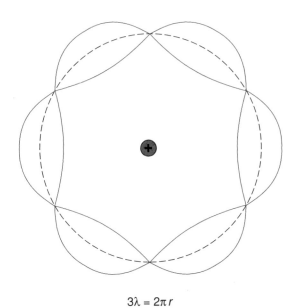

$$3\lambda = 2\pi r$$

Figure 16.22 If the electron wave was pictured as a standing wave wrapped around a circular orbit, DeBroglie showed that its wavelength could take on only certain values. These values yield the stable orbits predicted by Bohr.

which is the Bohr condition. (The angular momentum, L, is equal to the radius, r, times the ordinary momentum, p, for a particle moving in a circle.)

In other words, by assuming that particles had wavelike properties, and by picturing a standing particle-wave wrapping around a circular orbit, de Broglie could explain why only certain orbits might be stable. This was one of the fundamental questions left unanswered in Bohr's theory. Although the suggestion was radical, it demanded attention.

De Broglie's picture of a standing wave on a circular orbit should not be taken too literally. In fact, both the Bohr model and this standing-wave explanation predict the wrong value for the angular momentum of the various stable states in the hydrogen atom. The basic problem is that this picture is two-dimensional, and the atom itself is three-dimensional. We need a more sophisticated analysis to picture the standing waves properly.

The suggestion that particles had wavelike properties, however, was quickly borne out by experiment. It was known that X rays, which are electromagnetic waves, could be diffracted by a crystal lattice to form interference patterns that were characteristic of the crystal structure. Various workers soon showed that this could also be done with an electron beam. The interference patterns that resulted looked just like those obtained with X rays for the same crystal, and the wavelengths needed to explain these patterns were exactly those predicted by de Broglie's relationship, $\lambda = h/p$.

Schrödinger's Quantum Mechanics

Erwin Schrödinger had spent much of his professional life studying the mathematics of standing waves in two and three dimensions. He was therefore well prepared at the time of de Broglie's particle-wave hypothesis to explore the implications of standing electron waves in the atom. In the year following de Broglie's suggestion, Schrödinger developed a theory of the atom that utilized three-dimensional standing waves to describe the orbits of the electron about the nucleus.

Within the next five years, Schrödinger and other scientists pursuing the same problem from different approaches worked out the details of the theory that we now call *quantum mechanics*. This new theory gave a much more complete and satisfactory view of the hydrogen atom than that provided by the Bohr model. It predicted the same primary energy levels for the different orbits as the Bohr model, however, and therefore retained the picture established by Bohr of how the basic features of the hydrogen spectrum are produced.

The mathematics of quantum mechanics are quite complex, and the orbits are not simple curves as pictured in the Bohr model. Instead, they consist of three-dimensional probability distributions centered upon the nucleus. These probability distributions are predicted by the standing-wave features of the electrons, as already suggested. They describe the probability of finding electrons at various distances and orientations about the nucleus.

The probability distributions for a few stable orbits of the hydrogen atom appear in figure 16.23. The places where the electron is most likely to be found are represented by the darker or denser areas in the drawings. The average distances of the electron from the nucleus for different stable orbits are consistent with the orbital radii given by the Bohr model.

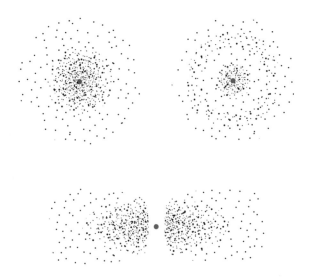

Figure 16.23 The probability of finding the electron at different distances from the nucleus is represented by these particle-density diagrams of a few of the stable hydrogen orbits predicted by quantum mechanics.

The Uncertainty Principle

The fact that we are dealing with probability distributions rather than well-defined orbital paths is a fundamental and necessary feature of quantum mechanics. The waves associated with electrons and other particles define a probability of finding the electron at various positions, but they cannot tell us exactly where the particle is located. In a similar manner, electromagnetic waves give us a probability of finding photons at various positions. In situations where the wave properties are dominant, we lose information about precise particle locations.

This limitation on what we can know about a particle's location is summarized in the famous *Heisenberg uncertainty principle*. This principle states that the position and momentum of a particle cannot both be known simultaneously with high precision. There will be an uncertainty in one that depends upon how precisely we have determined the other. In symbols, this limit takes the following form:

$$\Delta p \, \Delta x \geq h,$$

where h is Planck's constant, Δp is the uncertainty in the momentum, and Δx is the uncertainty in position. If the uncertainty in position is small, the uncertainty in momentum must be large, and vice versa.

Since momentum is related to the particle wavelength by the de Broglie relationship $\lambda = h / p$, what we are really saying is that if we know the wavelength accurately, we cannot know the position of the particle accurately. The converse is also true: If we know the position accurately, we cannot know the wavelength accurately. Thus some experiments tend to bring out the particle-like aspects of photons or electrons, whereas others tend to bring out the wavelike features.

Werner Heisenberg (1901–1976) was one of the people responsible for the development of quantum mechanics. His uncertainty principle is regarded as a fundamental limitation on what we can observe. It is not merely that we lack experimental capability; this limitation can be regarded as an inevitable feature of wave pulses. If we attempt to localize a wave by creating a brief pulse, the wavelength cannot be accurately defined. On the other hand, an extended wave, which permits accurate definition of the wavelength, gives us no precise information regarding position.

Atomic Structure

Quantum mechanics provides us with a means of answering most of the questions raised by the Bohr model regarding atomic structure. In particular, it is successful in predicting the structure and spectra of atoms with many electrons, although the computations are difficult. The Bohr model had managed to predict the spectra of single-electron atoms such as hydrogen or singly ionized helium. (Helium normally has two electrons, but if it is ionized by removing one electron, it becomes a single-electron atom.)

The picture that emerges from quantum mechanics had already been partially assembled from attempts to explain the regularities in the periodic table of the elements. The theory provides us with quantum numbers that describe the various possible stable orbits. One of these is n, the principle quantum number that was needed to compute the energies in the Bohr model. Quantum mechanics provides three others, however, associated with the magnitude and orientation of the angular momentum and the spin of the electron.

No two electrons in a particular atom can have the same set of quantum numbers. Thus, once a given orbit is filled, other electrons must take on new, and generally higher, values of at least one of the quantum numbers. The number of possible combinations increases rapidly as the principle quantum number, n, increases. Thus for $n = 1$, there are only two possible combinations corresponding to two different orientations of the electron-spin axis, but for $n = 2$, there are eight and for $n = 3$, there are eighteen and so on. These various possibilities involve different combinations of the orbital angular momentum magnitude and direction, each of which has two possible spin orientations. Once the two possible states for $n = 1$ are filled, the next electron added must go into an $n = 2$ level, or *shell.*

From this process, we can explain the regularities in the periodic table. The first two elements, hydrogen and helium, have one and two electrons respectively. Two electrons fill the $n = 1$ shell; thus the next element, lithium, which has three electrons, must have its third electron in the $n = 2$ shell. Since lithium has one electron beyond the closed-shell $n = 1$ orbit, its chemical properties are quite similar to those of hydrogen, which has just one electron. Likewise, the next element in that column of the periodic table, sodium, has one electron beyond the $n = 2$ closed shell. Its other ten electrons fill the $n = 1$ and $n = 2$ levels: two in the first shell and eight in the second shell. Figure 16.24 shows a schematic representation of this shell structure for a few atoms.

The element immediately preceding sodium in the periodic table is neon, which has ten electrons, two in the $n = 1$ shell and eight in the $n = 2$ shell. Like helium, therefore, it has a closed-shell arrangement and does not react very readily at all with other elements. Helium and neon are both *noble gases,* which are very nonreactive chemically. Fluorine, however, with nine electrons, one short of a closed shell, is very reactive. It forms compounds with elements such as hydrogen and sodium, which can contribute an electron to form a closed-shell arrangement.

Although the details become more complicated with higher numbers of electrons, the principles used in explaining the entire periodic table are the same as those just outlined. Not only does the theory explain the regularities of the periodic table, but it is highly successful in predicting specific features of the ways different elements combine to form chemical compounds. Thus quantum mechanics has become the fundamental theory of chemistry as well as of atomic, nuclear, and solid-state physics.

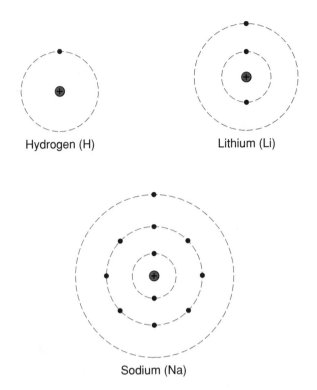

Hydrogen (H)   Lithium (Li)

Sodium (Na)

Figure 16.24 **The chemical properties of sodium, with one electron in the *n* = 3 shell, are similar to those of hydrogen and lithium, which also have a single electron in their outermost shell.**

Since the year 1900, we have gone from a state of knowing nothing at all about atomic structure to a state of having very detailed knowledge of that structure, and of how atoms interact with one another. This revolution began with the discoveries outlined near the beginning of this chapter and culminated in the development of quantum mechanics between 1925 and 1930. Since then, that theory has increased our understanding of an enormous range of phenomena in physics, chemistry, and even biology. The theory must be regarded as one of the major achievements of science, but does not represent the last word. As the topics in chapter 20 indicate, many mysteries remain to be unraveled.

SUMMARY

In a period of less than fifty years, we have progressed from knowing virtually nothing about the structure of the atom to a detailed knowledge of that structure. This chapter has outlined some of the important discoveries and theoretical developments that led to our knowledge of the atom, starting from chemical evidence for the existence of atoms and culminating in the theory called *quantum mechanics,* which explains atomic structure.

Chemical Evidence.   Recognition of the importance of weighing chemical reactants and products led to the statement of the law of definite proportions and the concept of atomic weight. If each element consisted of atoms that have the same mass, we could explain the mass ratios that were observed in chemical reactions.

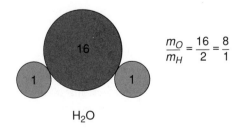

$$\frac{m_O}{m_H} = \frac{16}{2} = \frac{8}{1}$$

H_2O

Cathode Rays.   The study of cathode rays, produced by placing a high voltage across two electrodes in an evacuated tube, led to the discovery of the electron as well as the discovery of X rays. The electron is a negatively-charged particle with a mass much smaller than the smallest atom, so it was the first known subatomic particle available for designing atomic models.

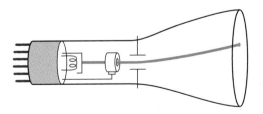

Radioactivity and the Nucleus. The discovery of X rays led to the discovery of another type of radiation emitted from uranium and other heavy elements. This radiation, called *natural radioactivity,* actually consisted of three types of emission, labeled α, β, and γ. The alpha rays (which are helium ions) turned out to be useful projectiles for scattering experiments. These experiments led to the discovery of the nucleus of the atom.

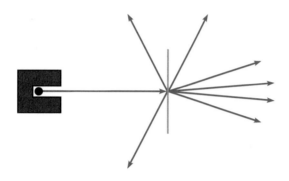

Spectra and the Bohr Model. Bohr explained the regularities in the observed spectrum of hydrogen (the colors of light emitted by excited hydrogen atoms) with a model that incorporated the new quantum ideas introduced by Planck and Einstein. Bohr pictured light as being emitted when an electron jumped from one stable orbit to another with lower energy. The energy difference explained the frequency and wavelength of the emitted photons.

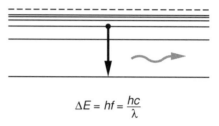

$$\Delta E = hf = \frac{hc}{\lambda}$$

Particle Waves and Quantum Mechanics. De Broglie's suggestion that particles such as electrons could have wave-like properties characterized by a wavelength related to the momentum of the particle, $\lambda = h / p,$ was one path that led to the development of quantum mechanics. The stable orbits of the electrons in atoms can be described in terms of three-dimensional standing waves in this theory. The resulting patterns predict the probability of finding the electron at various distances and orientations relative to the nucleus.

QUESTIONS

Q16.1 Is a chemical element the same thing as a chemical compound? Explain.

Q16.2 When a substance is burned, are all of the products of that reaction solid particles that can be easily weighed? Explain.

Q16.3 Is mass conserved in a chemical reaction? Explain.

Q16.4 Does an atom of hydrogen have approximately the same mass as one atom of oxygen? Explain.

Q16.5 Is it possible for any number of hydrogen atoms to combine with just one atom of oxygen? Explain.

Q16.6 Could the law of definite proportions be explained by a model in which different atoms of the same element have widely varying masses? Explain.

Q16.7 Do cathode rays consist of electromagnetic waves? Explain.

Q16.8 Do X rays consist of electromagnetic waves? Explain.

Q16.9 What characteristics of the negatively charged particles that make up a cathode-ray beam suggested that these particles might be contained within atoms? Explain.

Q16.10 Would you expect X rays to be produced by a television picture tube? Explain.

Q16.11 Following Roentgen's discovery of X rays, Becquerel discovered a seemingly similar type of radiation given off by phosphorescent materials containing uranium or thorium. Was this new radiation essentially the same thing as X rays? Explain.

Q16.12 Was it necessary for Becquerel's phosphorescent materials to be exposed to sunlight in order for them to exhibit natural radioactivity? Explain.

Q16.13 What are two important differences that distinguish alpha particles from beta particles when they are passed through a magnetic field? Explain.

Q16.14 When alpha particles are scattered from a thin piece of gold foil, do most of them go on through with very little deflection? Explain.

Q16.15 Does most of the mass of the atom reside inside or outside of the nucleus? Explain.

Q16.16 Would you expect electrons to be effective in deflecting an alpha particle beam? Explain.

Q16.17 How are the atomic spectra of hydrogen or other gaseous elements produced experimentally? Explain.

Q16.18 Does the spectrum of hydrogen consist of randomly spaced wavelengths, or is there a pattern to the spacing? Explain.

Q16.19 According to Planck's theory, can light of a given frequency be emitted from a blackbody radiator with any amount of energy selected from a continuously varying range of energies? Explain.

Q16.20 According to Bohr's theory of the hydrogen atom, is it possible for the electron to orbit the nucleus with continuously varying values for the radius of the orbit? Explain.

Q16.21 What happens to the excess energy when an electron jumps from a higher-energy orbit to a lower-energy orbit in the hydrogen atom? Explain.

Q16.22 Does an electron have a wavelength? Explain.

Q16.23 According to the theory of quantum mechanics, is it possible to pinpoint exactly where an electron is located in an atom? Explain.

Q16.24 The Bohr model of the hydrogen atom predicts a circular orbit for the electron about the nucleus; the theory of quantum mechanics predicts a three-dimensional probability distribution for locating the electron. Which of these views provides the more realistic picture of the hydrogen atom? Explain.

Q16.25 The chemical properties of sodium, with eleven electrons, are similar to those of hydrogen with just one electron. How do we explain this fact?

Q16.26 Does helium, with two electrons (one more than hydrogen), react chemically with other substances more readily or less readily than hydrogen? Explain.

EXERCISES

E16.1 If carbon, with an atomic weight of 12, combines with oxygen, with an atomic weight of 16, to form carbon dioxide (CO_2), what is the ratio of the mass of oxygen to that of carbon that you would expect to react completely in this reaction?

E16.2 If instead of forming carbon dioxide, carbon monoxide (CO) is formed in a reaction involving carbon and oxygen, how many grams of carbon would react with 32 grams of oxygen? (Use the atomic weights given in exercise 16.1.)

E16.3 If 38 grams of fluorine react completely with 2 grams of hydrogen to form the compound hydrogen fluoride (HF), what is the atomic weight of fluorine?

E16.4 If aluminum, with an atomic weight of 27, combines with oxygen, with an atomic weight of 16, to form the compound aluminum oxide (Al_2O_3), how much aluminum would be required to react completely with 48 grams of oxygen?

E16.5 How many electrons would be required to produce 1 coulomb of negative charge?

E16.6 If the mass of a hydrogen atom is 1.67×10^{-27} kg and that of an electron is 9.1×10^{-31} kg, how many electrons are needed to make a mass equivalent to that of one hydrogen atom?

E16.7 Using the Balmer formula, find the wavelength of the line in the hydrogen spectrum for $m = 10$. Would this line be visible to the unaided eye? Explain.

E16.8 Using the Rydberg formula, find the wavelength of the spectral line for which $m = 4$ and $n = 1$. Would this line be visible to the unaided eye? Explain.

E16.9 Using the Rydberg formula, find the wavelength of the spectral line for which $m = 5$ and $n = 3$. Would this line be visible to the unaided eye? Explain.

E16.10 Suppose that a photon has a wavelength of 540 nm (green).

a. What is the frequency of this photon?

b. What is the energy of this photon in joules?

E16.11 Suppose that a photon has an energy of 3.3×10^{-19} J.

a. What is the frequency of this photon?

b. What is the wavelength of this photon?

E16.12 An electron in a hydrogen atom jumps from an orbit in which the energy is 3.02×10^{-19} J higher than that of the final, lower-energy orbit.

a. What is the frequency of the photon emitted in this transition?

b. What is the wavelength of the emitted photon? Is this wavelength visible to the unaided eye? Explain.

CHALLENGE PROBLEMS

CP16.1 Suppose that an electron beam in a cathode-ray tube passes between two parallel plates that have a voltage difference of 100 volts across them as shown in the diagram. The distance between the plates is 4 cm, and the plates have a length of 10 cm parallel to the initial direction of the electron beam.

a. In what direction will the electron beam deflect as it passes between these plates? Explain.

b. Using the expression for a uniform field, $\Delta V = Ed$, find the value of the electric field in the region between the plates.

c. What is the magnitude of the force exerted upon individual electrons by this field? ($F = qE$)

d. What are the magnitude and direction of the acceleration of the electron?

e. What type of path will the electron follow as it passes through the region between the plates? Explain.

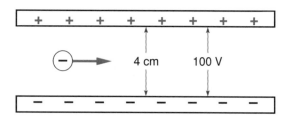

CP16.2 When an alpha particle approaches the nucleus of a gold atom head-on, it will stop and turn around at some point due to the repulsive electrostatic force between the alpha particle and the nucleus (both with a positive charge). This happens at the point where all of the original kinetic energy of the alpha particle has been converted to electrostatic potential energy. We call the distance d between the nucleus and the alpha particle at this point the *distance of closest approach*.

a. A typical kinetic energy of an alpha particle from a radioactive source might be 5.0 MeV, where 1 MeV is a million electron volts, and 1 electron volt equals 1.6×10^{-19} C joules. What is this kinetic energy in joules?

b. What is the electrostatic potential energy for this alpha particle at the distance of closest approach (in joules)?

c. The charge on an alpha particle is $+2e$ and that of the gold nucleus is $+79e$, where $e = 1.6 \times 10^{-19}$ C. Express each of these charges in coulombs.

d. The electrostatic potential energy between two charged particles separated by distance d is given by

$$PE = q_2 V_1 = \frac{kq_1 q_2}{d}$$

where k is the Coulomb constant. Using algebra, rearrange this equation to express d in terms of PE, k, and the two charges.

e. Find the distance of closest approach for this situation. Compare this value to the approximate diameter of an atom, 1.0×10^{-10} m.

CP16.3 Study the energy-level diagram shown in figure 16.20. In the Balmer series, all the spectral lines involve transitions to the $n = 2$ energy level, and the Lyman series involves transitions to the $n = 1$ level. The energies are all negative as a result of the negative potential energy for two charges of opposite sign.

a. Which transition in the Balmer series produces the smallest frequency for a photon (and the largest wavelength)?

b. What is the energy difference in joules for the two levels, involved in the transition of part (*a*)?

c. What are the frequency and wavelength of the photon emitted in this transition?

d. In a similar manner, find the frequency and wavelength of the photon with the longest wavelength in the Lyman series.

e. Compare the two wavelengths obtained in (*c*) and (*d*). Which of these, if either, will be visible to the unaided eye? Explain.

CP16.4 When an electron is removed completely from an atom, we say that the atom is *ionized*. An ionized atom has a net positive charge, since an electron has been removed.

a. From the energy-level diagram in figure 16.20, how much energy would be required to ionize a hydrogen atom when it is in its lowest energy level?

b. How much energy would be required to ionize the atom when it is in the first excited state above the lowest level?

c. If an electron with zero kinetic energy was captured by an ionized hydrogen atom and went immediately to the lowest energy level, what wavelength would you expect to observe for the photon emitted in this transition?

HOME EXPERIMENTS AND OBSERVATIONS

HE16.1 With the set *turned off,* take a close look at the screen of a color television set. A magnifying glass will help you to see the detail.

   a. Describe the pattern of lines that you observe. Produce a careful sketch showing the arrangement of the lines.

   b. If there is a black and white television set handy, compare the pattern on its screen to that of the color set. What differences can you describe?

17

The Nucleus and Nuclear Energy

In 1986 the newspapers were full of reports and comments on a serious nuclear accident at the Chernobyl nuclear power plant in Russia. Radioactivity was being dispersed across parts of Europe, several fire-fighters and reactor employees were killed, and the fears of the public regarding nuclear power were dramatically reawakened.

Closer to home, we have about seventy operating nuclear reactors producing electric power in the United States, as well as submarines that have nuclear reactors to run their engines, and many smaller reactors used for research and other purposes (fig. 17.1). Many of these have operated with only minor problems, and have had minimal impact on the surrounding environment. Still, the use of nuclear power, its environmental consequences, and its economics have been tremendously controversial issues over the past two decades.

What goes on inside a nuclear reactor? How is it that we can derive power from uranium, and what is the nature of the nuclear wastes that are generated? Need we fear those benign-looking clouds of steam that emerge from the cooling towers? Can a reactor explode in the same manner as a nuclear bomb? What is the difference between nuclear fission and nuclear fusion? An educated citizenry should know the answers to these questions. Otherwise, people are at the mercy of the extremists on either side of the issue who assert simple but misleading views.

The development of knowledge about the nucleus of the atom makes one of the most fasci-nating tales of twentieth-century science. The political consequences of nuclear weapons and nuclear power have been critical components of that story. These issues, more than any other, have thrust science and physics into the caldron of national and international policy. They appear to be there to stay; nuclear issues are a part of our common concern, if not our everyday experience.

Figure 17.1   The large cooling tower is often the most prominent feature of a modern nuclear power plant. What is the source of energy in such a plant?

Chapter Objectives

*This chapter explains how physi-
cists gained knowledge of the
nucleus and its structure. The story
includes the discovery of nuclear
fission just before World War II and
the wartime effort to develop the
atomic bomb. The postwar desire to
find peaceful uses for the atom led
to the development of commercial
power plants, but this same period
also saw the invention of the hydrogen (fusion) bomb and a rapid esca-
lation in the nuclear arsenals of the
major world powers.*

*The primary objective here is to pro-
vide a basis for understanding the
science that underlies these issues.
As the chapter outline indicates,
many questions can be answered in
a straightforward manner to lay this
foundation.*

*The answers to many of these ques-
tions are important to an understanding of what happened at
Chernobyl, which is described in
the "Everyday Phenomenon" box in
section 17.4. More important, this
understanding provides a basis for
addressing the nuclear issues that
face us as citizens of the world.*

Chapter Outline

1 *The structure of the nucleus.* What is
the nucleus made of, and how do the
pieces fit together? How have we
learned so much about something
that is far too small to be seen?
What is the distinction between different isotopes of the same element?

2 *Radioactive decay.* What is radioactive decay, and how is it related to
changes in the nucleus? What do
the terms *half-life* and *exponential
decay* mean? Why can radioactivity
be dangerous?

3 *Nuclear reactions and nuclear fission.* What is nuclear fission, and
how was it discovered? How does
fission lead to the possibility of a
chain reaction and the release of
large amounts of energy?

4 *Nuclear reactors.* How do nuclear
reactors work? What are the functions of the moderator, the control
rods, the coolant, and other reactor
components? What is the nature of
nuclear wastes?

5 *Nuclear weapons and nuclear
fusion.* How does an *atomic,* or
nuclear, bomb work? What is the difference between a hydrogen bomb
and the earlier fission bombs? What
is nuclear fusion, and how is this
process capable of releasing
energy?

17.1 THE STRUCTURE OF THE NUCLEUS

Our understanding of the structure of the nucleus is entirely a twentieth-century phenomenon. Indeed the existence of a nucleus in atoms was not suspected until Rutherford's famous scattering experiments using alpha particles, which he performed around 1909–1911 (see section 16.3 of chapter 16). The idea that this tiny but massive center of the atom, called the *nucleus,* also has a structure that we can decipher may seem amazing in itself. But such is the case, and we will examine the major steps that led to this understanding.

Just as in the case of atomic structure, before we can build a model of the nucleus we need to know something about its components. In other words, what are the building blocks from which the nucleus is constructed? Ernest Rutherford, whom we have credited with discovery of the nucleus, was also a major contributor to the answer to this question. The evidence came from more scattering experiments; in fact, scattering experiments of one sort or another are the major tool for probing the nucleus and other subatomic particles.

Discovery of the Proton

The experiment that provided the first clue to what the nucleus was made of was performed by Rutherford in 1919. Once again, he used alpha particles as his probe. Figure 17.2 shows a conceptual diagram of this scattering experiment. A beam of alpha particles was used to bombard a cell containing nitrogen gas. As expected, some alpha particles went through the sample without hitting anything, and others were deflected (scattered) by the nuclei of the nitrogen atoms. These can be detected readily with scintillation detectors.

The unexpected result of this experiment was the emergence of a different kind of particle within the cell containing the nitrogen. These particles were positively charged like the alpha particles, but were clearly different, judging by the nature of the scintillations and by how far the particles traveled in air. In fact, they behaved essentially like the nuclei of hydrogen atoms, which Rutherford had observed in earlier experiments in which he had bombarded hydrogen gas with alpha particles. The mass of a hydrogen atom is approximately one-fourth that of an alpha particle, which is the nucleus of a helium atom, as noted in chapter 16.

Finding hydrogen nuclei emitted from a cell in which none had been present initially suggested an exciting possibility: perhaps the hydrogen nucleus was a basic constituent of the nucleus of other elements. It was already known that the atomic masses of different elements were close to being integer multiples of the atomic mass of hydrogen. The atomic mass of nitrogen, for example, is approximately 14 times that of hydrogen. The atomic mass of carbon is approximately 12 times that of hydrogen, and the atomic mass of oxygen is 16 times that of hydrogen. These masses could be explained if we assumed that the nuclei of these elements were made up of 12, 14, and 16 hydrogen nuclei for carbon, nitrogen, and oxygen respectively. Further experiments by Rutherford and others showed that hydrogen nuclei could be ejected from sodium and other elements, as well as from nitrogen, by bombarding them with alpha particles.

We now call this particle a *proton.* It is the nucleus of the hydrogen atom as well as a constituent of the nuclei of other elements. It has a charge of +e, opposite in sign but identical in size to that of the electron. Its mass is much greater than the electron's, however; it is approximately equal to that of the hydrogen atom, which is 1835 times the mass of the electron. Thus the proton became the first candidate for a nuclear building block.

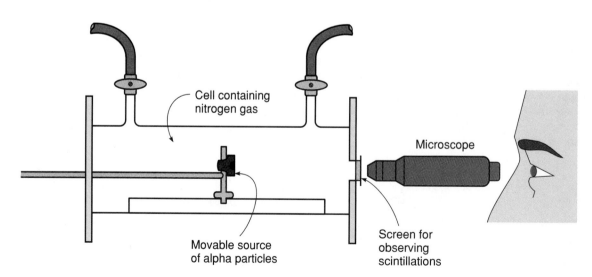

Figure 17.2   A drawing of the apparatus used in Rutherford's scattering experiment leading to the discovery of the proton. Nitrogen gas was the target for the alpha particles.

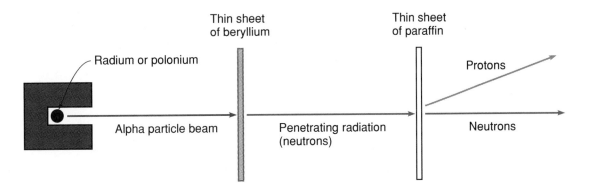

Figure 17.3 **A schematic diagram of Chadwick's experiment. Radiation coming from the beryllium target was used to bombard a paraffin target.**

Discovery of the Neutron

The hypothesis that the nuclei of different elements could be made of just protons had some serious problems, however. The most obvious problem was that of the charge of the nucleus. From its place in the periodic table, from scattering experiments with X rays, and from the X-ray spectrum of nitrogen, it was known that the charge of the nitrogen nucleus should be $+7e$ rather than $+14e$. If there were 14 protons in the nucleus of the nitrogen atom, the nuclear charge would be too large. Likewise, carbon and oxygen have nuclear charges of $+6e$ and $+8e$ respectively, rather than 12 or 16 times e.

For a while, physicists considered the possibility that there were also electrons within the nucleus, which partially neutralized the extra charge of the protons. This view also had some serious problems, however, particularly in light of the newly emerging insights from quantum mechanics. The energies of electrons confined to the very small region of the nucleus would have to be much larger than those emerging in the beta rays observed in the radioactive decay of nuclei. It did not seem likely, therefore that electrons could exist as separate particles within the nucleus.

It took several years to resolve this riddle. It was yet another scattering experiment, performed by W. Bothe and H. Becker in Germany around 1930, that provided the clue. Becker and Bothe bombarded thin beryllium samples with alpha particles and found that a very penetrating radiation was emitted. Since gamma rays were the only radiation known to be this penetrating, they originally assumed that gamma rays were involved. However, other experiments showed that this new emission had an even greater ability to pass through lead than gamma rays, and had other properties quite unlike those of gamma rays.

In 1932, the British physicist James Chadwick showed that this new emission from beryllium behaved like a neu-

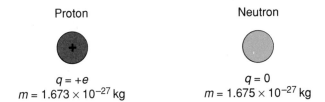

Figure 17.4 **The basic building blocks of the nucleus are the proton and the neutron.**

trally charged particle with a mass approximately equal to that of the proton. Chadwick's experiment used the penetrating emission coming from the alpha bombardment of beryllium to bombard a piece of paraffin (fig. 17.3). Paraffin is a compound of carbon and hydrogen, and hydrogen nuclei (protons) emerged from the paraffin when it was placed in the path of the penetrating radiation coming from the beryllium. Making the assumption that a neutral particle with a mass equal to that of the proton was colliding with protons in the paraffin neatly explained the energies of the protons emerging from the paraffin. This new particle was called a *neutron;* it has no charge, and its mass is essentially equal to that of the proton.

The discovery of the neutron answered the riddle of the basic building blocks of the nucleus (fig. 17.4). If the nucleus is made of neutrons and protons, we can explain both its charge and its mass. Nitrogen, for example, can have a nucleus made up of 7 protons and 7 neutrons for a total mass that is 14 times that of hydrogen and a nuclear charge 7 times that of hydrogen. You can easily deduce the required numbers of protons and neutrons for carbon and oxygen; the numbers for several elements are shown in figure 17.5.

Isotope	Symbol	Number of protons	Number of neutrons	Relative size
Helium-4	$_2\text{He}^4$	2	2	
Beryllium-9	$_4\text{Be}^9$	4	5	
Nitrogen-14	$_7\text{N}^{14}$	7	7	
Chlorine-37	$_{17}\text{Cl}^{37}$	17	20	
Iron-56	$_{26}\text{Fe}^{56}$	26	30	
Uranium-238	$_{92}\text{U}^{238}$	92	146	

Figure 17.5 **The proton and neutron numbers for the most common isotopes of several elements. The nucleus gets larger as the number of protons and neutrons increases.**

Isotopes

Another puzzle was also resolved with the discovery of the neutron. It had been known for some time that the atoms of many elements had different values of nuclear mass. Nuclear masses could be measured with high accuracy by passing nuclei of known velocity through a magnetic field and observing how much their paths were bent by the magnetic force on the positively charged nucleus. For example, chlorine was known from chemistry to have an average atomic mass 35.46 times that of hydrogen. When chlorine ions are passed through a magnetic field, however, it is clear that two different masses are actually present, one 35 times that of hydrogen and the other 37 times that of hydrogen. Their chemical properties are identical; both behave like chlorine.

These different-mass versions of the same element are called *isotopes*. Different isotopes have the same number of protons in the nucleus, but different numbers of neutrons. The two common isotopes of chlorine, for example, both have 17 protons in the nucleus, but one has 18 neutrons, for a total mass number of 35, and the other has 20 neutrons, for a total mass number of 37. (The mass number is just the sum of the proton and neutron numbers.) Table 17.1 provides other examples.

Table 17.1

Neutron and Proton Numbers for Different Isotopes of the Same Elements

Isotope Name	Symbol*	Protons	Neutrons
Hydrogen-1	$_1\text{H}^1$	1	0
Hydrogen-2 (deuterium)	$_1\text{H}^2$	1	1
Hydrogen-3 (tritium)	$_1\text{H}^3$	1	2
Carbon-12	$_6\text{C}^{12}$	6	6
Carbon-14	$_6\text{C}^{14}$	6	8
Chlorine-35	$_{17}\text{Cl}^{35}$	17	18
Chlorine-37	$_{17}\text{Cl}^{37}$	17	20
Uranium-235	$_{92}\text{U}^{235}$	92	143
Uranium-238	$_{92}\text{U}^{238}$	92	146

* See Section 17.2 for a discussion of the notation used here.

The chemical properties of an element are determined by the proton number, which is also called the *atomic number.* Since an atom is normally neutral, the atomic number also represents the number of electrons orbiting the nucleus. Nitrogen, for example, has an atomic number of 7; it has 7 protons in the nucleus and 7 electrons orbiting the nucleus. There are also 7 neutrons in the nucleus, but in general the neutron number is *not* equal to the atomic number. Chlorine-37, for example, has an atomic number of 17 and a neutron number of 20. Except for the lightest elements, the neutron number is generally larger than the atomic number.

With the discovery of the neutron in the year 1932, many pieces of the puzzle fell into place. Atomic masses as well as the chemical properties of atoms could now be explained. Physicists could begin constructing models of the nucleus and designing new experiments to test these models. Perhaps most important, however, the neutron provided a powerful new probe for exploring the structure of the nucleus. Since it has no charge, it is much easier to get a neutron to penetrate the nucleus and begin rearranging things. The proton and the alpha particle, on the other hand, are both positively charged and are therefore repelled by the positive charge of the nucleus. A host of new experiments were begun, some of which produced even more spectacular surprises than those we have been describing here.

17.2 RADIOACTIVE DECAY

The discovery of the neutron and the picture that this created of nuclear structure provided a new basis for understanding many already-known phenomena. Natural radioactivity, for example, had been discovered in 1896, and by 1910 Rutherford, Soddy, and others had demonstrated that one element was actually being changed into another in the process of radioactive decay. What exactly is happening here, and how can our new insight into nuclear structure help to clarify this phenomenon?

Alpha Decay

One of the first radioactive elements to be studied extensively was radium, which was isolated and identified at the turn of the century by Marie and Pierre Curie. Radium was found in the uranium ore, pitchblende, but was shown to be much more radioactive than uranium itself. In its radioactive decay, radium emitted alpha particles, which were shown to be the nuclei of helium atoms by Rutherford.

The dominant isotope of radium that is found in pitchblende contains a total of 226 *nucleons* (neutrons and protons) in its nucleus. Since the atomic number of radium is 88, this means that there are 88 protons and $226 - 88 = 138$ neutrons. We therefore call this isotope *radium-226,* and often write it as $_{88}Ra^{226}$. The subscript is the atomic number, and the superscript is the mass number.

If we know that radium-226 emits alpha particles, then we can figure out what element results from its decay. The process is quite simple; we know how many protons and neutrons are contained in both the radium and helium nuclei, so we know also how many of each are left over in the decay product, which is often referred to as the *daughter* element (fig. 17.6). A reaction equation is useful to keep track of these numbers:

$$_{88}Ra^{226} \Rightarrow {}_{86}X^{222} + {}_{2}He^{4}.$$

Here we have found the atomic number (86) of the unknown element by subtracting the atomic number of helium, 2, from that of radium, 88. The mass number is found in a similar manner: $222 = 226 - 4$. The unknown element X can then be identified by looking in a periodic table (see the inside backcover) to find which element has an atomic number of 86. It turns out to be the noble gas radon (Rn), so the daughter nucleus is radon-222 or $_{86}Rn^{222}$. (The number of neutrons can also be found by subtracting the proton number 86 from the total nucleon number 222, which yields 136.)

Since momentum is conserved in the decay, the alpha particle is shown in figure 17.6 to have a much larger velocity than the recoil velocity of the radon nucleus. The radon nucleus has a much larger mass than that of the alpha particle, so its recoil velocity will be much smaller than that of the alpha particle to give it the same magnitude of momentum ($\vec{p} = m\vec{v}$; see chapter 6).

We find that the alchemist's dream of turning one element into another actually happens in radioactive decay and other nuclear reactions (although the alchemists were usually trying to produce gold). It turns out that radon is itself radioactive. Radon-222 undergoes alpha decay to produce polonium-218, which in turn undergoes alpha decay to yield lead-214. Although lead-214 is not a stable isotope of lead, lead is often the end product of the radioactive decay of other heavier elements.

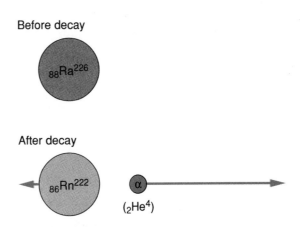

Figure 17.6 Alpha decay of radium-226. The daughter isotope is radon-222.

Beta and Gamma Decay

Lead-214 undergoes beta decay, which we have not yet considered in detail. The emitted particle in beta decay is either an electron or a *positron* (a positively charged version of the electron). In the case of lead-214, an ordinary (negatively charged) electron is emitted. The mass of an electron is so small that it can be ignored on the scale of nuclear masses; in other words, its mass number is zero. Its charge number, on the other hand, is –1, so the reaction equation takes the following form:

$$_{82}Pb^{214} \Rightarrow {}_{83}X^{214} + {}_{-1}e^0 + {}_0\bar{v}^0.$$

Here again we see that the mass numbers and charge numbers add up on either side of the reaction equation. The charge, or atomic, number of the resulting element is 83, since 83 – 1 = 82 is the original atomic number of the lead (Pb) isotope. Looking up this atomic number in the periodic table, we see that bismuth-214 is the daughter element in this decay (fig. 17.7). One of the neutrons inside the nucleus of

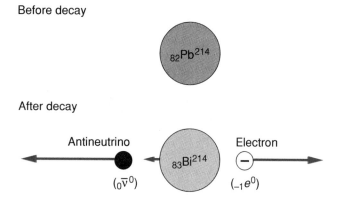

Figure 17.7   Beta decay of lead-214. The daughter isotope, bismuth-214, has a higher atomic number than lead.

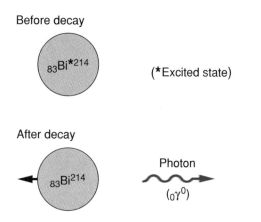

Figure 17.8   Gamma decay of bismuth-214. The daughter isotope is a more stable (lower energy) version of the original bismuth-214.

lead-214 has been changed to a proton in this process, yielding a nucleus with a higher atomic number. Thus we could substitute $_{83}Bi^{214}$ where the symbol *X* appears in the reaction equation.

Some comment should be made regarding the third particle appearing on the right-hand side of our beta-decay reaction equation. This particle is called an *antineutrino* and is represented by the Greek letter $\bar{v}$ pronounced "nu." The bar over the symbol indicates that it is an *antiparticle;* all elementary particles also have antiparticles. The antineutrino had not been directly observed in beta decay, but was included in order to conserve energy. Since the electrons in beta decay come out with a range of energies, physicists reasoned that something else must be involved to account for the remaining energy. Neutrinos were not actually observed until 1957, but physicists believed in their existence for many years because of their faith in the validity of the principle of conservation of energy. Neutrinos (and antineutrinos) have an extremely small mass and no charge, so they do not affect the numbers in the reaction equation.

Through a series of further beta and alpha decays, bismuth-214 decays finally to a stable isotope of lead: lead-206. Some of the isotopes involved in this decay chain also emit gamma rays, which are high-energy X rays. Since the emitted particle in this case is a photon, which has no charge or mass, neither the mass number nor the charge number changes in a gamma decay. We are left with a more stable version of the original isotope (fig. 17.8).

Half-Lives

How long does it take for these different decays to happen? All forms of radioactive decay are spontaneous events that occur in a random manner. There is no way to predict exactly when a specific unstable atom is going to throw out a particle and change to a different isotope. The decay of different isotopes, however, occurs over different average amounts of time. We usually use the concept of *half-life* to describe the characteristic time required for decay.

The concept of half-life is best explained through an example. The half-life of radon-222 is approximately 3.8 days. If we started with 20 000 atoms of radon-222, 3.8 days later we would have 10 000 remaining; the other half would have decayed to polonium-218 ($_{84}Po^{218}$). In two half-lives, or 7.6 days, half of these would have decayed, leaving only 5000 atoms of radon-222. In three half-lives we would be down to 2500, and in four half-lives this number would be halved again, yielding 1250.

The half-life is therefore the time required for half of the original number of atoms to decay. Since each time 3.8 days passes, the number of radon-222 atoms is reduced by half, it does not take very many half-lives to reduce the number of atoms of a given isotope to a tiny fraction of the original amount. After ten half-lives, or 38 days, for example, the number of radon-222 atoms remaining of the original 20 000 would be just 20 atoms, or one-thousandth of the original number!

This process is illustrated in the graph of figure 17.9. The curve that results is called an *exponential decay curve,* because mathematically it can be represented by a simple exponential function. Exponential decay and growth frequently occur in nature; they result whenever the number of decays or additional events are proportional to the original number of cases.

The half-lives of different radioactive isotopes vary over an enormous range. The half-life of radium-226, for example, is 1620 years, which is still relatively short compared to that of the common isotope of uranium, uranium-238, which is 4.5 billion years. At the other extreme, polonium-214 has a half-life of just 0.000164 seconds. It does not stick around very long!

The shorter the half-life, the greater the rate of radioactivity. With regard to the environment, however, it is isotopes with intermediate half-lives that pose the greatest problem. An isotope with a very short half-life is highly radioactive while it lasts, but decays very quickly and therefore does not remain dangerous. An isotope with an extremely long half-life, such as uranium-238, is not very radioactive, although it can be a hazard if there is enough of it. An isotope with a half-life of several years, however, is much more radioactive than uranium-238, and remains in the environment long enough to pose serious problems. These are the isotopes of concern in fallout from weapons tests or nuclear accidents.

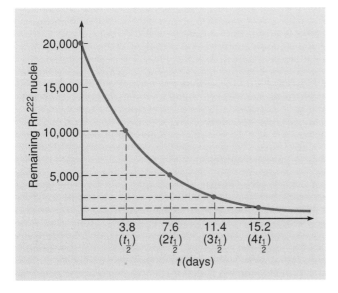

Figure 17.9 Decay curve for radon-222. The amount remaining decreases by one-half every 3.8 days, the half-life.

Box 17.1

Radiation Exposure

Although a variety of units are used for measuring the amount of ionizing radiation and its impact upon human tissue, the most commonly used unit for comparing doses of differing kinds of ionizing radiation (X rays, natural radioactivity, etc.) is the *rem*. This is an acronym standing for *r*oentgen equivalent in *m*an; the *roentgen* is a unit that describes the amount of ionization produced by radiation. The rem takes into account the differing effects of different types of radiation in human tissue.

A whole-body dose of 600 rem is usually lethal. Much smaller doses can also produce damage, however, and these are generally quoted in millirem (mrem); 1 mrem equals one-thousandth of a rem. On the average, people in the United States receive about 295 mrem per year from natural sources and another 64 mrem per year from human-produced sources. As the table below shows*, the largest human-produced source is that from medical use of diagnostic X rays and radioactive isotopes.

Natural sources	mrem/yr
inhaled (radon)	200
cosmic rays	27
terrestrial radioactivity	28
internal radioactivity	40
	295

Artificial (human-produced)	
medical	53
consumer products	10
other	1
	64

Both the natural sources and the human-produced sources can vary widely, depending on where you live and what medical procedures you receive. The average dose received by people in the United States from nuclear-power sources is not significant on this scale, but individuals working in the industry may be exposed to larger amounts. Current standards set a limit of 5 rem/yr (5000 mrem/yr) for nuclear workers, X-ray technicians, or other people exposed to radiation in their occupations.

* For further discussion see "Health Effects of Low-Level Ionizing Radiation" (*Physics Today,* August 1991, pp. 34–39) from which this table was adapted.

The reason that radioactivity is dangerous is that the emitted particles—alpha particles, beta rays (electrons), and gamma rays—can penetrate the body and cause alterations in the chemical compounds that make up our cells. These alterations can cause cancer and other forms of damage, including mutations in offspring. In high enough doses, these alterations produce radiation sickness and death. The effects at very low doses are still being debated: some scientists believe that any amount is potentially damaging; others hold that very low levels may have beneficial effects that counter the negative effects.

Because its half-life is similar to the estimated age of the earth, uranium-238 is present naturally in our environment. It appears in trace amounts in all rocks and soils and in more concentrated amounts in uranium ores. We are constantly exposed, therefore, to very low levels of radioactive emissions from trace amounts of uranium and from cosmic rays that impinge upon earth from space; we cannot escape that fact (see box 17.1). Only when levels of radioactivity become significantly larger than this natural background is there cause for concern. The invisible and seemingly mysterious nature of radioactivity, however, often produces fear that is out of proportion to the danger.

17.3 NUCLEAR REACTIONS AND NUCLEAR FISSION

We have seen that the nucleus can change spontaneously in the radioactive decay process. Is it possible, though, to initiate changes in the nucleus experimentally? The discovery of the neutron in 1932 provided a new tool for attempting this, and the results of such experiments became entangled with political developments in Europe and the world in a way that scientists could not have anticipated.

Nuclear Reactions

We have actually already considered changes in the nucleus that were produced by experiment prior to the discovery of the neutron: Rutherford's discovery that nitrogen nuclei emit protons upon bombardment by alpha particles is an example of such a change. If we consider what is going on in the scattering experiment pictured in figure 17.2, we can write a reaction equation of the following form:

$$_2He^4 + {_7}N^{14} \Rightarrow {_8}X^{17} + {_1}H^1.$$

Here the alpha particle is a helium nucleus, and the emitted proton is a hydrogen nucleus. The other product of the reaction is an element with an atomic number of 8, which turns out to be oxygen (fig. 17.10). We can write $_8O^{17}$ in place of the unknown X, therefore. Oxygen-17 is not the most common isotope of oxygen (that is oxygen-16), but it is found in nature as a stable isotope.

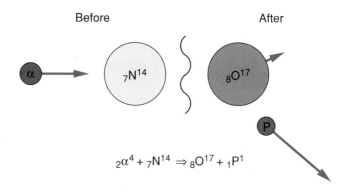

Before   **After**

$$_2\alpha^4 + {_7}N^{14} \Rightarrow {_8}O^{17} + {_1}P^1$$

Figure 17.10 The collision of an alpha particle and a nitrogen nucleus results in a proton being emitted and an oxygen-17 nucleus remaining in place of nitrogen.

This is an example of a *nuclear reaction*. Notice that the charge and mass numbers add up to the same total on either side of the equation: the total charge number is 9 and the total mass number is 18. (This was true also for the decay equations in section 17.2.) This fact was used to identify the other reaction product, oxygen-17, but we could also confirm this result by analyzing the gas found in the cell after performing the experiment. Oxygen would be found, even though none was present at the beginning of the experiment.

The emission of neutrons from beryllium that is bombarded by a beam of alpha particles also represents a nuclear reaction. The reaction equation in this case would be

$$_2He^4 + {_4}Be^9 \Rightarrow {_6}C^{12} + {_0}n^1.$$

The neutron has no charge, so its atomic or charge number is 0, and its mass number, like that of the proton, is 1. The other reaction product, carbon-12, is found by looking in the periodic table to find the element with an atomic number of 6. The resulting isotope of carbon turns out to be the most common one. Again, after the experiment has been performed, we will find small amounts of carbon present in the target, which was originally pure beryllium.

Energy and Mass

Another feature of nuclear reactions is that energy is often released in the reaction, although reactions occur in which energy is absorbed as well. The amount of energy released or absorbed can be predicted from knowledge of the masses of the isotopes involved and Einstein's famous $E = mc^2$ relationship. The meaning of this relationship (explained in more detail in chapter 18) is that mass and energy are equivalent: mass is energy, and energy is mass. The constant c^2, which is the velocity of light squared, is just a unit-conversion factor that allows us to convert mass units to energy units and vice versa. If the mass of the products is less than that of the reactants, then the energy represented by this mass difference shows up in other forms, usually as kinetic energy of the emerging particles.

This process is illustrated for the beryllium reaction in the computations shown in box 17.2. The masses for the isotopes and particles involved are given here in atomic mass units (or unified mass units, u). These are based upon the mass of the carbon-12 atom, so that the mass of carbon-12 is exactly 12.000 000 by definition of u. The mass difference is converted to kilograms and then multiplied by c^2 to find the energy released in joules. This energy is shared between the emerging neutron and the recoiling carbon-12 nucleus, but the neutron gets the larger share. This is because conservation of momentum requires a larger velocity for the smaller mass of the neutron (see the discussion of recoil in chapter 6).

Although the amount of energy released in a single reaction may seem small, it is actually roughly a million times larger than the typical energy released (per atom) in a chemical reaction. This type of mass change is the source of the

particle energies involved in radioactive decay, which were initially a mystery to physicists. Since the work of Einstein and others in the early part of the twentieth century, however, physicists had been aware of the potential for large quantities of energy to be released from nuclear reactions.

The Discovery of Fission

Prior to 1932, alpha particles or protons were the primary bullets available for attempting to rearrange the nucleus. The discovery of the neutron immediately suggested the availability of a powerful new probe, since its zero charge means that it is not repelled by the positive charge of the nucleus. Alpha particles and protons are positively charged and are repelled by the charge on the nucleus; thus, they require high initial energies and head-on collisions to produce a reaction. The neutron, on the other hand, can slip right into the nucleus at relatively low energies.

The Italian physicist Enrico Fermi was one of the first to explore the potential of the neutron for producing nuclear changes. He began a series of experiments in 1932–1934 in which he actually attempted to produce new elements. The element with the largest mass and atomic numbers then known was uranium, so Fermi decided to bombard uranium samples with neutrons produced from the beryllium reaction (fig. 17.11). He would then analyze the sample to see whether he could detect elements with atomic numbers higher than 92.

The results of these initial attempts were both confusing and disappointing. Fermi and his chemist co-workers were able to predict the likely chemical properties of the elements being sought by knowing what should come next in the periodic table, but attempts to isolate such elements were not successful. Since the expected quantities would be small and the exact chemical properties unknown, however, the lack of clear-cut results was perhaps not surprising.

Others took up the effort, and the first real breakthrough in this line of research occurred in 1938, when two German scientists, Otto Hahn and F. Strassmann, isolated the element barium from uranium samples that had been bombarded with low-energy neutrons. This was an astonishing result that Hahn and Strassmann carefully rechecked before announcing their finding. The reason such a result was totally unexpected was that barium is nowhere near uranium in the periodic table. Barium has an atomic number of 56, which is just a little more than half that of uranium.

What kind of reaction could produce barium from uranium? Two other German scientists, then working in Denmark and Sweden because of the growing persecution of Jews in Germany, provided a possible answer. Lise Meitner and her nephew, O. R. Frisch, speculated that the uranium atom might actually be splitting into two much smaller nuclei. If one of these nuclei was barium, for example, with an atomic number of 56, the other should have an atomic number of 36 in order to add up to 92, the atomic number of uranium. This element

---

Box 17.2

Sample Exercise: Mass-Energy Calculation

The nuclear masses for the reactant and products of the reaction

$$_2\text{He}^4 + {}_4\text{Be}^9 \Rightarrow {}_6\text{C}^{12} + {}_0\text{n}^1$$

are given below. Using these values and Einstein's $E = mc^2$ relationship, calculate the energy released in this reaction.

Reactants		Products	
Be^9	9.012 186 u	Neutron	1.008 665 u
He^4	+4.002 603 u	C^{12}	+12.000 000 u
	13.014 789 u		13.008 665 u

The difference in mass between the reactants and products is

$$13.014\ 789\ \text{u}$$
$$-13.008\ 665\ \text{u}$$
$$0.006\ 124\ \text{u} = \Delta m \text{ (mass difference)}$$

Since $1\ \text{u} = 1.661 \times 10^{-27}\ \text{kg}$,

$$\Delta m = (0.006\ 124\ \text{u})(1.661 \times 10^{-27}\ \text{kg/u})$$
$$= 1.017 \times 10^{-29}\ \text{kg}$$
$$E = \Delta mc^2 = (1.017 \times 10^{-29}\ \text{kg})(3.0 \times 10^8\ \text{m/s})^2$$
$$= \mathbf{9.15 \times 10^{-13}\ J}$$

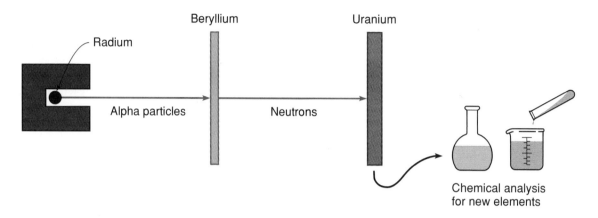

Figure 17.11   A schematic diagram of Fermi's experiments, in which he attempted to produce new elements by bombarding uranium with neutrons.

happens to be krypton, a noble gas (fig. 17.12). A possible reaction equation, therefore, might be

$$_0n^1 + _{92}U^{235} \Rightarrow _{56}Ba^{142} + _{36}Kr^{91} + 3_0n^1.$$

The equation is written with some extra neutrons being emitted because we will indeed end up with an excess of neutrons if we split a large nucleus into two smaller ones. This is because the ratio of neutrons to protons gets larger and larger as elements become heavier in the periodic table. The isotopes of barium and krypton that we have suggested here also contain an excess of neutrons. They are therefore unstable and undergo beta decay, which means that our reaction products are radioactive.

We have also jumped ahead of the story a bit in writing uranium-235 as the isotope of uranium involved. Naturally occurring uranium is mostly uranium-238; only 0.7% is uranium-235. It is uranium-235, however, that most readily undergoes *fission,* which is what we call this type of nucleus-splitting reaction. Low-energy neutrons are absorbed much more readily by uranium-235 than by uranium-238, and fission is also much more likely for uranium-235 than for uranium-238 when a neutron is absorbed.

The two elements that emerge from the fission reaction are called *fission fragments:* in this case they are barium and krypton. Many other elements can result, however, all having atomic numbers between 30 and 60 and therefore being near the middle of the list of known elements. They are generally radioactive because of the excess of neutrons. These fission fragments make up the bulk of nuclear wastes that result from fission applications.

The excitement generated in the scientific community by these discoveries and ideas is difficult to describe. Neils Bohr, who discussed these speculations with Meitner and Frisch in Denmark, was the first to suggest on theoretical grounds in 1939 that uranium-235 was probably the isotope involved. Bohr then traveled to the United States, where he

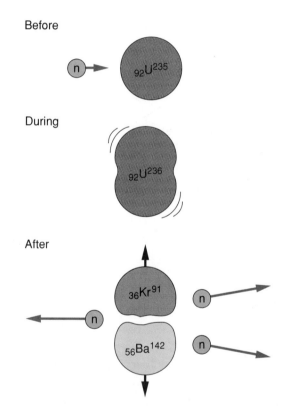

Figure 17.12   Barium-142 and krypton-91 are two possible fission fragments that can be produced when a neutron is absorbed by uranium-235 causing a fission reaction.

spread the word to a growing community of nuclear scientists. Many of these scientists were European refugees who had fled the unstable situation in Europe created by the beginning of World War II and the persecution of the Jews in Germany and other areas under Nazi control.

The excitement and concern were produced, in part, by the immediate recognition of the possibility of a chain reaction involving nuclear fission. Since the fission reaction is initiated by neutrons and several more neutrons are emitted in the reaction itself, a rapidly expanding chain reaction seemed a very real possibility (fig. 17.13). Such a chain reaction could release enormous quantities of energy, as predicted by the mass/energy equivalence relationship.

Among the European-born scientists working in the United States by 1939 were Enrico Fermi, Edward Teller, and Albert Einstein. Although Einstein was not primarily interested in nuclear physics, he was by then recognized throughout the world as a brilliant theoretician. Even politicians were familiar with his name, and for this reason some of his colleagues prevailed upon Einstein to write a letter to President Franklin D. Roosevelt suggesting the need for a crash research program to explore the military implications. The Manhattan project, which was begun shortly thereafter, led ultimately to the development of both nuclear reactors and nuclear weapons.

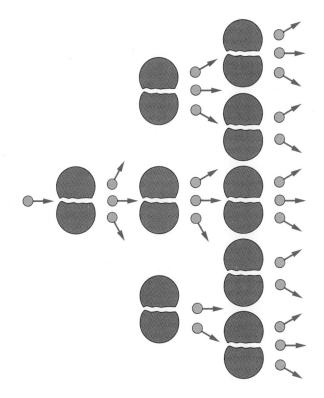

Figure 17.13   A chain reaction involving nuclear fission. Neutrons are produced in each fission of a uranium-235 nucleus, which in turn can initiate additional fission reactions.

17.4 NUCLEAR REACTORS

By 1940, the possibility of a chain reaction involving nuclear fission was apparent to physicists in both Europe and the United States. If the fission reaction is initiated by neutrons, and each reaction produces several additional neutrons, why don't chain reactions occur all the time in uranium samples? What conditions are necessary to produce a chain reaction? These questions became urgent because war had broken out in Asia as well as in Europe by then, and the military implications were ominous.

Achieving a Chain Reaction

Understanding the conditions necessary for a chain reaction is critical to a grasp of how nuclear reactors and nuclear bombs work. The key in either case lies in tracing what happens to the neutrons produced by the fission reaction. If enough of these neutrons are captured by other uranium-235 nuclei, new fissions will occur, and the reaction will be sustained. If too many of the neutrons produced are absorbed by other elements or escape from the reactor or bomb without colliding with other uranium-235 nuclei, a chain reaction will not occur.

It is important to remember that natural uranium consists largely of uranium-238 (99.3%); only 0.7% is uranium-235. Uranium-238 also absorbs neutrons, but this does not usually result in fission. The main reason, therefore, that a chain reaction does not occur in natural uranium is that the neutrons produced in one fission reaction are more likely to be absorbed by uranium-238 nuclei. Thus one way of increasing the likelihood of a chain reaction is to increase the proportion of uranium-235 in the sample.

Unfortunately, (or fortunately, depending on your point of view), it is notoriously difficult to separate uranium-235 from uranium-238. Different isotopes of the same element have identical chemical properties; thus, chemical separation techniques are useless. The very small difference in mass between the two isotopes must be used as the basis for separation. Various techniques were tested during the war years; the most promising seemed to be a gas-diffusion technique that was implemented in a plant at Oak Ridge, Tennessee. After enormous effort and expense, however, scientists succeeded in separating only enough uranium-235 (just a few kilograms) for one bomb by the end of the war.

A different strategy for achieving a chain reaction is used in nuclear reactors, which can be designed to run on either natural uranium or uranium that has been only slightly enriched in the amount of uranium-235 present. The trick is to slow the neutrons down between each fission reaction. Low-energy, or slow, neutrons have a much greater probability of being absorbed by uranium-235 than by uranium-238 nuclei, whereas the faster neutrons emitted in the fission reactions are almost equally likely to be absorbed if they encounter either a uranium-235 or uranium-238 nucleus.

Thus, if the neutrons can be slowed down before encountering additional uranium, we have a better chance of producing additional fission reactions.

Using this idea, the first controlled chain reaction was achieved by Enrico Fermi and co-workers at the University of Chicago in 1942. Fermi constructed a *nuclear pile* consisting of blocks of pure graphite between which were interspersed small pieces of natural uranium (fig. 17.14). Graphite is a solid form of the element carbon, which has a relatively small mass number (12). Carbon does not absorb neutrons readily; neutrons that collide with carbon-12 nuclei just bounce off. In each collision, however, the neutrons lose some energy, since the carbon nuclei gain some recoil energy in the process. The graphite, therefore, serves to slow down the neutrons without absorbing them; such a material is called a *moderator.*

In addition to the moderator material and the uranium fuel elements, one other feature is essential to any reactor. Because of the real possibility that a chain reaction will grow very rapidly under the proper conditions, we need some means of controlling the rate of reaction. This is done with *control rods,* which contain a material that absorbs neutrons readily and can be inserted or removed from the pile in order to maintain the desired level of reaction. Boron is the material most commonly used for this purpose, because of its great ability to absorb neutrons. In Fermi's pile, however, the control rods were made of cadmium, which is also a good neutron absorber.

On December 2, 1942, Fermi and his co-workers slowly removed some of the control rods from their carefully constructed pile. By monitoring the rate of neutron flow at different points in the pile, they were able to establish that a self-sustaining chain reaction was indeed achieved. The reactor had gone *critical;* that is, for each fission reaction produced, one of the new neutrons generated went on to be absorbed by another uranium-235 nucleus, thus producing

another fission reaction. If more than one additional fission reaction is produced for each initial reaction, we say that the reactor is *supercritical;* if less than one is produced, the reactor is said to be *subcritical.*

The idea in starting up a reactor is to let the reactor go just slightly supercritical initially, until the desired reaction level is reached, and then to reinsert the control rods slightly to maintain the reactor at a critical or steady-state level. Inserting the control rods farther decreases the reaction level; the control rods therefore are like the gas pedal in an automobile. Frequent adjustment of some of the control rod positions can be used to fine-tune the level of reaction. Other control rods are designed to be rapidly inserted to shut down the reactor.

Plutonium Production

After Fermi's successful demonstration of a chain reaction in his nuclear pile, the next major step in reactor development took place at Hanford, Washington during the closing years of World War II. At Hanford, several reactors were built as part of the effort to produce nuclear weapons. Up to this point, we have not considered what happens when uranium-238 absorbs a neutron, other than to indicate that this does not usually result in fission. It turns out that Fermi's original objective of producing new elements heavier than uranium occurs in this case. A series of nuclear reactions produces plutonium, which is now the primary source of weapons material for fission bombs.

The reactions that generate plutonium-239 from uranium-238 are summarized in box 17.3 and shown in figure 17.15. The first step in this series is the absorption of the neutron by a uranium-238 nucleus to produce uranium-239. This, in turn, undergoes beta decay (half-life, 23.5 minutes) to produce an isotope of neptunium (Np^{239}), a new element with an atomic number of 93, one higher than that of uranium. This also undergoes beta decay (half-life, 2.35 days) producing

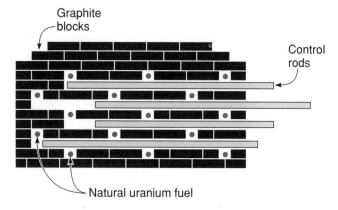

Graphite blocks

Control rods

Natural uranium fuel

Figure 17.14 A schematic diagram of Fermi's "pile," the first human-produced nuclear reactor. Small pieces of natural uranium were interspersed between bricks of graphite (carbon).

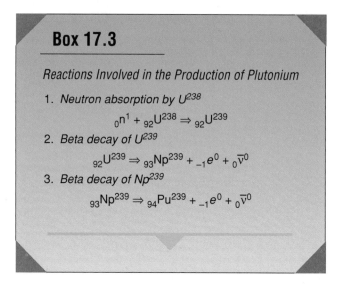

Box 17.3

Reactions Involved in the Production of Plutonium

1. *Neutron absorption by U^{238}*

$$_0n^1 + _{92}U^{238} \Rightarrow _{92}U^{239}$$

2. *Beta decay of U^{239}*

$$_{92}U^{239} \Rightarrow _{93}Np^{239} + _{-1}e^0 + _0\bar{\nu}^0$$

3. *Beta decay of Np^{239}*

$$_{93}Np^{239} \Rightarrow _{94}Pu^{239} + _{-1}e^0 + _0\bar{\nu}^0$$

Neutron
absorption

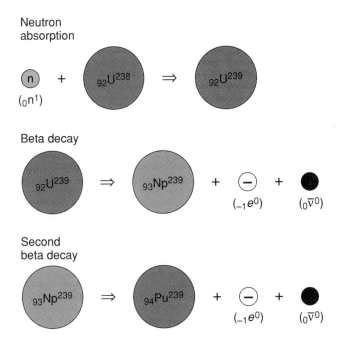

Beta decay

Second
beta decay

Figure 17.15 **Absorption of a neutron by uranium-238,
followed by two beta-decay reactions,
produces plutonium-239, which can also
be used as a fission fuel.**

yet another new element with an atomic number of 94, plutonium (Pu^{239}). Plutonium-239 is relatively stable; it has a half-life of approximately 24 000 years. Like uranium-235, however, it also readily undergoes nuclear fission when it absorbs a neutron.

Since nuclear reactors use either natural uranium or uranium only slightly enriched in uranium-235, the production of plutonium-239 is a natural by-product of the operation of a reactor. Also, it is a different element: it can be separated from uranium with chemical techniques because its chemical properties are different than those of uranium. Nuclear reactors can therefore be used to produce fissionable material for weapons. A crash program was begun at Hanford, Washington during the war years to build nuclear reactors for precisely this purpose. By the end of the war, enough plutonium had been produced to make two bombs.

Modern Power Reactors

Most of the modern power reactors that have been built since World War II do not use graphite as the moderator; instead, they use ordinary (*light*) water. Water (H_2O) contains nuclei of both hydrogen (H) and oxygen (O), but the hydrogen nuclei are most effective in slowing the neutrons. Unfortunately, hydrogen also absorbs neutrons to form the heavier isotopes of hydrogen, deuterium (H^2) and tritium (H^3). *Heavy* water is that made using deuterium in place of the ordinary lighter isotope of hydrogen. Heavy water absorbs

neutrons less readily than light water and is also sometimes used as a moderator.

The neutrons that ordinary hydrogen absorbs to form deuterium are removed from circulation; therefore, light water is not effective as a moderator when natural uranium is used as the fuel. Enriching the uranium-235 concentration slightly, however, from 0.7% to 2% or 3%, compensates for the neutrons lost to absorption by hydrogen, and a chain reaction can be achieved. Light water reactors must therefore use slightly enriched uranium as the fuel. Reactors that use heavy water as the moderator can use natural uranium as the fuel.

The advantage of using water as a moderator is that water can also be used as a *coolant,* removing excess energy in the form of heat from the reactor core. The kinetic energy of the neutrons and fission fragments released in the fission reactions shows up as heat as this energy is randomized through collisions with other atoms. Any large reactor must have some means of cooling the core, or the temperature will get so high that some of the reactor components may melt. The coolant circulates through the reactor carrying the energy generated in the fission reactions to the steam turbines, which produce electric power. The turbines themselves must be cooled in order to operate efficiently. It is heat from the water used to cool the turbines that is released to the atmosphere via the cooling towers; this water never passes through the reactor itself.

Any material that absorbs neutrons in a reactor, such as hydrogen, is referred to as a *poison*. Impurities in the moderator or fuel elements and even the fission products themselves can act as poisons to the reaction. The longer a reactor is operated, the more the poisons build up in the fuel elements and, of course, the more the uranium-235 fuel gets used up. The fuel rods must therefore be replaced from time to time (fig. 17.16). The spent fuel rods, containing uranium, plutonium, and the radioactive fission fragments must be stored or disposed of in some manner. Radioactive elements in spent fuel make up most of the nuclear wastes produced by nuclear reactors that generate electrical power.

Our current policy on waste disposal proposes burying these radioactive materials in solid rock formations, without separating the plutonium and remaining uranium from the fission fragments. (A mountain range in Nevada has been selected as the most suitable site, but this decision is still being challenged.) This proposal avoids the need for the expensive and environmentally hazardous processing involved in chemical separation. It has the disadvantage, however, of wasting fissionable material in the form of plutonium and uranium. It also requires that the disposal site remain stable for thousands of years because plutonium and uranium have much longer half-lives than most of the fission fragments. Plutonium-239 has a half-life of 24 000 years, whereas most of the fission fragments have half-lives of the order of several years or less. They decay quite rapidly, therefore, and would not have to be isolated from the environment nearly as long.

Figure 17.16 **An assembly containing fuel rods and control rods for use in a modern power reactor.**

Figure 17.17 shows a schematic diagram of a modern power reactor. The reactor core is contained within a thick-walled steel reactor vessel through which the coolant circulates. This, in turn, is contained within a heavily-reinforced concrete containment building that is designed to withstand strong pressure variations as well as to protect the reactor from external influences. The coolant passes through steam generators, and the steam passes through the steam turbines that turn the electrical generators. There is a lot of plumbing in a nuclear power plant.

Within the plant, but outside of the containment building, there is a control room where the controls for the pumps, control rods, and other equipment associated with the reactor are located. Here we also find temperature and radiation gauges and other monitoring equipment that tell the operators what is happening inside. Reactors in this country are designed with a lot of redundancy in safety equipment, and reactor operators are trained to deal with many contingencies. Because of the complexity of reactor operation, however, when accidents do occur, operator error is often a factor in either the initial event or the response to a problem. There is no such thing as a completely fail-safe reactor.

When nuclear power was first introduced in the late 1950s, it was regarded as a clean and inexpensive means of generating electrical power. Concerns regarding reactor safety and waste disposal have caused many people to modify that view, but it is certainly still true that nuclear reactors cause far less atmospheric pollution than burning of

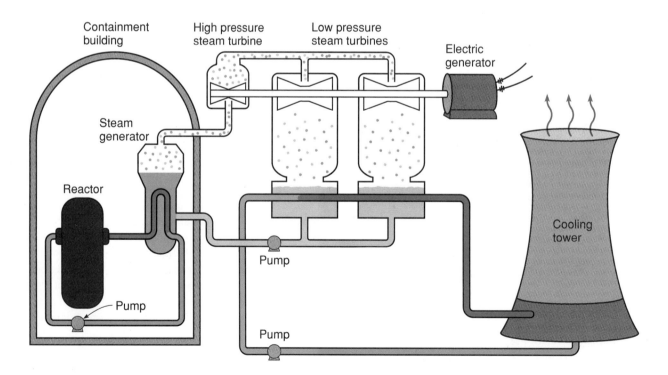

Figure 17.17 **A schematic diagram at a modern pressurized-water nuclear reactor. Hot water coming from the reactor is converted to steam when the pressure is reduced in the steam generators. The steam turns the turbines, which power the electric generators.**

fossil fuels such as coal or oil to generate power. Disposal of nuclear wastes has become a political issue, however, and reactor safety issues have also been a source of debate.

The development of nuclear power in the United States is at a virtual standstill now, primarily because of economic issues. The high costs and time required for initial construction and the concerns about accidents have caused utilities to back away from ordering new reactors. In Japan, Europe, and other parts of the world where fossil fuels are less available than in the United States, the use of nuclear power continues to expand, however. The future of nuclear power in the United States is uncertain as we approach the end of the century that saw its birth. The design of new, smaller reactors that are inherently stable may bring new life to the industry.

Box 17.4

What Happened at Chernobyl?

The Situation. Most of us can remember the publicity and concern associated with the 1986 reactor accident at Chernobyl in the USSR. In terms of loss of life and radiation releases to the environment, this was the worst nuclear reactor accident that had ever occurred anywhere in the world. The questions that concerned citizens wanted answered at the time were

1. What type of reactor was involved?
2. How did the accident occur?
3. Could a similar accident happen in the United States?

The Analysis. We will take each question in turn and briefly explain what is known about the reactor and the accident.

1. *The reactor.* The nuclear reactor at Chernobyl, like many others in the USSR, was a dual-purpose reactor designed both to generate electric power and to produce weapons-grade plutonium. It used graphite as the moderator, but also circulated water through the core as a coolant. The presence of ordinary water in the core requires some enrichment of uranium-235 (to about 2%), but not as much as would be required if water were also the moderator, as in the commercial reactors in this country.

   Because the water in the core was used only as a coolant, not as the moderator, the reactor had an unusual design characteristic: In the event of loss of the coolant or increase in temperature, the rate of reaction would actually increase, because the water used as the coolant absorbs neutrons. If the water is lost (or changes to steam becoming less dense), fewer neutrons are absorbed, and the chain reaction accelerates. This result is not possible in a reactor in which water also serves as the moderator, because loss of the moderator reduces the rate of reaction.

   In a reactor used to produce weapons-grade plutonium, the fuel rods must be removed after

The damaged reactor at Chernobyl. Rapid buildup of heat caused explosions that ignited the graphite moderator.

about 30 days. This is necessary to avoid consumption of the plutonium-239 through fission and the buildup of plutonium-240, another isotope of plutonium that does not undergo fission as readily. The Chernobyl-type reactors were therefore designed to allow easy access to the top of the reactor for replacement of fuel rods, as shown in the second photo. The building housing the reactor was not a containment building; that is, it was not designed to withstand strong pressure variations, as are the containment buildings housing U.S. reactors.

2. *The accident.* When the accident occurred, an experiment was being conducted to see whether the electrical generators could be used to provide power to the reactor pumps in case external power were lost. This required that the reactor be run at a low power level to simulate a condition in which it was being shut down. The person in charge of the experiment was an electrical engineer who was primarily interested in the genera-

Continued

Box 17.4 Continued

tor response and was not highly knowledgeable regarding the operation of the reactor.

The following description simplifies what in reality was a complex chain of events and errors. The reactor had been partially shut down in order to perform the experiment. In order to bring it back to the desired power level, most of the control rods were removed, and several other safety features were disengaged in order to reach the desired conditions. In the initial stages of the experiment, the water flow to the reactor core had been reduced, which caused a rapid increase in the fission reactions because of the characteristic mentioned earlier. The heat thus generated caused explosions that blew the top of the reactor building open and ignited the graphite moderator (which can burn much like coal).

The fire in the graphite had to be extinguished, so fire-fighters from nearby towns were called in. Of the thirty-one deaths directly caused by the accident, many were firemen who were exposed to high levels of radiation in fighting the fire. Radiation, in the form of fission fragments carried by the emissions from the fire, was dispersed over the surrounding countryside and, at decreasing levels, over much of western Europe. Some increase in cancer deaths among those exposed in the surrounding communities is expected to occur as a result, although these effects will not be seen for several years.

3. *U.S. comparisons.* A number of features of commercial reactors in the United States make an accident of the type that occurred at Chernobyl impossible. Most important, a loss of coolant slows the chain reaction because water is used as the moderator in our commercial reactors; an explosive increase in the chain reaction is therefore not possible. The partial meltdown that occurred at Three Mile Island in Pennsylvania was caused by heat generated from residual radioactivity of the fission fragments after a shutdown of the chain reaction. This can still be a serious problem, but does not result in an explosive buildup of the chain reaction.

(The N-reactor at Hanford, used for both plutonium production and power, is the U.S. reactor that is most similar to that at Chernobyl, but even it does not have the loss-of-coolant instability that is characteristic of Chernobyl-type reactors. The N-reactor is no longer in operation.)

The top of a Chernobyl-type reactor, showing a technician working on the square tops of the fuel-rod assemblies.

Secondly, reactor containment buildings in the United States (as well as in many other countries) are built with heavily-reinforced concrete, so that the fission fragments themselves are highly unlikely to escape from the containment building (or even from the reactor vessel itself) in the event of a partial meltdown or other accident. Despite its serious nature, the accident at Three Mile Island in Pennsylvania released very little radioactivity to the environment. The economic impact of the accident was considerable, however, due to the loss of the reactor and the cleanup costs.

Operator error was a major factor in the accident at Chernobyl. We would like to think that the training provided to U.S. reactor operators would preclude the serious errors in judgment that occurred at Chernobyl. Operator misjudgments have also been a problem, however, in accidents that have occurred in our own nuclear industry. The complexity of the details of reactor behavior are such that it is difficult to plan and train for all possible contingencies.

17.5 NUCLEAR WEAPONS AND NUCLEAR FUSION

In a nuclear reactor, the objective is to release energy from fission reactions in a controlled manner, never letting the chain reaction get out of hand. In a bomb, on the other hand, the objective is to release energy very quickly; a supercritical chain reaction is the desired state. How can this state be achieved? What are the necessary conditions for a nuclear explosion?

Critical Mass

The initial approach to this problem involved separating uranium-235 from uranium-238 in order to produce a highly enriched sample of uranium-235. If most of the uranium-238 could be removed, then neutrons emitted in the initial fission reactions would be much more likely to encounter other uranium-235 nuclei, and a rapidly increasing chain reaction should result. The other important consideration here is the size of the uranium sample, however. If the mass of uranium is too small, neutrons will escape through the surface of the sample before they have had sufficient opportunity to encounter other uranium-235 nuclei.

The concept of *critical mass* emerges from these considerations. A critical mass of uranium-235 is one just large enough for a chain reaction to be self-sustaining. For a mass smaller than the critical mass, too many of the neutrons generated in initial fission reactions escape through the surface of the uranium. Unless neutrons are provided from a separate neutron source, the reaction would die out under these conditions.

For a mass larger than the critical mass, though, there is a good chance that one or more of the neutrons produced in an initial fission reaction will be absorbed by other uranium-235 nuclei and produce additional fission reactions. If, on the average, at least one of the two to three neutrons produced goes on to generate an additional fission reaction, the chain reaction will be sustained. If, for each initial reaction, more than one neutron goes on to generate an additional fission, the chain reaction will grow very rapidly, because the time between reactions is very short. This is the *supercritical* state necessary for an explosion. Obviously, the larger the mass and volume of uranium available, the greater the probability of absorption by other uranium-235 nuclei.

The final problem in producing a nuclear explosion is that of creating a supercritical mass without having it blow apart prematurely. The energy released in a fission chain reaction causes rapid heating and expansion. If a supercritical mass were to be built up slowly from subcritical pieces, the mass would begin to come apart as soon as it became supercritical, and the bomb would fizzle. One approach to solving this problem is to bring two subcritical pieces of pure uranium-235 together very rapidly to produce a strongly supercritical mass. A gun arrangement such as that shown in figure 17.18

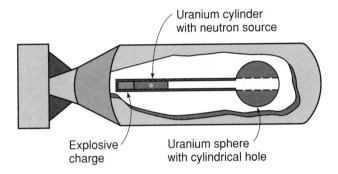

Figure 17.18 **The gun concept used in the Little Boy uranium bomb design. A subcritical-size cylinder of uranium-235 is fired into the hole in a subcritical sphere of uranium-235 to provide a supercritical mass of uranium-235.**

Uranium cylinder with neutron source

Explosive charge

Uranium sphere with cylindrical hole

was considered for the first uranium bomb. A subcritical cylinder of uranium-235 is fired into a subcritical sphere of uranium-235 containing a cylindrical hole, thus quickly assembling a supercritical mass. A neutron source must also be present in order to initiate the reaction.

The major problem in producing a uranium bomb is the extreme difficulty of separating uranium-235 from the much more abundant isotope, uranium-238. The gas diffusion plants at Oak Ridge, Tennessee, were only able to produce enough pure uranium-235 during the war years for one bomb. Building an arsenal at that rate would have been a very slow and expensive prospect. It became apparent that it might make more sense to build bombs using plutonium-239, which is a natural by-product of reactors that use natural uranium (or enriched uranium) for fuel. The reactors built at Hanford, Washington during World War II were designed to produce plutonium for weapons as noted in section 17.4.

The fission reaction of plutonium-239 is different from that of uranium-235, however, so that the gun design used for the uranium bomb would not work with plutonium. The primary difference is that plutonium-239 absorbs fast neutrons much more readily than uranium-235, so that we do not have to slow the neutrons down to increase the probability of absorption. This causes the chain reaction to grow even more rapidly than that for uranium, and two subcritical pieces cannot be brought together quickly enough to avoid the fizzle produced by premature disintegration.

The design used for plutonium bombs involves the concept of *implosion:* explosives are arranged around a subcritical mass of plutonium and fired in a synchronized fashion to create a tremendous inward pressure on the plutonium. This increases the density of the plutonium sample enough to make the mass become supercritical (fig. 17.19). The same number of atoms are now confined in a smaller volume, thus increasing the probability of the absorption of neutrons by other plutonium-239 nuclei.

By the end of World War II, enough weapons material had been assembled to produce just three nuclear bombs.

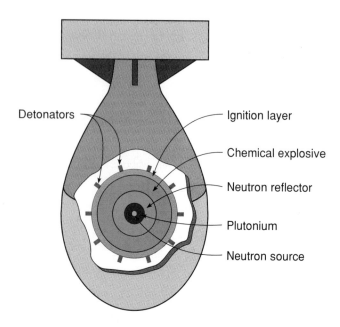

Detonators

Ignition layer

Chemical explosive

Neutron reflector

Plutonium

Neutron source

Figure 17.19 **The Fat Man plutonium bomb used chemical explosives arranged around a barely subcritical mass of plutonium-239. When imploded by the explosives, the increased density makes this mass supercritical.**

Two of them were plutonium bombs, dubbed *Fat Men* because of their round shape. The third bomb was a uranium bomb, called *Little Boy* because of its slim shape. In an historic test, one of the plutonium bombs was exploded at White Sands, New Mexico in the summer of 1945, producing the first of the awesome mushroom clouds associated with nuclear explosions. Shortly thereafter, the other two bombs were dropped on Hiroshima and Nagasaki, Japan.

Throughout the war years, the effort to build and test nuclear bombs was regarded as a race with Nazi Germany, where the fission reaction had been originally discovered. Many of the scientists working on the Manhattan Project were European refugees. The thought that Germany might acquire the bomb first was a horrifying possibility. As the war was winding down, however, and the success of the bomb-building effort was approaching, it became apparent that Germany no longer had the resources to produce a nuclear bomb. A debate began among the scientists involved in the project as to how the bomb should be used. Many scientists favored a demonstration of the bomb's effect without actually using it on a military target.

A demonstration involved serious problems, however, not the least of which was the fact that we had only three bombs, and the design of one, the uranium bomb, could not be tested beforehand because there was only one available. Ultimately the decision to drop the bombs on Japan was made by the top military authorities, including the President, Harry S. Truman. The wisdom of that decision has been debated ever since.

Following the war, weapons production continued as the Hanford reactors continued to generate more plutonium. Soon Russia had acquired the ability to manufacture fission bombs, and the pressure to proceed to the next step in the race increased. This step was the development of the hydrogen bomb, which was first successfully tested in 1952. A hydrogen bomb involves nuclear fusion rather than fission.

The Fusion Reaction

What happens in a fusion, or hydrogen, bomb? Nuclear fusion is another kind of nuclear reaction that also releases large quantities of energy. It is the energy source of the sun and other stars as well as that of the hydrogen bomb. We could not exist without the sun's energy, so fusion is obviously an important aspect of everyday phenomena, even though we may not be aware of the process. How does nuclear fusion differ from fission, and how can we generate a chain reaction involving fusion?

Nuclear fusion involves the combination of very small nuclei to form somewhat larger nuclei. In this sense it is the exact opposite of nuclear fission, which, of course, involves the splitting of large nuclei into smaller fission fragments. The fuel for fusion is therefore made up of very light elements, usually isotopes of hydrogen, helium, and lithium. (Lithium has an atomic number of 3.) The end product is often the particularly stable nucleus of the common isotope of helium (He^4), which we have already encountered as the alpha particle.

As long as the mass of the helium-4 nucleus and other reaction products is slightly less than the sum of the masses of the isotopes combining to form the helium nucleus, the mass difference will show up as other forms of energy, according to Einstein's $E = mc^2$ formula. One possible reaction, for example, is the combination of two isotopes of hydrogen, deuterium (H^2) and tritium (H^3) to form helium-4 plus a neutron.

$$_1H^2 + {}_1H^3 \Rightarrow {}_2He^4 + {}_0n^1$$

As shown in figure 17.20, the sum of the masses of the two particles on the right side of this equation is less than the sum of the masses of the two particles on the left side. The alpha particle and the neutron will therefore have a greater total kinetic energy than that of the initial two particles.

The problem in getting this to happen is that all nuclei are positively charged and therefore repel one another. Large initial kinetic energies are needed to overcome this repulsive force so that the two nuclei can combine. High densities of material are also necessary in order to increase the probability that the reactions will occur; thus, the material that is to undergo fusion must be confined in a very small space at very high temperature. These two requirements are usually incompatible.

The chain reaction that results under these conditions is called a *thermal chain reaction*. Very high temperatures are needed to initiate the reaction, and the energy released in the reaction raises the average kinetic energy of the reactants, and therefore the temperature. Chemical explosions are also thermal chain reactions, but the temperatures required for chemical reactions are much lower and more easily attainable. The temperatures required for nuclear thermal chain reactions are of the order of a million degrees centigrade or more!

It turns out that the easiest way to get both the high temperatures and the high densities required for a fusion chain reaction is to explode a fission bomb to initiate the fusion reaction. The high temperature produced by the fission bomb creates the high kinetic energies necessary for fusion. The fission explosion also compresses the fusion fuel momentarily; the time is very short, but it is still long enough for

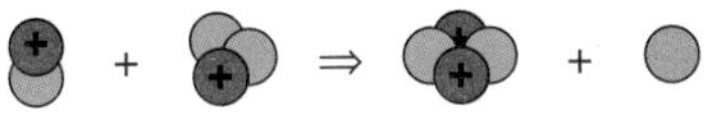

$${}_1H^2 + {}_1H^3 \rightarrow {}_2He^4 + {}_0n^1$$

Masses

${}_1H^2$	2.014102 u	${}_2He^4$	4.002603 u
${}_1H^3$	3.016050 u	${}_0n^1$	1.008665 u
	5.030152 u		5.011268 u

$\Delta m = 5.030152\ u - 5.011268\ u = .018884\ u \Rightarrow KE$

Figure 17.20 A deuterium nucleus and a tritium nucleus combine to form a helium-4 nucleus and a neutron. The difference in mass leads to an increase in the kinetic energy of the emerging particles.

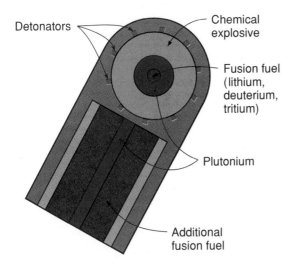

Figure 17.21 A fission bomb is exploded around the fusion fuel in order to produce the high temperatures and density required for a fusion chain reaction in a hydrogen (or thermonuclear) bomb.

considerable additional energy to be released from fusion reactions (fig. 17.21). This is essentially what happens in a hydrogen bomb (also called a *thermonuclear* bomb).

Hydrogen bombs can be made in various sizes, unlike fission bombs, which are restricted to the size associated with the critical mass of the fissionable material. Much larger energy yields are therefore possible with fusion bombs than with pure fission bombs. Even though the energy released per reaction is smaller for fusion reactions than for the typical fission reaction, pound for pound the fusion reactions pack more wallop because fusion fuels consist of very light elements. Hydrogen bombs can be made with an explosive power equivalent to that of 20 million tons of TNT or more. (Tons of TNT has become the standard basis for comparison in quoting bomb yields.) Both fission and fusion bombs now make up the arsenals of the nuclear powers.

Controlled Fusion

Producing controlled fusion reactions in a reactor that might be used as a commercial power source has not yet been accomplished. The problem again is that of confining the fuel at very high temperatures in a very small space for long enough to release a significant amount of energy from fusion reactions. Since any solid will melt at temperatures well below those required for fusion reactions, magnetic fields must be used to confine the fuel, or some other scheme must be devised for keeping things together (fig. 17.22). Another approach involves bombarding a small pellet of fusion fuel with laser beams (or particle beams) coming in from several directions to both heat and compress the pellet.

Research continues on this problem. Projections made in the 1970s that we would have working fusion reactors by the 1990s were clearly optimistic; it now seems that this goal may not be achieved until well into the 21st century. We can-

Figure 17.22 The Tokamak Fusion Test Reactor at Princeton, New Jersey, designed to confine and heat fusion fuels using magnetic fields.

not even be completely confident that an economically feasible reactor will ever be built, but we have already invested heavily in the effort, and we expect that, someday, the goal will be reached. Experimental reactors such as the Tokamak (fig. 17.22) have generated energy from fusion, but have not yet reached the break-even point, where as much energy is released as is required to initiate the reaction. We need to do much better than this, obviously, to make the process commercially viable.

Considerable excitement was generated a few years ago (1988–89) when scientists working in Utah claimed to have achieved *cold fusion* in a cell that did not require extraordinary temperatures or densities. Their cell involved a palladium electrode immersed in a beaker containing heavy water (in which deuterium replaces the ordinary isotope of hydrogen). Passing a current through the cell draws deuterium atoms into the spaces between the atoms in the palladium electrode. The Utah group claimed to have observed excess heat that could not be explained by normal chemical or physical processes occurring in the cell.

Although certain mechanisms might produce occasional fusion reactions under these circumstances, it is not expected that enough fusion could occur in this manner to produce usable quantities of energy. If it were possible, however, the commercial potential would be enormous, and so the claims of the Utah group generated a great deal of publicity and public interest. Many workers have attempted to reproduce their experiments, but so far the results have been disappointing. At the time of this writing, the prevailing opinion among physicists is that usable fusion energy is unlikely to be obtained from cold fusion.

Obviously, we do not yet know the best approach to achieving energy from fusion. Work is continuing on magnetic containment as well as the particle-beam, or laser, approaches. A number of scientists are working on the cold-fusion concept also. If a breakthrough occurs in the near future, some of today's students of science and engineering will be involved in a new expansion of applications of nuclear power.

SUMMARY

This chapter has traced how knowledge of the nuclear constituents, the proton and the neutron, led to an understanding of isotopes and of what was happening in radioactive decay. The discovery of the neutron also provided a new tool for probing and modifying nuclei, which led ultimately to the discovery of nuclear fission. Nuclear fission and fusion are nuclear reactions that are capable of releasing large amounts of energy, which can be utilized both in nuclear weapons and in nuclear reactors.

Nuclear Structure. Experiments in which alpha particles were scattered from various nuclei led to the discovery of both the proton and neutron as constituents of the nucleus. The number of protons in a nucleus is the atomic number of an element, and the total number of protons and neutrons (nucleons) is the mass number of a given isotope.

Radioactive Decay. Radioactive decay is a spontaneous nuclear reaction in which a particle or gamma ray is emitted, and the structure of the nucleus changes. One element is transformed into another in the processes of alpha or beta decay. The half-life is the time required for half of the original number of radioactive nuclei to undergo decay.

Lithium-7 $(_3Li^7)$

$\dfrac{\begin{array}{l}3 \text{ protons} \\ 4 \text{ neutrons}\end{array}}{7 \text{ nucleons}}$

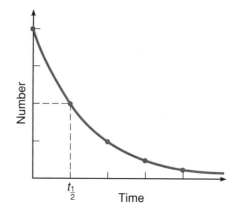

Nuclear Reactions and Fission. In any nuclear reaction, one element may change into another, but the total charge number (atomic number) and mass number (nucleon number) are conserved. The fission reaction was discovered by bombarding uranium samples with neutrons, leading to the splitting of the uranium nucleus into fission fragments and the emission of additional neutrons.

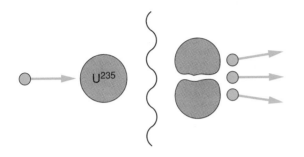

Nuclear Reactors. Nuclear reactors generate energy by allowing a controlled nuclear-fission chain reaction to take place. The moderator slows down the neutrons in order to increase the probability of absorption by uranium-235, the fissionable isotope of uranium. Control rods absorb neutrons to control the reaction level, and the coolant carries off the energy generated.

Weapons and Fusion. The primary material used in fission weapons is plutonium, which is produced in reactors when neutrons are absorbed by uranium-238, the most common isotope of uranium. Nuclear fusion involves the combination of small nuclei to form larger nuclei, which also releases energy. Hydrogen bombs use a fission trigger to produce the high temperatures and densities required for much larger fusion explosions.

QUESTIONS

Q17.1 In 1919, Rutherford bombarded a sample of nitrogen gas with a beam of alpha particles.

  a. In addition to alpha particles, what other particle emerged from the nitrogen gas in this experiment?

  b. What conclusion did Rutherford draw from this experiment?

Q17.2 Is it possible for two atoms of the same chemical element to have different masses? Explain.

Q17.3 Is it possible for atoms of the same chemical element to have different chemical properties? Explain.

Q17.4 Which number, the mass number or the atomic number, determines the chemical properties of an element? Explain.

Q17.5 Is it possible for two different atoms of the same chemical element to have different numbers of neutrons in their nuclei? Explain.

Q17.6 In a nuclear reaction, is it possible for the total mass of the products of the reaction to be less than the total mass of the reactants? Explain.

Q17.7 In alpha decay, do we expect the atomic number of the daughter nucleus to be equal to, greater than, or less than the atomic number of the isotope that decays? Explain.

Q17.8 In beta decay, do we expect the atomic number of the daughter nucleus to be equal to, greater than, or less than the atomic number of the decaying isotope? Explain.

Q17.9 In gamma decay, do we expect the atomic number of the daughter nucleus to be equal to, greater than, or less than the atomic number of the decaying isotope? Explain.

Q17.10 Do all radioactive substances decay at the same rate? Explain.

Q17.11 In a time equal to two half-lives for a certain radioactive isotope, would you expect that all of that isotope will have decayed? Explain.

Q17.12 In a chemical reaction, the number of atoms of any given element present in the reactants is the same as that in the products of the reaction. Is this true also in a nuclear reaction? Explain.

Q17.13 Chemical reactions and nuclear reactions can both release energy. On the average, would you expect the energy released in a chemical reaction to be greater than, equal to, or less than that released in a nuclear reaction? Explain.

Q17.14 Why do we expect fission fragments to have higher neutron numbers than stable isotopes of the same element, and therefore to be radioactive? Explain.

Q17.15 The most common isotope of uranium is U^{238}. Is this the isotope that is most likely to undergo fission? Explain.

Q17.16 What property of the fission reaction leads to the possibility of a chain reaction? Explain.

Q17.17 What is the function of the moderator in a nuclear reactor? Explain why the moderator is needed in order to obtain a chain reaction using natural uranium.

Q17.18 Do the control rods in a nuclear reactor absorb or emit neutrons? Explain.

Q17.19 If you wanted to slow down the chain reaction in a nuclear reactor, would you remove or insert the control rods? Explain.

Q17.20 Will a reactor using ordinary water as the moderator be able to use natural, unenriched uranium as a fuel? Explain.

Q17.21 If plutonium and uranium are removed from the spent fuel of a nuclear reactor, will the remaining nuclear wastes need to be stored for thousands of years before they will be nonradioactive? Explain.

Q17.22 What was the purpose of the nuclear reactors that were built at Hanford, Washington during World War II? Explain.

Q17.23 Is nuclear fusion essentially the same process as fission? Explain.

Q17.24 Is nuclear fission the primary process involved in the energy generated in the sun? Explain.

Q17.25 Do we currently have commercial nuclear reactors that use nuclear fusion as their energy source? Explain.

Q17.26 Which can produce larger yields of energy, a fission weapon or a fusion weapon? Explain.

EXERCISES

E17.1 Sodium has an atomic number of 11 and an atomic weight of 22.99. How many neutrons would you expect to find in the most common isotope of sodium?

E17.2 $_{94}Pu^{239}$ is an isotope of plutonium produced in nuclear reactors.

  a. How many protons are in the nucleus of this isotope?

  b. How many neutrons are in the nucleus of this isotope?

E17.3 Uranium-238 undergoes alpha decay. Complete the reaction equation for this decay and identify the daughter nucleus.

$$_{92}U^{238} \Rightarrow ? + \alpha$$

E17.4 The fission fragment barium-142 undergoes beta decay. Complete the reaction equation and identify the daughter nucleus.

$$_{56}Ba^{142} \Rightarrow ? + _{-1}e^0 + _0\bar{v}^0$$

E17.5 Oxygen-15 is a radioactive isotope of oxygen that undergoes beta-positive decay, in which a positive electron (or positron) is emitted. Complete the reaction equation and identify the daughter nucleus.

$$_8O^{15} \Rightarrow ? + _{+1}e^0 + _0v^0$$

E17.6 Suppose that you have 1000 atoms of a radioactive substance with a half-life of 2 hours.

  a. How many atoms of that element remain after 4 hours?

  b. How many atoms remain after 8 hours?

E17.7 When you measure the rate of radioactivity of a given isotope 12 days after making an initial measurement, you discover that the rate has dropped to one-fourth of its initial value. What is the half-life of this isotope?

E17.8 How many half-lives must go by in order for the radioactivity of a given isotope to drop to

  a. One-sixteenth of its original value?

  b. One sixty-fourth of its original value?

E17.9 Suppose that we discover that one of the fission fragments for a given fission reaction of uranium-235 is iodine-133 and that four neutrons are emitted in this reaction. Complete the reaction equation and identify the other fission fragment.

$$_0n^1 + _{92}U^{235} \Rightarrow ? + _{53}I^{133} + 4_0n^1$$

E17.10 Suppose that two deuterium nuclei ($_1H^2$) combine in a fusion reaction in which a neutron is emitted. Complete the reaction equation and identify the resulting nucleus.

$$_1H^2 + _1H^2 \Rightarrow ? + _0n^1$$

CHALLENGE PROBLEMS

CP17.1 Using the periodic table found on the inside back cover, we can get some idea of how the number of neutrons increases relative to the number of protons as we look at heavier elements. By rounding off the atomic weight to the nearest whole number, we can estimate the total number of nucleons (neutrons and protons), and the atomic number tells us the number of protons in the nucleus.

   a. What are the neutron and proton numbers for carbon (C), nitrogen (N), and oxygen (O)?

   b. What is the ratio of neutrons to protons for the stable isotopes of these three elements? (Ratio = N_n / N_p)

   c. Taking three elements near the middle of the table, silver (Ag), cadmium (Cd), and Indium (In), find the number of neutrons and protons for each in the same manner.

   d. Compute the ratio of neutrons to protons for each of these elements and find their average ratio (add the three ratios together and divide by 3).

   e. Repeat the process of steps *c* and *d* for thorium (Th), protactinium (Pa), and uranium (U).

   f. Compare the ratios of parts *b, d,* and *e.* Can you see why there are extra neutrons when uranium or thorium undergo fission?

CP17.2 Uranium and thorium are the radioactive isotopes found in some abundance in the earth's crust. As each isotope of these elements decays, new radioactive elements are created that have much shorter half-lives than uranium or thorium themselves. A whole series of alpha and beta decays occurs until a stable isotope of lead (Pb) is reached. One such series begins with the isotope thorium-232. The elements in the series are as follows:

Th $\Rightarrow$ Ra $\Rightarrow$ Ac $\Rightarrow$ Th $\Rightarrow$ Ra $\Rightarrow$ Rn $\Rightarrow$ Po $\Rightarrow$ Pb $\Rightarrow$ Bi
$\Rightarrow$ Po $\Rightarrow$ Pb

   a. Using the periodic table, find the atomic numbers for all of these elements.

   b. Identify which of the reactions in this series involve alpha decay and which involve beta decay. (The change in atomic number provides all the information you need.)

   c. Write the reaction equations for the first three decays in this series.

   d. Fill in the mass numbers for all of the isotopes in this series.

CP17.3 Consider the following fusion reaction: $_1H^2 + _1H^2 \Rightarrow _2He^3 + _0n^1$. From tables of nuclear masses, we can find the following masses for the reactants and products in this reaction:

   $H^2 = 2.014102$ u $He^3 = 3.016029$ u   n = 1.008665 u

   a. Find the mass difference, Δm between the reactants and the products in this reaction.

   b. Following the procedure used in Box 17.2, convert this mass difference to energy units.

   c. Is there energy released in this reaction? If so, where does it go? Explain.

CP17.4 Nuclear power has been a source of constant controversy over the last few decades. Although its use has grown during that time, we still get over half of our electric power by burning fossil fuels. Because the environmental and economic considerations are very different for these different energy sources, people often have a hard time keeping them in perspective.

   a. Burning fossil fuels produces carbon dioxide as a natural by-product. Carbon dioxide is one of the gases that contributes to the greenhouse effect and global warming, as discussed in chapter 9. Is this a problem with nuclear power also? Explain.

   b. What environmental problems are associated with nuclear power that are not present in the burning of fossil fuels? Explain.

   c. What environmental problems are associated with fossil fuels that are not present in the use of nuclear power? Explain.

   d. What environmental problems are common to both of these sources of power? Explain. (Check the discussion of heat engines in chapter 10.)

   e. In balance, which of these power sources would you choose to develop further if other options were not seen as being feasible? Explain. (Reasonable people may differ here!)

HOME EXPERIMENTS AND OBSERVATIONS

HE17.1 A better understanding of the concept of half-life and the associated exponential decay curve may be gained by using piles of pennies (or other suitable stackable objects) to represent atoms.

a. Collect as many pennies as you can find. Fifty to one hundred would make a good starting point.

b. Divide your initial pile into two stacks, placed side by side. The left-hand pile represents the number of atoms remaining after one half-life has passed.

c. Divide the right-hand pile in half. Place the left-hand stack for this division next to the original left-hand stack. This represents the number of atoms remaining after two half-lives have passed.

d. Continue this process, always dividing the remaining right-hand stack in half and placing the left-hand stack obtained from this division next to those stacks already accumulated. The resulting row of stacks, each one smaller than the preceding one, has the form of an exponential decay curve. Can you identify what each stack in your collection represents?

HE17.2 If you have access to a dart board, you can perform an experiment that may give you some feeling for what happens in a chain reaction and for the concept of critical mass.

a. On a piece of paper, draw a regular array of circles all the same size (say 2 cm in diameter) and separated by equal distances (say 4 cm center-to-center).

b. With this piece of paper tacked to your dart board, start with three darts and throw them at the board without aiming for any particular circle. For each dart that hits a circle, award yourself three more darts. (The idea is to throw the darts somewhat randomly, so that they hit the board, but are not grouped at any one spot.)

c. If you find yourself with no darts remaining after a few tries, your array is subcritical. If you find yourself with a continually increasing number of additional chances, your array is supercritical. If roughly one out of three darts thrown hits a circle, the array is just critical and the game can be barely sustained. Which is true for your array? Does it, depend upon who is throwing the darts?

d. Try different-sized circles with different spacings. What size and spacing comes the closest to producing a critical array?

e. Compare the dart-board array to the situation involved in defining a critical mass for uranium or plutonium. What are the similarities and differences?

Part 6

SUPPLEMENTARY

Topics

18 *Einstein and Relativity*

Have you ever had the experience of sitting in a stationary bus and peering out the window at another bus sitting alongside of yours? Suddenly the other bus moves forward, but you have the distinct sensation that your own bus is moving backwards (fig. 18.1). The sensation lasts briefly until the other bus moves out of your view and you realize that you are not moving.

Your senses have deceived you because of the motion of your *frame of reference.* We normally measure our own motion with respect to objects that we expect to remain at rest. If these objects are fixed to the surface of the earth somehow, then our frame of reference is the earth. Position, velocity, and acceleration are measured by reference to that frame. If objects that we have identified with that supposedly fixed frame of reference suddenly move, our senses interpret that as though we are moving. Most of us have experienced that sensation.

All motion must be measured with respect to some frame of reference, and that frame of reference may also be moving. The earth, of course, rotates on its axis and also undergoes orbital motion about the sun. The sun, in turn, is moving with respect to other stars, and so on. Something that is at rest in one frame of reference may be moving with respect to some other frame of reference. We need to define our frame of reference in order to provide a complete description of any motion.

The problem of defining a frame of reference, and of describing how a given motion might look as seen from different frames of reference, was discussed by both Galileo and Newton. This is not a difficult problem if the velocities we are dealing with are not unusually large; relative motion in this sense is part of our everyday experience. The motion of a boat relative to a moving stream, for example, is familiar to many of us.

If, however, we imagine ourselves moving along with a light beam, as Einstein did as a boy, some very interesting possibilities arise. This, of course, is *not* part of our everyday experience. With a little imagination, though, we should be able to extrapolate from ordinary velocities to velocities near the extremely high velocity of light. This is exactly what Einstein did. The unusual consequences of this extrapolation are described by his theory of relativity.

The theory of special relativity, introduced by Einstein in 1905, deals with the special case in which different frames of reference are moving at constant velocity with respect to one another. The theory of general relativity, which Einstein developed roughly ten years later, deals with cases of accelerated frames of reference. Together these theories have revolutionized our view of the universe.

Figure 18.1   The forward motion of an adjacent bus can give you the impression that your own bus is moving backwards.

Chapter Objectives

After examining how relative motion is handled in classical physics, this chapter introduces Einstein's postulates of special relativity and explores their consequences for our views of space and time. We then consider how Newton's laws of motion must be modified to be valid for very high velocities, and explore the idea of mass/energy equivalence. Finally we briefly consider the theory of general relativity.

The primary objective is to give you a sense of what Einstein's theories of relativity are about. They do have consequences for space travel, nuclear energy, and other phenomena that are becoming a part of our modern reality. The ideas may stretch your mind more than many of those we have already introduced, but that is part of the fun of playing with relativity.

Chapter Outline

❶ *Relative motion in classical physics.* How did Galileo and Newton handle ideas involving relative motion? How do velocities add when the frame of reference itself is moving?

❷ *The velocity of light and Einstein's postulates.* What is an appropriate frame of reference for measuring the velocity of light? How did experimental evidence lead to the radical new view suggested by Einstein's postulates of the theory of special relativity?

❸ *Space and time effects.* How do Einstein's postulates lead us to some apparently strange conclusions regarding the nature of time and space? What are the effects that we call *time dilation* and *length contraction*?

❹ *Relativistic dynamics.* How must Newton's second law of motion be modified in order to make it valid for very high velocities? What does the concept of mass/energy equivalence mean, and how does it arise?

❺ *General relativity.* How does the theory of general relativity differ from that of special relativity? What new consequences emerge from the general theory?

18.1 RELATIVE MOTION IN CLASSICAL PHYSICS

Imagine that you have dropped a twig in the water of a moving stream and are observing its motion. The twig moves with the stream, of course, because the water is moving (fig. 18.2). What is the velocity of the twig with respect to the bank of the stream? What is its velocity with respect to the water, and how are these two velocities related? How does this picture change if we consider a rowboat or motor boat moving on a moving stream? These simple questions are easily handled within the framework of the classical mechanics developed by Galileo and Newton.

Velocity Addition

When you drop a twig in the water, it quickly reaches the velocity of the stream. Once that happens, its velocity with respect to the bank is the same as that of the water with respect to the bank. Its velocity with respect to the water, on the other hand, is zero. If you are watching the twig from a boat that is also floating with the current, the twig does not appear to be moving.

A more interesting situation arises if your boat is moving (with the aid of oars, a motor, or a sail) with respect to the water. In this case, the velocity of the boat relative to the water is not zero; the boat moves relative to the water and the water moves relative to the stream bank. This situation is illustrated in figure 18.3.

If the boat and the stream are moving in the same direction, intuition suggests that the velocity of the boat relative to the water and that of the water relative to the earth should add to yield the overall velocity of the boat relative to the earth. In symbols, this idea takes the following form:

$$\vec{v}_{be} = \vec{v}_{bw} + \vec{v}_{we},$$

where $\vec{v}_{be}$ represents the velocity of the boat relative to the earth, $\vec{v}_{bw}$ the velocity of the boat relative to the water, and $\vec{v}_{we}$ the velocity of the water relative to the earth. If we are careful with the subscripts, we see that the shared subscript, w, on the right side of this equation drops out, leaving the two outer subscripts, b and e, on the left side of the equation.

From figure 18.3, we can justify our expectation that these velocities should add by considering the distances that the water and the boat travel in a given time interval. In the time t, a piece of wood floating with the stream moves a distance d_{we}, which is the distance that the water has moved relative to the earth. In that same time, however, the boat has moved a distance d_{bw} relative to the water, so that it is now that much ahead of the piece of wood. The total distance that the boat has moved relative to the earth in this time, d_{be}, is obviously the sum of the other two distances:

$$d_{be} = d_{bw} + d_{we}.$$

If we divide both sides of this equation by the time interval t, we get

$$\frac{d_{be}}{t} = \frac{d_{bw}}{t} + \frac{d_{we}}{t}.$$

Since speed is distance divided by time, this is just

$$v_{be} = v_{bw} + v_{we}.$$

In other words the speed of the boat relative to the earth (the total distance divided by time) is equal to the speed of the boat relative to the water plus the speed of the water relative to the earth. Since speed is just the magnitude of the velocity in this case, and all three velocities have the same direction (downstream), we can also say that the velocities add.

Although we have illustrated this idea for the simple case in which all three velocities are in the same direction, the

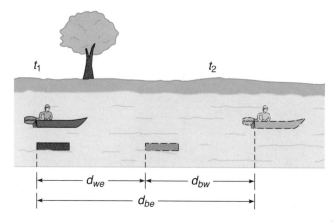

Figure 18.3 A motorboat and a piece of wood move downstream. In a given time, the boat moves a distance d_{be} relative to the earth, and the piece of wood moves a distance d_{we}.

Figure 18.2 A floating stick moves with the current. What is its velocity with respect to the stream bank?

velocity addition result is valid more generally. For example, if the boat is moving upstream, we can still write

$$\vec{v}_{be} = \vec{v}_{bw} + \vec{v}_{we}.$$

In a one-dimensional problem like this, vector directions can be indicated with plus or minus signs. If we choose the upstream direction to be positive, we have

$$v_{be} = v_{bw} + (-v_{we}) = v_{bw} - v_{we},$$

since the water is moving downstream (fig. 18.4).

If the boat can travel at a speed of 10 m/s relative to the water, and the water is moving at a speed of 4 m/s relative to the earth, then the magnitude of the boat's velocity relative to the earth is

$$v_{be} = 10 \text{ m/s} - 4 \text{ m/s} = 6 \text{ m/s}.$$

Since this result is positive, the boat is moving upstream relative to the earth. To do so, of course, the boat must be able to travel with a speed relative to the water that is larger than the speed of the current. If it could not, then v_{be} would be negative, and the boat would be carried downstream, as indicated in figure 18.4.

Velocity Addition in Two Dimensions

The principle of addition of relative velocities can also be extended to two or three dimensions. Suppose, for example, that we are traveling across the stream, as shown in figure 18.5. If we point the motorboat directly towards a point straight across the stream, will the boat end up at that point? Not if the stream is moving.

We can analyze this situation just as we did in the one-dimensional case. At the same time that the boat is moving in a direction straight across the stream relative to the water, the water is moving downstream relative to the earth. The displacement of the boat relative to the water would have to be added to the displacement of the water relative to the earth to get the overall displacement of the boat relative to the earth. Since these displacements are not in the same direction, they must be added as vectors:

$$\vec{D}_{be} = \vec{D}_{bw} + \vec{D}_{we}.$$

Since velocity is displacement divided by time, dividing both sides of this equation by time yields the familiar velocity-addition formula:

$$\vec{v}_{be} = \vec{v}_{bw} + \vec{v}_{we}.$$

The addition process is most easily handled by the head-to-toe vector-addition method introduced in chapter 2 and discussed in appendix C. This process is illustrated in figure 18.6 for both the displacement vectors and the velocity vectors. Notice that the velocity of the boat relative to the earth is not equal to the simple numerical sum of the other two velocities. Since it is the hypotenuse of a right triangle in this case, it is equal to the square root of the sum of the squares of the other two sides:

$$v_{be} = \sqrt{v_{bw}^2 + v_{we}^2}.$$

If we want to hit a point on the bank that is directly across the stream, we must point the boat at an angle to the line drawn perpendicular to the bank, as shown in figure 18.7. When the two velocities add in this case, they yield a velocity for the boat relative to the earth that is straight across the stream. The magnitude of this velocity will be smaller than that of the boat relative to the water, however, as you can see from the vector diagram.

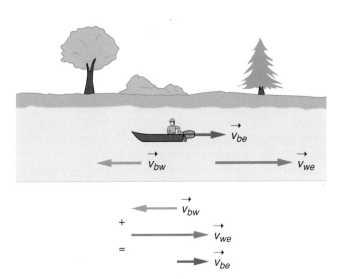

Figure 18.4 A motorboat moving upstream loses ground to the stream current if it cannot travel fast enough relative to the water (v_{bw}).

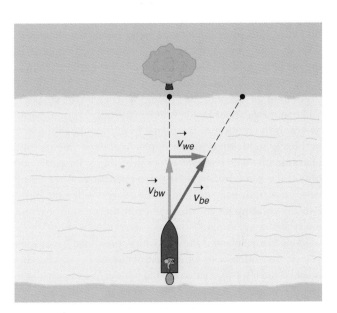

Figure 18.5 A motorboat pointed straight across the stream ends up at a point on the opposite bank that is somewhat downstream.

Displacement

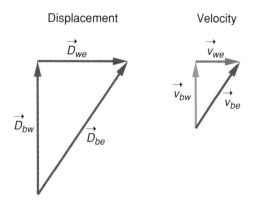

Velocity

Figure 18.6 **The displacement vectors and the velocity vectors must be added head-to-toe to obtain the total displacement and velocity for the case pictured in figure 18.5.**

We can apply the same type of analysis to an airplane. In this case, the velocity of the plane relative to the air adds to that of the air relative to the earth (the wind velocity) to yield the velocity of the plane relative to the earth:

$$\vec{v}_{pe} = \vec{v}_{pa} + \vec{v}_{ae}.$$

Just as before, a tail wind has a different effect than a head wind or a cross wind.

The Principle of Relativity

The ideas involving relative velocity that we have just been discussing can also be applied to events within moving vehicles. Suppose, for example, that you are walking up the aisle of a large airliner that is traveling with a constant velocity relative to the earth. Your velocity relative to the plane must be added to that of the plane relative to the earth in order to find your velocity relative to the earth. In practice, however, you are usually much more aware of your velocity relative to the plane than your velocity relative to the earth: the plane is your frame of reference.

As long as the plane is moving with constant velocity, you can move about inside it quite naturally without being aware of the motion of the plane. In fact, you can throw a ball back and forth or perform physical experiments on a plane moving with constant velocity relative to the earth, and obtain the same results as if these operations were being done in a stationary building. (Of course, since the earth itself is rotating on its axis and orbiting the sun, it is not really stationary either.)

Even when a plane is moving with approximately constant velocity, we often have some sense of motion because air turbulence causes it to bounce around a little. We can also look out the window and watch the clouds or earth go by. In smooth air and with the window shades closed, though, we often lose the impression that we are moving at all. This is even more striking in an elevator that is moving up or down

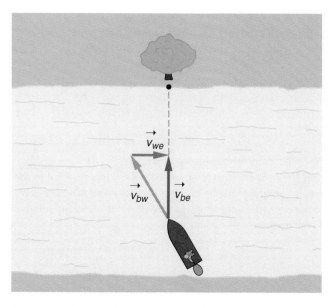

Figure 18.7 **In order to travel straight across the stream, the boat must be pointed somewhat upstream.**

with constant velocity. Since an elevator usually has no windows and can move very smoothly, it becomes difficult to tell whether or not it is actually moving.

These ideas were discussed by both Galileo and Newton and are often summarized in the *principle of relativity:*

The laws of physics are the same in any inertial frame of reference.

This means that we cannot tell whether our frame of reference is in motion or not by performing physical experiments. As long as our frame of reference is moving with constant velocity with respect to other inertial frames of reference, the results of our experiments are the same.

But what is an *inertial* frame of reference? This is where the logical difficulties arise, and Newton was well aware of this problem. Newton's second law of motion is valid only in inertial frames of reference, which can be defined as any frame of reference that is moving with constant velocity with respect to some other inertial frame of reference. If our frame of reference is accelerated with respect to a valid inertial frame, then it is *not* a valid inertial frame of reference.

If, for example, your airplane lurches up or down due to accelerations produced by air turbulence, your experimental results (and your ability to walk a straight line) are modified. Likewise, if an elevator is accelerating up or down, your apparent weight changes. If you were standing on a bathroom scale, this change would be registered on the dial of the scale. You can also feel it in the pit of your stomach.

If we want to apply Newton's second law in these situations, we have to modify the law by adding imaginary, or inertial, forces that arise as a result of the acceleration. The *centrifugal force* that we sometimes talk about feeling in a rotating frame of reference is an example of such an imaginary force. A rotating frame has a centripetal acceleration and is therefore not a valid inertial frame. We feel like we are being pulled outward by the centrifugal force, but from a valid inertial frame, we see that this is really just our own inertia at work, our tendency to continue moving in a straight line.

The centrifugal force that appears to be present in a rotating frame of reference is not a valid force in the Newtonian sense, because it is not due to the interaction of the affected body with any other body. In other words, it does not obey Newton's third law, which is part of Newton's definition of force. It arises solely because of the acceleration of the frame of reference. In this sense, it is like the increase or decrease in apparent weight that is observed in an accelerating elevator.

It looks easy enough, then, to define an inertial frame of reference: it is one that is not accelerated. But with respect to what? For most purposes, we can treat the surface of the earth as an inertial frame of reference, since its acceleration is small. But as we have noted before, the earth is rotating about its axis and is also moving (in a curved path) about the sun, so it is accelerated with respect to the sun. The sun itself is accelerated with respect to other stars, so it cannot be used as a completely valid inertial frame of reference either.

The problem, then, lies in the apparent impossibility of establishing a frame of reference that is absolutely at rest, or at least not accelerated in any sense. Maxwell's prediction and description of electromagnetic waves brought new attention to this problem in the latter half of the nineteenth century. The possibility that measuring the velocity of light could help to establish an absolute inertial frame of reference was an exciting idea. Questions involving the appropriate reference frame for the measurement of the velocity of light (treated in the next section), led ultimately to Einstein's theory of special relativity.

18.2 THE VELOCITY OF LIGHT AND EINSTEIN'S POSTULATES

Chapter 14 introduced the idea that light is an electromagnetic wave, as predicted by Maxwell's theory of electromagnetism. An electromagnetic wave involves time-varying electric and magnetic fields that propagate through empty space, as well as through air, glass, and other transparent materials. Thus, to understand this view of the nature of light, we need to understand the concepts of electric and magnetic fields, and to picture these fields varying as the wave travels.

Is there a medium through which light waves travel even when they are traveling in a vacuum? All other known types of wave motion travel through some medium or material; sound waves through air (and other materials), water waves

in water, waves on a rope on the rope itself, and so on. What about light; does it also have a medium? During the latter part of the nineteenth century (and in some sense even now), this was one of the fundamental questions of physics.

The Luminiferous Ether

When Maxwell invented the concepts of electric and magnetic fields, he used a mechanical model to help him visualize these ideas. An electric field can exist in otherwise empty space. It represents the force per unit charge that would be experienced by a charge at the point in space where the field exists. There does not have to be a charge there (or anything else), however, in order to define the field. The field is therefore a property of space.

The presence of a field must somehow modify or distort space in order to produce its effect on a charge. Maxwell's mechanical model of this process involved imagining that empty space consisted of an infinite array of tiny, massless, interconnected springs (fig. 18.8). An electric field could be viewed as a distortion of this array of springs. Although Maxwell did not claim that this was an accurate picture of the nature of empty space, his model helped him to think about fields and the process of wave propagation.

Even if the spring picture was not to be taken literally, it seemed necessary to attribute elastic properties to empty space in order to explain the propagation of a wave through space. This invisible, elastic, and apparently massless medium that could exist in a vacuum was called the *luminiferous ether*. It was the medium through which light waves

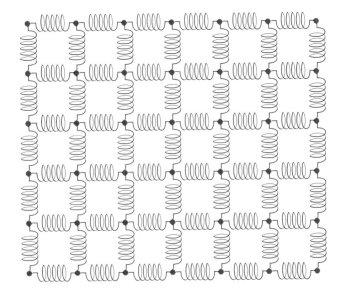

Figure 18.8 Maxwell imagined empty space as consisting of an array of massless, invisible, interconnected springs that could be distorted by electric or magnetic fields.

and other electromagnetic waves traveled. Whether or not it was really necessary to assume its existence in order to explain the propagation of electromagnetic waves was a matter of some debate. Maxwell himself was apparently not totally convinced of its necessity.

The very existence of electromagnetic waves, however, suggested the existence of some medium through which they were propagating. The possible existence of such a medium, which we will call the *ether,* opened an exciting possibility for solving the problem regarding inertial frames of reference mentioned in the preceding section. Perhaps the ether could serve as an absolute reference frame. Any other valid inertial frame of reference would then have to be traveling with constant velocity relative to the ether. The ether itself could be pictured as being embedded and fixed somehow in empty space.

How could we measure motion relative to the ether? Simply by measuring the velocity of light. If the earth is moving relative to the ether, then this motion should affect the velocity of light. The velocity-addition formula introduced in the previous section should apply. Unfortunately, as indicated in chapter 14, measuring the velocity of light with high accuracy is difficult.

These ideas are best illustrated by considering a wave that is easier to visualize than an electromagnetic wave. Suppose, for example, that we consider a water wave traveling on a moving stream, as shown in figure 18.9. If the wave is moving with velocity v_{ws} relative to the stream, and the stream is moving in the same direction with velocity v_{se} relative to the earth, then the velocity of the wave relative to the earth, v_{we}, should be equal to

$$v_{we} = v_{ws} + v_{se},$$

just as in the case of the boat going downstream. The velocity of the wave in the medium (the stream in this case) adds to that of the medium relative to the earth to give the overall velocity of the wave relative to the earth.

It is not difficult to extend these ideas to light waves traveling in the ether. If the earth is moving through the ether, it is as if the ether were streaming by the earth. The velocity of light that we measure should then be the vector sum of the velocity of light relative to the ether and the velocity of the ether relative to the earth. Accurate measurement of the velocity of light relative to the earth would then allow us to determine whether or not the earth is moving in a certain direction relative to the ether.

The Michelson-Morley Experiment

The most famous experiment designed to detect the possible motion of the earth relative to the ether was performed by Albert Michelson (1852–1931) and Edward Morley (1838–1923) during the 1880s. Michelson and Morley were working at what is now Case Western Reserve University in Cleveland. In order to detect small differences in the velocity of light, they used a special instrument designed by Michelson, now called the *Michelson interferometer.* As the name suggests, this instrument uses interference phenomena to detect small differences in the velocity of light or in the distance that the light travels.

A Michelson interferometer is shown in figure 18.10. The instrument has two perpendicular arms along which light can travel. Light coming in from the source on the left is split into two beams by a partially silvered mirror, or beam splitter. When light strikes this mirror, approximately half of it passes through the mirror, and half is reflected, producing two beams of equal intensity. These two beams travel along perpendicular paths, as shown, and are reflected by fully silvered mirrors, thus returning through the beam splitter. Here again, the beams are partially transmitted and partially reflected.

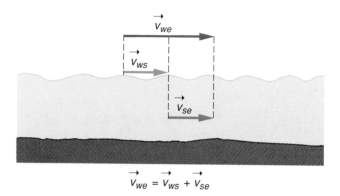

Figure 18.9 **The velocity of the wave relative to the stream adds to the velocity of the stream to yield the overall velocity of the wave relative to the earth.**

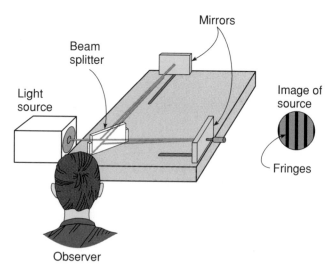

Figure 18.10 **A Michelson interferometer. Light waves traveling along the two perpendicular arms interfere to form a pattern of light and dark fringes.**

The observer looks at images of the source viewed along a path on the fourth side of the interferometer at the bottom of the diagram. Because the light coming to the observer has come from the same source but has traveled along different paths, it may be either in or out of phase for different points on the image of the source. If one of the end mirrors is slightly tilted so that it is not exactly perpendicular to the light beam, these phase differences will create a pattern of dark and light fringes, as shown. The dark fringes represent destructive interference, the light fringes represent constructive interference.

If anything happens to change the time required for either one of the light beams to make its trip to the end mirror and back, the fringe pattern will shift. Michelson and Morley reasoned that if the ether was moving in a direction parallel to one of the arms, then the time of flight for the beam moving parallel to the ether stream would be slightly different from that of the beam moving perpendicular to the ether stream. These time differences could be computed by computing the effective wave velocities for light traveling along each arm. The computation is exactly like those done in section 18.1 for a boat moving parallel or perpendicular to the current of a moving stream.

In order to see the expected shift of fringes, then, the interferometer would have to be rotated through 90° (along with the observer), so that the arm that had been parallel to the ether stream was now perpendicular to the ether stream, and vice versa. Michelson and Morley mounted the interferometer on a rock slab and floated the whole thing in a vat of mercury in order to avoid vibrations and to accomplish the rotation of the interferometer smoothly. (Mercury was the only obtainable fluid dense enough to allow a heavy rock slab to float.)

The computations done by Michelson and Morley were based on the assumption that the velocity of the earth relative to the ether would be due, in part, to the orbital motion of the earth around the sun. Since they could not assume that the ether was fixed with respect to the sun, however, they had to do the experiment at different times of the year. Since the orbit of the earth about the sun is approximately circular, there should be some time during the year when the motion of the earth is parallel to a component of the ether's motion. Six months later the earth should be moving in the opposite direction to that of the ether (fig. 18.11). Thus they used the velocity of the orbital motion of the earth as the minimum velocity of relative motion of the earth to the ether. (It would have this value if the ether was fixed with respect to the sun; if the ether were moving relative to the sun, the relative velocity of the ether to the earth would be even larger at some time during the year.)

In terms of its objectives, the result of the Michelson-Morley experiment was disappointing. No fringe shift was observed at any time during the year. Although the shift was

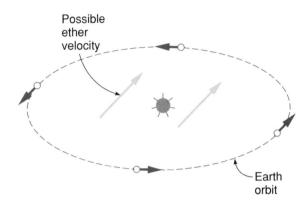

Figure 18.11 **Regardless of the direction of motion of the ether relative to the sun, at some time during the year the earth should be moving relative to the ether.**

expected to be small (of the order of half a fringe width), it should have been observable according to the assumptions upon which the experiment was based. The experiment failed to detect any motion of the earth relative to the ether. Often in science, though, the failure to find what is expected can be a very important result with far-reaching consequences.

Einstein's Postulates

The failure of the Michelson-Morley experiment to detect any motion of the earth relative to the ether raised new questions regarding the nature of the ether. Why could we not detect its motion? Perhaps the ether was being dragged along with the earth, much like the atmosphere, so that an experiment performed on the surface of the earth would not detect any motion. This assumption seemed to be precluded, however, by other observations involving shifts in the apparent positions of stars viewed at different times during the year.

Although Einstein was just a child at the time the Michelson-Morley experiment was performed and may not have been very familiar with its details, he was certainly aware of the debate that followed regarding the nature of the ether. His solution to the dilemma was both simple and radical. He merely assumed as a basic postulate what seemed to be the case experimentally, namely that the velocity of light is not affected by motion of the source or of the frame of reference.

Einstein actually stated two postulates in his introductory paper on special relativity, published in 1905. The first was a reaffirmation of the principle of relativity stated in similar form over two hundred years earlier by Galileo and Newton and discussed in the first section of this chapter:

Postulate 1: The laws of physics are the same in any inertial frame of reference.

The second postulate involved the velocity of light:

Postulate 2: The velocity of light in a vacuum is the same in any inertial frame of reference, regardless of the relative motion of the source and observer.

Although both postulates are important to Einstein's theory, it is the second one that really requires a radical change in our thinking. What he was saying, in essence, is that light (or any electromagnetic wave) does not behave like any other known wave motion (or particle motion). If we throw a ball on a moving airplane, the velocity of the ball relative to the earth is the vector sum of the velocity of the ball relative to the plane plus that of the plane relative to the earth. If the steward or stewardess speaks on the sound system, the sound wave travels with a velocity relative to the earth that is the vector sum of the velocity of sound in the air (inside the plane) plus the velocity of the plane relative to the earth (fig. 18.12).

If, however, we shine a flashlight on the plane, the velocity of light measured on the plane must be the same, according to Einstein's second postulate, as that measured (for the same flashlight beam) by an observer who is at rest with respect to the earth. The classical velocity-addition formula does *not* hold for light. This was not an easy idea for physicists to accept in 1905.

In fact, a closer examination of this second postulate requires us to rethink the very nature of space and time themselves. It is this aspect of relativity that really challenges our minds. We begin to explore some of these consequences in the next section.

18.3 SPACE AND TIME EFFECTS

Any velocity is a ratio of distance (a measure of space) divided by time, meters/second or miles/hour for example. The law of velocity addition discussed in the first section of this chapter follows from this basic nature of velocity. It also depends, however, upon the implicit assumption that space and time can be measured by different observers in the same manner and with the same results, regardless of whether these observers are moving relative to one another. This assumption is certainly consistent with our ordinary experience.

If the velocity of light does not obey the velocity-addition formula, then our measures of space and time must be different as measured by different observers. In other words, if we accept Einstein's second postulate that the velocity of light is the same for all observers, we must give up the idea that space and time are the same for all observers. This goes

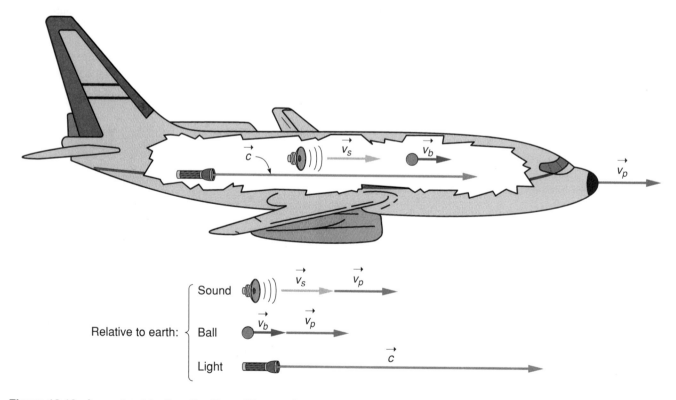

Figure 18.12 **In contrast to the situation with sound waves or a ball, the velocity of the plane does not add to the velocity of the flashlight beam to yield the velocity of light relative to the earth. The velocity of light is the same for all observers. (Velocity vectors not drawn to scale!)**

against our intuition, or common sense; it requires us to throw out some ideas that seem to be inherently true.

In order to approach these questions, Einstein devised what he called *thought experiments:* experiments that are impractical to actually perform because of the tremendous velocities involved, but which can be readily imagined and their consequences explored. They allow us to see how the concepts of space and time must be altered in order to accept the absolute nature of the velocity of light in a vacuum. Anyone can do the thought experiments; no physical equipment is required!

Time Dilation

Suppose that you want to measure time using the velocity of light as your standard of measurement. Suppose also that you are riding in a spaceship that is moving at a large velocity with respect to the earth. The spaceship has a large glass window on one side, so that an observer on earth can also observe your experiment.

How might you go about using the velocity of light as a standard for a time measurement? One way would be to direct a light beam at a mirror directly overhead and use the time required for the beam to go up to the mirror and back as a basic unit of time. This arrangement would be a light-clock; it uses the velocity of light to establish a time standard.

If the distance from the light source to the mirror is d, as shown in figure 18.13, then the velocity of light, c, should be related to this distance as follows:

$$c = \frac{2d}{t_0},$$

where t_0 is the total time of flight of the light beam. Using algebra to rearrange this expression, we get $2d = ct_0$, or

$$t_0 = \frac{2d}{c}.$$

Thus the basic unit of time obtained using this light-clock is $t_0 = 2d/c$. If we know c (our standard) and the distance d, we can compute the elapsed time.

The observer standing on the earth as you flash by in your glass-walled spaceship can also observe the time taken for the light beam to make its trip to the mirror and back. She sees the events somewhat differently, however. If the spaceship is moving with velocity of magnitude v, the mirror moves at that velocity also. In order for the light beam to be reflected from the mirror and return to the source (which meanwhile has moved), it must travel along the diagonal path shown in figure 18.14.

If the observer on earth uses this same light-clock to establish a measure of time, her basic measure, t, will be larger than t_0. This is because she sees the light beam traveling a longer distance at the *same* velocity, c, as that measured by the observer on the spaceship. We assume that she can measure the vertical distance d in the same manner as the observer on the spaceship because this distance is perpendicular to the direction of relative motion and should be unaffected by the motion. The longer path for the light beam then must yield a larger measure of time.

The difference in the times measured by the two observers can be found by considering the geometry of the diagram in figure 18.14. By the symmetry of the situation, the time taken to travel from the source to the mirror should be the same as that taken to travel from the mirror back to the source. If the total time of flight is t, then the time for the first half of the trip is just $t/2$. During this time, the spaceship has traveled a distance $v(t/2)$ and the light travels a distance $c(t/2)$, the velocity times the time in both cases. Since the distance that the light travels forms the hypotenuse of a right triangle, the Pythagorean theorem yields

$$c^2\left(\frac{t}{2}\right)^2 = v^2\left(\frac{t}{2}\right)^2 + d^2.$$

The square of the hypotenuse equals the sum of the squares of the two sides of the triangle.

Using algebra, we can solve this expression for t to yield

$$t = \frac{\left(\frac{2d}{c}\right)}{\rule{2cm}{0.4pt}}.$$

(If you are comfortable with algebra, you might accept the challenge of showing that this is true. Several steps are required.) However, as we saw earlier, $2d/c$ is equal to t_0, the time measured by the observer on the spaceship. The relationship between the two times is therefore

$$t = \frac{t_0}{\sqrt{1 - \left(\frac{v}{c}\right)^2}}.$$

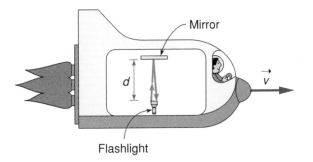

Figure 18.13 **In a light-clock, the time taken for light to travel the distance *d* to the overhead mirror and back becomes the basic measure of time, *t_0*.**

Mirror

d

$\vec{v}$

Flashlight

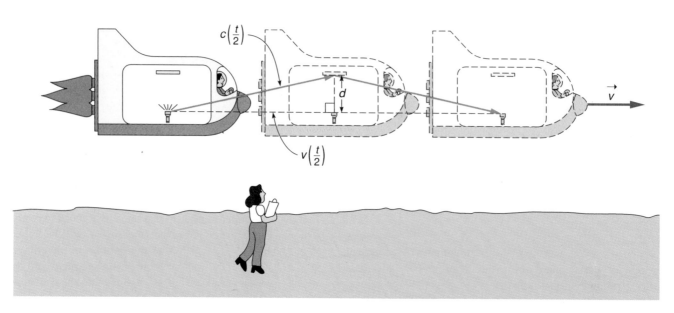

Figure 18.14 **To an observer standing on earth, the light appears to take the diagonal path to the mirror and back to the source. This yields a longer time as measured by the light clock.**

This is the time-dilation formula; since the number in the denominator will always be less than one, t will always be larger than t_0.

The time t_0 is often called the *proper time;* it is the time measured in the spaceship, in this case, where the light starts and finishes its trip at the same point in space. In general, this is what defines a proper time interval:

A proper time interval is the elapsed time between two events measured in a frame of reference in which the two events occur at the same point in space.

This is true for the time measured by the person on the spaceship, but not for the observer standing on earth. She sees the light as leaving the source at one point in space and returning to the source at a different point in space. She therefore measures a *dilated,* or longer elapsed time between the start and finish of the beam's flight.

It is important to step back and try to see what we have actually done with this thought experiment. We have used the velocity of light as a standard for measuring time in a light-clock. By insisting that the two observers moving relative to one another observe the same value for c, the velocity of light, we find that they arrive at different measures of time using the same clock. If they agree upon the value of c, they cannot agree on the time of flight of the light beam! Time is not viewed as passing at the same rate in the two frames of reference.

It is also important to recognize that for ordinary velocities, the difference in these two times would be extremely small. For $v = c / 100 = 0.01c$, for example, which is still the enormous velocity of 3 million meters per second, the quantity v / c in the time-dilation formula is just 0.01 and $(v / c)^2 = 0.0001$. The dilated time t is then

$$t = \frac{t_0}{\sqrt{1 - \left(\frac{v}{c}\right)^2}} = \frac{t_0}{\sqrt{(1 - .0001)}} = \frac{t_0}{\sqrt{.9999}}.$$

This yields $t = 1.00005\, t_0$, so the difference between t and t_0 is only 5 thousandths of a percent in this case. The relative velocity v must be almost as large as the velocity of light in order for the difference in these times to be noticeable.

Length Contraction

Not only do our two observers disagree on the elapsed time for the flight of the light beam, they will also disagree on the distance that the spaceship and the mirror have traveled during this time. We can see this by extending our thought experiment to make a measurement of the distance that the spaceship travels in the time required for one round trip of the light beam.

This distance can be measured most readily by the observer who is standing still on the surface of the earth. With the help of assistants suitably spaced along the path of the spaceship, she can mark where the spaceship was when the light pulse was emitted and where it was when the pulse returned to the source. With a tape measure, then, she can measure the distance between these two points at her leisure, since this distance is fixed on the earth's surface (fig. 18.15).

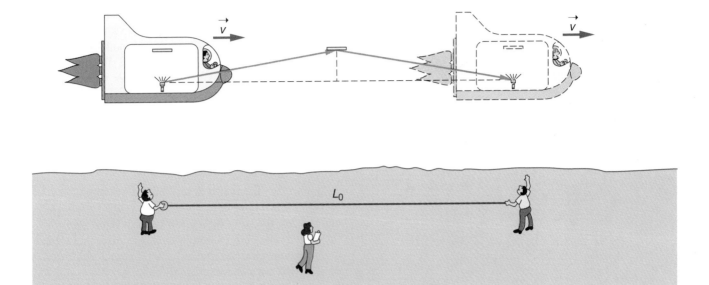

Figure 18.15 With the help of an assistant, the observer on earth can mark the positions of the spaceship for the times when the light pulse is emitted and when it returns. The distance between these positions, L_0 can then be easily measured.

The observer in the spaceship has a somewhat more difficult task. He sees the earth as moving past him and must somehow locate the endpoints of the desired distance simultaneously. Alternatively, if he could measure the velocity of the spaceship independently of this distance, he could compute the distance by multiplying the velocity v times the time of flight of the light beam. This latter quantity is the time t_0, the proper time, since that is the time measured by our astronaut. Using this method, he would measure a length $L = vt_0$, for the distance covered during the flight of the light beam.

Using the same reasoning, the observer on earth could also compute this distance. She finds $L_0 = vt$, where t is the time that she measures for the flight of the light beam. We have used the symbol L_0 for this quantity because this is the *rest length*, the length measured by the observer who is at rest relative to the distance being measured. Since we have already discovered that t is greater than t_0, we see that the rest length L_0 must be larger than the length L measured by the astronaut. The person in the spaceship measures a *contracted*, or shorter, length than the rest length.

The relationship between these lengths can be found by substituting our expression for t in the expression for L_0:

$$L_0 = vt = \frac{vt_0}{\sqrt{1 - \dfrac{v^2}{c^2}}} = \frac{L}{\sqrt{1 - \dfrac{v^2}{c^2}}},$$

since $vt_0 = L$. Multiplying both sides of this equation by the expression involving the square root yields

$$L = L_0 \sqrt{1 - \frac{v^2}{c^2}}.$$

This is the length-contraction formula; since the expression under the square root is always less than 1, L is always less than L_0, the rest length. Again, in order for the effect to be noticeable, v must be very large, as illustrated in the example of box 18.1.

As shown by the result of the sample exercise, the spaceship pilot measures a contracted length (720 km) for the trip, but also measures a shorter time (4 milliseconds). The observer on earth measures the longer distance of 900 km (the rest length) and also a longer (dilated) time (5 ms). The ratio of the distance to the time has the same value for both observers:

$$v = \frac{L}{t_0} = \frac{L_0}{t} = 1.8 \times 10^8 \text{ m/s},$$

which is the relative velocity of the spaceship with respect to the earth (or of the earth with respect to the spaceship).

The fact that both observers agree on the value of the relative velocity was an assumption made in our original

Box 18.1

Sample Exercise

Suppose that a spaceship traveling at a velocity of 1.8×10^8 m/s (180 million m/s) covers a distance of 900 km as measured by an observer on the earth.

  a. What is the distance traveled in this time as measured by the pilot of the spaceship?
  b. How long does it take to cover this distance as measured by the observer on earth and as measured by the pilot?

a. $v = 1.8 \times 10^8$ m/s.
  $c = 3.0 \times 10^8$ m/s.
  $L_0 = 900$ km
    $= 9.0 \times 10^5$ m.

$$\frac{v}{c} = \frac{1.8}{3.0} = 0.6.$$

$$\sqrt{1 - \frac{v^2}{c^2}} = \sqrt{1 - (0.6)^2}$$

$$= \sqrt{1 - 0.36}$$

$$= \sqrt{0.64} = 0.80.$$

$$L = L_0 \sqrt{1 - \frac{v^2}{c^2}}$$

$$= 0.80 \, L_0 = 0.80 \, (900 \text{ km}) = \textbf{720 km.}$$

b. As seen by the observer on earth, $L_0 = vt$.

$$t = \frac{L_0}{v} = \frac{(9.0 \times 10^5 \text{ m})}{}$$

$$= 5.0 \times 10^{-3} \text{ s} = \textbf{5.0 m/s.}$$

As seen by the spaceship pilot, $L = vt_0$.

$$t_0 = \frac{L}{v} = \frac{(7.20 \times 10^5 \text{ m/s})}{(1.8 \times 10^8 \text{ m/s})}$$

$$= 4.0 \times 10^{-3} \text{ s} = \textbf{4.0 m/s.}$$

Notice that t is related to the proper time t_0 by the time-dilation formula:

$$t = \frac{t_0}{\sqrt{1 - \frac{v^2}{c^2}}} = \frac{4.0 \text{ m/s}}{0.80} = 5.0 \text{ m/s.}$$

thought experiment. It is really dictated by Einstein's first postulate, that the laws of physics take the same form in any inertial frame of reference. Since there is nothing to indicate which frame of reference is in motion, the situation is completely symmetrical between the two frames; the observer on earth sees the spaceship moving by with velocity v, and the pilot sees the earth as moving by with velocity v.

Other Examples and Simultaneity Disagreement

Although the effects just described seem strange, they are real effects that have been observed in a wide variety of circumstances. It is very difficult to get ordinary objects moving at velocities large enough to show noticeable effects, but subatomic particles routinely travel at such velocities. A particle that has a characteristic lifetime when it is at rest in the lab seems to have a longer (dilated) lifetime when it is moving at velocities near the velocity of light. It therefore travels farther before decaying, as seen from an earthbound observer's perspective.

As seen from the perspective of an observer traveling with the particle, however, the particle has its proper lifetime, t_0, and travels a contracted distance, L, shorter than the rest length measured by the earthbound observer. The situation is essentially the same as that described for the spaceship. All of these observations are consistent if we treat them according to Einstein's theory.

The problem of measuring distances from a moving spaceship, if examined closely, reduces to a problem of locating simultaneously the end points of the length being measured, as mentioned earlier. It turns out that not only do the two observers disagree about the elapsed time, they also disagree about whether certain events are simultaneous or not. Two events, separated in space, may be seen as simultaneous (occurring at the same time) by one observer, but as occurring at different times by an observer who is moving relative to the first observer.

These space and time effects are explored further in box 18.2. Here we treat the famous *twin paradox*, which examines the difference in aging rates of two twins, one who makes a space trip to a distant star and back, and the other who remains on earth. The fact that the traveling twin ages less than the one remaining on earth can be understood using the time-dilation idea. It is not just science fiction!

Box 18.2

Future Everyday Phenomenon:
The Twin Paradox

The Situation. One of the most often-discussed phenomena of relativity is the so-called twin paradox. One of a pair of identical twins, Adele, travels at a very large velocity to a distant star and then returns to earth. The second twin, Bertha, remains on earth the entire time that her twin is traveling.

Since one of the twins is traveling at a velocity approaching the speed of light, the two twins should measure time as passing at different rates due to the time-dilation effect. When the traveling twin returns, will she find that she is older or younger than the twin who has remained at home? And since each twin can regard the other as moving while she herself is standing still (as long as the velocity is constant), should not the situation be symmetrical for the two observers? This latter question lies at the heart of the apparent paradox.

The Analysis. Let us suppose that Adele's entire trip is made at the speed $v = 0.6c$. The factor that appears in the time-dilation formula is then

$$\gamma = \frac{1}{\sqrt{1 - \frac{v^2}{c^2}}} = \frac{1}{\sqrt{1 - (.6)^2}} = \frac{1}{0.8} = 1.25.$$

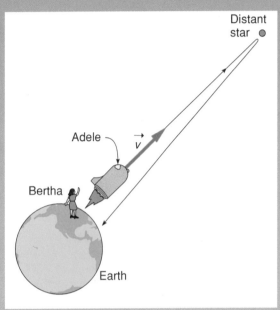

Twin Adele flies to a distant star and back while her twin sister, Bertha, remains at home.

Suppose that for Adele the trip takes 12 years; this is a proper time in her frame of reference, the spaceship. She has lived 12 years and experienced an appropriate number of heart beats (or other suitable biological ticks of the clock) during that time. In other words, as far as she is concerned she is 12 years older than when she left.

Her twin, Bertha, on the other hand, sees a dilated time for this same time interval:

$$t = \frac{t_0}{\sqrt{1 - \frac{v^2}{c^2}}} = 1.25\, t_0 = (1.25)(12 \text{ years}) = 15 \text{ years.}$$

Thus, as far as she is concerned, a time of 15 years has passed since her twin sister left, and Bertha has aged 15 years in waiting for her sister to return. Adele is therefore 3 years younger than her identical twin at the end of the trip!

Suppose, however, that we regard the spaceship as fixed, while spaceship earth takes off and returns with Bertha aboard. Would we not arrive at the reverse conclusion? Would not such an analysis sug-

gest that Bertha should be 3 years younger than Adele? Surely we cannot hold both of those results to be true. Therein lies the paradox.

The resolution of the paradox lies in realizing that we have ignored an asymmetry involving accelerations. In order to make a trip such as this, Adele's spaceship must accelerate away from earth until it reaches the enormous velocity of $0.6c$ that we have assumed. When it reaches the distant star, it must turn around, which involves a deceleration and then an acceleration in the opposite direction. It must decelerate again when it reaches the earth.

Because the earth has a much larger mass than the spaceship, it is not reasonable to picture the earth as making these accelerations. It requires an enormous expenditure of energy to accelerate even the spaceship to such a velocity. Thus our situation is not really completely symmetrical; the spaceship exists in two different frames of reference, each having velocity v relative to the earth, but in opposite directions.

We could assume rates of acceleration for the spaceship and use general relativity to analyze the

Continued

Box 18.2 Continued

passage of time during the periods of acceleration. This is not really necessary, however. If we assume that the accelerations take place in times that are small compared to the overall time of flight, then the computation done above using special relativity produces the correct result; Adele does age less than Bertha. We can confirm this by doing a thought experiment utilizing the basic assumptions of special relativity and carefully treating the behavior of clocks in either frame of reference. We must assume, however, that it is the spaceship that changes frames of reference, not the earth.

This difference in the passage of time and the resulting difference in aging of the two twins is a real effect. An experiment done using very accurate clocks and the much slower speeds of a jet plane has obtained results consistent with this effect, although the results are marginal given the very small difference in time that is involved. If it were feasible to obtain velocities as high as 0.6c or even 0.995c, then the difference in aging of the two twins would be quite striking. At a velocity of 0.995c, the time-dilation factor is approximately 10 rather than 1.25. A trip that took 10 years as seen by Adele, would take 100 years as seen by Bertha. Adele would return to earth 100 earth years later having aged only 10 years herself!

Thus it is theoretically possible for us to travel into the future, while aging much less than people who

At a speed of 0.995 c for v, the difference in aging of Adele and Bertha could be quite striking.

remain on earth. It would be extremely difficult, however, to reach the velocities required to make the differences in aging noticeable. Other practical problems are associated with space travel at such speeds, including the damaging effects of high-speed particle bombardment. Nor is it possible, even in theory, to travel backwards in time without violating some fundamental ideas involving cause and effect. We can fast-forward time in this sense, but not reverse it.

18.4 RELATIVISTIC DYNAMICS

We have seen that accepting Einstein's postulates requires some major changes in how we think about space and time. Since space and time measurements are involved in the concepts of velocity and acceleration, and acceleration plays a key role in Newton's laws of motion, we might suspect that Newton's laws must also be modified in order to be consistent with Einstein's postulates. Does Newton's second law of motion, for example, still apply when objects are moving at very large velocities?

In addressing these questions, Einstein discovered that it was necessary to modify Newton's second law by redefining the concept of momentum. As he explored the consequences of this new approach to dynamics in his early papers on relativity, he was led also to a striking conclusion involv-

ing the relationship between mass and energy. This is the idea summarized in the often-quoted equation, $E = mc^2$.

Velocity Addition

Before we consider Newton's second law, we should return to the question of how velocities add in relativistic mechanics. Suppose, for example, that our spaceship pilot fires a projectile inside his spaceship. Will an observer on earth measure the velocity for this projectile predicted by the classical expression for velocity addition? We know that this expression is not valid for light, but what about objects traveling at velocities near the speed of light?

To simplify the notation used earlier for velocity addition, we can label our two observers: let observer A be the spaceship pilot and observer B be the earthbound person.

Then v_A will represent the velocity of the projectile as measured by observer *A*, and v_B that measured by observer *B*, as indicated in figure 18.16. If, as before, the velocity of the spaceship relative to the earth is represented by *v*, then classically we would predict that

$$v_B = v + v_A,$$

provided that the projectile is fired in the direction of the spaceship's motion.

If we fired a beam of light instead of a projectile, both observers would measure $c = 3.0 \times 10^8$ m/s, according to Einstein's second postulate, so the simple classical expression for addition of relative velocities does not hold for light. For objects traveling near the speed of light, we also find that this simple expression cannot be used. Einstein was able to show that the expression for velocity addition must be modified for large velocities to take the following form:

$$v_B = \frac{v + v_A}{1 + \frac{v v_A}{c^2}}.$$

Although this expression is more complicated than the simple one used by Galileo and Newton, there is an obvious similarity between the two expressions. If either *v* or v_A is small compared to the velocity of light, then the second term in the denominator of the relativistic expression is very small (much less than 1). When this is the case, we can often ignore this term. What is left is just the ordinary classical expression for velocity addition, $v_B = v + v_A$. For ordinary

velocities, therefore, the relativistic expression predicts the same result as the classical expression.

Another interesting feature of this expression emerges if we let the velocity of the projectile in the spaceship, v_A, be equal to the velocity of light. This yields

$$v_B = \frac{v + c}{1 + \frac{vc}{c^2}}.$$

Multiplying both the numerator and denominator by *c* produces

$$v_B = \frac{c(v + c)}{(c + v)} = c.$$

In other words, if the projectile happens to be a beam of light, the relativistic velocity-addition formula yields *c* for both observers, consistent with Einstein's second postulate. For any velocity, v_A, less than *c*, however, the velocity v_B will always be less than *c* also, as you can see from the example in box 18.3.

Notice that the classical velocity-addition formula with the values given in box 18.3 would yield a velocity of 1.4*c*, which is greater than the speed of light. The relativistic expression, on the other hand, will always yield a value less than or equal to *c*. No observer will see any object as having a velocity greater than the speed of light. The speed of light is apparently an upper limit for the velocity of any object.

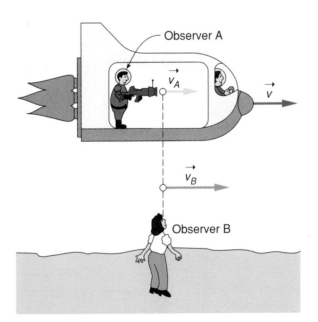

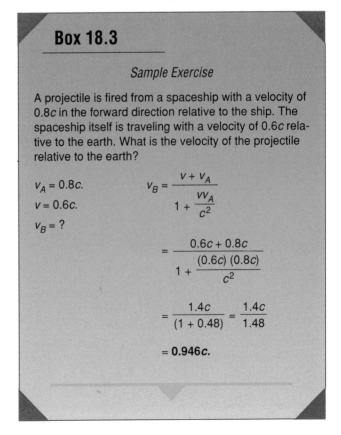

Box 18.3

Sample Exercise

A projectile is fired from a spaceship with a velocity of 0.8*c* in the forward direction relative to the ship. The spaceship itself is traveling with a velocity of 0.6*c* relative to the earth. What is the velocity of the projectile relative to the earth?

$v_A = 0.8c.$

$v = 0.6c.$

$v_B = ?$

$$v_B = \frac{v + v_A}{1 + \frac{v v_A}{c^2}}$$

$$= \frac{0.6c + 0.8c}{1 + \frac{(0.6c)(0.8c)}{c^2}}$$

$$= \frac{1.4c}{(1 + 0.48)} = \frac{1.4c}{1.48}$$

$$= 0.946c.$$

Figure 18.16 **A projectile fired within the spaceship with a velocity, v_A relative to the spaceship is observed to have a velocity v_B relative to the earth. How are these velocities related?**

Newton's Second Law

When Einstein examined Newton's second law in the light of his postulates and his new formula for the addition of relative velocities, he discovered that there were problems there also. An acceleration measured in one frame of reference was not the same as that of the same object measured in some other frame of reference; if the spaceship pilot fires a projectile with an acceleration of a_A, the observer on earth would measure a different acceleration, a_B.

If we use the classical expression for velocity addition (valid for speeds much less than c) to compute the acceleration, we get the same acceleration for both observers. This assumes that the two observers agree upon their measurement of time. At ordinary velocities, we can therefore apply Newton's second law in the form $\Sigma \vec{F} = m\vec{a}$ in either frame of reference, using the same forces to explain the acceleration of the object. This apparently will not work when the velocities are very large, however, because the accelerations are no longer equal in different frames of reference. Essentially, then, we have a violation of Einstein's first postulate: Newton's second law does not seem to take the same form in any inertial frame of reference.

As discussed in chapter 6, the most general form of Newton's second law is stated in terms of momentum rather than acceleration:

$$\Sigma \vec{F} = \frac{\Delta \vec{p}}{\Delta t} \; .$$

The net force is equal to the rate of change of momentum, where momentum is defined as the product of the mass and the velocity, $\vec{p} = m\vec{v}$. We also encounter problems at large velocities in using Newton's second law in this form. Even the law of conservation of momentum does not seem to take the same form when viewed by observers in different frames of reference.

Einstein discovered that in order to salvage the law of conservation of momentum at relativistic (large) velocities, he had to redefine momentum. This new definition of momentum took the following form:

$$\vec{p} = \frac{m\vec{v}}{\sqrt{1 - \dfrac{v^2}{c^2}}} \; ,$$

where v is the velocity of an object in a given frame of reference. Using this new definition of momentum, Einstein was able to show that different observers could agree that momentum was conserved in a collision, even though these different observers would measure different values for the velocities and momenta. At low velocities, of course, this revised definition of momentum reduces to the ordinary definition of momentum, $\vec{p} = m\vec{v}$.

Since the law of conservation of momentum follows directly from Newton's second law, this revised definition of momentum must be used there also. In other words, Ein-stein found that he could make Newton's second law conform to his postulates by using this new definition of momentum in the general form of Newton's second law, $\Sigma \vec{F} = \Delta \vec{p} / \Delta t$. At ordinary velocities, Newton's second law works in the usual manner because the relativistic momentum reduces to the classical definition, $\vec{p} = m\vec{v}$. At very large velocities, however, we are forced to use the more complex relativistic definition of momentum. Thus Einstein's theory of special relativity represents a significant revision of Newton's theory of mechanics.

Mass and Energy

As Einstein began to explore the consequences of revising Newton's second law, he discovered that the concept of mechanical energy also took on a new meaning. In classical physics, the kinetic energy of an object is found by computing the work done to accelerate the object to a given speed, resulting in the familiar expression, $KE = \frac{1}{2}mv^2$. (See chapter 8.) Using the same procedure, we can compute the kinetic energy for an object that is accelerated to a very large velocity (fig. 18.17). In this case, however, we must use the modified version of Newton's second law to describe the force.

The kinetic energy found by using the relativistic modification of Newton's second law to describe the required force can be written as follows:

$$KE = mc^2 \left(\frac{1}{\sqrt{1 - \dfrac{v^2}{c^2}}} - 1 \right).$$

Because the expression $1 / \sqrt{1 - v^2/c^2}$ appears so frequently in relativistic expressions, the Greek letter γ (gamma) is often used to represent this quantity. With this substitution, the expression for kinetic energy takes a simpler form:

$$KE = mc^2(\gamma - 1) = \gamma mc^2 - mc^2.$$

Notice that only the first term in this expression depends upon the velocity because γ depends upon velocity.

Arriving at this result was a simple process for Einstein; the computation is not difficult for someone experienced in using calculus. Interpreting the result, however, offered a significant challenge. Since γ is always greater than or equal to 1, the expression for kinetic energy appears to be the differ-

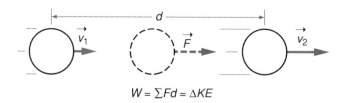

$$W = \Sigma Fd = \Delta KE$$

Figure 18.17   As before, the work done by the net force used to accelerate an object equals the increase in kinetic energy of the object.

ence between two terms, only one of which depends upon velocity. Apparently, accelerating an object increases its energy *above* an energy that it already possesses by virtue of its mass, mc^2.

We often refer to the mc^2 term in this expression as the *rest energy* and give it the symbol E_0; thus, the expression becomes

$$KE = \gamma E_0 - E_0.$$

Rearranging this expression, we have

$$\gamma E_0 = E_0 + KE = E.$$

Here we have let $E = \gamma E_0$ represent the total energy, the sum of the rest energy plus the kinetic energy. When an object is accelerated, the total energy increases because the factor γ increases as the velocity gets larger.

The rest-energy term was the most striking new feature of Einstein's computation of kinetic energy. Since c is a constant of nature, multiplying the mass of an object by c^2 to obtain an energy value ($E_0 = mc^2$) is simply multiplying the mass by a constant. All of this indicates that mass *is* energy; if we increase the mass of an object or system, we increase its energy, or if we increase the energy of a system, we increase its mass! This is the essence of the $E_0 = mc^2$ relationship, Einstein's principle of the equivalence of mass and energy.

The equivalence principle is illustrated in figure 18.18, which shows a Bunsen burner heating a flask of water. Since heat flow represents a flow of energy, we are increasing the internal energy of the water in this process. This means that

we are actually increasing the mass of the water, since energy is mass. The amount by which the mass is increased in this example, however, would be very small and extremely difficult (if not impossible) to measure. If, for example, we added 1000 joules of heat energy, the increase in mass would be given by

$$\Delta m = \frac{E}{c^2} = \frac{1000 \text{ J}}{} = 1.1 \times 10^{-14} \text{ kg}.$$

Since a flask of water would normally contain a few tenths of a kilogram of water, a change of 10^{-14} kg would be utterly negligible.

Because we are so used to thinking of mass and energy as very different quantities, we often find the principle of mass-energy equivalence used or stated in a confusing manner. The principle itself has been thoroughly confirmed; it correctly predicts the amount of energy released in nuclear reactions such as fusion or fission, as described in chapter 17. We often read that mass is converted to energy in such reactions. It would be better to say that rest-mass energy has been transformed to kinetic energy in these reactions. In other words, mass cannot be converted to energy because it already is energy; we are merely transforming one type of energy into another.

The principle of mass-energy equivalence, like the other effects that we have been describing, can be regarded as just another result of applying Einstein's postulates in a careful and consistent manner to our basic understanding of mechanics. The surprising nature of the results, however, has required fundamental revisions in our understanding of the concepts of energy and mass, as well as of space and time.

18.5 GENERAL RELATIVITY

Our discussion so far has been restricted to cases involving inertial frames of reference, that is, reference frames moving at constant velocity relative to one another. This is the province of the theory of special relativity.

What happens if our frame of reference is accelerating? Can we extend the type of thinking used with special relativity to treat cases in which our frame of reference is accelerated? This was a question that Einstein addressed shortly after his development of the theory of special relativity. He did not publish the resulting theory of general relativity until 1915, however, about ten years after his first paper on special relativity. Once again, physicists were required to make some radical adjustments in their customary view of the universe.

The Principle of Equivalence

We have discussed accelerating frames of reference earlier, at the beginning of this chapter as well as in chapter 4, when we considered how things appear to someone inside an accelerating elevator. If the elevator is moving with con-

Figure 18.18   A Bunsen burner adds mass to a flask of water by increasing the internal energy of the water. Energy and mass are equivalent.

stant velocity, then Einstein's first postulate (the principle of relativity) tells us that the laws of physics will apply exactly as they would if the elevator were at rest. In other words, no experiment done inside the elevator can establish whether we are moving or not with respect to the earth.

If, however, the elevator is accelerating, we do expect to see some differences from what we would expect if the elevator were at rest or moving with constant velocity. In particular, as discussed in chapter 4, if you are standing on a bathroom scale while the elevator is accelerating upward, the scale will indicate a greater weight than if the elevator were not accelerating (fig. 18.19). From the standpoint of Newton's second law, this greater apparent weight is recorded because the scale must exert a larger upward force on your feet than your actual weight, which is the downward pull of gravity. This is necessary in order to provide a net force acting on you in the upward direction, so that you will accelerate upwards along with the elevator.

This change in the reading of the scale could be used as an indication that the elevator is accelerating. If the elevator is accelerating upwards, the reading will be greater than normal; if the elevator is accelerating downwards, the reading will be less than normal. If the cable of the elevator is cut and the elevator accelerates downward with an acceleration g

(free fall), then the scale will read zero; you are then in a condition of apparent weightlessness. Until things come to a crashing halt at the bottom of the shaft, you can float around the inside of the elevator much like an astronaut in an orbiting space shuttle.

Other experiments will also lead to results that you would not expect if the elevator were not accelerated. For example, a dropped ball will approach the floor of the elevator with an apparent acceleration that is different from g. If the elevator is accelerating upwards, the apparent acceleration of the ball will be larger than g; if it is accelerating downwards, the apparent acceleration of the ball will be less than g, as illustrated in figure 18.20. The period of a swinging pendulum will also have a different value than the one you would observe if the elevator were at rest.

All of these experiments have one thing in common: we can interpret all of them in terms of an apparent acceleration of gravity that is different from $g = 9.8$ m/s². Since weight is equal to mass times the acceleration of gravity (mg), we can regard the change in apparent weight as being due to a change in the apparent value of the acceleration of gravity. Similarly, we can explain the changes in the acceleration of the ball or the period of the pendulum in the same way. In other words, although we can detect the acceleration of the elevator, we cannot distinguish these effects from those that would occur if the acceleration of gravity were being increased or decreased.

Figure 18.19 If the elevator is accelerating upward, the scale reads a value F_N, which is higher than the normal weight of the person.

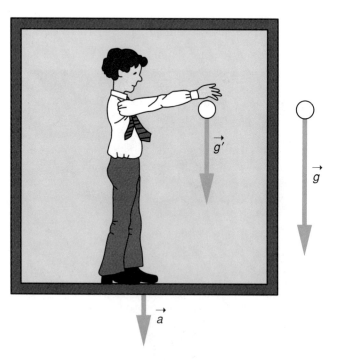

Figure 18.20 A ball dropped in an elevator that is accelerating downward approaches the floor with an apparent acceleration g' that is less than g.

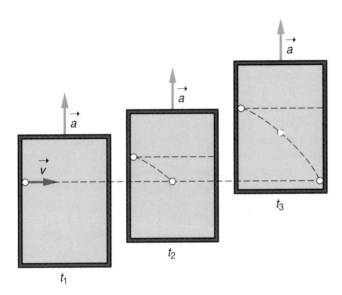

Figure 18.21 A ball thrown horizontally within an accelerating elevator in outer space (where the earth's gravitational pull is negligible) falls towards the floor in the same manner as a projectile near the earth's surface.

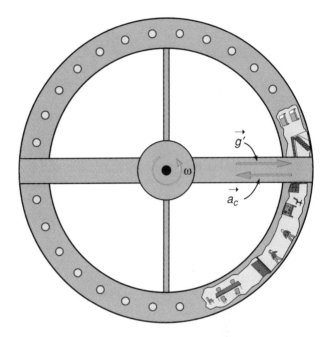

Figure 18.22 The centripetal acceleration a_c of a rotating, wheel-like space station can produce an artificial gravity for the astronauts.

This is the essential idea underlying Einstein's basic postulate of general relativity, the principle of equivalence:

It is impossible to distinguish an acceleration of a frame of reference from the effects of a gravitational field.

From inside the elevator you cannot tell whether the elevator is accelerating or whether the gravitational field is increasing or decreasing. Since we do not expect the gravitational field to change, we would usually interpret the effects as being due to an acceleration of our frame of reference.

Suppose that our elevator has been moved to outer space, where gravitational effects would be much smaller than near the surface of the earth. If the elevator is not accelerating, we would be in a weightless condition. If the elevator is accelerating upwards, as shown in figure 18.21, then the results of experiments performed in the elevator will be the same as if there were a gravitational field acting in the direction opposite to the acceleration. In fact, the direction of acceleration is what defines the upward direction in this situation.

If a ball is thrown horizontally in the elevator, its trajectory will be the same as if it were thrown on the surface of the earth. From the perspective of someone inside the elevator, the upward acceleration is equivalent to a downward acceleration of gravity of the same magnitude as the acceleration of the elevator. (This is the principle of equivalence at work.) The ball therefore "falls" towards the floor of the elevator, and we can predict its motion by the same methods used in chapter 3 to describe projectile motion.

If the acceleration of the elevator is equal to 9.8 m/s², everything that we experience inside the elevator will be identical to what we would experience on the surface of the earth. In fact, it has often been proposed that a space station should be given a constant acceleration in order to mimic the effects of gravity on the earth. Since a linear (straight-line) acceleration is not consistent with the mission of a space station, we generally picture the space station as having a centripetal acceleration provided by a constant rotational velocity. Since the direction of a centripetal acceleration is toward the center of rotation, that direction would be up, as indicated in figure 18.22.

Gravitation Bending of Light

The principle of equivalence also has implications for the propagation of light. Imagine performing an experiment similar to that pictured in figure 18.21, this time using a beam of light rather than a ball. If the elevator is not accelerating, the beam will trace a horizontal line across it. From the theory of special relativity, we know that this is true whether or not the elevator is moving with constant velocity relative to any other inertial frame of reference.

If the elevator is accelerating upwards, however, a different result is observed. If the acceleration is large enough, the path of the light beam will be curved, as viewed from within the elevator, in the same manner as that of the projectile in figure 18.21. We can visualize this by superimposing the positions of the accelerated elevator upon the straight-line light beam observed from outside of the elevator. As shown in figure 18.23, the path traced by the beam relative to the elevator is curved.

By the principle of equivalence, however, we cannot distinguish the acceleration of our frame of reference from the presence of a gravitational field. We should expect, then, that the path of a light ray would also be bent in passing through a strong gravitational field. Naturally, it takes a very large acceleration of the frame of reference or a very strong gravitational field to produce a noticeable bending because of the extremely large velocity of light. The relatively puny field of the earth is not enough to produce much of an effect.

When light from a distant star passes near our sun, however, the gravitational field of the sun should be large enough to produce a noticeable effect. Einstein was able to predict how much bending would be produced by the gravitational field of the sun, and thus how the true positions of stars would be distorted when their light passed near the sun. The effect is small, but it was expected to be measurable.

Unfortunately, it is very difficult to observe stars from the surface of the earth during the daytime. Light from the sun is scattered in the earth's atmosphere and completely washes out the much more feeble light from the stars. Such observations are only feasible during a total eclipse of the sun, when the light from the sun is blocked by the moon. Einstein suggested that such measurements be attempted during a total eclipse, and this has since been done almost every time that the opportunity has occurred. The results of these measurements have confirmed Einstein's predictions.

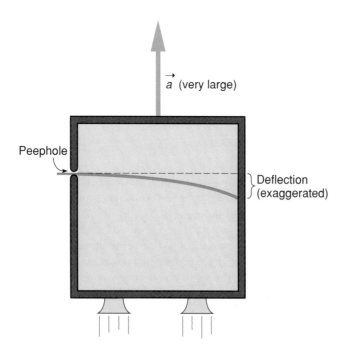

Figure 18.23   The path traced by a beam of light relative to a rapidly accelerating elevator is curved due to the motion of the elevator.

Time Measurements, Space Curvature, and Black Holes

Special relativity tells us that moving clocks appear to run more slowly than stationary clocks. This is the essence of the time-dilation effect; the time interval measured by observers who see the start and finish of some event occurring at the same point in their frame (the proper time) is shorter than that measured by observers who are moving with respect to that frame. The moving observers see a dilated (or longer) time, so their clock appears to run more slowly.

By extending these ideas to accelerated frames of reference, we find that an accelerated clock will also run slowly compared to a nonaccelerated clock. By the principle of equivalence, then, we also expect that a clock in a strong gravitational field will run more slowly than one in a weaker gravitational field. This is the time effect predicted by general relativity; it is often referred to as the *gravitational red shift.* If the period, T, (the time for one cycle) of a light wave is increased, the frequency, which is $1 / T$, is decreased. A lowered frequency shifts the light to the red portion of the visible spectrum.

We can see that the general theory of relativity is, in many ways, a theory regarding the nature of gravity. Gravity affects the nature of a straight-line path and also the nature of time; it has an impact, in other words, upon how we measure both space and time. In order to develop a self-consistent mathematical framework for handling these effects, Einstein found it necessary to resort to a non-Euclidean, or curved, space-time geometry. The complexity of the mathematics of curved space-time is often somewhat forbidding to novices, but the basic idea can be understood without too much difficulty.

In Euclidean, or ordinary, geometry, two parallel lines never meet. In non-Euclidean geometry, however, two parallel lines can meet. An example is provided by parallel lines drawn on the surface of a sphere, such as the lines of longitude that we use on maps. Here the parallel lines drawn perpendicular to the equator meet at the poles because the surface upon which they are drawn is a sphere (fig. 18.24). It is all a matter of how we define the rules of our geometry.

Einstein's special theory of relativity showed that measurements of time can depend upon spatial measures, and that measurements of length, in turn, depend upon time measurements. Therefore, we can no longer regard space and time as independent of one another. To fully represent these ideas using geometry, we must use four dimensions, or coordinates: three perpendicular spatial coordinates and a fourth one representing time. To describe a motion or event, we need to locate its path in this *space-time continuum.*

Although the space-time continuum is four-dimensional and therefore difficult to visualize, diagrams such as that shown in figure 18.25 are often used to illustrate how space might be curved in the vicinity of a very strong gravitational field. The diagram shows only two dimensions on a curved surface, but its drainlike aspect does suggest how things

Figure 18.24 The parallel lines of longitude drawn on a globe meet at the poles of the sphere.

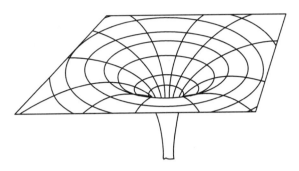

Figure 18.25 The gravitational effects of a black hole can be represented by a strong curvature of space in the vicinity of the black hole.

might be pulled into the center of the field. Since light rays are bent by strong gravitational fields, they can be pulled into the center of the field as well as particles having some mass.

Figure 18.25 is actually a two-dimensional representation of a *black hole*. Black holes are thought to be produced by very massive collapsed stars, which generate an extremely strong gravitational field and therefore a strong curvature of space in their vicinity. This field is so strong that light rays coming in at certain angles are bent into the center in such a manner that they do not emerge. Thus, light can get in but cannot get out; a black hole is a perfect absorber of light and therefore appears black.

Although black holes cannot be directly observed because they do not emit or reflect light, their presence might be inferred from the effects of their gravitational fields upon nearby stars and other matter. If, for example, a binary star consists of two stars, one visible and the other a black hole, the motion of the visible star can indicate the presence of its partner. Astronomers have at least one good candidate for this type of suspected black hole in Cygnus X-1. Several other observations suggest the presence of black holes, but their existence cannot be regarded as proven.

Einstein's theories of special and general relativity have had an enormous impact upon twentieth century physics. Most of the predicted effects are well confirmed, from the energy released in nuclear reactions to astronomical effects such as the bending of starlight. Our fundamental concepts of space and time have been modified and intermixed by these ideas. Although removed from everyday experience, these ideas can certainly excite the imagination and perhaps whet your appetite to read more on this subject.

SUMMARY

Accepting the idea that the velocity of light is not added to the velocity of the source or frame of reference in the way that much smaller velocities are in classical mechanics led Einstein to a radical new way of looking at the nature of space and time. This chapter has described the basic postulates of Einstein's theories of special and general relativity, and has examined some of the consequences of accepting those postulates.

Relative Velocities. If an object is moving relative to a frame of reference (such as a stream) that is itself moving, classical mechanics predicts that these velocities will add as vectors. If the velocity of a boat relative to the water is $\vec{v}_{bw}$, and that of the water relative to the earth is $\vec{v}_{we}$, then that of the boat relative to the earth is $\vec{v}_{be} = \vec{v}_{bw} + \vec{v}_{we}$.

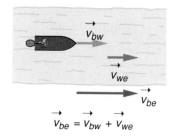

$$\vec{v}_{be} = \vec{v}_{bw} + \vec{v}_{we}$$

Einstein's postulates. Experiments designed to find the velocity of the ether, the assumed medium for electromagnetic waves, failed to detect any motion of the ether relative to the earth. This failure contributed to Einstein's formulation of the two basic postulates of special relativity: (1) The laws of physics have the same form in any inertial frame of reference (the principle of relativity), and (2) The velocity of light is the same in any inertial frame of reference, regardless of the motion of the source. This second postulate was a radical departure from the expectations of classical physics.

Time and Space Effects. Application of Einstein's postulates to the measurement of time and length leads to the conclusion that observers in different frames of reference will not agree upon these measurements. A person observing a moving clock will see a longer (or dilated) time than the proper time measured by an observer for whom the clock is at rest, and a person measuring a moving length will observe a shorter (contracted) length than that measured by an observer who is at rest with respect to the length being measured.

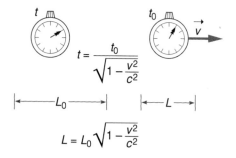

$$t = \frac{t_0}{\sqrt{1 - \dfrac{v^2}{c^2}}}$$

$$L = L_0 \sqrt{1 - \frac{v^2}{c^2}}$$

Relativistic dynamics: Newton's Second Law and Energy. Extending his thinking to dynamics, Einstein found that Newton's second law could be preserved only if we redefined momentum and used the general form of Newton's second law written in terms of momentum. A computation of kinetic energy then led to the recognition that mass is energy, which is expressed in the relationship $E_0 = mc^2$.

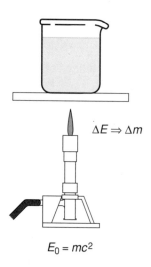

$\Delta E \Rightarrow \Delta m$

$$E_0 = mc^2$$

General Relativity. Special relativity deals with frames of reference that are moving with constant velocity relative to one another. If we consider accelerated frames of reference, we are in the realm of general relativity, for which the basic postulate is the principle of equivalence: An acceleration of a frame of reference cannot be distinguished from the presence of a gravitational field. This additional postulate leads to new effects involving the bending of light and the modification of time by gravitational fields, and also requires the use of non-Euclidean, curved-space geometries to describe the space-time continuum.

QUESTIONS

Q18.1 If a boat is moving upstream, is its velocity relative to the bank of the stream greater than, less than, or equal to its velocity relative to the water? Explain.

Q18.2 If an airplane is flying in the same direction as the wind, is its velocity relative to the ground greater than, less than, or equal to its velocity relative to the air? Explain.

Q18.3 If a boat is moving across a stream, will the magnitude of its velocity relative to the bank equal the numerical sum of the magnitudes of its velocity relative to the water plus the velocity of the water relative to the bank? Explain.

Q18.4 Does the addition of the velocities of things like airplanes and wind speed necessarily involve the theory of special relativity? Explain.

Q18.5 Would it be appropriate, from the perspective of special relativity, to add the velocity of light relative to the earth to the velocity of the earth relative to the sun in order to obtain the velocity of light relative to the sun? Explain.

Q18.6 Was the ether (the assumed medium for light waves) something that was presumed to exist in a vacuum? Explain.

Q18.7 Was the Michelson-Morley experiment designed to detect the motion of the earth relative to the sun? Explain.

Q18.8 Did the Michelson-Morley experiment succeed in meeting its objective? Explain.

Q18.9 Which of Einstein's postulates deals most directly with the failure to detect motion of the earth relative to the ether? Explain.

Q18.10 Do either of Einstein's postulates contradict the classical assumptions regarding the addition of relative velocities? Explain.

Q18.11 A chess game taking place on earth is observed by observer *A*, who is passing by in a spaceship. Observer *B* is standing on earth looking over the shoulder of the players. Which of these two observers measures the longer time for the time between moves in the game? Explain.

Q18.12 A radioactive isotope with a certain half-life is moving with a high speed in a particle accelerator. Does an observer at rest in the lab measure the proper time for the half-life of this isotope? Explain.

Q18.13 A spaceship is moving with a large velocity past observer *A*, who is standing on earth. Observer *B* is a mission specialist aboard the spaceship. Which of these observers measures the longer length for the length of the space ship? Explain.

Q18.14 Is it possible for a father to be younger (to have aged less) than his child? Explain.

Q18.15 Is it possible for an astronaut to leave on a space trip and return before his twin brother was born? Explain.

Q18.16 Is Newton's second law, written in the form $\sum \vec{F} = m\vec{a}$, valid for objects traveling at velocities near the velocity of light? Explain.

Q18.17 If we compress a spring and lock it into its newly compressed configuration, have we changed the mass of the spring? Explain.

Q18.18 Is it completely correct to say that mass is converted into energy in a nuclear reaction such as a fission reaction? Explain.

Q18.19 Is the increase in kinetic energy of an object equal to the work done to accelerate the object when that object is moving at very large velocities? Explain.

Q18.20 If the velocity of an object is reduced to zero, does all of its energy disappear? Explain.

Q18.21 If you are on an elevator that is accelerating downward, will your apparent weight (as measured by a bathroom scale) be greater than, less than, or equal to your weight measured when the elevator is not accelerating? Explain.

Q18.22 When you are inside a closed space vehicle, is it possible for you to tell whether the vehicle is accelerating, or whether you are simply moving closer to (or farther away from) some massive body such as the sun or the earth? Explain.

Q18.23 Would your experiences inside a freely falling elevator be similar in any way to those inside a spaceship moving with constant velocity when it is a long distance away from any planet or star? Explain.

Q18.24 Would a clock located on the surface of the sun measure time at the same rate as a clock located a long distance away from any planet or star? Explain.

Q18.25 Is a black hole just a hole in space that contains no mass? Explain.

EXERCISES

E18.1 A boat that can travel with a velocity of 15 m/s in still water is moving at maximum speed *against* the current of a stream that flows with a velocity of 4 m/s relative to the earth. What is the velocity of the boat relative to the bank of the stream?

E18.2 A plane that can travel at 180 MPH in still air is flying with a tail wind of 20 MPH. How long does it take for the plane to travel a distance of 400 miles (relative to the earth)?

E18.3 A ball is thrown with a velocity of 60 MPH down the aisle (towards the tail of the plane) of a jetliner that is traveling with a velocity of 300 MPH relative to the earth. What is the velocity of the ball relative to the earth?

E18.4 An astronaut aims a flashlight towards the tail of her spaceship, that is traveling with a velocity of 0.5c relative to the earth. What is the velocity of the light beam relative to the earth?

E18.5 The factor $\gamma = 1 / \sqrt{1 - v^2/c^2}$ appears in many expressions derived from the special theory of relativity. Show the following:

  a. $\gamma = 2.294$ when $v = 0.9c$ (or 9/10 the speed of light).

  b. $\gamma = 1.155$ when $v = 0.5c$ (half the speed of light).

  c. $\gamma = 1.005$ when $v = 0.1c$ (one-tenth the speed of light).

E18.6 An astronaut cooks a three-minute egg in his spaceship, which is whizzing past earth at a speed of 0.9c. How long has the egg cooked as measured by an observer on earth? (Use the results of exercise 18.5; the time-dilation formula written in terms of γ is $t = \gamma t_0$.)

E18.7 An observer on earth notes that an astronaut on a spaceship puts in a 4-hour shift at the controls. How long is this shift as measured by the astronaut himself, if the spaceship is moving with a velocity of 0.5c relative to the earth? (Use the results of exercise 18.5. Be careful, though; which observer measures the proper time?)

E18.8 A spaceship that is 40 m long as measured by its occupants is traveling at a speed of 0.1c relative to the earth. How long is the spaceship as measured by mission control in Houston? (Use the results of exercise 18.5; the length-contraction formula takes the form $L = (1 / \gamma) L_0$ when written in terms of γ.)

E18.9 The crew of a spaceship that is traveling with a velocity of 0.5c relative to the earth measures the distance between two cities on earth (in a direction parallel to their motion) as 150 km. What is the distance between these two cities as measured by people on earth? (Use the results of exercise 18.5; who measures the proper length?)

E18.10 A spaceship is traveling with a velocity of 0.9c relative to the earth. What is the momentum of the spaceship if its mass is 2000 kg? ($p = \gamma mv$ and $c = 3 \times 10^8$ m/s; use the results of exercise 18.5.)

E18.11 Suppose that a particle has a mass-energy of 10 joules when it is at rest.

  a. What is its total energy when it is moving with a velocity of 0.9c ? ($E = \gamma E_0$; use the results of exercise 18.5.)

  b. What is the kinetic energy of the particle at this speed? ($KE = E - E_0$)

CHALLENGE PROBLEMS

CP18.1 A boat capable of moving with a velocity of 3 m/s is pointed straight across a stream that is flowing with a velocity of 1 m/s. The width of the stream is 30 m.

  a. Draw a vector diagram to show how the velocity of the stream adds to that of the boat relative to the water to obtain the velocity of the boat relative to the earth.

  b. Use the Pythagorean theorem to find the magnitude of the velocity of the boat relative to the earth.

  c. How long does it take for the boat to cross the stream? (Hint: We need to consider only the component of the velocity of the boat that is straight across the stream if we use the stream width for the distance.)

  d. How far downstream from its starting point does the boat hit the opposite bank?

  e. How far does the boat actually travel in reaching the opposite bank?

CP18.2 Suppose that a beam of π-mesons (or *pions*) is moving with a velocity of $0.99c$ with respect to the laboratory. When the pions are at rest, they decay with a half-life of 1.77×10^{-8} s.

a. Calculate the factor $\gamma = 1 / \sqrt{1 - v^2/c^2}$ for the velocity of the pions relative to the laboratory.

b. What is the half-life of the moving pions as observed by an observer in the laboratory?

c. How far do the pions travel, as measured in the laboratory, before half of them have decayed?

d. As measured in a frame of reference that moves with the pions, how far do the pions travel before half of them have decayed?

e. Which observer, one moving with the pions or one at rest in the laboratory, measures the following:

   (1) The proper time for the half-life?

   (2) The rest length for the distance traveled?

f. Does the distance traveled divided by the half-life equal the velocity, v, for both frames of reference?

CP18.3 Suppose that an astronaut flies to a distant star and then returns to earth. Except for brief intervals of time during which she is accelerating or decelerating, her spaceship travels at the incredible velocity of $v = 0.995c$ relative to the earth. The star is 20 light years away. (A light-year is the distance light travels in one year.)

a. Show that the factor γ for this velocity is approximately equal to 10.

b. How long does the trip to the star and back take as seen by an observer on earth?

c. How long does the trip take as measured by the astronaut?

d. What is the distance traveled as measured by the astronaut?

e. If the astronaut left a twin sister at home on earth while she made this trip, how much younger is the astronaut than her twin when she returns?

CP18.4 Suppose that a beaker contains 500 g of water. Heat is added to raise the temperature of the water from $0°$ to $100°$ C.

a. How much heat energy in joules must be added to the water to raise its temperature this much? ($c = 1.0$ cal/g·C° and 1 cal = 4.186 J)

b. By how much does the mass of the water increase in this process. ($E_0 = mc^2$)

c. Compare this mass increase to the original mass of the water. Would this increase in mass be measurable?

d. If it were somehow possible to convert the original mass of the water into kinetic energy, how many joules of kinetic energy could be produced?

HOME EXPERIMENTS AND OBSERVATIONS

HE18.1 If you live close to a small stream, you can test the velocity addition ideas contained in the first section of this chapter. A small battery-powered or wind-up boat is also necessary. This could be borrowed or purchased cheaply at a variety store.

a. Test your boat first in the bathtub or a pond and estimate how fast it can move in still water.

b. Find a place in the stream where the current is smooth (no eddies) and you have easy access to the water. Drop a twig in the stream and, with the help of a watch, estimate the velocity of the current. Is this velocity larger or smaller than that of your boat in still water?

c. Place your boat, with its motor running, in the stream and point it downstream. Estimate its velocity relative to the bank. Does the result agree with what you would predict based upon the addition of relative velocities?

d. Repeat this process with the boat pointed upstream. (If the current is too strong, it may be difficult to keep it headed in this direction.)

19

The Behavior of Fluids

Boats hold a special fascination for many of us. As a child, you probably floated twigs in puddles and streams. You noticed that some things float and some do not; stones sink quickly to the bottom of the stream. A balloon filled with helium pulls upward on the string, but a balloon filled with air drifts downward to the floor. What makes the difference?

As you grew older, you may have wondered why a steel boat floats, but a piece of metal dropped in the water quickly sinks (fig. 19.1). Does the shape of the material have something to do with whether it floats or not? Could you make a boat out of concrete?

The behavior of things that float (or sink) in either water or air is one aspect of the behavior of fluids. Water (a liquid) and air (a gas) are both examples of fluids. They flow readily and conform to the shape of their containers, unlike solids, which have a shape of their own. Although liquids are usually much denser than gases, many of the principles that apply to liquids apply also to gases; therefore, it makes sense to consider them together under the common heading of fluids.

The concept of pressure, introduced in the discussion of thermodynamics in chapter 9, plays a central role in describing the behavior of fluids. We will explore that concept much more thoroughly in this chapter. Pressure is involved in Archimedes' principle, which explains why things float, as well as in most of the other phenomena that we will consider. What exactly is pressure, and how can this concept be used to explain the behavior of floating objects, hydraulic systems, and why a plane flies, among other things?

Figure 19.1   A steel boat floats, but a piece of metal dropped in the water quickly sinks. How do we explain this?

Chapter Objectives

The first objective of this chapter is to explore thoroughly the meaning of the concept of pressure. We will then consider atmospheric pressure and describe some ways in which gases and liquids are both similar and different. Those ideas will prepare us to explore the behavior of floating objects, as well as effects that occur when fluids are in motion. Moving fluids are involved in Bernoulli's principle, which helps to explain why an airplane can fly or why a curve ball curves.

Chapter Outline

1 *Pressure and Pascal's principle.* What is pressure, and how is it transmitted from one part of a system to another? How does a hydraulic jack or press work?

2 *Atmospheric pressure and gas behavior.* How do we measure atmospheric pressure, and how and why does it vary? How do gases differ from liquids, and what is an ideal gas?

3 *Archimedes' principle.* What does Archimedes' principle say, and how does it depend upon variations in pressure with depth? Why does a steel boat float, but a lump of steel sink?

4 *Fluids in motion.* What special characteristics apply to moving fluids, and what is viscosity? How does the velocity of a moving fluid vary if we change the width of its pipe or channel?

5 *Bernoulli's principle.* What does Bernoulli's principle say, and how is it related to the conservation of energy? How can it be used to explain why an airplane flies, and other phenomena?

19.1 PRESSURE AND PASCAL'S PRINCIPLE

A small woman wearing high-heel shoes sinks into soft ground, whereas a large man wearing size 13 shoes walks across the same ground without difficulty (fig. 19.2). Why is this so? The man weighs much more than the woman, so he must exert a larger force on the ground. But the woman's high heels leave much deeper indentations in the ground.

Clearly weight alone is not the determining factor here, it is more a matter of how that weight is distributed across the area of contact between the shoes and the earth. The woman's shoes have a small area of contact, whereas the man's shoes have a much larger area of contact. The force that the man's feet exert upon the ground because of his weight is therefore distributed over a larger area.

Definition of Pressure

The quantity that determines whether or not the shoes will sink into the ground is actually the pressure exerted upon the partially-fluid soil by the shoes. The concept of pressure was introduced in chapter 9:

Pressure is force per unit area; $P = \dfrac{F}{A}$.

Its units, therefore, are newtons per meter squared (N/m^2), the metric unit of force divided by the metric unit of area. This unit is also called a *pascal* ($1 \text{ Pa} = 1 \text{ N/m}^2$).

The heel of a woman's high-heel shoe can have an area as small as 1 or 2 square centimeters (cm^2). When you walk, there are times when almost all of your weight is supported by your heel. (The weight shifts from the heel to the toe as you move forward with the other foot off the ground; get out of your chair and try it!) The woman's weight divided by the small area of her heel thus produces a large pressure on the ground.

The man's shoe, on the other hand, may have a heel area of as much as 100 cm^2 (for a size 13 shoe). Since the heel area of his shoe can be one hundred times larger than the woman's, he can weigh two or three times more than the woman, but still exert a pressure on the ground ($P = F / A$) that is a small fraction of that exerted by the woman. His smaller pressure leads to a smaller distortion of the ground.

These ideas are illustrated schematically in figure 19.3. On the left side of the diagram, a force is applied to a piston with a small area, while on the right side a somewhat larger force is applied to a much larger piston. If the cylinders contain a fluid (gas or liquid), the pressure exerted on the fluid in the left cylinder will be greater than that exerted on the fluid in the right cylinder because of the difference in areas.

Given the definition of pressure ($P = F / A$), a force distributed over a small area will produce a larger pressure than the same force distributed over a larger area; the greater the area the smaller the pressure. As the linear dimensions of a surface increase, the area increases even more rapidly. For a circle, for example, the area is equal to π (3.14) times the square of the radius of the circle (πr^2). A circle with a radius of 10 cm, then, has an area that is 100 times larger $(10)^2$ than a circle with a radius of 1 cm. These ideas are illustrated in the sample exercise in box 19.1 where a force of 50 N exerts a smaller pressure than one of 10 N because of the difference in the areas over which they are applied.

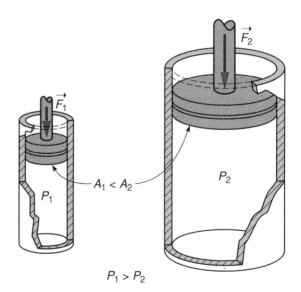

Figure 19.3 The pressure exerted upon the left-hand piston, P_1, is greater than that exerted upon the right-hand, P_2, even though the force F_2 is larger than F_1. $P = F / A$, and A_1 is smaller than A_2.

Figure 19.2 A small woman's high heels sink into the soft ground, but the larger shoes of the much larger man do not.

Box 19.1

Sample Exercise

A force of 10 N is applied to a circular piston with a radius of 1 cm. On a second cylinder, a force of 50 N is applied to a piston with a radius of 10 cm. What is the pressure exerted upon the fluid in each cylinder?

$F_1 = 10$ N. $A_1 = \pi r^2 = (3.14)(1 \text{ cm})^2$

$r_1 = 1.0$ cm. $= 3.14 \text{ cm}^2 = 3.14 \times 10^{-4} \text{ m}^2$.

$P_1 = ?$ $(1 \text{ m}^2 = (100 \text{ cm})^2 = 10^4 \text{ cm}^2.)$

$$P_1 = \frac{F_1}{A_1} = \frac{10 \text{ N}}{3.14 \times 10^{-4} \text{ m}^2}$$

$$= 3.18 \times 10^4 \text{ N/m}^2 = \textbf{31.8 kPa.}$$

(A kilopascal, kPa, is 1000 Pa.)

$F_2 = 50$ N. $A_2 = \pi r^2 = (3.14)(10 \text{ cm})^2$

$r_2 = 10$ cm. $= 314 \text{ cm}^2 = 3.14 \times 10^{-2} \text{ m}^2$.

$P_2 = ?$

$$P_2 = \frac{F_2}{A_2} = \frac{50 \text{ N}}{3.14 \times 10^{-2} \text{ m}^2}$$

$$= 1.59 \times 10^3 \text{ N/m}^2 = \textbf{1.59 kPa.}$$

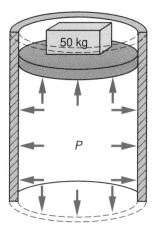

Figure 19.4 The pressure exerted by the piston on the fluid is transmitted uniformly throughout the fluid, causing it to push outward with equal force per unit area upon the walls and bottom of the cylinder.

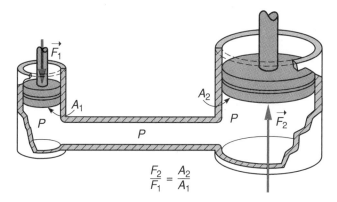

Figure 19.5 A small force, F_1, applied to a piston with a small area can produce a much larger force, F_2, on the larger piston. This allows a hydraulic jack to lift heavy objects.

Pascal's Principle and Hydraulics

What happens inside the fluid when a pressure is exerted on it? Does pressure have direction? Does it transmit a force to the walls or bottom of the cylinder in the examples we have been considering? These questions point to another important feature of the nature of pressure in fluids.

When we apply a force (push down) on the piston in a cylinder, as pictured in figure 19.4, the piston exerts a force on the fluid. By Newton's third law, of course, the fluid also exerts a force (in the opposite direction) on the piston. The fluid inside the cylinder will be squeezed and may decrease in volume somewhat; we say that it has been *compressed*. It is easy to imagine that compression of the fluid will cause it to push outwards on the walls and bottom of the cylinder as well as on the piston.

The fluid behaves like a spring, but it is an unusual spring; it pushes outward uniformly in all directions when it is compressed. Any increase in pressure is thus transmitted uniformly throughout the fluid, as is indicated in figure 19.4. If we ignore variations in pressure due to the weight of the fluid itself, the same pressure that pushes upward upon the piston also pushes outward on the walls and downward on the bottom of the cylinder.

This ability of a fluid to transmit the effects of pressure uniformly is the essential feature of Pascal's principle and the basis of the operation of a hydraulic jack and other hydraulic devices. Blaise Pascal (1623–62) was a French scientist whose primary contributions were in the areas of fluid statics and probability theory. Pascal's principle is usually stated as follows:

Any change in the pressure of a fluid is transmitted uniformly, in all directions, throughout the fluid.

How does a hydraulic jack work? Hydraulic systems are the most common area of application of Pascal's principle. Their operation depends on the uniform transmission of pressure, as well as on the relationship between pressure, force, and area (the definition of pressure). The basic idea is illustrated in figure 19.5.

A force applied to a piston with a small area can produce a large increase in pressure of the fluid because of the small area of the piston. This increase in pressure is transmitted through the fluid to the piston on the right, which has a much larger area. Since pressure is force per unit area ($P = F / A$), the force exerted on the larger piston by this pressure is proportional to the area of the piston ($F = PA$, if we multiply both sides of the definition by A).

Equating the pressures for the two cylinders leads to a simple expression describing the effect of the jack:

$$P_1 = \frac{F_1}{A_1} = \frac{F_2}{A_2} = P_2.$$

Rearranging this expression using algebra shows that the ratio of the two forces is equal to the ratio of the areas:

$$\frac{F_2}{F_1} = \frac{A_2}{A_1}.$$

The force exerted upon the second piston can be much larger than that applied to the first piston if the area A_2 is much larger than A_1.

Since it is quite feasible to build a system in which the area of the second piston is over one hundred times larger than that of the first piston, we can produce a force on the second piston that is over one hundred times larger than that exerted on the first piston. The mechanical advantage (F_2 / F_1, as defined in chapter 8) of a hydraulic system is usually large, therefore, in comparison to that of a lever or other simple machine. The force applied to the lever handle of a hydraulic jack can easily be large enough to lift a car (fig. 19.6).

Of course, as in our discussion of simple machines in chapter 8, we pay a price for getting this larger output force. The work done by the larger piston in lifting the car cannot be any greater than the work input to the handle of the jack. Since work is equal to force times distance ($W = Fd$), the dis-

tance that the larger piston moves will be quite small compared to the distance moved by the smaller piston. Conservation of energy, in other words, tells us that

$$W = F_2 d_2 \leq F_1 d_1.$$

If F_2 is much larger than F_1, then d_2 must be much smaller than d_1 because the output cannot exceed the amount of work that we put into the system.

With a hand-pumped jack, you move the smaller piston several times, allowing its chamber to refill with fluid after each stroke. The total distance moved by the smaller piston is then the sum of the distances moved on each stroke. The larger piston inches upward during this process.

The fluid in a hydraulic jack is usually an oil of some kind. Hydraulic systems are also used in the brake systems of cars and in many other applications. Oil is more effective than water as a hydraulic fluid because it is not corrosive and also serves as a lubricant to assure smooth operation of the system. The fluid used is almost always a liquid rather than a gas because liquids are much less compressible than gases. We generally do not want the fluid volume to change significantly during the operation of a hydraulic system.

Hydraulic systems take advantage of the ability of fluids to transmit changes in pressure, as described by Pascal's principle. They also usually use the multiplying effect produced by pistons of different areas. Thinking about how a hydraulic system works should reinforce your understanding of the concept of pressure itself.

19.2 ATMOSPHERIC PRESSURE AND GAS BEHAVIOR

Living on the surface of the earth, we live at the bottom of a sea of air. Except for the presence of smog or haze, that air is usually invisible, and we seldom give it a second thought. How do we know that it is there? What measurable effects does it have?

We do feel the presence of the air, of course, if we are riding a bike or walking on a gusty day. Skiers, bicycle racers, and race-car designers are all very conscious of the need to reduce their resistance to the flow of air past themselves and their vehicles. The labored breathing of a mountain climber reflects the thinning of the atmosphere near the top of a high mountain. And, of course, modern barometers are capable of accurately measuring atmospheric pressure and its variations with weather or altitude.

Measurement of Atmospheric Pressure

It was during the seventeenth century that we first learned how to measure atmospheric pressure and became aware of its variation with altitude. Galileo had noticed that the water pumps he designed were capable of pumping water to a height of only 32 feet, but never adequately explained why.

Figure 19.6 A hydraulic jack can easily lift a car.

His disciple, Evangelista Torricelli (1608–47), is generally credited with inventing the barometer.

Torricelli was actually most interested in the concept of a vacuum. He reasoned that Galileo's pumps were creating a partial vacuum and that the pressure of the air pushing down on the water reservoir was actually responsible for lifting the water. In thinking about how to test this hypothesis, he was struck by the idea of using a much denser fluid than water. Mercury, or *quicksilver,* was the logical choice; it is fluid at room temperature and is approximately 13 times as dense as water.

Torricelli's early experiments employed a glass tube, approximately 1 m (39–40 inches) in length, that was sealed at one end and open at the other. He filled the tube with mercury and then, holding his finger over the open end, inverted the tube and placed its open end in an open container of mercury (fig. 19.7). The mercury would flow from the tube into the container until an equilibrium was reached, leaving a column of mercury approximately 76 cm (30 inches) in height in the tube. The pressure of air pushing down on the surface of the mercury in the open container was apparently strong enough to support a column of mercury 76 cm in height.

Torricelli was careful to demonstrate experimentally that there was actually a vacuum in the space at the top of the tube, above the column of mercury. The reason that the mercury column does not fall is that the pressure at the top of the column is zero, whereas that at the bottom of the column is equal to atmospheric pressure. We still often quote atmospheric pressure in either centimeters of mercury or inches of mercury (the commonly used unit in the United States). How are these units related to the pascal?

This relationship is easy to establish if we know the density of mercury ($d = 13.55$ g/cm^3 = 13.55×10^3 kg/m^3 at 20° C). The weight of the column of mercury is equal to its mass times the acceleration of gravity:

$$W = mg = dVg.$$

(Multiplying the density, d, by the volume, V, yields the mass of the column.) The volume of the column, however, is just its height times its cross-sectional area, A ($V = Ah$). The weight of the column can therefore be expressed as $W = d(Ah)g$.

The pressure at the bottom of the column is just the weight of the column divided by the cross-sectional area of the column:

$$P = \frac{F}{A} = \frac{W}{A} = \frac{d(Ah)g}{A} = dhg.$$

The pressure is equal, in other words, to the density of mercury times the height of the column times the acceleration of gravity (fig. 19.8). Putting in the numerical values, we see that

$$P_a = dhg = (13.55 \times 10^3 \text{ kg/m}^3)(0.76 \text{ m})(9.8 \text{ m/s}^2)$$
$$= 1.01 \times 10^5 \text{ N/m}^2 = 101 \text{ kPa}.$$

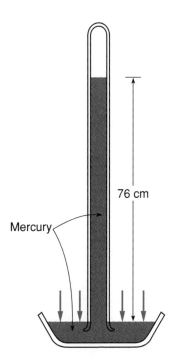

Figure 19.7   Torricelli filled a glass tube with mercury and inverted it into an open container of mercury. Air pressure acting upon the mercury in the dish could support a column of mercury 76 cm in height.

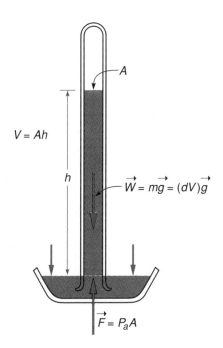

Figure 19.8   The weight of the mercury column ($W = dVg$) is supported by atmospheric pressure P_a. The force acting upon the bottom of the column ($F = P_aA$) must equal the weight of the column.

Figure 19.9 Two teams of eight horses were unable to separate Von Guericke's evacuated metal hemispheres. What force pushes the two hemispheres together?

Atmospheric pressure, P_a, is thus approximately 101 kilopascals when measured at sea level. This converts to 14.7 pounds per square inch (psi), which is probably a more familiar unit. Living at the bottom of this sea of air, you have 14.7 pounds pushing on every square inch of your body. Why do you not notice that? The reason is that air permeates your body and pushes back out. If you were suddenly thrown into a vacuum, you would explode!

A very famous experiment designed to demonstrate the effects of air pressure was performed by Otto von Guericke (1602–86) during the seventeenth century. Von Guericke had two bronze hemispheres produced that could be smoothly joined together at their rims. He then pumped the air out of the sphere formed in this manner, using a crude vacuum pump that he had developed. As shown in figure 19.9, two teams of eight horses each were unable to pull the spheres apart. When the stopcock was opened to let air back into the evacuated sphere, however, the two hemispheres could be easily separated.

Variations in Atmospheric Pressure

If we live at the bottom of a sea of air, we should expect that the pressure will decrease as we go up from sea level. The reason is simple; the pressure that we experience is due to the weight of the fluid above us. As we go up from the earth's surface, there is less atmosphere above us, so the pressure should decrease.

It was just such reasoning that led Blaise Pascal to attempt to measure atmospheric pressure at different altitudes shortly after Torricelli's invention of the mercury barometer. Although Pascal himself was in poor health through most of

Figure 19.10 A balloon that was partially inflated near sea level expanded as the experimenters climbed the mountain.

his adult life, and therefore not up to climbing mountains, he sent his brother-in-law to the top of the Puy-de-Dome mountain in France with a barometer similar to Torricelli's. His brother-in-law found that the height of the mercury column that could be supported by the atmosphere was significantly lower at the top of the mountain than near sea level; it was about 7 cm lower at the top of the 1460 m (4800 ft) mountain than at the bottom.

Pascal also had his brother-in-law take a partially inflated balloon to the top of the mountain. As Pascal had predicted, the balloon expanded as the climbers gained elevation, indicating a decrease in the external atmospheric pressure (fig. 19.10). Pascal was even able to show a decrease in pressure within the city of Clermont, between the low point in town and the top of the cathedral tower. The decrease was small but measurable.

It was also observed that pressure (measured by a barometer) seemed to vary in relation to changes in the weather. Lower column heights were observed on stormy days than on clear days. This pressure variation has been used ever since as an indication of changes in weather; a falling atmospheric pressure points to stormy weather ahead. The readings announced are usually corrected to sea level, so that variations in altitude will not mask the changes related to the weather.

Boyle's Law and Ideal-Gas Behavior

You might think that we could calculate the variation in atmospheric pressure with altitude in much the same way that we computed the pressure due to a column of mercury earlier in this section. We would merely need to compute the weight of the column of air that is above us. However, even though mercury and air are both fluids, there is a significant difference (besides the difference in density) between the behavior of a column of mercury and that of a column of air.

Mercury, like most liquids, is not very compressible; if we increase the pressure acting upon a given volume of mercury, its volume does not change very much. Thus the density of the mercury (mass per unit volume) is essentially the same near the bottom of the column as it is near the top. A gas such as air, on the other hand, is quite compressible. As the pressure changes, the volume changes, and so does the density. Therefore we cannot use a single value of density to compute the weight of our column of air; the density decreases as we rise in the atmosphere (fig. 19.11).

This variation of the volume and density of a gas with pressure was studied by Robert Boyle (1627–91) in England, as well as by Edme Mariotte (1620–84) in France. Boyle's results were first reported in 1660, but went unnoticed on the European continent, where Marriotte published very similar conclusions in 1676. Both of them were interested in the springiness, or compressibility, of air.

Both experimenters used a bent glass tube sealed on one side and open on the other, as shown in figure 19.12. In Boyle's experiment, the tube was partially filled with mercury, so that air was trapped in the closed portion of the tube. He allowed air to pass back and forth initially so that the pressure in the closed side of the tube was equal to atmo-

spheric pressure, and the columns of mercury on both sides were therefore at the same height (fig. 19.12a).

As he then added mercury to the open end of the tube, the volume of the air trapped on the closed side decreased. He found that when he added enough mercury to increase the pressure to twice atmospheric pressure, the height of the air column on the closed side decreased to one half of its initial value (fig. 19.12b). In other words, doubling the pressure caused the volume of air to decrease by one half. More generally, he discovered that the volume was inversely proportional to the pressure.

In symbols this relationship can be expressed as follows:

$$PV = \text{constant}.$$

If the pressure increases, the volume must decrease in proportion in order to keep the product of the pressure and the volume constant. We often write Boyle's law (known as *Mariotte's law* in Europe) as follows:

$$P_1 V_1 = P_2 V_2,$$

where P_1 and V_1 are the initial pressure and volume, and P_2 and V_2 are the final pressure and volume. The expression assumes a fixed mass or quantity of gas, at a constant temperature.

Thus, as we gain altitude, the atmospheric pressure decreases, and the volume of a given mass of air increases. Since density is mass per unit volume ($d = m/V$), the density must decrease when the volume increases. In computing the weight of a column of air, we would have to take this change in density into account, and our computation becomes much more complex than that for the column of mercury.

A further complication is that the density of a gas also depends on temperature, and the temperature of the atmosphere usually decreases as we gain altitude. As we saw in

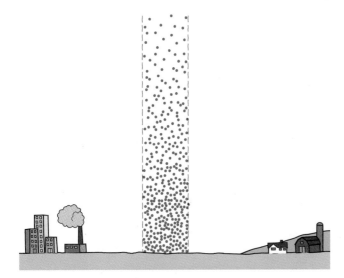

Figure 19.11 The density of a column of air decreases as altitude increases because the air expands as the pressure decreases.

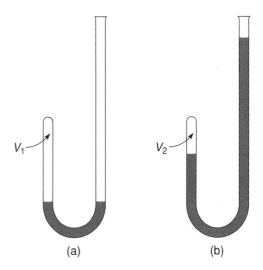

Figure 19.12 In Boyle's experiment, adding mercury to the open side of the bent tube caused a decrease in the volume of the trapped air in the closed side.

chapter 9 when we discussed the concept of absolute temperature, the pressure of a fixed volume of gas increases in proportion to the absolute temperature of the gas; conversely, the volume of the gas increases if the pressure remains constant. The constant in Boyle's law is related, then, to the temperature of the gas.

All of these effects can be summarized in the equation of state for an ideal gas, which was introduced briefly in chapter 9 and is usually written as follows:

$$PV = nRT,$$

where P and V are the pressure and volume, n is the mole number (a measure of the quantity of gas present), R is the ideal-gas constant, and T is the absolute temperature. (R has the value 8.314 J/mole K.) From this expression, we see that increasing either the quantity of the gas or the temperature of the gas increases the product of the pressure and the volume involved in Boyle's law. The ideal-gas equation approximately describes the behavior of any gas provided that the temperature is well above the condensation point for that gas.

If the quantity of gas is held constant, we can see from the ideal-gas equation that the product of the pressure and the volume *divided* by the absolute temperature must be constant, and we can write.

$$\frac{P_1 V_1}{T_1} = \frac{P_2 V_2}{T_2} = \text{constant}.$$

Boyle's law is valid when the temperature of the gas does not change; it is a special case of this more general expression. The ideal-gas law (which includes Boyle's law as a special case) can be stated in words as follows:

The product of the pressure and volume of an ideal gas divided by the absolute temperature is equal to a constant, which is the product of the mole number, *n*, and the ideal-gas constant, *R*.

The application of Boyle's law, as well as the more general ideal-gas law, is illustrated in the sample exercise of box 19.2.

In the first part of the sample exercise, the pressure increases by a factor of 3, and the volume decreases by the same factor: the final volume is one-third of the initial volume. In the second part, the absolute temperature increases by a factor of $4/3$ and so does the volume.

Notice that we can use any units for pressure and volume in a problem like that in box 19.2, provided that they are consistent. In (*a*), for example, we expressed the pressure in atmospheres, which is commonly done. Since this unit appeared in both the numerator and denominator of our final expression, the unit cancels and does not appear in the result. Temperature, however, must be expressed in kelvins, since the proportionality holds only for absolute temperatures; Celsius or Fahrenheit temperatures would result in a different (and incorrect) ratio.

Box 19.2

Sample Exercise

Suppose that a fixed quantity of gas is held in a cylinder capped at one end by a movable piston. The pressure of the gas is originally 1 atmosphere (101 kPa), and the volume is originally 0.3 m³. What is the volume of the gas under the following conditions?

a. The pressure is increased to 3 atmospheres while the temperature is held constant.
b. The pressure is held constant at 1 atmosphere, but the temperature is increased from 27°C to 127°C.

a. $P_1 = 1$ atm.
  $V_1 = 0.3$ m³.
  $P_2 = 3$ atm,
  $V_2 = ?$

$$P_1 V_1 = P_2 V_2.$$
$$V_2 = \frac{P_1 V_1}{P_2}$$
$$= \frac{(1\ \text{atm})(0.3\ \text{m}^3)}{3\ \text{atm}}$$
$$= \left(\frac{1}{3}\right)(0.3\ \text{m}^3) = \mathbf{0.1\ m^3}.$$

b. $P_1 = P_2 = 1$ atm.
  $V_1 = 0.3$ m³.
  $T_1 = 27$°C.
  $T_2 = 127$°C.
  $V_2 = ?$

$T_1 = 273 + 27$°C $= 300$ K.
$T_2 = 273 + 127$°C $= 400$ K.

$$\frac{P_1 V_1}{T_1} = \frac{P_2 V_2}{T_2}.$$
$$\frac{V_1}{T_1} = \frac{V_2}{T_2}.$$
$$V_2 = \frac{T_2}{T_1} V_1$$
$$= \left(\frac{400\ \text{K}}{300\ \text{K}}\right)(0.3\ \text{m}^3)$$
$$= \left(\frac{4}{3}\right)(0.3\ \text{m}^3) = \mathbf{0.4\ m^3}.$$

As we can see, gases are springy. They can be readily compressed to a small fraction of their initial volume. They can also expand if the pressure is reduced, as did the gas in Pascal's balloon. Changes in temperature also affect the volume or pressure of a gas much more than those of a liquid.

This major difference between gases and liquids can be explained by differences in the atomic or molecular *packing* of these two types of fluid. The atoms or molecules in a gas

Liquid	Gas

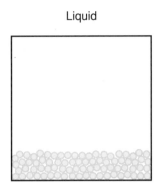
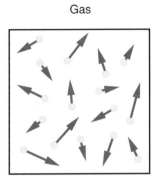

Figure 19.13 **The atoms in a liquid are closely packed, whereas those in a gas are separated by much larger distances.**

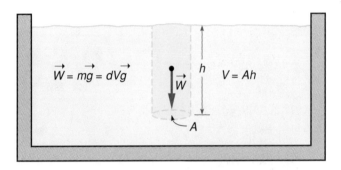

Figure 19.14 **The weight of a column of water within a swimming pool is equal to the mass of the column times the acceleration of gravity($m\vec{g}$). The mass is equal to the density times the volume (dV).**

are separated by distances that are large compared to the size of the atoms themselves, as shown in figure 19.13. The atoms in a liquid, on the other hand, are *closely packed,* much like those in a solid. They cannot be easily squeezed.

19.3 ARCHIMEDES' PRINCIPLE

Why do some things float and others sink? Is it a matter of how heavy the object is? A large ocean liner made mostly of steel floats, but a small pebble quickly sinks. Clearly it is not just a matter of total weight; as you may know, it is more directly related to the density of the object. Objects that are more dense than the fluid in which they are immersed will sink; those that are less dense will float.

The complete answer to the question of why things sink or float is found in Archimedes' principle, which describes the buoyant force that is exerted on any object that is fully or partly immersed in a fluid. This buoyant force, in turn, depends directly upon the increase in fluid pressure that occurs with increasing depth.

Variation of Pressure with Depth in a Liquid
When you swim to the bottom of a swimming pool at the deep end, you can feel the pressure building on your ears. The pressure at the bottom of the pool is greater than that at the surface for the same reason that the pressure of the atmosphere is greater near the surface of the earth than at higher altitudes. The weight of the fluid above us partly determines the pressure that we experience.

We have already considered the weight of a column of liquid in the context of the mercury barometer in section 19.2. If we consider such a column within a swimming pool, as pictured in figure 19.14, it is easy enough to compute the weight of that column; it is equal to the mass times the acceleration of gravity (*mg*), and the mass is just the density times the volume (*dV*). In a liquid, in contrast to the atmosphere, the density does not change significantly with depth because

liquids are not usually very compressible.

As was shown for the mercury column in section 19.2, the weight of the column of water is

$$W = mg = (dV)g = d(Ah)g,$$

where *d* is the density, *A* is the cross-sectional area of the column, and *h* is the height of the column. The increase in pressure produced by this weight is equal to the weight (the additional force) divided by the area, since pressure is force per unit area:

$$\Delta P = \frac{W}{A} = \frac{dAhg}{A} = dhg.$$

In words, the increase in pressure with depth is equal to the density of the fluid times the depth times the acceleration of gravity.

The pressure resulting from the weight of the fluid is written as an increase in pressure (ΔP) because the pressure at the top of the pool is atmospheric pressure rather than zero. By Pascal's principle, the atmospheric pressure pushing on the surface of the pool is transmitted uniformly throughout the fluid. The total pressure at some depth *h* under the water is therefore

$$P = P_a + \Delta P = P_a + dgh.$$

For many purposes, however, it is the increase in pressure beyond atmospheric pressure that is of most interest. Since our bodies have internal pressures essentially equal to the atmospheric pressure, for example, it is the increase in pressure beyond atmospheric pressure (ΔP) to which our eardrums are sensitive.

A large can filled with water is often used to demonstrate the variation of pressure with depth. If you punch holes in the can at different depths, the water emerges from a hole near the bottom of the can with a much greater horizontal velocity than from a hole punched near the top (fig. 19.15). Since the can itself is submerged in the atmosphere, it is once again the difference in pressure with depth that is significant. A larger ΔP provides a larger accelerating force for the emerging water.

Figure 19.15 Water emerging from a hole near the bottom of a can filled with water has a larger horizontal velocity than water emerging from a hole near the top.

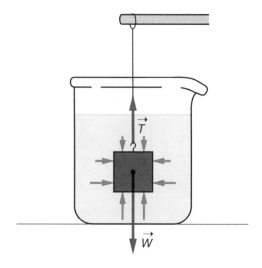

Figure 19.16 The pressure acting upon the bottom of the suspended metal block is greater than that acting on the top of the block due to the increase of pressure with depth.

The Principle

We are now in a position to understand Archimedes' principle. Suppose that we choose an object that we know will sink in water, such as a steel block. If this block is suspended from a string in a beaker of water, as shown in figure 19.16, there is a buoyant force, $\vec{F}_B$, acting upwards on the block. The presence of this force can be verified by weighing the block while it is submerged. If the string is attached to the arm of a balance, the apparent weight of the block will be less than its actual weight. The difference is equal to the buoyant force.

This buoyant force is present because the pressure is greater at the bottom of the block than at the top. The bottom of the block is at a greater depth in the fluid; therefore, the force acting on the bottom of the block is greater than the force acting on the top. The buoyant force itself, $\vec{F}_B$, is equal to the difference in these two forces.

The pressure increases with depth, as already noted, by an amount $\Delta P = dgh$. If we use Δh to represent the difference in depth from the top of the block to its bottom, then the increase in pressure from the top to the bottom of the block is $\Delta P_{block} = dg\Delta h$. Since pressure is force per unit area, the buoyant force is equal to this pressure difference times the cross-sectional area:

$$F_B = (\Delta P_{block})A = dg\Delta hA.$$

Note that the product ΔhA is equal to the volume, V, of the block—its height times its cross-sectional area. Notice also that d here is the density of the fluid in which the block is immersed, not the density of the block. We can rewrite this result, then, as follows:

$$F_B = d_f gV.$$

The buoyant force is equal to the density of the fluid, d_f, times the acceleration of gravity, g, times V, the volume of the block.

If we examine this expression for the buoyant force, we see that it is actually equal to the weight of the fluid displaced by the volume of the block. The mass of the fluid displaced is equal to the density of the fluid times the volume of the block ($d_f V$), and multiplying that mass by g yields the weight of the fluid. This was the insight that we now know as Archimedes' principle:

The buoyant force acting on an object that is fully or partially submerged in a fluid is equal to the weight of the fluid that is displaced by the object.

Archimedes' principle readily explains why an object sinks or floats. If the object is more dense than the fluid in which it is submerged, then the weight of the object ($W_o = d_o Vg$) is greater than the weight of the fluid displaced when the object is fully submerged ($F_B = d_f Vg$). Since the weight, which acts downward, is greater than the buoyant force, which acts upward, the net force acting on the object is downward, and the object sinks.

If the object is less dense than the fluid, the buoyant force is larger than the weight of the object when it is fully submerged. The net force acting on the object is then upward, and the object floats. At the surface of the fluid, just enough of the object will remain submerged so that the weight of the fluid it displaces is equal to the weight of the object. The net force is then zero, and the object is in equilibrium; it has no acceleration. Table 19.1 summarizes these ideas.

Table 19.1
Summary of Conditions for Sinking and Floating

Condition	Effect
$d_o > d_f$ (density of object greater than that of the fluid).	The weight of the object is greater than that of the fluid displaced by the object, or $$W_o > F_B,$$ and the object sinks.
$d_o < d_f$ (density of object less than that of the fluid).	The weight of the object is less than that of the fluid displaced when the object is submerged, or $$W_o < F_B,$$ and the object rises to the surface.
$d_o = d_f$ (density of object equal to that of the fluid).	The weight of the object equals that of the fluid displaced when the object is submerged, or $$W_o = F_B,$$ and the object floats when fully submerged. (Such an object can rise or sink in fluid if its average density changes slightly, as it can in a fish or a submarine.)

Figure 19.17 **A fully loaded tanker will ride much lower in the water than an empty tanker.**

Boats and Balloons

Why does a boat made of steel float? The answer, of course, is that the boat is not made of solid steel all the way through; there are open spaces within the boat filled with air and with other materials. Steel itself is much denser than water, and a solid piece of steel will quickly sink. As long as the average density of the boat is less than that of water, though, a steel boat will certainly float.

According to Archimedes' principle, the buoyant force that acts upon the boat must be equal to the weight of the water that its hull displaces. For the boat to be in equilibrium (with a net force of zero), the buoyant force must equal the weight of the boat. As the boat takes on cargo, its total weight increases and so must the buoyant force. This means that the amount of water displaced by the hull must increase, so the boat sinks lower in the water. There is a limit to how much cargo can be added to a boat (often expressed in terms of tons of displacement). A fully loaded oil tanker rides much lower in the water than an unloaded tanker, as shown in figure 19.17.

Other important considerations in ship design dictate the shape of the hull and the manner in which the boat must be loaded. If the center of gravity is too high, or if the boat is unevenly loaded, it may tip over. Wave action and winds can add to this danger, of course, so some margin of safety must be included in the design. Once water enters the boat, the overall weight of the boat and its average density increase. When the average density becomes greater than that of water, down she goes.

An amusing exercise is sometimes performed in elementary classrooms, but it can provide significant challenges to college students as well. It consists of a contest to see who can build the best boat out of clay. Each contestant is provided with the same amount of clay and challenged to build a boat that will hold the maximum number of steel washers (or other suitable weights) without capsizing. What principles would you use in your boat design?

Buoyant forces do not act only on objects floating in liquids; they also act on objects submerged in a gas such as air. If a balloon is filled with a gas that is less dense than air, then its average density can be less than that of air, and the balloon rises. Helium and hydrogen are the two common gases with densities less than that of air. Helium is more commonly used, even though its density is somewhat greater than that of hydrogen, because hydrogen can combine explosively with the oxygen in air. It is therefore dangerous to use in balloons or blimps.

The average density of a balloon is determined, of course, by the material of which it is made as well as by the density of the gas that fills it. The best materials for making balloons can be stretched very thin without losing strength. They should also remain impermeable to the flow of gas, so that the helium or other gas will not be lost rapidly through the skin of the balloon. Balloons made of mylar (which is often aluminum coated) are much better in this latter respect than ordinary latex balloons.

Hot-air balloons take advantage of the fact that any gas expands when it is heated. If the volume of the gas increases, the density must decrease. As long as the air inside the balloon is much hotter than the air surrounding the balloon, there is an upward buoyant force. The beauty of a hot-air balloon is that we can readily adjust the density of the air within the balloon by turning the gas-powered heater on or off. This gives us some control over whether the balloon will ascend or descend.

Buoyant forces and Archimedes' principle are useful in a variety of applications in addition to explaining the behavior of boats and balloons. We can use Archimedes' principle to determine the density of objects, or of the fluid in which they are submerged. This, in fact, was Archimedes' original application; he is said to have used the idea to determine the density of the king's crown in order to see whether it was truly pure gold. Buoyant force can also be a nuisance at times; we need to correct for the effect of the buoyant forces33 of the air, for example, when we want an accurate weight for a low-density object.

19.4 FLUIDS IN MOTION

If we return to the stream bank mentioned in the introduction to this chapter, we can observe some other features besides whether our boats will sink or float. We might notice, for example, that the velocity of the current varies from point to point in the stream. Where the stream narrows, the velocity increases. The velocity is also usually greater near the middle of the stream than close to the banks. Eddies and other features of turbulent flow may also be observed.

These are all characteristics of the flow of fluids. The velocity of flow is affected by the width of the stream and by the *viscosity* of the fluid, which is a measure of the frictional effects within the fluid. Some of these features are easy to understand, whereas others, particularly the behavior of turbulent flow, are still areas of active research.

Continuous Flow

The change in the velocity of the stream as its width changes is one of the easier features to understand. As long as no tributaries add water to the stream, the flow of water over a limited length of the stream is a good example of the continuous flow of a fluid. In a given time, the same amount of water that enters the stream at some upper point must leave the stream at some lower point; otherwise water would be piling up or being lost somewhere within that segment of the stream.

This may be easier to visualize if we imagine water flowing in a pipe, as shown in figure 19.18. How can we describe the *rate* of flow of water through the pipe? We might describe it as being so many gallons per minute, which is a volume per unit of time. In the metric system, we might use liters per second or cubic meters per second as our units. As figure 19.18 shows, the volume of a slug of water of length L flowing past some point in the pipe equals the product of the length and the cross-sectional area, L times A. The rate at which this volume moves is obviously proportional to its velocity.

The rate of flow turns out to be equal to the cross-sectional area of the pipe times the velocity of the current (vA). This is clear if we simply write the definition of rate of flow:

$$\text{Rate of flow} = \text{volume / time} = \frac{V}{t} = \frac{(LA)}{t} = vA.$$

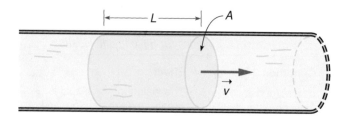

$$\text{Rate of flow} = \frac{V}{t} = \frac{LA}{t} = vA$$

Figure 19.18   The rate at which water moves through a pipe is defined by the volume per unit time that passes by a given point. This is equal to the velocity of the water times the cross-sectional area of the pipe.

The last step in this process follows from recognizing that L is just the distance the water travels in the time t. This distance divided by the time equals the velocity of flow ($v = L / t$). This expression for the rate of flow (vA) is valid for any fluid and should make intuitive sense to you. The greater the velocity, the greater the rate of flow; and the larger the cross section of the pipe or stream, the greater the rate of flow.

If the pipe gets either wider or narrower, then the velocity of flow must change if the rate of flow is to stay the same. This is the underlying justification for what is often called the *equation of continuity*. As long as the rate of flow is the same throughout the pipe or stream, then

$$\text{Rate of flow} = v_1 A_1 = v_2 A_2.$$

If the density of the fluid does not change, then a decrease in the cross-sectional area of the pipe requires an increase in the velocity of flow in order to maintain the same rate of flow.

The same principle applies to a stream; where the stream is narrow, its cross-sectional area is generally smaller than at a wider point in the stream. (We say *generally* because the depth of the stream can also vary. It may be deeper at the narrower places, but usually not enough to keep the cross-sectional area the same as at the wider places.) As long as the cross-sectional area has decreased, the velocity must increase to maintain the rate of flow.

Viscosity

Up to this point, we have ignored the variation in the velocity of the fluid across its cross-sectional area. We mentioned earlier that the velocity is usually greatest near the middle of the stream. The reason for this is that there are frictional, or *viscous*, effects between layers of the fluid itself. There are also frictional effects between the fluid and the walls of the pipe or the banks of the stream.

If we imagine that the fluid is made up of layers, we can see why its velocity will be greatest near the center. Figure 19.19 shows different layers of a fluid moving through a

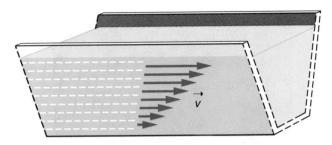

Figure 19.19 Because of frictional or viscous forces between layers, each layer of a fluid flowing in a trough moves more slowly than the layer immediately above.

Low viscosity

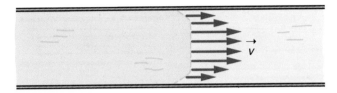

High viscosity

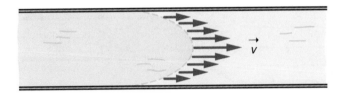

Figure 19.20 The velocity increases rapidly from the wall inward for a low-viscosity fluid but changes more gradually for a high-viscosity fluid.

trough. Since the bottom wall of the trough is not moving, it exerts a frictional force on the bottom layer of fluid, which therefore moves more slowly than the layer immediately above it. This next layer exerts a frictional drag, in turn, upon the layer above it, which therefore flows more slowly than the next one above it, and so on.

The viscosity of the fluid determines the strength of the frictional forces between its layers. The magnitude of the frictional forces also depends upon the area of contact between layers and the rate at which the velocity is changing across the layers. If these other factors are the same, however, a fluid with a high viscosity, such as molasses, will experience a larger frictional force between layers than a fluid with a low viscosity, such as water.

There is usually a thin layer of fluid lying next to the walls of the pipe or trough that does not move at all. The fluid velocity then increases with distance from the wall. The distance over which this variation takes place is determined by the fluid's viscosity as well as by its overall velocity through the pipe. For a fluid with a low viscosity, the transition to the maximum velocity occurs over a relatively short distance from the wall. For a fluid with a high viscosity, the transition takes place over a larger distance, and the fluid's velocity may vary throughout the pipe or trough. These ideas are illustrated in figure 19.20.

The viscosity of different fluids varies enormously; molasses, thick oils, and syrup all have higher viscosities than water or alcohol. Most liquids have much higher viscosities than gases. The viscosity of a given fluid can also change substantially if its temperature changes; an increase in temperature usually produces a decrease in viscosity. Heating a bottle of syrup, for example, makes it less viscous, so that it flows more readily.

Even glass, which is a noncrystalline solid, can be thought of as an extremely viscous liquid. The windows of very old buildings, such as some of the cathedrals in Europe, are thicker near the bottom than near the top. It is possible that the glass has gradually flowed, over the centuries, towards the bottom of the pane. (Another possibility is that the windows were originally manufactured thicker at the bottom.) The viscosity of glass is so great that at ordinary tempera-

tures it appears to be completely solid. Heating it will cause it to flow, however, as any glassblower knows.

Laminar and Turbulent Flow

One of the most fascinating aspects of the flow of fluids is the question of why the flow is smooth, or *laminar,* under some conditions, but turbulent under other conditions. Both modes of flow can be observed in most streams. Some sections of the stream will have a smooth flow, with no eddies or other similar disturbances. The flow of the stream can be described in this case by *streamlines* that indicate the direction of flow at any point. The streamlines for laminar flow will be approximately parallel to one another, as shown in figure 19.21. The velocities of different layers may vary, but one layer moves smoothly past another.

As the velocity of the stream increases, however, this simple pattern disappears. Ropelike twists in the stream-flow lines appear and then whorls and eddies; the flow becomes *turbulent.* For most applications, turbulent flow is undesirable because it greatly increases the resistance to the flow of the fluid through a pipe or past other surfaces. It does make river rafting much more exciting, however!

If the density of the fluid and the width of the pipe or stream do not vary, then the transition from laminar to turbulent flow is predicted by two quantities; the average velocity of flow and the viscosity. Higher velocities are more likely to produce turbulent flow, whereas higher viscosities inhibit turbulent flow. Larger fluid densities and pipe widths are also more conducive to turbulent flow. From experimental evidence, scientists have been able to use these quantities to predict with some accuracy the velocity at which the transition to turbulent flow will begin.

Laminar flow

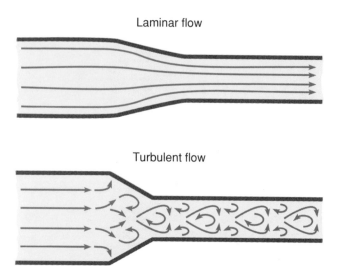

Turbulent flow

Figure 19.21 **In laminar, or smooth, flow, the stream-lines are approximately parallel to one another. In turbulent flow, the stream-flow patterns are much more complicated.**

You can observe the transition from laminar to turbulent flow quite easily yourself. The higher velocities associated with the narrowing of a stream often produce turbulent flow in a stream. The transition is also apparent in the flow of water from a spigot. A small flow rate usually produces laminar flow, but as the rate increases, the flow becomes turbulent. It may even be smooth near the top of the water column, but turbulent lower down as the water is accelerated by gravity. Try it the next time you are near a sink.

You can also observe this phenomenon in the smoke rising from a cigarette or candle. Near the source, the upward flow of the smoke column is usually laminar. As the smoke accelerates upward (due to the buoyant force) and the column widens, however, the flow becomes turbulent (fig. 19.22). Whorls and eddies appear that are similar to those that we see in a stream.

The conditions that produce turbulence are well understood, but until recently, scientists have not been able to explain why the flow patterns develop as they do. Some surprising regularities emerge from the seemingly chaotic behavior of turbulent flow in very different situations, however. Recent theoretical advances in the study of *chaos* have produced a much better understanding of the reasons for these regularities.

The study of chaos and the regularities that appear in turbulent flow have provided new insights into global weather patterns and various other phenomena. Perhaps the most striking example of planetary flow patterns has been provided by photographs sent back from the Voyager flyby of the planet Jupiter. Whorls and eddies can be seen in the flow of the atmospheric gases on Jupiter. These include the famous red spot, which is now considered to be a giant and very stable atmospheric eddy (fig. 19.23).

Figure 19.22 **Smoke rising from a cigarette first exhibits laminar flow. As the velocity increases and the column widens, turbulent flow appears.**

Figure 19.23 **Whorls and eddies, including the giant red spot, can be observed in the atmosphere of the planet Jupiter.**

19.5 BERNOULLI'S PRINCIPLE

Have you ever wondered how a large passenger jet gets off the ground? You know that those things fly, but somehow it still seems improbable. Why does an airplane fly? What force keeps it in the air?

Although the forces that act upon an airplane wing, or *airfoil,* can be analyzed in different ways, the simplest explanation lies in a principle regarding the flow of fluids that was introduced by Daniel Bernoulli (1700–82) in a treatise on hydrodynamics published in 1738. As we will see, Bernoulli's principle is a direct result of applying the conservation of energy to the flow of fluids.

The Principle

From our discussion of energy concepts in chapter 8, you should expect that if we do work on a fluid, its energy will increase. This increase could show up as an increase in the velocity and hence the kinetic energy of the fluid, or as an increase in potential energy, accomplished either by squeezing the fluid (elastic potential energy) or by lifting it (gravitational potential energy).

It is best to take these possibilities one at a time. Consider a fluid that is not compressible (not squeezable) and is flowing in a level pipe of constant cross-sectional area. If we were to do work on this fluid, the increase in energy should show up as kinetic energy. In order to accelerate the fluid and increase its kinetic energy, a net force must act on the fluid. This implies that a pressure difference must exist between one point and another within the pipe, as shown in figure 19.24.

The amount of work done in a fixed time would equal the net force times the distance that the fluid moves in that time. Since force is pressure times area, the net force would be equal to the difference in pressure across a given volume of the fluid times the cross-sectional area of the pipe. When

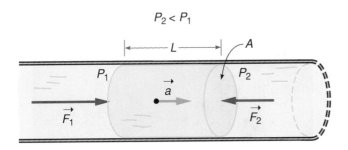

$$W = (F_1 - F_2)L = \Delta KE$$

Figure 19.24 A pressure difference ($P_1 - P_2$) acting upon a portion of fluid within a level pipe produces an increase in the kinetic energy of the fluid.

we multiply this times the distance that the fluid moves, we have

$$W = (\Sigma F)L = (P_1 - P_2)AL = (P_1 - P_2)V,$$

where $V = AL$ is the volume of fluid that has moved past a given point in this time. The work done on this volume of fluid equals the pressure difference times the volume.

Kinetic energy is defined as $(1/2)mv^2$, so the corresponding increase in kinetic energy would be

$$\Delta KE = \left(\frac{1}{2}\right)mv_2^2 - \left(\frac{1}{2}\right)mv_1^2 = \left(\frac{1}{2}\right)m(v_2^2 - v_1^2).$$

If we substitute the density times the volume for the mass ($m = dV$) and set the work done equal to the increase in kinetic energy, we have

Work done = Increase in *KE;*

$$(P_1 - P_2)V = \left(\frac{1}{2}\right)(dV)(v_2^2 - v_1^2).$$

Dividing both sides by the volume and rearranging this expression yields

$$P_1 + \left(\frac{1}{2}\right)dv_1^2 = P_2 + \left(\frac{1}{2}\right)dv_2^2.$$

This last expression is Bernoulli's principle for the case in which no changes of potential energy are taking place. What it says is that the sum of the pressure plus the kinetic energy per unit volume must remain constant. (The expression $(1/2)dv^2$ is the kinetic energy per unit volume, since d, the density, is equal to the mass per unit volume of the fluid.) Thus if the pressure decreases as the fluid moves through the pipe, the kinetic energy and velocity must increase. Likewise, if the velocity increases, then the pressure must decrease.

The more general expression of Bernoulli's principle includes the possibility of changes in potential energy also. Since gravitational potential energy is equal to *mgh,* the potential energy per unit volume is *mgh / V = dgh.* (As in the expression for kinetic energy, the mass is here replaced by the density, *d,* in order to get the energy per unit volume.) Bernoulli's principle then can be more generally written as follows:

$$P + \left(\frac{1}{2}\right)dv^2 + dgh = \text{constant.}$$

In words it can be stated as follows:

The sum of the pressure plus the kinetic energy and potential energy (per unit volume) of a flowing fluid must remain constant. If the energy terms increase, the pressure must decrease.

Because of the assumptions involved in justifying Bernoulli's principle, the principle is valid only for noncompressible fluids undergoing smooth (laminar) flow. According to this more general statement of Bernoulli's principle, if the height, *h,* of the fluid above some reference level increases, the pressure of the fluid decreases. The pressure will be greater in the lower portions of the pipe than in the higher portions. Looking at this in another way, we are just restating the fact that pressure increases with depth in the fluid, as established earlier in this chapter. The kinetic energy term that involves motion of the fluid is our primary interest in the discussion that follows.

Pipes and Hoses

Consider a pipe with a constriction in its center section, as shown in figure 19.25. Would you expect that the pressure of water flowing in the pipe would be greatest in the area of the constriction, or in the wider sections of the pipe? Intuition might lead you to suspect that the pressure would be greater in the constricted section, but this is not the case.

We know from the continuity principle introduced in section 19.4 that the velocity of the water will be greater in the constricted section (where the cross-sectional area is smaller) than in the wider portions of the pipe. By Bernoulli's principle, then, the pressure must be smaller in the constricted section in order to keep the sum of $P + (½)dv^2$ constant. If the velocity increases, the pressure must decrease. If we place open tubes at different places (see fig. 19.25), the water level in these tubes will rise higher above the main pipe at its wider portions, indicating a higher pressure there. (For the water to rise in these open tubes, the pressure must be greater than atmospheric pressure. These open tubes serve, then, as simple pressure gauges.)

This result goes against our intuition because we tend to incorrectly associate higher pressure with higher velocity. Another example is provided by the nozzle on a hose. The nozzle constricts the flow and thus increases the velocity of flow. The pressure of the water is therefore smaller at the

narrow end of the nozzle than it is farther back in the hose, contrary to what intuition might suggest.

If you place your hand in front of the nozzle, you will feel a large force as the water strikes it. This force results from the change in the velocity and momentum of the water as it strikes your hand. By Newton's second law, a large force is required to produce this change in momentum, and by Newton's third law, the force exerted on the water by your hand is equal in magnitude to the force exerted on your hand by the water. That force is not directly associated with the fluid pressure in the hose, however; the pressure is actually much greater farther back in the hose where the water is not moving very fast.

Airplane Wings

Although Bernoulli's principle, as stated above, is only strictly valid for fluids whose density does not change (noncompressible fluids), we can still use it in a qualitative sense to describe effects due to the motion of air and other gases that are compressible. Even for compressible fluids, a higher fluid velocity is usually associated with a lower fluid pressure. The lift on an airplane wing and a number of other easily demonstrated phenomena can be explained in this way.

A simple demonstration may make a believer of you. Take a half sheet of tablet paper (or even a facial tissue) and hold it in front of your mouth, as shown in figure 19.26. It should hang down limply in front of your face. If you blow across the top of it, the paper rises, and if you blow hard enough, the paper may even stand out horizontally. What is happening here?

Blowing across the top of the paper causes the air to flow across the top with a greater velocity than underneath the paper. (The air underneath is presumably not moving much

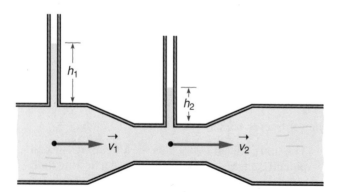

Figure 19.25 Vertical open pipes can serve as pressure gauges; the height of the column of water is proportional to the pressure. The pressure of a moving fluid is greater where the velocity is smaller.

Figure 19.26 Blowing across the top of a limp piece of paper causes the paper to rise, demonstrating Bernoulli's principle.

at all.) This greater velocity causes a reduction in pressure, by Bernoulli's principle. Since the pressure is then greater on the bottom of the paper than on the top, (and pressure is force per unit area), the upward force due to air pressure on the bottom of the paper is larger than the downward force on the top. The paper therefore rises. Contrary to our expectation, blowing between two strips of paper causes them to move closer together rather than farther apart for the same reason.

Similar forces are at work on an airplane wing. In this case, however, the greater velocity of air flow on the top of the wing is produced either by the shape of the wing or by its tilt as it moves through the air. An airplane wing usually has a cross section shaped like the one shown in figure 19.27. As the wing moves through the air, the air moving across its top travels f0arther and thus faster relative to the wing than that flowing underneath. (Although this seems simple enough, a full description of the patterns of air flow past the wing and their causes is actually quite complex.)

By Bernoulli's principle, again, this higher velocity of air flow past the top of the wing is associated with a lower pressure on the top of the wing than on the bottom. There is therefore a net upward, or *lift,* force acting upon the wing. A full analysis of this lift force requires that the different directions and magnitudes of the forces acting at different points on the wing be taken into account. (Given the shape of the airfoil and its angle of attack, the airfoil is deflecting the air flow downward. There is therefore a reaction force on the wing in the upward direction. This provides another method for analyzing the lift force.) The simple argument using Bernoulli's principle produces the correct conclusion, however.

The design of airplane wings and the flow of air past wings has been extensively studied in wind tunnels, where the wing is stationary, and air is blown past it. The effects of the angle at which the wing is set and the use of flaps to change the curvature have been studied in this manner. Under some conditions the air flow over the top of the wing can become turbulent—an undesirable effect that tends to reduce the lift force. Fluid-flow considerations are therefore extremely important to the design and operation of all kinds of aircraft.

Another demonstration of Bernoulli's principle involving air flow is often seen in department stores where vacuum cleaners are being advertised. A ball can be suspended in an upward-moving column of air produced by a vacuum cleaner. Figure 19.28 demonstrates this effect, using a hair dryer as the blower. As the air moves upward, the velocity of air flow is greatest in the center of the stream and falls off to zero farther from the center. Once again, Bernoulli's principle requires the pressure to be smallest in the center, where the velocity is greatest.

The pressure increases as you move away from the center of the air column towards regions where the air is moving less rapidly. Thus if the ball moves out of the center of the column a larger pressure, and therefore a larger force, acts on the outer side of the ball than on the side nearer the center of the column. The ball is thus moved back into the center. The upward force of air hitting its bottom holds it up, while the low pressure in the center of the column keeps it near the center.

All of these phenomena reveal the effects of a reduction in fluid pressure associated with an increase in fluid velocity, as predicted by Bernoulli's principle. Intuitively, we do not expect this. We are tempted to think that blowing between two pieces of paper will push them apart, for example, but a simple trial shows that the opposite effect is produced. Understanding such surprises is part of the fun of physics.

Figure 19.28 **A ball is suspended in an upward-moving column of air produced by a hair dryer. The air pressure is smallest in the center of the column, where the air is moving with the greatest velocity.**

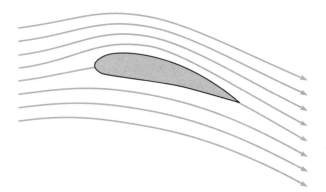

Figure 19.27 **The shape of an airplane wing causes an air-flow pattern in which the air flowing across the top of the wing moves with a greater velocity than that flowing past the bottom.**

Box 19.3

The Situation. Baseball players know that they can be badly fooled by a well-thrown curveball. The idea that a fast-moving curveball can be deflected by as much as a foot on its way to the plate has been difficult for some people to accept, however. Many people have insisted, over the years, that it is just an illusion. Does the path of the curveball actually curve? If so, how can we explain this fact? What is it about the motion of the curveball that could produce such an effect?

The Analysis. There is no secret to throwing a curveball. A right-handed pitcher throws a curveball with a counterclockwise spin (as viewed from above), so that it curves away from a right-handed batter. It is most effective when it starts out appearing to be headed towards the inside of the plate, and then curves over the plate and down and away from the batter.

The usual explanation given for the deflection of the ball's path involves Bernoulli's principle. Because the surface of the spinning ball is rough, it drags a layer of air around with it creating a whirlpool of air in the vicinity of the ball. But the ball is moving towards the plate, which produces an additional flow of air past the ball in the direction opposite to its velocity. The whirlpool created by the spin of the ball causes the air to move more rapidly on the side opposite the right-handed batter than on the side nearer the batter, as shown in the drawing.

A batter is badly fooled by a curveball. Does the path of the ball actually curve?

By Bernoulli's principle, a greater velocity of air flow is associated with a lower pressure; therefore, the air pressure is lower on the side of the ball opposite the right-handed batter than on the nearer side. This pressure difference produces a deflecting force on the ball that pushes it away from the right-handed batter.

Although Bernoulli's principle does yield the proper conclusion for the direction of the deflecting force and thus the direction of the curve, it cannot really be used to make accurate quantitative predictions. Air is a compressible fluid, and (as already indicated) the usual form of Bernoulli's principle is valid only for non-compressible fluids such as water or other liquids. Bernoulli's principle is also valid only for laminar flow, and the air-flow patterns around a rapidly spinning baseball are likely to be turbulent. More accurate methods of treating the effects of air flow past the ball must be used to generate quantitative predictions of the degree of curvature to be expected.

Both theoretical computations and experimental measurements have confirmed that there is a deflecting force on the ball and that the path of the ball does indeed curve. The degree of curvature depends upon the rate of spin of the ball and the roughness of its surface, as we might expect from Bernoulli's principle. There is a continuing debate as to whether the orientation of the seams of the baseball also has an effect. The pitcher's grip on the ball is probably an important factor in how much spin he can put on the ball. Once the ball is released, however, experimental evidence suggests that the orientation of the seams is not important in determining the strength of the deflecting force.

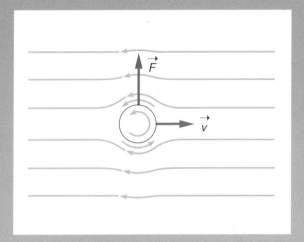

The whirlpool of air created by the spin of the ball causes the air to move more rapidly on one side of the ball than on the other. This produces a deflecting force, as predicted by Bernoulli's principle.

A good discussion of the theory and the experimental evidence is found in an article by Robert Watts and Ricardo Ferrer in the January 1987 issue of the *American Journal of Physics,* pages 40–44. The curved motion of spinning balls is also important in other sports, such as golf and soccer. A good athlete needs to be able to recognize and utilize the effects of these curves.

SUMMARY

The concept of fluid pressure is central to understanding the behavior of both liquids and gases. This chapter has focused on the definition and measurement of pressure, and on the effects of pressure in both stationary and moving fluids.

Pressure and Pascal's Principle. Pressure is defined as the force per unit area exerted on or by a fluid. Pressure is exerted uniformly in all directions in a fluid, according to Pascal's principle, which can be used to explain the operation of hydraulic systems.

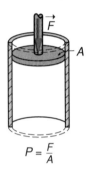

$$P = \frac{F}{A}$$

Atmospheric Pressure and Gas Laws. We can measure atmospheric pressure by determining the height of a column of mercury that can be supported by the atmosphere. Atmospheric pressure decreases with increasing altitude. The density of air also varies with altitude because a lower pressure leads to a larger volume. This is Boyle's law, which is a special case of the more general ideal-gas law, $PV = nRT$.

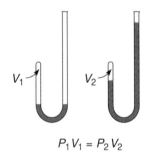

$$P_1 V_1 = P_2 V_2$$

Archimedes' Principle. The increase of pressure with increasing depth in a fluid produces a buoyant force on objects immersed in the fluid. Archimedes' principle states that this force is equal to the weight of the fluid displaced

by the object. If the buoyant force is less than the weight of the object itself, the object sinks; otherwise it floats.

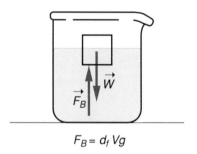

$$F_B = d_f V g$$

Fluids in Motion. The rate of flow of a fluid is equal to the velocity times the cross-sectional area, vA. In steady-state flow, the velocity must increase if the area decreases. Fluids with high viscosity have a greater resistance to flow than those with low viscosity. As the velocity of flow increases, the flow may change from laminar (smooth) to turbulent.

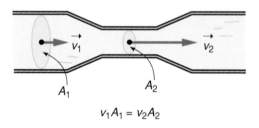

$$v_1 A_1 = v_2 A_2$$

Bernoulli's Principle. Energy considerations require that the sum of the kinetic or potential energies (per unit volume) of a fluid increase as the pressure decreases (Bernoulli's principle). Higher fluid velocities (and kinetic energies) are therefore associated with lower pressures. These ideas can be used to explain the lift on an airfoil, the curve of a curveball, and many other phenomena.

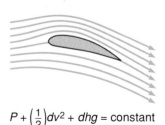

$$P + \left(\frac{1}{2}\right) d v^2 + d h g = \text{constant}$$

QUESTIONS

Q19.1 Is it possible for a 100-lb woman to exert a greater pressure on the ground than a 250-lb man? Explain.

Q19.2 The same force is applied to two different cylinders containing air; one with a piston of large area and the other with a piston of small area. In which cylinder will the pressure be greater? Explain.

Q19.3 A penny and a quarter are embedded in the concrete bottom of a swimming pool filled with water. Which of these coins experiences the greater downward force due to water pressure acting upon it? Explain.

Q19.4 Why are bicycle tires often inflated to a significantly higher pressure than automobile tires, even though the automobile tires must support a much larger weight? Explain.

Q19.5 The fluid in a hydraulic system pushes against two pistons, one with a large area and the other with a small area.

a. Which piston experiences the greater force due to fluid pressure acting upon it? Explain.

b. If the small-area piston moves, does the large-area piston move through the same distance, a greater distance, or a smaller distance than the smaller piston? Explain.

Q19.6 When a mercury barometer is used to measure atmospheric pressure, is there usually air in the closed end of the tube above the mercury column? Explain.

Q19.7 Could water be used in place of mercury in making a barometer? Explain.

Q19.8 If you climbed a mountain carrying a mercury barometer, would the level of the mercury column in the barometer increase or decrease as you went up? Explain.

Q19.9 If you carried a partially-filled, airtight balloon to the top of a mountain, would the balloon expand or contract as you went up? Explain.

Q19.10 The plunger of a sealed hypodermic syringe containing air is slowly pulled out. Does the air pressure inside the syringe increase or decrease when this happens? Explain.

Q19.11 Helium is sealed inside a balloon that is impermeable to the flow of gas. If a storm suddenly comes up, would you expect the balloon to expand or contract? Explain. (Assume that there is no change in temperature.)

Q19.12 Is it possible for a solid metal ball to float in mercury? Explain.

Q19.13 A rectangular metal block is suspended by a string in a beaker of water so that the block is completely surrounded by water. Is the water pressure pushing on the bottom of the block equal to, greater than, or less than the water pressure pushing on the top of the block? Explain.

Q19.14 Suppose that you have a lump of clay that is more dense than water. Is it possible to shape the clay so that it will float? Explain.

Q19.15 A block of wood is floating in a pool of water.

a. Is the buoyant force acting upon the block greater than, less than, or equal to the weight of the block? Explain.

b. Is the volume of the fluid displaced by the block greater than, less than, or equal to the volume of the block? Explain.

Q19.16 A rowboat is floating in a swimming pool when the anchor is dropped over the side. When the anchor is dropped, will the water level in the swimming pool increase, decrease, or remain the same? Explain.

Q19.17 Is it possible that some objects might float in salt water but sink in fresh water? Explain.

Q19.18 A steady stream of water flowing in a narrow pipe reaches a point where the pipe widens. Does the velocity of the water increase, decrease, or remain the same when the pipe widens? Explain.

Q19.19 Why does the stream of water flowing from a faucet often get narrower as the water falls? Explain.

Q19.20 Does a stream of liquid with a high viscosity flow more rapidly, under the same conditions, than a stream with low viscosity? Explain.

Q19.21 If the velocity of flow in a stream decreases, is the flow likely to change from laminar to turbulent flow? Explain.

Q19.22 If you blow between two limp pieces of paper that are held hanging down a few inches apart, will they come closer together or move farther apart? Explain.

Q19.23 A steady stream of water flowing in a pipe comes to a point where the pipe narrows. Will the water pressure in the narrow portion of the pipe be greater than, less than, or equal to that in the wider portion of the pipe? Explain.

Q19.24 When an airplane is flying, the pressure on the bottom of the wing must be greater than that on the top. Is the velocity of the air flowing past the bottom greater than, less than, or equal to that flowing past the top? Explain.

Q19.25 A hair dryer can be used to create a stream of air. Is the air pressure in the center of the stream greater than, less than, or equal to that at some distance from the center of the stream? Explain.

EXERCISES

E19.1 A force of 60 N pushes down on the movable piston of a closed cylinder containing a gas. The piston has an area of 0.5 m^2. What is the pressure in the gas?

E19.2 The pressure of a gas contained in a cylinder with a movable piston is 200 Pa (200 N/m^2). The area of the piston is 0.25 m^2. What is the force exerted on the piston by the gas?

E19.3 A woman with a weight of 100 lb puts all of her weight on one heel of her high-heel shoe. The heel has an area of 0.40 in^2. What is the pressure that her heel exerts on the ground in pounds per square inch (psi)?

E19.4 In a hydraulic system, a force of 200 N is exerted upon a piston with an area of 0.001 m^2. The load-bearing piston in the system has an area of 0.20 m^2.

  a. What is the pressure in the hydraulic fluid?

  b. What is the force exerted on the load-bearing piston by the hydraulic fluid?

E19.5 The load-bearing piston in a hydraulic system has an area 12 times as large as that of the input piston. If the larger piston supports a load of 3600 N, how large a force must be applied to the input piston?

E19.6 With the temperature held constant, the pressure of an ideal gas in a cylinder with a movable piston is increased from 30 kPa to 90 kPa. The original volume of the gas in the cylinder was 0.6 m^3. What is the final volume of the gas after the pressure has been increased?

E19.7 With the temperature held constant, the piston of a cylinder containing an ideal gas is pulled out so that the volume increases from 0.1 m^3 to 0.4 m^3. If the original pressure of the gas was 80 kPa, what is the final pressure?

E19.8 With the pressure held constant, the temperature of an ideal gas increases from 300 K to 500 K. If the original volume of the gas was 0.30 m^3, what is the final volume?

E19.9 With the volume held constant, the temperature of an ideal gas is increased from 27° C to 77° C. The original pressure of the gas was 100 kPa.

  a. What are the initial and final absolute temperatures (in kelvins)?

  b. What is the final pressure of the gas?

E19.10 A certain boat displaces a volume of 5 m^3 of water. (The density of water is 1000 kg/m^3.)

  a. What is the buoyant force acting upon the boat?

  b. What is the weight of the boat? Explain.

E19.11 A block of wood of uniform density floats so that exactly half of its volume is under water. What is the density of the block?

E19.12 A stream moving with a velocity of 0.5 m/s reaches a point where its cross-sectional area decreases to one-third of the original area. What is the velocity of the flow in this narrowed portion of the stream?

E19.13 Water emerges from a faucet with a velocity of 2.0 m/s. After falling a short distance, its velocity increases to 3.0 m/s as a result of the acceleration due to gravity.

  a. Does the cross-sectional area of the stream of water increase or decrease in this process? Explain.

  b. Express the final cross-sectional area of the stream as a fraction of the original area.

E19.14 An airplane wing with an average cross-sectional area of 10 m^2 experiences a lift force of 4000 N.

  a. What is the difference in air pressure, on the average, between the bottom and top of the wing?

  b. Is the air velocity relative to the wing larger on the bottom or on the top of the wing? Explain.

CHALLENGE PROBLEMS

CP19.1 Suppose that the input piston of a hydraulic jack has a diameter of 3 cm, and the load piston has a diameter of 30 cm. The jack is being used to lift a car with a mass of 1200 kg.

 a. What are the radii of the input and load pistons in meters?

 b. What are the areas of the input and load pistons in m^2? ($A = \pi r^2$)

 c. What is the ratio of the area of the load piston to that of the input piston?

 d. What is the weight of the car in newtons? ($W = mg$)

 e. What force must be applied to the input piston to support the car?

CP19.2 Water has a density of 1000 kg/m^3. The depth of a swimming pool at the deep end is often about 3 m.

 a. What is the volume of a column of water 3 m deep with a cross-sectional area of 0.5 m^2?

 b. What is the mass of this column of water?

 c. What is the weight of this column of water in newtons?

 d. What is the additional pressure (above atmospheric pressure) exerted by this column of water on the bottom of the pool?

 e. How does this value compare to atmospheric pressure? Since atmospheric pressure is transmitted uniformly from the air above through the water to the bottom of the pool, what is the total pressure at the bottom of the pool?

CP19.3 A steel block with a density of 7800 kg/m^3 is suspended from a string in a beaker of water so that it is completely submerged but not resting on the bottom. The block is a cube of 5 cm (0.05 m) on a side.

 a. What is the volume of the block in m^3?

 b. What is the mass of the block?

 c. What is the weight of the block?

 d. What is the buoyant force acting on the block? ($d_w = 1000$ kg/m^3)

 e. What tension in the string is needed to hold the block in place?

CP19.4 A flat-bottomed rectangular boat is 6 m long and 2 m wide. The sides of the boat are 1 m high. The mass of the boat is 200 kg, and it carries 5 people with a total mass of 450 kg.

 a. What is the total weight of the boat plus the people in newtons?

 b. What is the buoyant force required to keep the boat and its load afloat?

 c. What volume of water must be displaced in order to support the boat and its load? ($d_w = 1000$ kg/m^3)

 d. How much of the boat (to what depth) is under water? (What height of the sides is required to produce a volume equal to that in (c)?)

 e. How much of the boat would be under water if it were empty?

CP19.5 A pipe of circular cross section has a diameter of 10 cm. It narrows at one point to a diameter of 5 cm. A steady stream of water completely fills the pipe and is moving with a velocity of 1.5 m/s in the wider portion.

 a. What are the cross-sectional areas of the wide and narrow portions of the pipe? ($A = \pi r^2$, and the radius is half the diameter.)

 b. What is the velocity of the water in the narrow portion of the pipe?

 c. Is the velocity of the water in the narrow portion more than twice as large as that in the wider portion? Explain.

 d. Is the pressure in the narrow portion of the pipe greater than, less than, or equal to the pressure in the wider portion? Explain.

HOME EXPERIMENTS AND OBSERVATIONS

HE19.1 Fill a plastic milk jug with water and then, using an awl or other round pointed device, punch three similar holes in the jug at different heights. (You may wish to punch the holes before filling the jug. You can use your fingers to block the flow of water while filling the jug.)

    a. Lift the jug to the edge of the sink, and let the water flow out of the holes. Which hole produces the greatest initial velocity for the emerging water?

    b. Which stream travels the greatest horizontal distance? What factors determine how far the water will go horizontally?

    c. Observe the shapes of the water streams as they fall into the sink. Do they become narrower as the water falls? If so, how can this be explained?

HE19.2 If you have some clay available, try building a boat.

    a. Does the clay sink when it is rolled into a ball? What does this indicate about the density of the clay?

    b. Is a flat boat more effective than a canoe-shaped boat in carrying the maximum load of steel washers or other weights? Try them, both. What are the problems with each?

HE19.3 If you are a swimmer, try floating with your lungs full of air, and then again with the air largely expelled from your lungs.

    a. How much of you is above water in each case? Is there a significant difference? Can you float in the latter case?

    b. With your lungs full of air, can you estimate what percentage of the volume of your body is above water? What does this indicate about the average density of your body in this condition?

HE19.4 If you have a hair dryer handy, try supporting a ping-pong ball in a vertical column of air coming from the dryer.

    a. Can you get the ping-pong ball to stay in place? How far out of the center can the ball wander and still return?

    b. Will this work with a tennis ball or a small balloon? What differences do you note in each?

    c. Blow the hair dryer between two pieces of paper. What effect do you observe? How far apart should the pieces of paper be in order to produce the maximum effect?

20 *Beyond Everyday Phenomena*

The general aim of this book has been to build your understanding of some of the basic ideas of physics by stressing their origin in and application to everyday phenomena—things that happen all the time around us. To be sure, we have strayed at times from such everyday observations, for example, in our discussion of special relativity or the structure of the atom and its nucleus. Even these ideas have their origins in simple experiments, however, that can be described without too much difficulty.

Many of the basic ideas that we have discussed have emerged in the twentieth century. This is particularly true of those involving the structure of the atom and its nucleus. Even the discoveries of fission and fusion occurred about fifty years ago, though, during the 1930s and 1940s. Where have things gone since then? What can we expect to hear from physics in the future?

One of the most intriguing aspects of science, of course, is that we can never be sure just where it will lead us. Like a good mystery, there are clues and indications, but the answers are elusive. Unlike a mystery novel, however, there is no final resolution. Successes in science increase our understanding and often lead to advances in technology, but they always raise new questions.

Physics does make the news from time to time, and will continue to do so. Questions are raised regarding the desirability of spending several billion dollars to build a new particle accelerator or space station. Ordinary citizens are sometimes called upon to make their own judgments on such issues. The proposed construction of the Superconducting Supercollider (a particle accelerator) in Texas has been a political issue for several years (fig. 20.1).

Our everyday lives are affected in more ways than we realize by advances in physics. Most of us use personal computers and other electronic devices such as video recorders that have micro-computers built into them. Computers have made enormous changes in the way we do things. The invention of the transistor, which made the modern computer feasible, came about through advances in solid-state physics involving our understanding of the nature of semiconductors.

Modern physics is active on many fronts. Some research areas are driven by expected contributions to better technology, while others are motivated simply by a desire to better understand the universe in which we live. Although we cannot hope to touch upon all of these areas, we will describe a few that have received attention in the popular press, and which are likely to continue to do so. Ideas that seem well removed from common experience today may someday become common.

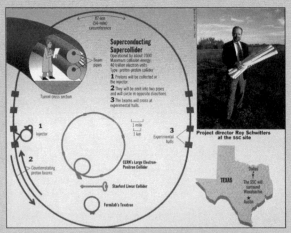

Figure 20.1 Graphics for a *Time* magazine article discussing the new Superconducting Supercollider (SSC) to be built in Texas. Do its expected benefits to science justify its enormous cost? *Copyright 1990 The Time Inc. Magazine Company. Reprinted by permission.*

Chapter Objectives

This chapter describes a few broad areas of research in physics that have either excited the popular imagination or are likely to have a significant impact on developing technologies. The descriptions are necessarily brief, but they attempt to bring out some of the fundamental ideas and issues. The ideas discussed include elementary particles, the origins of the universe, semiconductor electronics, supercomputers, and superconductors and other exotic materials.

Chapter Outline

1 *Quarks and other elementary particles.* What are the basic building blocks of the universe, and how do we gain information about them? What are the relationships between the many particles that have been discovered?

2 *Cosmology and the beginning of time.* How was the universe formed, and how is it changing? How do our studies of elementary particles shed light on the origins of the universe?

3 *Semiconductors and microelectronics.* What are transistors, and how do they operate? Why have semiconductor devices revolutionized electronics, and how are they made?

4 *Computers and artificial intelligence.* What do computers do, and how are they constructed? How have analogies between the brain and computers led to new ways of designing and programming computers? What lies ahead in computer technology?

5 *Superconductors and other new materials.* What are superconductors, and why are they important? What other new materials are emerging from the study of solid-state physics?

20.1 QUARKS AND OTHER ELEMENTARY PARTICLES

One of the most enduring quests in science has been the search for the ultimate building blocks of nature, the particles or entities from which everything else is constructed. Until the twentieth century these were thought to be atoms, a view that was strengthened by the advances in chemistry during the nineteenth century (see chapter 16).

The discovery of the electron by J. J. Thomson in 1897 provided us with the first subatomic particle; it was apparently present in all atoms. This was followed in 1911 by the discovery of the nucleus of the atom, and then by the recognition in 1932 that nuclei are themselves made up of protons and neutrons. Physicists are now aware that neutrons and protons also have a substructure and can be regarded as being composed of quarks (fig. 20.2).

Where will this all end? What are quarks, and why do we believe in them? Will physicists someday discover that quarks also have a substructure? This final question cannot be answered with certainty, but recent theoretical advances in high-energy physics have succeeded in bringing order to what had seemed like a bewildering array of new particles. We will consider just a few features of this new *standard model.*

Discovering New Particles

The electron, proton, and neutron, the basic constituents of the atom, were just the first in a long parade of subatomic particles that have been discovered in the twentieth century. The positron, for example, was discovered in 1932 shortly after its existence was suggested on theoretical grounds by Paul A. M. Dirac. This discovery was followed by the discovery of the muon and pion. The list grew rapidly during the 1950s and 1960s as work in high-energy physics intensified.

How are these discoveries made? Most of them involve scattering experiments, similar in basic concept to the original scattering experiment performed by Rutherford and his associates, which led to the discovery of the nucleus. Targets are bombarded with high-energy particles, and various types of particle detectors are used to study what emerges from the collisions. The emerging particles leave tracks in photographic emulsions, cloud chambers, bubble chambers, or other more sophisticated detectors (fig. 20.3). Cloud chambers and bubble chambers utilize the fact that a rapidly moving charged particle produces water droplets or bubbles in a supersaturated vapor or in a superheated fluid.

Analyses of these tracks, in conjunction with other measurements, allow us to deduce the mass, kinetic energy, and charge of the created particles. The path of a positively charged particle, for example, bends one way in a magnetic field, and that of a negatively charged particle bends in the opposite direction. The degree of curvature of the path is related to the mass of the particle. The mass is particularly important; it is one of the major identifying characteristics of the new particles.

The source of the high-energy particles used to bombard targets in these scattering experiments is usually a particle accelerator of some kind. Rutherford, of course, used alpha particles obtained from radioactive substances, but these are limited in their available energy. Other early workers used the high-energy particles in the cosmic rays that stream in from outer space. Particle accelerators, however, are capable of producing both high-energy and high-density beams of particles, which yield a better probability of interesting collisions.

Since mass *is* energy, producing particles with masses much greater than the proton or neutron requires higher and higher collision energies, which is the stimulus for building ever-bigger particle accelerators. A modern particle accelerator uses electric and magnetic fields to accelerate and shape the beam. The beam itself is contained in a long evacuated tube, which can either be straight (as in a linear accelerator) or bent into a large ring. Either design makes it possible to accelerate two separate beams of particles so that they collide head-on at the point where the reactions will be studied. Collisions of this nature provide a larger collision energy than that which results from colliding a particle beam with a stationary target.

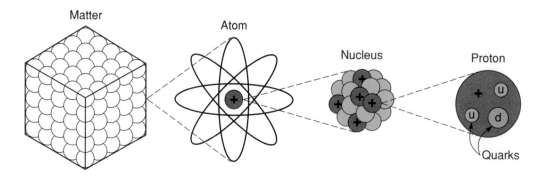

Figure 20.2 **Atoms, once thought to be the basic building blocks of all matter, are now known to consist of electrons, protons, and neutrons. Protons and neutrons also have a substructure consisting of quarks.**

Figure 20.3   Particle tracks in a bubble chamber provide information on the new particles produced in collisions or decays.

Beam energies are measured typically in electron volts (eV), which is a unit of energy representing the energy gained by an electron when it is accelerated through a potential difference of one volt. Modern accelerators are capable of reaching collision energies up to 100 GeV or more, where $1 \text{ GeV} = 10^9 \text{ eV} = 1$ billion eV. These energies are over a thousand times those available from radioactive decay. The largest modern accelerators now operating include the Stanford Linear Accelerator Center (SLAC) in California, the CERN electron-positron collider in Switzerland, and the Fermilab proton-antiproton collider near Chicago.

The Particle Zoo

As more and more particles were discovered, attempts were made to organize and classify them. These attempts were guided by theoretical considerations, and although they provided incomplete models, they were often successful in predicting the existence of particles that were later discovered in experimental work. The original classification schemes were based primarily upon the masses of the particles.

The particles were grouped into three primary groups: leptons, mesons, and baryons. The leptons are the lightest particles and include electrons, positrons, and the neutrinos that are involved in beta decay. Mesons are intermediate in mass and include the pion (originally called the π *meson*) and the kaon. (Muons were also originally grouped with the mesons because their mass was also intermediate between that of the electron and the proton, but they were later classified with the lepton group by virtue of their other properties.) Baryons are the heaviest; they include the neutron and proton as well as many heavier particles. A partial listing of these particles appears in table 20.1.

Each particle has an antiparticle, which has the same mass as the particle itself, but opposite values of other properties

such as charge. The positron, for example, is the antiparticle of the electron and has a positive charge instead of a negative charge. A proton is positively charged and an antiproton has a negative charge. When a particle runs into its antiparticle, the two can annihilate each other, producing high-energy photons or other particles. The antiparticles are not shown in table 20.1

The spin is listed in table 20.1 because this is one property that distinguishes mesons from leptons and baryons. As you can see from the table, all of the mesons have zero spin, whereas all of the leptons and baryons have a spin of $\frac{1}{2}$. Spin is a quantum property that is related to the angular momentum of the particle. If the particle is charged, the spin also generates a magnetic dipole moment, which affects how the particle interacts with other particles.

The theory that describes the interactions between these particles is called *quantum electrodynamics,* which is based upon both quantum mechanics and relativity. Advances in this theory during the early 1970s began to point to a more fundamental organization scheme for all of these particles. This is where quarks come into the picture. The mesons and baryons (which together are called *hadrons*) could all be viewed as being made up of quarks, new particles suggested by the theory. Mesons consisted of two quarks each, a quark and an antiquark, and baryons consisted of three-quark groupings.

As the theory developed, it became apparent that six types of quarks were necessary (not counting the antiparticles) in order to account for all of the baryons and mesons. These have been dubbed the *up, down, charmed, strange, top,* and *bottom* quarks. Different combinations of these six quarks (and their antiparticles) can account for all of the observed particles in the meson and baryon groups. This has left the leptons as the only particles (in addition to the quarks) that are still elementary, that is, not made up of any more fundamental particles as far as we know.

The proton, for example, consists of two up quarks, each of charge $+(\frac{2}{3})e$, and one down quark of charge $-(\frac{1}{3})e$. Likewise, a neutron consists of two down quarks and one up quark, which produce a total charge of zero (fig. 20.4). Scattering experiments, in which extremely high-energy electrons are collided with protons, provide strong evidence of this substructure. These experiments indicate the presence of hard scattering centers within the proton, with the appropriate charges for two up quarks and one down quark. Such experiments are similar in concept to Rutherford's original scattering experiment, which led to the discovery of the nucleus of the atom.

The theory also suggests similarities across the groups of leptons and quarks. Physicists now group these particles into three families with similar properties. Each family consists of two leptons and two quarks, with one of the leptons in each family being a neutrino. Table 20.2 shows the particles that belong to each family. There are just twelve elementary particles (three families of 4 particles each) in this scheme; twenty-four if we count the antiparticles.

Particle	Mass (MeV)	Charge	Spin	Lifetime
Leptons				
Electron neutrino	0 ?	0	½	
Muon neutrino	0 ?	0	½	
Electron	0.511	$-e$	½	
Muon	105.7	$-e$	½	2.2×10^{-6} s
Mesons				
Pion	139.6	$+e$	0	2.6×10^{-8} s
Neutral pion	135.0	0	0	8.3×10^{-15} s
Kaon	493.7	$+e$	0	1.2×10^{-8} s
Neutral kaon	497.7	0	0	?
Eta	548.8	0	0	7×10^{-19} s
Baryons				
Proton	938.3	$+e$	½	
Neutron	939.6	0	½	920 s
Lambda	1115.6	0	½	2.5×10^{-10} s
Sigma	1189.4	$+e$	½	8.0×10^{-11} s
Neutral sigma	1192.5	0	½	?
Xi	1321.3	$-e$	½	1.7×10^{-10} s
Neutral Xi	1314.9	0	½	3.0×10^{-10} s
Omega	1672	$-e$	½	1.3×10^{-10} s

Table 20.1

Leptons, Mesons, and Baryons

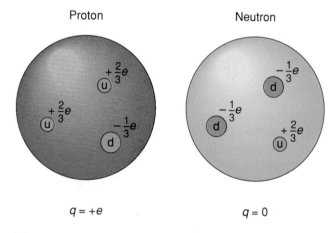

Proton

Neutron

$q = +e$

$q = 0$

Figure 20.4 **A proton consists of two up quarks and one down quark. A neutron consists of two down quarks and one up quark.**

At the time of this writing, the existence of these twelve particles has been confirmed experimentally, except for the tau neutrino and the top quark. Neutrinos are notoriously difficult to detect, and the tau neutrinos are expected to be much rarer than the electron or muon neutrinos. The success of the overall model, however, creates a strong confidence in the existence of the tau neutrino.

The quarks are never present individually but always in combination with other quarks. For this reason, they cannot be directly observed as tracks in a bubble chamber or by similar devices. Their existence can be inferred, however, by scattering experiments and by the observation of reactions that are predicted based upon the quark model. Higher energies and beam densities than those available in existing particle accelerators are expected to be necessary to observe reactions related to the top quark. Current efforts to boost the collision energies of the CERN or Fermilab accelerators may do the trick. If not, the supercollider to be built in Texas should reach the required energies and beam densities.

Table 20.2		
The Three Families of Elementary Particles		
First family	**Second family**	**Third family**
Electron	Muon	Tau particle
Electron neutrino	Muon neutrino	Tau neutrino
Up quark	Charmed quark	Top quark
Down quark	Strange quark	Bottom quark

Fundamental Forces

What holds all of these particles together? The primary force responsible for binding the quarks within neutrons, protons, and other baryons (as well as mesons) is called the *strong nuclear interaction*. This is also the force responsible for holding the neutrons and protons together inside the nucleus of an atom. The strong nuclear force obviously must be stronger than the electrostatic repulsion of the positively charged protons; otherwise the nucleus would fly apart. The strong force has a very short range, however, and falls off in strength very rapidly when the distance between particles increases beyond nuclear dimensions.

In addition to the strong nuclear interaction, physicists have traditionally recognized three other fundamental forces: the electromagnetic force, the gravitational force, and the weak nuclear force. The weak nuclear force is involved in the interactions of leptons, including the process of beta decay, which involves electrons and neutrinos. One of the goals of modern theoretical physics has been to unify all of these forces within a single theory. Since these forces are usually described in terms of their fields, as we have done with electric and magnetic fields, such a theory is referred to as a *unified field theory*.

One of the major successes of the standard model that has created order within the particle world is that it unifies the weak nuclear force with the electromagnetic force. These two forces can now be viewed as different manifestations of the same fundamental force, called the *electroweak* interaction. James Maxwell's theory of electromagnetism had earlier unified what had been seen as two independent forces, the electrostatic force and the magnetic force, into a single electromagnetic force. Now that force has been joined by theory with the weak nuclear force.

We should perhaps say, then, that there are only three fundamental forces, the strong nuclear interaction, the electroweak force, and the gravitational force. This statement might also be misleading, however, since substantial progress has been made towards unifying the strong nuclear interaction with the electroweak interaction in extensions of the standard model (fig. 20.5). These theories are now referred to as *grand unified theories,* or *GUTs* for short.

One force has thus far resisted incorporation into an overall unified field theory: the gravitational force. Its theoretical basis is found in Einstein's theory of general relativity. The mathematics of general relativity seem to be incompatible, in some ways, with those of quantum mechanics and the standard model. Fame and fortune await those who may succeed in unifying the gravitational force with the other fundamental forces.

When the superconducting supercollider is completed and operational, we expect to see new particles and phenomena at the higher energies that will then be available. The top quark should be observed, if that has not already been achieved with other accelerators. We may also find entirely new particles or phenomena, however, that have not been anticipated by the standard model. Such new findings might point the way to new theoretical insights. This, of course, is what science is all about, and this is where the real rewards of the supercollider may be found.

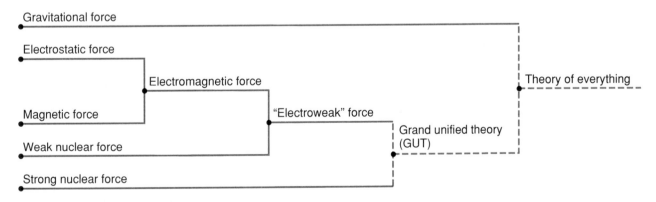

Figure 20.5 Fundamental forces of nature, which once were viewed as independent forces, have been unified into fewer fundamental forces by advances in theoretical understanding. Complete unification of all fundamental forces may lie ahead.

20.2 COSMOLOGY AND THE BEGINNING OF TIME

The previous section has provided a glimpse of advances in physics involving the world of the very small; substructures of substructures of substructures. The quark is a building block of protons and neutrons; they in turn are building blocks of the nucleus, which in turn is a part of the atom. Atoms make up molecules and the ordinary matter of which our world is made.

What happens if we focus on the very large? Our world, the earth, is a part of the solar system, as discussed in chapter 5. Our sun is just one of a seemingly infinite number of stars, as you probably know. Stars, in turn, are grouped into galaxies, which themselves seem to come in clusters. How do we know all of this? What is the ultimate structure of the universe, and how is it changing? The answers to some of these questions depend upon our understanding of atoms, nuclei, and quarks.

The Expanding Universe

Humans have long been fascinated with the night sky and with questions regarding the nature of the universe. The invention of the telescope around 1600 (in Galileo's time) provided a new aid for viewing the planets and stars. In fact, it was Galileo's discovery of the moons of Jupiter, using a crude telescope, that helped to turn the tide in favor of the Copernican heliocentric model of the solar system—here were objects whose orbits were clearly not centered upon the earth.

As telescopes were improved, however, we also became aware that there were many more objects out there than could be seen by the unaided eye. Not all of these objects appeared to be pointlike stars; some of them had a fuzzy appearance. As the resolution of telescopes was increased, it became apparent that some of these objects were not stars at all but collections of stars, which we now call *galaxies*. Many galaxies have a spiral structure like the one shown in figure 20.6.

The galaxy that we can see most readily with unaided eyes is our own, the Milky Way. On a clear night, the Milky Way is visible in what appears to be a continuous cloud of stars making a band across the sky (fig. 20.7). We are actually looking at the opposite side of our own spiral galaxy. The other stars that appear to us to be larger and brighter lie on the same side of the spiral as our sun, and are therefore much closer to us. The sun is but one of millions of stars that make up this galaxy.

We also know now that the universe is expanding; the other galaxies are receding from us. This realization emerged from the spectrographic studies of Edwin Hubble in the 1920s. Hubble was engaged in trying to estimate the distance to various stars and galaxies by measuring their relative brightness. In order to do this, he needed some assurance that he was looking at the same type of star. It was already

Figure 20.6 **A spiral galaxy viewed against a foreground of nearer stars. Our own Milky Way galaxy has a similar form.**

Figure 20.7 **The Milky Way appears as a continuous cloud of stars that can be seen as a band across the sky on a clear night.**

known that different types of stars had characteristic colors or spectra; red giants were very different from white dwarfs, and so on. Measuring the range of wavelengths (or spectra) coming from different stars gave him a basis for comparing the size and temperature of the stars he was viewing.

When Hubble applied these techniques to galaxies, however, he noticed a startling feature. Specific absorption lines in the spectra of stars in these other galaxies were all shifted substantially in wavelength and frequency towards the red portion of the spectrum. (Absorption lines make good reference points in the otherwise continuous spectrum of stars. They are produced by light being absorbed by gases in the outer atmosphere of a star.)

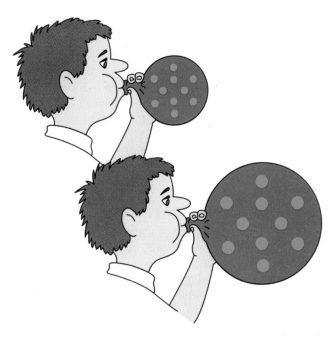

Figure 20.8   As the balloon is inflated, spots on the surface recede from one another. Spots that are initially farther away from some chosen point recede more rapidly from that point than nearer spots.

The only reasonable explanation for this shift was that these stars must be moving away from us, thereby producing a Doppler shift in the frequency of the light. The Doppler shift is the same phenomenon discussed in chapter 14 for sound waves; the frequency of a car horn is shifted to a lower value when it is moving away from us. For light, a lower frequency represents a shift towards the red end of the visible spectrum.

The farther away a given galaxy, the more rapidly it seemed to be receding from us. This is consistent with the hypothesis that the entire universe is expanding. From our knowledge of the curvature of space-time introduced in Einstein's theory of general relativity, we know that it is not necessary to be at the center of the universe to see things in this way. The analogy often used is that of points on an expanding balloon. From any point on the surface of the balloon, all other points are seen to recede as the balloon expands. Points farther away from the given point recede at a greater rate than points that are closer (fig. 20.8).

The Big Bang

If the universe is expanding, then it seems that at some point in time the entire mass of the universe was much more compressed than it is now. In other words, if we could run a moving picture of the expanding universe in reverse, it should revert to a very small volume. The beginning of the expansion (and perhaps the beginning of time) can then be viewed

as an explosion from which the universe has been expanding ever since. This initial rapid expansion or explosion is referred to as the Big Bang.

When a large quantity of matter is confined in a very small space, it no longer consists of individual atoms and molecules. The electrons get stripped from the atoms, and what is left is a dense plasma of electrons, protons, and neutrons. At even higher densities, the protons and electrons combine to form neutrons. This is thought to occur in the gravitational collapse of stars that have used up their fusion fuel, resulting in a very small and dense *neutron star.* If such a star had sufficient mass, it might collapse still further to form a black hole (see section 18.5).

At even higher densities, the matter would exist as a sea of quarks, in which individual quarks could not be identified as belonging to specific neutrons or protons. In the early stages of the Big Bang (just a microsecond or so after the beginning), all of the matter of the universe is viewed as having been just a sea of quarks. As the expansion continued, the quarks condensed into mesons and baryons, including neutrons and protons. At approximately three minutes after the beginning, the protons and neutrons began to fuse into nuclei, primarily isotopes of hydrogen and helium.

In these initial stages of the Big Bang, the temperature of the universe must have been extremely high. As the expansion continued, the universe cooled down and, of course, became less dense. At a much later point (roughly half a million years), things should have cooled down enough for electrons to begin to orbit about the nuclei to form atoms. Presumably during this same time, gravitational attraction began to produce clumps of matter that became galaxies, and matter within these galaxies began to condense into individual stars. The synthesis of larger nuclei via fusion reactions then began to take place within the stars.

The standard model of high-energy physics has been able to predict how many of these steps might have occurred. The model has had considerable success in explaining things that have been confirmed by astronomical observations. One of these is the ratio of helium to hydrogen that is observed in stars and galaxies. Another is the uniform background of microwave radiation that has been observed in the universe. In fact, this microwave radiation can be regarded as a residual effect of the Big Bang itself. Many physicists consider it to be one of the strongest pieces of evidence confirming the Big Bang hypothesis.

Thus we see that our success in describing the world of the very small (nuclei and quarks) plays a large role in our understanding of the universe. Much of this success has been achieved in just the last twenty years or so. Obviously, however, more remains to be done. Some physicists feel that good progress has been made on a completely unified field theory that will successfully incorporate gravity into a "theory of everything." Such advances are quickly applied to models of the universe to test their implications.

There are still many unanswered questions. Since we still have not achieved a theory of everything, we cannot model the very earliest stages of the Big Bang. We therefore cannot describe with assurance the initial conditions of the universe. Perhaps there are many universes, some of which have evolved in different ways than our own!

We do not expect that we will ever answer all of these questions. They hold a tremendous fascination, however, for physicists, astronomers, philosophers, and ordinary people. The nature of time itself, and of our own role in the universe, are parts of the puzzle.

20.3 SEMICONDUCTORS AND MICRO-ELECTRONICS

Most of us own a hand-held calculator. In our homes, we may have microwave ovens, video recorders, and perhaps a home computer. We probably also possess a number of other electronic devices, including television sets, radios, digital watches, stereo systems, and even electronic ignition systems in automobiles. All of these devices use solid-state electronics, and many of them incorporate microcomputers.

What do we mean by *solid-state electronics?* What led to this ongoing revolution in technology? Although these devices are part of our everyday experience, their internal operation is invisible to us. Despite their immense importance to our economy, most people have very little understanding of how they function.

Semiconductors

In chapter 11, we discussed the distinction between electrical conductors and insulators. Good electrical conductors, mostly metals, permit a relatively free flow of electrons or other charge-carrying particles, whereas good insulators do not. There is an enormous difference in the values of electrical resistivity between these two types of materials.

Table 11.1 in chapter 11 listed a few examples of conductors and insulators, as well as a few members of a third category called *semiconductors*. These materials have a much lower resistivity than good insulators, but a considerably higher resistivity than good conductors. What causes these differences in electrical conduction? Can we predict which materials will be good conductors, insulators, or semiconductors?

If you examine the periodic table in the inside back cover of this book, you see that all the metals lie on the left-hand side of the table or in the transition regions between the two sides. As we discussed in chapter 16, these elements have just one or two (or sometimes three) electrons outside of a closed shell of electron states. These outer electrons are responsible for the chemical properties of the metals. They are less tightly bound to the nucleus of the atom than the other electrons, and are therefore relatively free to migrate within the material as conduction electrons.

Elements which make good insulators, on the other hand, lie on the right-hand side of the periodic table. These elements are lacking one, two, or three electrons needed to complete a closed shell. They therefore readily accept electrons from other elements when they combine to form chemical compounds. In pure form, however, when they bond together to form solids or liquids, there are no loosely-bound electrons that can contribute to electrical conduction.

The elements that we commonly list as semiconductors (carbon, germanium, and silicon) are all found in column IV of the periodic table. These elements have four outer electrons (beyond a closed shell). When they bond to form solids, these electrons are shared with neighboring atoms, as shown in the two-dimensional representation in figure 20.9. (The actual crystal structure is three-dimensional, of course, which is more difficult to picture.) These electrons are more closely tied to their corresponding nuclei than those in a metal, but they are freer to migrate through the substance than those in a good insulator. Not surprisingly, then, the conducting properties of these materials are intermediate between those of metals and good insulators.

Although carbon, germanium, and silicon are called *semiconductors,* they certainly are not good conductors in their pure form. In fact, carbon is often used to make resistors for circuit applications. A short piece of carbon has a much higher resistance than a much longer and thinner piece of metal wire. The real value of semiconductors in electronics is a result of our ability to modify the resistivity of these materials by *doping* them with small amounts of impurity.

Suppose, for example, that we add a small amount of phosphorus or arsenic to silicon. These elements lie in column V of the periodic table and therefore have five outer electrons. Four of these five electrons will participate in the bonding of the phosphorus or arsenic atom with neighbor-

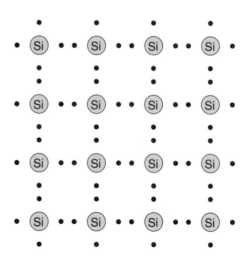

Figure 20.9 A two-dimensional representation of the sharing of the four outer electrons of silicon with neighboring atoms in solid silicon.

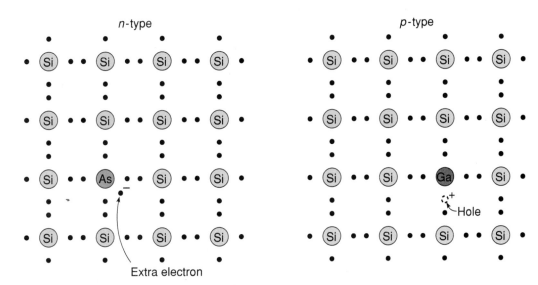

Figure 20.10 **Doping silicon with phosphorous or arsenic provides an extra electron, making an *n*-type material. Doping with boron or gallium leaves holes and produces *p*-type material.**

ing silicon atoms. The fifth electron, however, is not needed in these bonds and is therefore relatively free to migrate through the material. This doping, then, has the effect of introducing conduction electrons into the material, making it a considerably better conductor than pure silicon.

Doping with phosphorus, arsenic, or antimony produces what is called an *n-type* semiconductor, because the charge carriers are negatively charged electrons. *P-type* doping is also possible; in this case the impurities added are elements from column III in the periodic table, most commonly boron, gallium, or indium. Since atoms of these elements have just three outer electrons, they leave a *hole* in one of the bonds between the impurity atom and the neighboring silicon atoms (fig. 20.10).

A hole is just the absence of an electron, but it turns out that these holes are also somewhat free to migrate through the material. A moving hole acts as a positive charge carrier because it leaves an excess positive charge (associated with the charge on the nucleus of the silicon atom) wherever it goes. What happens is that electrons from neighboring silicon atoms move in to fill the hole, thus leaving a hole (and an excess positive charge) somewhere else in the material.

Diodes and Transistors

Besides improving the conducting properties of semiconductors by amounts that can be carefully controlled, doping introduces some other very important possibilities. The boundaries, or *junctions,* between *p-* and *n*-type materials have properties that have proved to be enormously useful in electronics; these junctions allow us to make diodes, transistors, and related devices.

A diode is a device that allows electric current to flow in one direction but not in another; it is essentially a one-way

valve for electric current. The diagrams in figure 20.11 illustrate why a diode behaves in this manner. The essential feature of a semiconductor diode is the junction between the *n*-type and *p*-type materials.

When the positive terminal of a battery is connected to the *p*-type material and the negative terminal to the *n*-type side of the diode, as shown in figure 20.11a, electrons are introduced from the battery into the *n*-type side. These electrons flow through the *n*-type material to the junction between the *n*-type and *p*-type materials. Here the electrons attract holes in the *p*-type material to the junction, and the holes are eliminated as electrons move across the junction to fill them. The positively-charged holes move through the *p*-type material from the positive side of the battery, and a continuous current flows. This manner of connection is referred to as a *forward bias* of the diode.

If we reverse the connections of the battery to the diode, as shown in figure 20.11b, a different condition exists. Holes are now pulled away from the junction by the negative charges from the negative terminal of the battery (which is now connected to the *p*-type side of the diode). Likewise, electrons in the *n*-type material are attracted toward the positive terminal of the battery. Since the holes and electrons are both pulled away from the junction, no recombination of holes and electrons occurs there. Thus there is no flow of current across the junction in this *reverse bias* condition.

Diodes have many uses in electronic circuits. One of the easiest to understand is what is called *rectification*. This is the process by which an alternating current is converted to a direct current. Since a diode allows current to flow in just one direction, the simplest type of rectifier would be just a single diode. Combinations of diodes can produce a steadier flow of current, however.

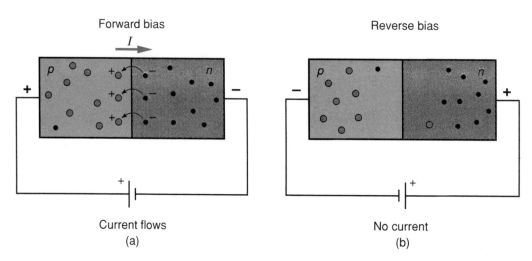

Current flows
(a)

No current
(b)

Figure 20.11   A forward bias of a diode allows electrons and holes to recombine at the junction, and an electric current thus flows across the junction. Reverse bias produces no recombination and no current.

The transistor is probably the most important semiconductor device, and they can take various forms. For many years, the most commonly used type was a *bipolar* transistor made up of two pieces of semiconductor material, heavily doped in the same manner, separated by a thin piece of oppositely-doped material. Depending on which type of doping is used in the two outer pieces of the sandwich, either *p-n-p* transistors or *n-p-n* transistors can be made. The diagram in figure 20.12 illustrates the operation of a *p-n-p* transistor.

The transistor can be regarded as a combination of two diodes that share the middle portion, which is called the base of the transistor. When connected in the usual manner, holes are introduced in the *emitter* of the *p-n-p* transistor from the positive terminal of a battery. The junction between the emitter and the base behaves then as a forward-biased diode, and the holes can flow into the thin base layer. Because the base layer is very thin and only lightly doped compared to the emitter and *collector,* these holes can flow across the base and into the collector, provided that not too many of them recombine with electrons in the base layer. The number of electrons in the base layer is therefore a critical property in determining how many holes get through. This number can be controlled by the current that is permitted to flow between the base and the emitter.

The important feature of this process is that a small change in the current from the base to the emitter can produce a large change in the current flowing between the collector and the emitter. This is why a transistor is so effective as an amplifier. Small variations in the signal applied to the base can produce large variations in the current that flows through the collector. A weak signal picked up by a radio antenna, for example, can be turned into a stronger signal by using transistor amplifiers. This is done routinely in radios, television sets, and stereo amplifiers.

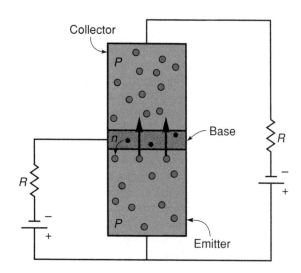

Figure 20.12   The rate of flow of holes from the emitter to the collector of a *p-n-p* transistor depends upon how much current is allowed to flow to the *n*-type base of the transistor.

Another important application of transistors is as voltage-controlled switches. One value of the voltage applied across the base and emitter, for example, can cause a large flow of current through the collector, while another value produces a very small flow. Thus the transistor is either on or off, depending upon the value of voltage applied to the base. This is the feature of transistor operation that is most useful in computers, which we will discuss in the next section.

A second type of transistor, called a *field-effect* transistor (*FET* for short), is often used in computer circuitry. In a field-effect transistor, the current flowing through a thin channel of *n*-type material, for example, is controlled by the voltage applied across two pieces of *p*-type material on either side of the channel. The strength of the electric field produced by this voltage strongly determines how much current will flow through the channel.

Integrated Circuits

The transistor was invented in 1947–48 by scientists at Bell laboratories in New Jersey. These included William Shockley, who invented the bipolar junction transistor, and J. Bardeen and W. H. Brattain, who first demonstrated transistor action in a simpler but less effective *point-contact* transistor. The invention followed directly from the growing understanding of the solid-state physics of semiconductors gained throughout the previous half century. By 1960, transistors were being routinely used in many electronic and switching applications.

Prior to that time, electronic amplification and switching was accomplished with vacuum tubes, which required much larger voltages and generated much more heat than semiconductor diodes and transistors. They were also considerably larger than transistors, so they took up more space, resulting in bulky instruments. Throughout the 1950s and 1960s, vacuum-tube technology was replaced by solid-state electronics in all kinds of electronic devices.

Another major revolution in technology took place during the 1960s, however. This was the development and rapid growth of miniaturized *integrated circuits*. An integrated circuit consists of several transistors, diodes, resistors, and electrical connections all constructed on a single tiny *chip* of semiconductor material, usually silicon. This development permitted the miniaturization of circuitry to much smaller sizes than those introduced by transistors themselves. A vacuum-tube computer that might have filled a large room could now be reduced to the size of a hand-held calculator!

The usual process of producing integrated circuits begins with the growth of a large cylindrical crystal of doped silicon. This crystal is then sliced into *wafers,* which are generally several centimeters in diameter but just a few millimeters thick (fig. 20.13). These wafers are polished and then run through a long series of processing steps in which insulating oxide layers are produced on the wafer, and circuit patterns are overlaid on it by photographic methods. Some regions are masked, and the unmasked portions are doped in the opposite manner to the original silicon crystal in order to produce diodes and transistors. Metal strips are overlaid to provide conducting connections between elements.

Several identical circuits are usually produced on a single silicon wafer. Near the end of the process, the wafer is cut into individual chips, each containing a miniature circuit. A single wafer may yield a hundred or more chips (fig. 20.14). The final steps involve making electrical connections to the chip, packaging it in a sealed plastic enclosure, and testing the resulting circuit (fig. 20.15). Producing integrated circuits (or *ICs*) has become a major industry in this country, as well as in Japan and Europe.

Figure 20.13 **The starting point in producing integrated circuits is a polished wafer of single-crystal silicon. The wafer shown here has been processed to produce tiny circuits on its surface.**

Figure 20.14 **A magnified view of the circuit on a single integrated-circuit chip. Thousands of circuit elements may be contained on such a chip, and many chips can be produced from a single silicon wafer.**

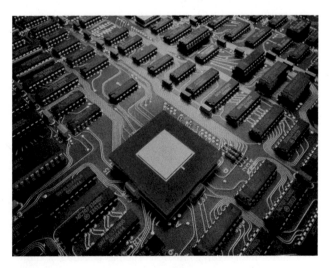

Figure 20.15 **Rows of packaged microchips arranged on the circuit board of a computer.**

Competition to produce ever smaller and faster circuitry for computers and other applications continues to push the technology forward. Physics and chemistry are central to the invention and improvement of new processing techniques. For some applications, silicon is being replaced by the semiconducting compound, gallium arsenide, formed by combining gallium and arsenic. Research in the solid-state physics of semiconducting elements and compounds has become one of the largest areas of activity in modern physics. The revolution in electronics technology is still going on.

20.4 COMPUTERS AND ARTIFICIAL INTELLIGENCE

The revolution in microelectronics discussed in the previous section has gone hand-in-hand with the development of computers and a revolution in the processing of information. Computers play an ever-increasing role in our everyday lives. Some people use them directly for computation or word processing; others use them less directly as components of appliances such as microwave ovens or video recorders. Computers have produced enormous changes in the way we do things.

What is a digital computer? How has the development of computers been tied to the development of integrated circuits? Can computers think? These are some of the questions that we explore in this section.

Digital Computers

The first true electronic computers were built during the 1940s; they used vacuum tubes as basic switching devices. Univac 1, which some people consider to be the first fully-operational digital computer, filled an entire room and required a large air-conditioning system to remove the heat generated by the tubes. A crew of technicians was needed to replace vacuum tubes as they burned out. It did use digital logic, however, and could be programmed in a manner that was fundamentally similar to the programming of most modern computers.

Computers that are much more powerful than Univac 1 and its immediate descendants now fit into small consoles that sit on a desk (fig. 20.16). They use transistors contained within integrated circuits as their basic switching devices. They still use digital logic, though, and must be programmed to achieve whatever function they are to serve.

What is digital logic? In a digital computer, all information is represented in terms of individual binary numbers. In other words, each discrete *bit* of information can have only two possible values, usually represented by 0 or 1. Logical operations that perform functions such as addition or multiplication are done by changing the values of each bit step-by-step in a manner that accomplishes the desired task.

Figure 20.16   A modern microcomputer is self-contained and may easily sit on a desk or on the user's lap.

Numbers are therefore represented in the computer in binary or *base two* form, unlike ordinary numbers, which are decimal (base ten) numbers. In a decimal number, each digit represents a different power of ten. The number 238, for example, is essentially:

$$(2 \times 10^2) + (3 \times 10^1) + (8 \times 10^0) = 200 + 30 + 8.$$

(Any number raised to the zero power is equal to 1.)

In a binary number, each digit represents a different power of two rather than a different power of ten. For example, the binary number 1011 is equal to:

$$(1 \times 2^3) + (0 \times 2^2) + (1 \times 2) + 1 = 8 + 0 + 2 + 1 = 11.$$

(Can you show, then, that the binary number 11010 is the decimal number 26? How would you write the number 5 in binary form?)

This might seem like a cumbersome way to write the number 11, but its advantage is that only the individual values 0 and 1 are needed to represent the number. Such a representation is ideally suited for simple switches, such as transistors, which have just two states, off or on, 0 or 1. Transistors can also be combined to produce memory elements that can "remember" which state they were last in. These memory elements allow us to store numbers and program codes. This is essentially what happens in a digital computer; information is stored and manipulated in binary form.

Of course, programming such a digital computer would be very tedious if you had to do it all in binary codes. This is what was done originally, in the early days of computer development. Before too long, however, programming languages were developed that allowed a computer user to code instructions in a less cumbersome manner. Special programs translate, or *compile,* these instructions from the programming language to binary code.

The development of such *higher-level* programming languages made the task of programming much easier, but it still requires careful attention to detail to get the instructions in exactly the right form to meet the desired objectives. Computers are very stupid in one sense: they generally do exactly what you tell them to, even if those instructions make no sense. "Garbage in, garbage out" has become a common expression among computer users.

The author took his first computer-programming course as a first-year graduate student in physics in 1963. He used an IBM 1620, which was a standard early computer, and the FORTRAN programming language. The computer was then used extensively in his research to model the kinetics of phase transitions and to perform complicated analyses of experimental data. This sort of use has been a tremendous boon to science. Computations that would have been impossibly tedious if done by hand could be readily performed on the computer. Computers are very good at doing repetitive computations quickly, without ever losing patience.

Later the author learned BASIC, which was commonly used on minicomputers in the late 1960s and 1970s. He bought his first home microcomputer in the early 1980s, and now has three of them around the house, one of which was initially purchased for his children and is used almost exclusively for playing computer games. Another is devoted to word processing and was heavily used in the composition of this textbook. In that type of use, we do not program at all, but use sophisticated *application programs* (or software) that are produced for the personal computer. These application programs are developed in a *user-friendly* fashion, so that they are more forgiving of errors in input than traditional computer programs.

During the 1960s, the input of data and programs was usually accomplished via punched cards or paper tape. These, in turn, had to be prepared on card-punch machines or teletypes, which were often quite noisy. Output was handled in the same manner and then fed into a separate printer, which was also generally noisy. Today most data input and output is handled via magnetic disks, which are much faster, more quiet, and easier to use. A printer must still be used to produce a hard copy, but modern laser printers produce excellent letter-quality output in almost noiseless operation.

Artificial Intelligence

What can a computer do that we cannot? Perhaps a question of greater interest is: What can we do as thinkers that computers cannot? Is there some sense in which a computer can actually think, or must it always just follow instructions? These questions are part of the new area of study called *artificial intelligence.* They have extensive implications for the future of computers and their applications.

In most uses, a computer does not think in the usual sense of the word. It merely computes and manipulates binary information following a set of instructions provided by the user or programmer. The output is only as good as the data that were entered and the particular programmed set of manipulations that were performed. People provide both the data and the programs. For this reason, you should always be wary of someone who tells you that "the computer did such and such." They are really telling you that they do not feel competent to question the program or the input data.

What computers can do very well is to execute a complicated set of computations or instructions at high speed and with great accuracy. They can also store vast amounts of information in various forms of memory, so that it can be readily retrieved. In the speed and accuracy with which they can do these things, modern computers far exceed human capabilities. All of these things are done in a step-by-step fashion, however, following instructions provided by human programmers. An ordinary computer makes no leaps of intuition.

Although the speed and accuracy of computers are wonderful assets, thinking surely involves more than this. Among other things, thinking involves the ability to organize information, to recognize patterns in events, to learn from experience, and to create new ways of treating problems. The human brain can do step-by-step computations but is also capable of these other aspects of thinking. (There are individual differences, of course, in how each of us handles these different modes of thinking.)

As scientists have studied how the human brain functions, they have come to recognize that our brains are organized in a very different manner than the normal computer. Instead of transistors, we have neurons as basic signal processors. Unlike transistors that can handle just one or two input signals, individual neurons may receive and transmit information in interaction with many other neurons. They make up a vast, interconnected network called a *neural network* (fig. 20.17).

The manner in which individual neurons transmit signals is determined by what that neuron has experienced before as well as by the strength of the signal and the nature of the interconnections of that particular neuron. It is an incredibly complex system. As we interact with our environment, certain pathways within the neural network develop more than others. In a greatly oversimplified sense, this is how concepts are learned.

Certain input, for example, from nerve impulses transmitted from our eyes stimulates a set of neurons that, when fired simultaneously, transmit signals along established pathways. These signals create a state within the brain that might correspond to the concept of *chair* (or whatever recognizable object we are looking at). These pathways and the response of the brain have been developed through our previous experience with chairs. In some sense, the same sort of process occurs when we recognize physical concepts such as acceleration or torque.

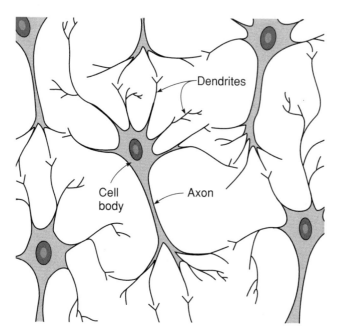

Figure 20.17 The human brain consists of a vast, interconnected network of neurons that transmit and process electrical signals.

One of the exciting advances in computer science in recent years has been the development of computers that mimic this neural-network mode of processing information. This can be done either by special programs (software), or by designing electrical elements that are actually connected to many other elements within the circuitry (hardware). Although the models now in use are very crude approximations of an actual neural network, they do learn from prior experience in a manner analogous to the learning of the brain. Such computers pose interesting new problems in how to go about programming, or teaching, the network.

Although we do not anticipate that neural-network computers will ever be able to fully model the complexity of our brains, they have added an aspect of intelligence that is not present in an ordinary digital computer. Of equal importance, perhaps, is the fact that in thinking about how to program neural-network computers we gain insights into the nature of teaching and learning that may carry over to humans. The entire field of artificial intelligence has caused us to examine the fundamental aspects of thinking and intelligence in new ways.

The field of artificial intelligence would not exist without computers. The continued development of computers, in turn, depends upon our ingenuity in inventing new ways of making and connecting the semiconductor components upon which the technology is based. The revolution in information processing is still unfolding, and may bring even greater changes into our lives than those that have already occurred through our extensive dependence upon computers.

20.5 SUPERCONDUCTORS AND OTHER NEW MATERIALS

A major scientific news item during the late 1980s was the discovery of so-called high-temperature superconductors. For a year or so, there were a series of news announcements, stories about Japan being ahead of us in the superconductor race, and speculations about exotic applications of superconductors. The news flap was almost as great as that produced by the supposed discovery of cold fusion a year or so later.

What was all the excitement about? What is superconductivity, and what does temperature have to do with it? What other exotic materials are in the works? These questions involve a field, which we now call *materials science,* that has emerged from a combination of metallurgy, chemistry, geology, and solid-state physics. *Materials research* is another physics-related discipline that has already had a major impact on technology and the way we live.

Superconductivity

Physicists have been aware of the phenomenon of superconductivity for some time; it was originally discovered in 1911 by a Dutch physicist, Heike Kamerlingh Onnes. Onnes found that if he cooled mercury to a temperature of about 4 K (4 degrees above absolute zero), the electrical resistance of his sample completely disappeared. An electric current, once started, would flow indefinitely with no continuing source of power.

The electric resistance of most materials decreases with decreasing temperature. In the case of Onnes' mercury sample, however, the resistance completely disappeared at the temperature of 4.2 K, which is called the *critical temperature, T_c*. This behavior is illustrated in the graph of figure 20.18. The resistance drops abruptly to zero at the critical temperature and is zero for any temperature below T_c.

Further research showed that many metals became superconducting if cooled to a low enough temperature. The metal niobium has one of the highest critical temperatures of a pure substance at 9.2 K. Some alloys have even higher critical temperatures; in 1973, an alloy of niobium and the semiconductor germanium was discovered to have a critical temperature of 23 K. That is still a very low temperature, though; 23 K is equal to –250°C (–418°F).

A theoretical explanation for the phenomenon of superconductivity was developed in 1957. The theory involves applying quantum mechanics to the behavior of electrons in a low-temperature metal. It is similar to a theory developed to explain the behavior of superfluids, which also exist at very low temperatures, primarily in liquid helium. A superfluid loses it viscosity (or resistance to the flow of the fluid) at a critical temperature, just as a superconductor loses its electrical resistance. Both superconductivity and superfluidity are regarded as quantum phenomena because quantum mechanics is required to explain their characteristics.

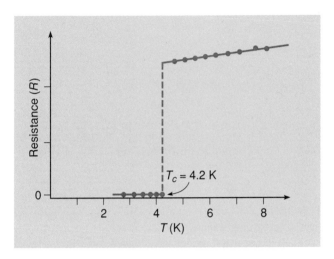

Figure 20.18 The electric resistance of mercury decreases as the temperature is decreased. It drops abruptly to zero, however, at the critical temperature of 4.2 K.

Figure 20.19 A small magnet levitates above a superconducting disk that is being cooled with liquid nitrogen. This is a simple test to establish the presence of superconductivity.

High-Temperature Superconductors

In 1986, a new type of superconducting compound was discovered. This was a ceramic material, a metal oxide containing various other elements. The original ceramic superconductor had a critical temperature of 28 K, not much higher than that obtained with metal alloys. The discovery of ceramic superconductors provoked a flurry of experimental activity, however, in which other combinations of elements were tried. By 1987, ceramic superconductors had been developed with critical temperatures first of 57 K and then of 90 K. Finally, in 1988 there were reports of materials with critical temperatures of over 100 K, and perhaps as high as 125 K.

These new ceramic superconductors are what we have called *high-temperature* superconductors, although these critical temperatures are still not what we would normally regard as high; 100 K is equal to −173°C, for example—still rather cold by most standards. The development of materials with critical temperatures in the vicinity of 90 K was a real breakthrough, however, because these temperatures can be reached using liquid nitrogen. Liquid nitrogen is readily available for industrial and scientific uses; it boils at 77 K (−196°C), so a bath of liquid nitrogen can cool samples to that temperature.

Ceramic superconductors can be produced using a variety of combinations of materials, most of which involve copper oxides as one of the components. The superconducting ceramic material that has become most commonly available is a combination of yttrium (Y), barium (Ba), copper (Cu), and oxygen (O) in the proportions $Y_1Ba_2Cu_3O_{(7)}$. (The number of oxygen atoms in the structure varies depending on how the material is prepared.) This material can be prepared in undergraduate laboratories and has a critical temperature of approximately 90 K.

One of the striking properties of a superconductor (called the *Meisner effect*) is the fact that it will completely exclude magnetic field lines produced by an external magnet or electric current. A magnet brought near to a superconducting material will therefore be repelled. This is commonly demonstrated by levitating a small magnet above a disk of superconducting material. The materials for performing this demonstration have been widely distributed among science teachers. A small amount of liquid nitrogen placed in the bottom of a styrofoam cup is sufficient to cool the superconducting disk (fig. 20.19).

The theory that successfully explained superconductivity in pure materials has not been particularly enlightening in explaining the superconductivity of these new ceramic materials. A common structural feature of many of these materials is that they seem to contain layers of copper or copper oxide sandwiched between atoms of the other elements (fig. 20.20). It is thought that the superconduction occurs through these layers, but exactly why this occurs is not completely understood. A successful theory might point the way to designing materials with even higher critical temperatures. This is the goal, of course.

There are many potential applications of high-temperature superconductors, especially involving the use of electromagnets. A strong electromagnet requires large currents flowing in tightly wound coils of wire. With ordinary conductors, these large currents generate a great deal of heat and thus limit the amount of current that can be established (and also the resulting strength of the electromagnet). Magnets already exist that use superconducting coils, but they must be maintained at a temperature below the critical temperature of the superconducting material. If we had materials that were superconducting near room temperature, such electromagnets would become much more feasible for general use.

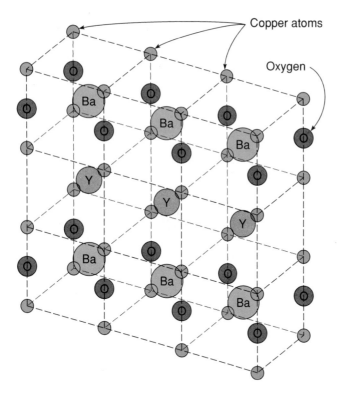

Copper atoms

Oxygen

Figure 20.20 **The atomic structure of many superconducting ceramics involves layers of copper atoms sandwiched between other elements, such as oxygen, yttrium, and barium.**

Figure 20.21 **The long molecules in some liquid crystals line up in layers; this allows the liquid to flow along the layers, but not in the perpendicular direction.**

The superconducting supercollider to be built in Texas, mentioned in the first section of this chapter, is designed to use superconducting magnets to control the particle beams. Other applications that have been suggested include the magnetic levitation of trains or other vehicles to reduce friction and thus attain higher speeds. Superconducting cables might also be used in power transmission to reduce losses that result from electric resistance.

Most of these applications will require superconducting materials with even higher critical temperatures than those available now and which can be readily formed into wires and cables. Many of the ceramic superconductors are quite brittle and not suitable for cables or coils. Success in developing more usable superconducting materials awaits a new generation of scientists, engineers, and dreamers.

Other Exotic Materials

The discovery of the new superconductors was a direct result of research into ceramic metal oxides that were already known to have some interesting electrical properties. A compound of barium, titanium, and oxygen, barium titanate (BaTiO$_3$), for example, had been used for many years to convert changes in

pressure into electrical signals or vice versa. This allows a crystal of barium titanate to be used as a tiny microphone or speaker. Other metal oxides have played important roles in various aspects of integrated-circuit processing.

The search for new materials builds upon our growing knowledge of how atoms interact in the solid state. As this understanding increases, we become more able to design materials to meet specific needs: Special optical or electronic properties for use in electronics and communications, for example, or perhaps high-strength, lightweight materials for use in aircraft. Different elements can be combined to make new materials in an infinite number of ways; the results cannot always be predicted.

Liquid crystals are one of the new materials that have found extensive application. Liquid crystals have a crystal-like organization in one direction, but are disordered and free to flow along other directions in the material. They therefore have some properties of both liquids and solids. Their usefulness has resulted primarily from the degree to which electric fields affect the flow of light through the material. This has led to their use in the display screens of hand-held calculators and to the development of very thin, flat television screens that do not require a bulky cathode-ray tube.

Liquid crystals are often made up of long organic molecules that can line up in layers, as shown in figure 20.21. These layers can slide along one another, so that the material can flow in the direction parallel to the layers. The regular spacing perpendicular to these layers provides the crystal-like properties. The author has conducted research on another class of substances known as "plastic crystals," in which the molecules are generally globular in shape and are partially free to rotate within the solid crystal. These also have many interesting properties, but so far they have not led to the extensive applications (and financial rewards) that have grown out of research on liquid crystals.

It is hard to know just where all of this research will lead. Two things are certain, however. One is that new materials will have some unexpected and exciting properties. The other is that some of this work will lead to new products that you will encounter in your everyday activities.

Box 20.1

Everyday Phenomenon:
Holograms

The Situation. You may have seen holograms on cereal boxes, toys, credit cards, and perhaps on simple jewelry such as pendants as shown in the photo. Although the credit cards or other surfaces are clearly two-dimensional, the image that you see when you view a hologram is three-dimensional. You can move your head from side to side and view these images from different aspects, just as you can with a three-dimensional object.

How are these three-dimensional images produced? What is a hologram, and how do we go about making them? Could holograms be used to develop three-dimensional television or movies? Holography seems like magic or science fiction to many people. What can actually be achieved with holograms?

The Analysis. Although the concept of holography had been suggested earlier, the first good holograms were produced in the early 1960s following the invention of the laser. A hologram is essentially an interference pattern produced by combining light waves reflected from some object with a wave coming directly from the laser. Lasers are highly coherent light sources; they produce much longer wave trains than ordinary light sources that produce short, uncorrelated pulses of light. This high coherence of the laser is necessary in order to produce interference patterns involving light scattered from reasonable-sized objects.

A common arrangement for producing a hologram is shown in the drawing on the next page. Light coming from the laser is split by a partially silvered mirror, or *beam-splitter* into two beams: one called the *object* beam, and the other the *reference* beam. The object beam is directed towards the object, which scatters and reflects the light back towards the photographic plate. The reference beam is directed to the photographic plate at some angle to the object beam. These two beams of light then interfere with one another to produce an interference pattern on the photographic plate.

When the photographic plate is developed, the resulting picture of the interference pattern becomes the hologram. If light similar to the original laser light is directed through the hologram, it is modified by the interference pattern in such a way that one of the light waves that is transmitted is identical in form to the original light wave that was reflected from the object. This is the light wave that we view when we look at a

A hologram on a pendant viewed from two different angles. How is this three-dimensional image produced?

hologram; it is diverging from a three-dimensional image of the original object. We are therefore observing a three-dimensional virtual image. (See chapter 15 for a discussion of image formation.)

The holograms that are most familiar are reflection holograms, which are designed to be viewed in light that is reflected from the hologram rather than transmitted through it. Reflection holograms have the additional advantage of not requiring a laser or other monochromatic light source for viewing; the reflection process selects out only certain wavelengths of light. In the reflection holograms seen on cereal boxes or credit cards, the interference pattern that forms the hologram is embossed onto a thin reflecting film.

Originally the process of producing holograms required that the object be held very still, with no vibrations, in order to produce an accurate interference pattern. As more powerful lasers and better films have been developed, however, shorter exposure times have become possible, and the objects do not have to be kept quite so motionless. It is also possible to generate the interference patterns from mathematical computations on a computer, so that we can produce computer-generated holograms of nonexistent objects. A single hologram contains an enormous amount of information, however, so the development of moving holograms that could be transmitted via television signals is not yet feasible.

Continued

Box 20.1 Continued

The invention of the laser has produced a tremendous growth in the field of optics over the last few decades. The development of holography is just one of the applications that this amazing light source has made possible. Holography is now used in a variety of technical applications as well as in art, special displays, and novelty items. New applications are being developed continually; it is difficult to know just where they will pop up next.

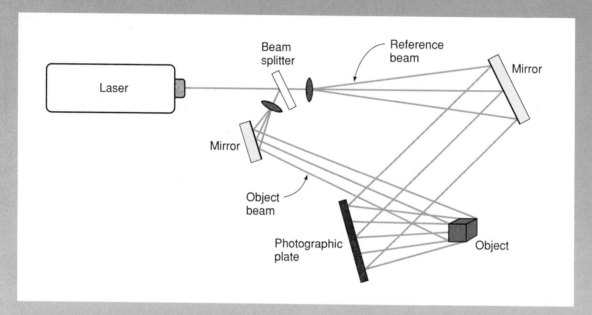

Light from the laser is split into two beams, one of which is reflected from the object and interferes with the second (reference) beam to form the hologram.

SUMMARY

This chapter goes beyond everyday phenomena to briefly describe some of the more exciting developments in the continually advancing research areas of physics. The ideas explored included quarks and other elementary particles, cosmology and the Big Bang, integrated circuits, digital computers and neural-network computers, and superconductors and other "designer" materials.

Elementary Particles. The standard model of high-energy physics can now describe all of the known particles as combinations of twenty-four elementary particles; six leptons, six quarks, and their antiparticles. Good progress has been made in bringing the fundamental forces of nature into a single unified field theory as part of this process.

Proton

Quarks

Cosmology. Astronomical measurements have shown that our sun is just one star in a large galaxy of stars and that there are many other galaxies, all receding from one another in an expanding universe. Knowledge of quarks and the fundamental forces are necessary to model the earliest moments of the universe following the Big Bang that started the current expansion.

Semiconductors and Integrated Circuits. Our understanding of the nature of semiconductors, and the manner in which their resistivity can be modified by doping with impurity atoms, led to the invention of the transistor in the late 1940s. Since then, integrated circuits have been developed by combining hundreds of transistors and other circuit elements on tiny silicon chips. A tremendous industry has grown from this ability to miniaturize electronic and computer circuitry.

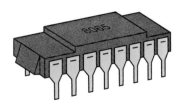

Computers and Neural Networks. The inventions of the transistor and integrated circuits have led to rapid growth in the use of digital computers. Ordinary computers perform operations in a step-by-step fashion, with high speed and accuracy, following the instructions provided by programs. New neural-network computers are organized more like the interconnected neurons in our brains, and can learn from experience to recognize patterns in the data that are provided.

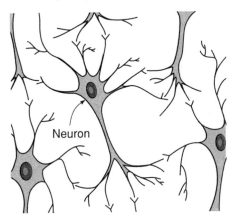

Neuron

Superconductors and Other New Materials. Superconductors are materials that lose all of their electrical resistance below a certain critical temperature. Recently, ceramic superconductors have been discovered with critical temperatures around 100 K, much higher than those previously known, but still well below room temperature. Research in the physics of solids and liquids has produced many other useful materials, such as liquid crystals, that are used in calculator displays and other similar applications.

QUESTIONS

Q20.1 Are leptons generally heavier than protons or neutrons? Explain.

Q20.2 Do we now consider protons to be elementary particles that do not have any underlying structure? Explain.

Q20.3 Are quarks constituents of electrons? Explain.

Q20.4 Are baryons and mesons made up of the same number of quarks? Explain.

Q20.5 Are higher energies in new particle accelerators required in order to produce particles with larger masses than those now known or particles with smaller masses? Explain.

Q20.6 Which fundamental force of nature is the most difficult to incorporate into a unified field theory? Explain.

Q20.7 How do we know that the universe is expanding? Explain.

Q20.8 Is our own sun part of a galaxy? Explain.

Q20.9 Does the term *Big Bang* refer to explosions of individual stars? Explain.

Q20.10 Is it necessary to know anything about very small entities such as quarks in order to produce successful models of very large-scale phenomena such as the beginning of the universe? Explain.

Q20.11 Is the electrical resistance of a semiconductor such as silicon increased or decreased when we dope it with impurity atoms of arsenic? Explain.

Q20.12 Does the doping of silicon with gallium make the resulting semiconductor an *n*-type or *p*-type semiconductor? Explain.

Q20.13 Does the direction in which the battery is connected to a diode make any difference in the amount of electric current that will flow through the diode? Explain.

Q20.14 Are diodes usually made from material that is doped with just one type of impurity atom? Explain.

Q20.15 What property of the behavior of transistors makes them useful for amplifying an electrical signal? Explain.

Q20.16 In making ever-smaller electronic instruments, do integrated circuits have an advantage over separate transistors and diodes? Explain.

Q20.17 Are decimal numbers better suited than binary numbers for manipulation by circuits containing transistors? Explain.

Q20.18 Did the original electronic computers use transistors in their circuitry? Explain.

Q20.19 Can an ordinary digital computer program itself? Explain.

Q20.20 Can an ordinary digital computer think? Explain.

Q20.21 Are the transistors in an ordinary digital computer connected together in a manner similar to the neurons in our brains? Explain.

Q20.22 Does a superconductor have zero resistance only *above* a certain critical temperature? Explain.

Q20.23 With the current high-temperature superconductors, can we now build superconducting magnets that will operate in a superconducting mode at room temperature? Explain.

Q20.24 Are liquid crystals fluids or solids? Explain.

Q20.25 Does the production of a hologram involve the interference of light waves? Explain.

EXERCISES

E20.1 The average distance from the sun to the earth is approximately 1.5×10^8 km. How many seconds are required for light to travel from the sun to the earth? ($c = 3 \times 10^8$ m/s, 1 km = 1000 m)

E20.2 Convert the following binary numbers to ordinary decimal numbers:

  a. 111

  b. 11001

E20.3 Write the following ordinary decimal numbers in binary form:

  a. 6

  b. 19

CHALLENGE PROBLEMS

CP20.1 The nearest star to our sun is about 4 light-years away, and a light-year is the distance that light travels in one year.

    a. How many seconds are there in a year?

    b. Since light travels at the rate of 3.0×10^8 m/s, how many meters are there in one light-year? (How many years as measured by an earth-based observer?)

    c. How far is it to the nearest star in meters?

    d. How many years would it take to travel to the nearest star if we were able to travel at one-tenth of the speed of light?

HOME EXERCISES AND OBSERVATIONS

HE20.1 On a clear night, go outside (preferably away from city lights) and study the night sky.

    a. Can you see the Milky Way? (It will usually appear to be a faint cloud of stars making an irregular band across the sky.) Make a sketch of its orientation.

    b. What are the brightest objects in the sky (other than the moon)? Are some of these objects planets? (Planets will produce a steadier-appearing light than stars and are quite bright.)

    c. Can you pick out the Big Dipper and other constellations? Make a sketch of the more prominent groupings of stars that you observe.

HE20.2 If you have a hand-held calculator (preferably one that is no longer working), carefully remove the back cover and examine the circuit board.

    a. Can you identify the ICs (integrated circuits)? How many of them are there?

    b. Can you identify the other electronic components attached to the circuit board? Make a sketch of the layout that you observe.

HE20.3 Find a hologram on a credit card or cereal box and examine it closely.

    a. As you move your head from side to side, can you see different features of the object? Is the image that you see clearly three-dimensional?

    b. Move your head up and down as you observe the hologram. Do the colors change? What sequence of colors do you observe? Is there any three-dimensional character to the hologram in the vertical (up-and-down) direction?

APPENDIX A
Using Simple Algebra

In chapter 1 we discussed the idea that mathematics is the language of physics. It is a shorthand for expressing relationships between physical quantities, which makes manipulating these relationships much easier than if they were expressed in words and sentences. People who are conversant in the language of mathematics are very comfortable in interpreting and using such relationships.

Many people, however, are not comfortable using even simple mathematics. For this reason, mathematics is used very sparingly in this textbook. Most of what is used is simple algebra, which almost all college students have been introduced to in high school. Very frequently that introduction does not produce a firm understanding of the basic principles underlying algebraic manipulations, though, and this leaves students feeling uneasy regarding the use of algebra. This appendix presents the basic concepts that underlie simple algebraic manipulations and provides illustrations of their use.

BASIC CONCEPTS

Three simple but fundamental concepts form the basis for most algebraic manipulations. If you grasp these concepts, you should feel no mystery in the ordinary use of simple algebra. These principles are the following:

1. The letters used in algebra represent numbers. Any operation that can be performed with numbers (addition, subtraction, multiplication, division, etc.) can also be performed with these symbols.

In mathematics courses, the letters x and y are often used to represent unknown numbers, and other letters are used to represent constants or known numbers. In physics, specific letters are used to represent specific quantities: t for time, m for mass, d for distance, s for speed, and so on. They all represent numerical quantities, however; some may be known, and others may be initially unknown. The relationship $s = d / t$, for example, tells us that we can find the numerical value of speed by dividing a numerical value for distance by a numerical value for time. This is a general relationship that holds for any possible numerical values of d and t.

2. If the same operation is performed on both sides of an equation, the equality expressed by that equation does not change.

This principle is the basis for all algebraic manipulations performed to express a relationship in a different form. For example, if we multiply both sides of the equation $s = d / t$ by the quantity t, the equality still holds. This operation yields

$$st = \frac{dt}{t} = d,$$

since t / t equals 1. (Any number divided by itself equals 1.) Performing this operation expresses the original equation in a new form, $d = st$, which tells us that distance is equal to speed multiplied by time. We can multiply both sides of an equation by the same quantity, divide both sides by the same quantity, add or subtract the same quantity from both sides, and the equation is still valid. We can also square both sides of the equation and perform various other operations, but the simple operations just listed are those most commonly used.

3. When we solve an algebraic equation, we are simply rearranging the equation using principle 2 so that the quantity that we are interested in knowing is expressed, by itself, on one side of the equation.

In the previous paragraph, in other words, we solved the equation $s = d / t$ for the quantity d, thus expressing the distance in terms of the other two quantities, speed and time. If we wanted to express the time of travel in terms of the speed and the distance, we could divide both sides of the equation $d = st$ by the quantity s:

$$\frac{d}{s} = \frac{st}{s} = t\left(\frac{s}{s}\right) = t,$$

or, $t = d / s$, the distance divided by the speed. Thus we see that the original equation, $s = d / t$, can be expressed in two other forms, $d = st$ and $t = d / s$, that merely restate the original equality in forms suitable for computing a specific quantity when the other two quantities are known. This is an extremely useful thing to be able to do.

Since the letters represent numbers (principle 1), we can always check the validity of the operations we perform by inventing numbers for the quantities and checking to see that the equalities still hold in the new form. For example, using the original equation if, $d = 6$ cm and $t = 2$ s, then

$$s = \frac{d}{t} = \frac{6 \text{ cm}}{2 \text{ s}} = 3 \text{ cm/s}.$$

If we put the same numbers in the final equation, we have

$$t = \frac{d}{s} = \frac{6 \text{ cm}}{3 \text{ cm/s}} = 2 \text{ s},$$

or 2 s = 2 s, which is obviously an equality.

These concepts are simple, and their application is not difficult once the basic ideas are grasped. A little practice, obtained by following the additional examples given below and performing the exercises at the end of this appendix, should help to build confidence in their use. For most people who have trouble with mathematics, lack of confidence is the fundamental problem. Often they have never fully accepted the idea that letters can represent numbers, and thus the manipulations and rules of algebra seem arbitrary and mysterious.

MORE EXAMPLES

1. Solve the equation $a = b + c$ for the quantity c.

Solution: We want to obtain a result in which c is by itself on one side of the equation, and the other two quantities are on the other side. This can be accomplished by subtracting the quantity b from both sides of the equation, since doing so will leave c by itself on the right side:

$$a - b = b + c - b = c$$

$b - b = 0$, of course, and $c + 0 = c$. Thus we see that $c = a - b$, since $a - b = c$. (It does not matter which side of an equality is stated first; the equality is the same in either case.) Subtracting b from the right side of the original equation leaves c by itself, and we have therefore solved the equation for c.

2. Solve the equation $v = v_0 + at$ for the quantity t.

Solution: This is best done in two steps. The first step is to subtract the quantity v_0 from both sides of the equation in order to isolate the product at.

$$v - v_0 = v_0 + at - v_0 = at$$
$$at = v - v_0$$

Then we divide both sides of this equation by a to get t by itself:

$$\frac{at}{a} = \frac{v - v_0}{a}$$
$$t = \frac{v - v_0}{a}$$

If you can understand why each of these operations was performed (What was the objective?), then you are well on your way to being able to do any of the algebraic operations used in this textbook.

3. Solve the equation $b = c + d / t$ for the quantity t.

Solution: Again, we first subtract the quantity c from both sides of the equation in order to isolate the term containing t:

$$b - c = c + \frac{d}{t} - c = \frac{d}{t}$$

The quantity t is in the denominator, however, so we multiply both sides of the equation by t:

$$(b - c)\, t = \frac{d}{t}\, t = d$$

Next, we divide both sides of the equation by $(b - c)$ in order to leave t by itself on the left side of the equation:

$$\frac{(b - c)\, t}{b - c} = \frac{d}{b - c}$$
$$t = \frac{d}{b - c}$$

Each of these steps has a specific objective. The first step isolates the quantity d / t, the second step removes t from the denominator so that we can more readily solve for t, and the final step leaves t by itself. You must recognize these objectives in order to gain confidence in performing such operations yourself. Even people who are familiar with algebra often forget just what they are trying to accomplish, or get careless about making sure that they are doing the same thing to both sides of an equation.

EXERCISES

(Answers to odd-numbered exercises are found in appendix D.)

EA.1 Solve the equation $F = ma$ for the quantity a.

EA.2 Solve the equation $b = c + d$ for the quantity c.

EA.3 Solve the equation $h = g - f$ for the quantity g.

EA.4 Solve the equation $a = b + cd$ for the quantity b.

EA.5 Solve the equation in exercise EA.4 for the quantity d.

EA.6 Solve the equation $a = b\,(c - d)$ for the quantity b.

EA.7 Solve the equation in exercise EA.6 for the quantity c. (Hint: first rewrite the equation as $a = bc - bd$, multiplying both terms inside the parentheses by b. This does not change the equality.)

EA.8 Solve the equation $a + b = c - d$ for the quantity c.

EA.9 Solve the equation in exercise EA.8 for the quantity d.

EA.10 Solve the equation $b\,(a + c) = dt$ for the quantity b.

EA.11 Solve the equation in exercise EA.10 for the quantity a.

EA.12 Solve the equation $x = v_0 t + (1/2)at^2$ for the quantity v_0.

EA.13 Solve the equation in exercise EA.12 for the quantity a.

APPENDIX B
Decimal Fractions, Percentages, and Scientific Notation

In physics, and many other fields in which numbers are important, we usually express fractions as decimal fractions and often use percentages as a means of expressing fractions or ratios. Because we need to deal with very large and very small numbers at times, we also use a means of expressing these numbers involving powers of ten, or *scientific notation,* to avoid writing out all of the zeros. Although scientific notation is used sparingly in this book, there are times when its use is highly desirable, if not essential, and it is important that you understand its meaning. It is part of the language of science.

DECIMAL FRACTIONS

Although most college students are familiar with decimal fractions and percentages, they are not always completely comfortable with their use or sure of their meaning. Fractions involve ratios or proportions, and research has shown that many people are not adept at using these concepts. One of the benefits of taking a course in physics is that it can strengthen your ability to think in terms of ratios or proportions, and to understand the mathematics used to describe them.

A decimal fraction is just a fraction for which the number in the denominator is some multiple of the number 10 (10, 100, 1000, etc.); the appropriate multiple is indicated by the location of the decimal point. For example, if we start with the fraction $\frac{1}{2}$ and divide 1 by 2 as the fraction indicates, a calculator will display the result as 0.5. The decimal point in front of the 5 is a shorthand notation for expressing the fraction $\frac{5}{10}$. The number 5 is half of 10, and so the fraction $\frac{5}{10}$ is the same thing as the fraction $\frac{1}{2}$ (one-half). In other words, the *ratio* of 5 to 10 is the same as the ratio of 1 to 2.

If the fraction $\frac{3}{4}$ is evaluated on a calculator by dividing 3 by 4, the calculator expresses the result as 0.75, which is equivalent to the fraction $\frac{75}{100}$, or seventy-five hundredths. Thus the first place, or number, after the decimal point represents tenths, the second place hundredths, the third place thousandths, and so on. The fraction $\frac{343}{1000}$ (343-thousandths) is expressed as 0.343, for example. We could read this as 3 tenths plus 4 hundredths plus 3 thousandths. Obviously, some space is saved by writing the fraction in decimal form.

Decimal fractions are very commonly used, although we may not always stop to think about their meaning. In base-

ball, for example, we express a batter's hitting efficiency as a decimal fraction. A batter who has produced 35 hits in 100 official at-bats is said to be hitting 350. This is really $\frac{350}{1000}$ or 0.350, but the decimal point is often omitted. Most people understand that it should be there, however, and that we are merely expressing the fraction $\frac{35}{100}$ in decimal form and including three figures to the right of the decimal point.

PERCENTAGES

Another common way of expressing decimal fractions is to write them as percentages. The word *percent* means "per one hundred," so a percentage is just a decimal fraction in which the denominator is 100. The fraction $\frac{1}{2}$, for example, is $\frac{5}{10}$, or $\frac{50}{100}$, and can be expressed as 50%; it is fifty hundredths. The fraction $\frac{3}{4}$ is 0.75 or 75% (seventy-five hundredths), and the fraction $\frac{343}{1000}$ is 0.343 or 34.3% (34.3 hundredths). Thus moving the decimal point two places to the right converts a decimal fraction to a percentage. This is equivalent to multiplying the fraction by 100.

The use of percentages is even more common than the direct use of decimal fractions. Interest rates and tax rates are usually expressed as percentages, for example, and we should all have some understanding of their meaning. An interest rate of 7% means that you will receive or pay $7 each year for every $100 that you have invested or borrowed, $\frac{7}{100}$ of the total amount. (We are ignoring here the possible effects of compounded interest.) A tax rate of 28% means that we will owe to the government $28 of every $100 that we earn (after deductions are subtracted). A percentage is always per one hundred, by definition.

Although it is easy enough to understand how a percentage is calculated (compute the decimal fraction and multiply by 100), having a good feeling for the proportions represented by different percentages is another matter. Pie charts, such as that shown in figure B.1 are often used to provide a visual representation of these proportions. The slices of the pie should be in proportion to the percentages or fractions being represented. If, as sometimes happens, the graphic artist does not understand this, the resulting pie chart may be very misleading.

The pie chart shown in figure B.1 represents the average monthly expenditures of someone who takes home $2000 a month (after taxes and other deductions). If she spends $500

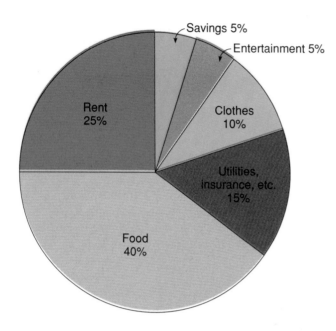

Figure B.1 **A pie chart showing the fractions of total take-home pay spent in different categories. The slices of the pie are in proportion to the percentages being represented.**

a month on rent, this is $^{500}\!/_{2000}$ or 0.25 (one-fourth) of her total income. Since 0.25 equals 25%, this is shown as 25% on the pie chart, and it takes up one-quarter of the total pie or circle; thus, the size of the slice is in proportion to the percentage being represented. Likewise, if she spends $800 a month on food, this is $^{800}\!/_{2000}$ or 0.40, which is 40% of her total take-home pay. The other slices represent smaller percentages and are correspondingly smaller. If we have taken into account all of her normal expenses, the sum of the percentages represented on the chart should add to 100%.

SCIENTIFIC NOTATION

Very large numbers or very tiny fractions often require many zeros in order to properly locate the decimal point. For example, 1.2 trillion dollars (corresponding roughly to the size of our accumulated national debt a few years ago) can be written as $1 200 000 000 000. The zeros are there only to locate the decimal point; they do not imply that all of the other numbers to the right of the 1 and 2 are exactly zero. If we count the digits to the right of the 1, we see that there are twelve (eleven zeros and the digit 2).

Another way of expressing this number would be to say that it is 1.2 times 1 trillion, where 1 trillion is the number 1 followed by twelve zeros. One trillion is also the result of multiplying 10 by itself twelve times. (1 000 000 000 000 = $10 \times 10 \times 10 \times 10 \times 10 \times 10 \times 10 \times 10 \times 10 \times 10 \times 10 \times 10$)

The shorthand notation for a number multiplied by itself twelve times is to say that it has been raised to the *power* 12, which we write as 10^{12}. The superscript represents the power to which the number has been raised, which is the number of times it has been multiplied by itself. We read this number as "ten to the twelfth power," or often just "ten to the twelfth."

Thus we can write the number 1.2 trillion as follows:

$$1.2 \times 10^{12}.$$

It is this notation, which expresses the number as some value times a power of 10, that we call *scientific notation*. The number 12 (the power) simply tells us how many places to the right we would have to move the indicated decimal point if we wrote out all of the zeros. Scientific notation has several advantages. It saves space, it properly indicates the accuracy, or precision, of the number being represented by eliminating the zeros, and it makes the number easier to manipulate in calculations involving very large or small numbers.

Some examples involving smaller numbers may help to make the concept clear. The number 586, to choose a much smaller number than 1.2 trillion, can be expressed as 5.86 times 100, or 5.86×10^2 since $10 \times 10 = 100$ (10 squared). The number 6180 can be expressed as 6.18×10^3, since 10 cubed is 1000. The number 5 400 000 (5.4 million) can be expressed as 5.4×10^6 since 10 to the sixth power is 1 million. Several other examples are provided in table B.1. (The last number listed under the positive powers of 10 is the approximate mass of the earth in kilograms.)

Table B.1 also shows several decimal fractions written in scientific notation. A fraction always has a negative exponent (negative power of 10) if the value of the fraction is less than 1.0. For example, the fraction 0.000 000 000 001 2 is 1.2 trillionths. It can be expressed as 1.2×10^{-12}, since one-trillionth is 10 raised to the power −12. This is the same thing as the number obtained by dividing the number 1.2 by 10^{12} (1 trillion). The superscript −12 tells you to move the decimal point twelve places to the *left* in order to express the number in normal decimal form.

Taking simpler examples again, the decimal fraction 0.15 can be written as 1.5 times 0.1 (one-tenth), or 1.5×10^{-1}. The fraction 0.0343 is 3.43 hundredths or 3.43×10^{-2}. Moving the decimal point two places to the left, as indicated by the power of 10, yields the original decimal fraction. The fraction 0.0079 is 7.9 thousandths, or 7.9×10^{-3}. Studying the other examples in table B.1 should make the pattern clear. (The last number in table B.1 is the value of the charge on the electron in coulombs, a quantity that arises frequently in modern physics.)

Another aid to expressing very large or very small numbers is the set of prefixes used in the metric system of units discussed in chapter 1. Since the prefix *mega* stands for

Table B.1
Examples of Scientific Notation

Positive Powers of Ten

5460	= 5.46 times one thousand	$= 5.46 \times 10^3$
23 400	= 2.34 times ten thousand	$= 2.34 \times 10^4$
6 700 000	= 6.7 times one million	$= 6.7 \times 10^6$
11 000 000	= 1.1 times ten million	$= 1.1 \times 10^7$
9 400 000 000	= 9.4 times one billion	$= 9.4 \times 10^9$
5 980 000 000 000 000 000 000 000		$= 5.98 \times 10^{24}$

Negative Powers of Ten (Fractions)

0.62	= 6.2 times one tenth	$= 6.2 \times 10^{-1}$
0.0523	= 5.23 times one hundredth	$= 5.23 \times 10^{-2}$
0.0082	= 8.2 times one thousandth	$= 8.2 \times 10^{-3}$
0.000 047	= 4.7 times one hundred-thousandth	$= 4.7 \times 10^{-5}$
0.000 002 4	= 2.4 times one millionth	$= 2.4 \times 10^{-6}$
0.000 000 007 9	= 7.9 times one billionth	$= 7.9 \times 10^{-9}$
0.000 000 000 000 000 000 16		$= 1.6 \times 10^{-19}$

1 million, the quantity 1.35 Mg (megagrams) is the same as 1.35×10^6 g (10^6 is 1 million). Likewise, 780 nm (nanometers) is the same as 780×10^{-9} m, since the prefix *nano* means one-billionth or 10^{-9}. The values of the commonly used metric prefixes are given in table 1.1 in chapter 1. These metric prefixes and the power-of-10 scientific notation are both forms of scientific shorthand used to shorten the representation of numbers.

MULTIPLYING AND DIVIDING POWERS OF 10

The process of multiplying or dividing numbers written in scientific notation is quite simple if you understand what is involved. (It is even easier if you have a calculator that handles scientific notation; you just punch the numbers in and push the appropriate function key.) Understanding the process can be very useful, however, for checking your results.

Suppose, for example, that we multiply the number 3.4×10^3 (3400) by 100 (10^2). Multiplying by 100 adds two zeros to the original number yielding 340 000, as you can quickly check by doing this operation on a calculator or by direct multiplication. Thus

$$(3.4 \times 10^3) \times (10^2) = 3.4 \times 10^5.$$

In other words, the *exponents,* or powers of 10, add ($3 + 2 = 5$). If we divided by 100, we would remove two zeros:

$$\frac{3.4 \times 10^3}{10^2} = 3.4 \times 10^1 = 34.$$

In this case, the exponent of the denominator is subtracted from the exponent of the number being divided ($3 - 2 = 1$).

The rules are therefore very simple:

1. **When numbers are multiplied, the powers of 10 add.**
2. **When numbers are divided, the power of the denominator is subtracted from the power of the numerator.**

These rules hold regardless of whether the powers themselves are positive or negative. Thus

$$(3.0 \times 10^6) \times (2.0 \times 10^{-4}) = 6.0 \times 10^2 = 600,$$

since $6 + (-4) = 2$. This should make sense to you, since multiplying by a fraction (a number with a negative power of 10) should result in a smaller number than the number being multiplied.

EXERCISES

If any of these ideas are unfamiliar, or if they are familiar but you are rusty in using them, working some or all of the following exercises will help to build confidence in their use. The answers to the odd-numbered exercises appear in appendix D.

EB.1–EB.4, express the numbers given as decimal fractions.

EB.1 a. $\dfrac{7}{10}$ b. $\dfrac{37}{100}$ c. $\dfrac{567}{1000}$ d. $\dfrac{4}{10\,000}$

EB.2 a. $\dfrac{52}{100}$ b. $\dfrac{3}{10}$ c. $\dfrac{67}{10\,000}$ d. $\dfrac{23}{1000}$

EB.3 a. $\dfrac{3}{4}$ b. $\dfrac{5}{8}$ c. $\dfrac{37}{52}$ d. $\dfrac{512}{914}$ (Use a calculator.)

EB.4 a. $\dfrac{2}{7}$ b. $\dfrac{13}{15}$ c. $\dfrac{212}{654}$ d. $\dfrac{34}{150}$ (Use a calculator.)

EB.5 Express the fractions in exercise B.3 as percentages.

EB.6 Express the fractions in exercise B.4 as percentages.

EB.7 Find the following:
  a. 50% of 96
  b. 75% of 124
  c. 65% of 150
  d. 85.2% of 644

EB.8 Find the following:
  a. 40% of 80
  b. 90% of 200
  c. 33.3% of 90
  d. 72% of 540

EB.9–EB.12, express the numbers given in scientific notation (power-of-10 notation).

EB.9 a. 6435 b. 5 200 000 c. 43 000 d. 790 000 000 000

EB.10 a. 2345 b. 67 800 c. 344 000 d. 78 000 000

EB.11 a. 0.0024 b. 0.000 577 c. 0.000 001 2 d. 0.000 000 007

EB.12 a. 0.025 b. 0.000 86 c. 0.000 000 32 d. 0.000 044

EB.13 Express the following numbers as decimal fractions:
  a. 2.5×10^{-3}
  b. 4.8×10^{-2}
  c. 5.77×10^{-5}
  d. 3.22×10^{-4}

EB.14 Perform the following operations:
  a. $(2.0 \times 10^4) \times (4.3 \times 10^5)$
  b. $(2.5 \times 10^3) \times (5.0 \times 10^4)$
  c. $(4.0 \times 10^8) \times (5.4 \times 10^{-5})$
  d. $\dfrac{8.0 \times 10^8}{2.0 \times 10^6}$

EB.15 Perform the following operations:
  a. $(3.0 \times 10^5) \times (1.5 \times 10^4)$
  b. $(4.0 \times 10^7) \times (5.0 \times 10^{-5})$
  c. $\dfrac{8.4 \times 10^8}{2.0 \times 10^6}$
  d. $\dfrac{1.2 \times 10^{12}}{2.0 \times 10^{-6}}$

APPENDIX C
Vectors and Vector Addition

Many of the quantities that we encounter in the study of physics are vector quantities, which is to say that their *direction* is important as well as their size or magnitude. Examples of vector quantities include displacement, velocity, and acceleration, which are introduced in chapter 2, as well as force, momentum, electric field, and many others that are encountered in later chapters. Direction is an essential feature of these quantities; the result of traveling with a velocity of 20 m/s due north is different from that of traveling with a velocity of 20 m/s due east.

Quantities for which direction is not an essential feature (or for which direction has no meaning at all) are called *scalar* quantities. Mass, volume, and temperature are examples of scalar quantities; for most purposes it would make no sense to talk about the direction of a volume or a mass. Specifying the magnitude (the numerical value with appropriate units) of a scalar quantity is sufficient; no other information is needed. Vectors, on the other hand, require at least two pieces of information to be completely specified.

DESCRIBING A VECTOR

Suppose that we want to describe the velocity of an airplane flying in a direction somewhat north of due east. The magnitude of the airplane's velocity can be specified by stating its speed as, for example, 400 km/hr. The direction of the airplane's velocity can be specified in a number of ways, but the simplest would be to specify an angle to some reference direction, 20° north of east, for example. These two numbers, 400 km/hr and 20° north of east, are sufficient to describe the airplane's velocity, provided that its motion is two-dimensional (in a horizontal plane, not climbing or descending). If the plane were climbing or descending, thus involving altitude as a third dimension, a second angle (the angle of ascent or descent) would have to be specified.

A picture or graph is often a more satisfying way of describing such a vector. Figure C.1 pictures the velocity of the airplane as an arrow pointed in a direction 20° north of east. We can represent the magnitude of the velocity by the length of the arrow if we choose an appropriate scale factor when drawing the diagram. For example, if 2 cm represent 100 km/hr, then we would draw the arrow with a length of 8 cm (4 times 2 cm) to represent the speed of 400 km/hr. A smaller speed would be represented by a shorter arrow and a larger speed by a longer arrow.

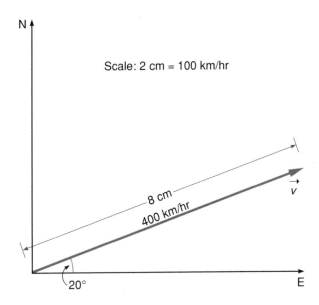

Figure C.1 **The velocity vector of 400 km/hr in the direction 20° north of east is represented by drawing the arrow to scale (2 cm = 100 km/hr) and at the appropriate angle (20°).**

The arrow is the universal symbol for representing vectors on diagrams: it clearly indicates direction, and it can also be drawn to different lengths to indicate magnitude. We also use the arrow symbol in writing expressions involving vector quantities; the symbol $\vec{v}$, with the arrow drawn above the letter, tells us that we are dealing with a vector. The symbol v, without the arrow, represents the scalar quantity, speed. Another method of describing a vector is discussed later in this appendix when vector components are introduced.

ADDING VECTORS

We are often interested in the net result of combining two or more vectors. In chapter 4, for example, the *net force* determines the acceleration of an object. This net force is the vector sum of whatever forces are acting on the object, perhaps as many as four or five forces. As a second example, an airplane's velocity relative to the ground is determined by the vector sum of its velocity relative to the air and the velocity of the air relative to the ground (the wind velocity), as discussed in chapter 18.

One of the most readily visualized examples of vector addition involves the displacement of a moving object. Suppose, for example, that a student wants to travel to an apartment complex that is located on north Main Street, a few blocks north and west of campus. She might get there from a starting point on the south side of campus by walking three blocks due west along Pacific Avenue, and then six blocks due north along Main Street, as indicated in figure C.2. The result of these two motions can be represented by a displacement vector; the first one displaces her three blocks due west, and the second one six blocks due north.

These two displacements are drawn to scale and in the appropriate direction on the vector diagram in figure C.2. (1 cm = 1 block.) Their net result is indicated by drawing displacement $\vec{C}$, which is the vector drawn from the starting point to the final destination. Vector $\vec{C}$ is thus the vector sum of vectors $\vec{A}$ and $\vec{B}$:

$$\vec{C} = \vec{A} + \vec{B}.$$

It combines their individual effects into a single net displacement. The length of vector $\vec{C}$ on the diagram is approximately 6.7 cm, which represents a distance of 6.7 blocks, given the scale factor used in drawing the diagram. Measuring the angle with a protractor yields an angle of approximately 27° west of north for the total displacement $\vec{C}$.

The process of vector addition that we have just described can be accomplished with any vectors. It is often referred to as the *graphical method* of adding vectors, or (more descriptively) as the *toe-to-head* technique. Its steps are as follows:

1. Draw the first vector to scale (1cm equals so many units of the vector quantity) and in the appropriate direction, using a ruler and a protractor.
2. Start the second vector with its toe at the head of the first vector, and draw it to the same scale and in the appropriate direction.
3. If more than two vectors are involved, draw the succeeding vectors to scale and in the appropriate direction, starting each one with its toe at the head of the previous vector.
4. To obtain the vector sum, draw a vector from the toe of the first vector to the head of the final vector. Measure the angle this vector makes to some reference direction with the protractor, and measure its length with a ruler. These two measurements represent the direction and magnitude of the vector sum. (The measured length must be multiplied by the scale factor used in drawing the original vectors to obtain the appropriate units.)

As another example, two velocity vectors are added in figure C.3. The first vector ($\vec{A}$) is a velocity of 20 m/s at an angle of 15° north of east. The second vector ($\vec{B}$) is a velocity of 40 m/s at an angle of 55° north of east. Each has been drawn to a scale of 1 cm = 10 m/s, so vector $\vec{A}$ is 2 cm long, and vector $\vec{B}$ is 4 cm long. The toe of vector $\vec{B}$ is placed at the

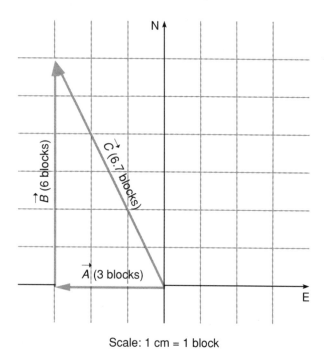

Scale: 1 cm = 1 block

Figure C.2 The net result of adding displacement $\vec{A}$ (three blocks due west) and displacement $\vec{B}$ (six blocks due north) is the displacement $\vec{C}$, obtained by drawing a vector from the starting point to the final destination.

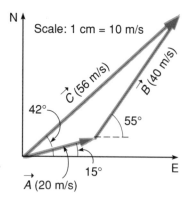

Figure C.3 The velocity vectors $\vec{A}$ and $\vec{B}$ are added to obtain the vector sum $\vec{C}$. A scale factor of 1 cm = 10 m/s is used, and the toe of vector $\vec{B}$ is placed at the head of the vector $\vec{A}$.

head of vector $\vec{A}$ in order to add the vectors, and the resulting sum, vector $\vec{C}$, is approximately 5.6 cm long, determined by measuring with a ruler. Using the scale factor of 1 cm = 10 m/s, we have

$$5.6 \text{ cm} \times \frac{10 \text{ m/s}}{1 \text{ cm}} = 56 \text{ m/s}.$$

Measuring the angle that $\vec{C}$ makes to the horizontal axis (east), we find that $\vec{C}$ is approximately 42° north of east. Thus the vector sum of vectors $\vec{A}$ and $\vec{B}$ is equal to 56 m/s at an angle of 42° north of east.

Notice that in both this example involving velocities and the previous example involving displacements, the magnitude of the vector sum is not equal to the sum of the magnitudes of the two vectors being added. In the first case, the vector sum $\vec{C}$ had a magnitude of 6.7 blocks, which is less than the 9 blocks (3 + 6) that the person actually walked. In the velocity example, the vector sum had an approximate magnitude of 56 m/s, which is less than the sum of 60 m/s obtained by adding the magnitudes of vectors $\vec{A}$ and $\vec{B}$. This is a general feature of the process of vector addition; the only case in which the magnitude of the vector sum equals the sum of the magnitudes of the vectors being added ($A + B$) is when these vectors are in the same direction.

VECTOR SUBTRACTION

How do we go about subtracting two vectors? Once you have mastered the concept of vector addition, subtraction represents a simple extension of these ideas. Subtraction can always be represented as the process of adding to the original quantity the *minus* value (the negative of) the quantity being subtracted. Thus the process of subtracting 2 from 6 is the same as adding –2 to 6. If we want to subtract vector $\vec{A}$ from vector $\vec{B}$ to get the vector difference $\vec{B} - \vec{A}$, we merely add $-\vec{A}$ to $\vec{B}$. (To get the negative of a vector, we reverse its direction.)

To illustrate this process using the same vectors as in the previous examples, velocity $\vec{A}$ is subtracted from velocity $\vec{B}$ in figure C.4. First we draw vector $\vec{B}$ to scale and then add to it the negative of vector $\vec{A}$. Notice that we have reversed the direction of $\vec{A}$ to get $-\vec{A}$: the negative vector is 15° below the horizontal instead of 15° above. The difference vector, $\vec{D}$, is obtained by drawing the vector from the toe of the first vector ($\vec{B}$) to the head of the second ($-\vec{A}$). Vector $\vec{D}$ has a length of approximately 5.6 cm and makes an angle of approximately 7° to the vertical axis (north). The length of 5.6 cm represents a velocity of 28 m/s (5.6 cm × 5 m/s per cm), since that is the scale factor used in figure C.4.

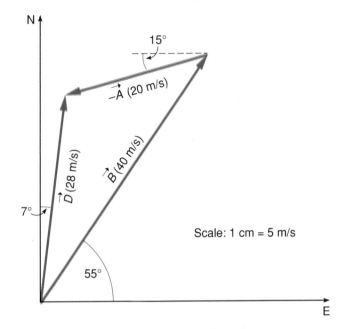

Figure C.4 The velocity vector $\vec{A}$ is subtracted from the velocity vector $\vec{B}$ to obtain the difference vector $\vec{D}$. The scale is 1 cm = 5 m/s.

VECTOR COMPONENTS

We often find it useful to describe vectors in terms of their horizontal and vertical *components*, rather than handling the entire vector directly. This is especially true when we are discussing projectile motion, as in chapter 3, but it is also useful in computing work (chapter 8) and in many other applications.

The components of a vector are any two (or more) vectors that yield the vector of interest when added together.

Usually it is most productive to define these components as perpendicular to one another, generally in the horizontal and vertical directions.

We can use graphic techniques to find the components of a vector, as well as to add or subtract vectors. The process is illustrated for a force vector in figure C.5. The force vector $\vec{A}$ has a magnitude of 8 N (the newton [N] is the metric unit of force) and a direction of 30° above the horizontal. Our first step in finding the horizontal and vertical components of this vector is to draw the vector to scale (1 cm = 1 N) and in the appropriate direction (30° above the horizontal) using a ruler and a protractor, as before.

The horizontal component of the vector $\vec{A}$ is then found by drawing a dashed line from the tip of $\vec{A}$ to the horizontal (x)

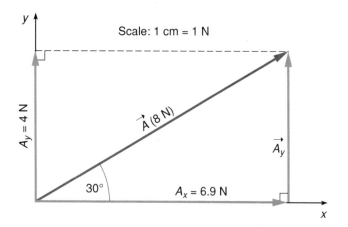

Figure C.5 **The components of the force vector $\vec{A}$ are found by drawing the vector to scale and then drawing lines from the tip of the vector to the *x* and *y* axes, so that these lines make right angles (90°) with the axes.**

the dashed line is drawn from the tip of $\vec{A}$ to the vertical (y) axis, making a right angle with the vertical axis. Measuring the lengths of these components with a ruler yields magnitudes of 6.9 N (6.9 cm on the graph) and 4 N (4 cm on the graph) for A_x and A_y respectively.

If we treat these two components of $\vec{A}$ as vectors and add them in the usual toe-to-head manner, we obtain the original vector $\vec{A}$, as is also shown in figure C.5. We can therefore use these two components to represent the vector, since their sum is identical to the original vector. Often, however, we are interested only in the horizontal effect or the vertical effect of the vector, and then we may use just one of the components. In the case of a force vector, for example, the effect of the force in moving an object in the horizontal direction is determined by the horizontal component of the vector rather than by the total vector. In projectile motion, the horizontal component of the velocity determines how far the object will travel horizontally in a given time, and so on.

The components of vectors can also be used in adding or subtracting vectors, as well as for many other purposes. In this book, however, the primary use of vector components is in breaking a vector down into its horizontal and vertical portions, in order to analyze the horizontal and vertical motions separately. Knowing that this can be done is important to your understanding of projectile motion and many other physical processes.

axis, so that the line makes a right angle with (is perpendicular to) the *x*-axis. The distance along the *x*-axis, measured from the toe of $\vec{A}$ (the origin) to the point where the perpendicular dashed line meets the axis, represents the magnitude of A_x, the horizontal component of $\vec{A}$. A_x represents the portion of $\vec{A}$ that is in the horizontal direction. A similar process yields A_y, the vertical component of $\vec{A}$, but in this case

---

EXERCISES

(Answers to the odd-numbered exercises appear in appendix D.)

C.1–through C.4, use the graphic toe-to-head technique to find the vector sums of the indicated vectors.

EC.1 Vector $\vec{A}$ = a displacement of 10 m due east.

Vector $\vec{B}$ = a displacement of 30 m due north.

EC.2 Vector $\vec{A}$ = a velocity of 30 m/s at 30° north of east.

Vector $\vec{B}$ = a velocity of 50 m/s at 45° north of east.

EC.3 Vector $\vec{A}$ = an acceleration of 4 m/s^2 due east.

Vector $\vec{B}$ = an acceleration of 6 m/s^2 at 40° north of east.

EC.4 Vector $\vec{A}$ = a force of 20 N at 45° above the horizontal (to the right).

Vector $\vec{B}$ = a force of 35 N at 20° to the left of vertical.

EC.5 Find the magnitude and direction of the difference vector $\vec{B} - \vec{A}$ in exercise C.1.

EC.6 Find the magnitude and direction of the difference vector $\vec{A} - \vec{B}$ in exercise C.1.

EC.7 Find the magnitude and direction of the difference vector $\vec{A} - \vec{B}$ in exercise C.2.

EC.8 Find the magnitude and direction of the difference vector $\vec{A} - \vec{B}$ in exercise C.3.

EC.9 Find the east and north components (x and y) of vector $\vec{A}$ in exercise C.2.

EC.10 Find the horizontal and vertical components of vector $\vec{A}$ in exercise C.4.

EC.11 Find the horizontal and vertical components of vector $\vec{B}$ in exercise C.4.

APPENDIX D
Answers to Odd-Numbered Exercises and Challenge Problems

CHAPTER 1

E1.1 1000 ml **E1.3** 131 cm = 1.31 m
E1.5 1250 kg = 1 250 000 g **E1.7** 1610 m = 1.61 km

CHAPTER 2

E2.1 50 MPH **E2.3** 2.5 hr
E2.5 a. 12 cm/min **b.** 0.12 m/min **c.** 4.72 in/min
E2.7 a. 0.015 km/s **b.** 54 km/hr **E2.9** 1.5 m/s^2
E2.11 −4.0 m/s^2

CP2.1 a. 2.0 m/s^2 **b.** 0.667 m/s^2 **c.** 1.20 m/s^2 **d.** No
CP2.3 c. 720 m **d.** 48 m/min **e.** 66.7 m/min (larger than the average velocity)

CHAPTER 3

E3.1 a. 0.24 m/s **b.** 3.20 m/s **c.** 9.87 m/s^2
E3.3 a. 11.0 m **b.** 44.1 m
E3.5 a. +5.2 m/s **b.** −14.4 m/s **E3.7 a.** 1.53 s **b.** 11.5 m
E3.9 a. 18 m/s **b.** 72 m **E3.11** 3.0 m

CP3.1 a. 0 **b.** 1.22 s **c.** 7.35 m **d.** 4.4 m **e.** Down
CP3.3 a. Car A: 2.5 m, 10 m, 22.5 m, 40 m;
car B: 9 m, 18 m, 27 m, 36 m **b.** Between 3 and 4 seconds
CP3.5 a. 10.2 s **b.** 883 m **c.** 17.7 s, 883 m (the same)

CHAPTER 4

E4.1 1.5 m/s^2 **E4.3** 5 kg
E4.5 a. 20 N (to the right) **b.** 10 m/s^2
E4.7 a. 668 N **b.** 68.1 kg **E4.9** 5.8 m/s^2 (downward)
E4.11 7.9 N

CP4.1 a. 2.4 m/s^2 (to the right) **b.** 4.8 m/s **c.** 4.8 m **d.** Yes
CP4.3 a. 36 N **b.** 6 m/s^2 (to the right) **c.** 18 N
d. Yes, $a = 6$ m/s^2

CHAPTER 5

E5.1 a. 4.2 m/s **b.** 2.1 m/s^2
E5.3 a. 13.3 m/s^2 **b.** 10 700 N = 10.7 kN
E5.5 a. 3.2 m/s^2 **b.** 224 N **c.** Gravity
E5.7 a. 150 N **b.** 67 N **E5.9** 30 lb **E5.11** 8.87 m/s^2

CP5.1 a. 12.6 m/s **b.** 5.29 m/s^2 **c.** 212 N, yes
d. The rider flies off if not constrained by a belt or bar.
e. 604 N
CP5.3 a. 3.53×10^{22} N **b.** 2.01×10^{20} N **c.** 176, no
d. 4.34×10^{20} N **e.** 0.463, yes

CHAPTER 6

E.61 a. 12.5 N·s **b.** 12.5 kg·m/s **E6.3** 12 kg·m/s
E6.5 a. 9 kg·m/s **b.** 9 N·s
E6.7 a. 360 kg·m/s, −480 kg·m/s **b.** (−) 120 kg·m/s
c. In the original direction of the defensive back (due west).
E6.9 a. 1.75 kg·m/s **b.** (−)1.75 kg·m/s **c.** (−)0.70 m/s
E6.11 a. 150 000 kg·m/s **b.** 4.29 m/s

CP6.1 a. 15 kg·m/s **b.** Yes **c.** 15 N·s **d.** 1500 N
CP6.3 a. (−)2400 kg·m/s **b.** 2400 N·s, 9600 N
c. The time will be larger with the use of the seat belt, and therefore the average force will be smaller.

CHAPTER 7

E7.1 a. 0.167 rev/s **b.** 1.05 rad/s
E7.3 a. 0.555 rev/s **b.** 3.49 rad/s **c.** 5.55 rev
E7.5 a. 0.050 rad/s^2 **b.** 12.5 rad
E7.7 a. 200 N·m **b.** (−)125 N·m **c.** 75 N·m
E7.9 12.5 N **E7.11 a.** 30 N·m **b.** 15 kg·m^2
E7.13 a. 0.0338 kg·m^2 **b.** 0.405 kg·m^2/s
E7.15 8.0 rev/s

CP7.1 a. 160 N·m **b.** 0.107 rad/s^2 **c.** 3.20 rad/s
d. 7.04 m/s, no **e.** −0.0107 rad/s^2, 300 s = 5 min
CP7.3 a. 1200 kg·m^2, 2700 kg·m^2 **b.** 1575 kg·m^2
c. 2.06 rad/s **d.** Yes

CHAPTER 8

E8.1 30 J   **E8.3 a.** 150 J **b.** 150 J
E8.5 a. 88.2 J **b.** 88.2 J   **E8.7 a.** 20 J **b.** 4000 N/m
E8.9 a. 90 J **b.** 30 m/s   **E8.11** 550 000 J
E8.13 a. 882 J **b.** 600 J **c.** 6.32 m/s

CP8.1 a. 16 J **b.** 10 J **c.** 10 J **d.** Yes. Some is converted to kinetic energy; some is lost as heat. **e.** 6.32 m/s
CP8.3 a. 54 J **b.** 54 J **c.** 46.5 m/s **d.** No. The rubber strap retains some kinetic energy.
CP8.5 a. Yes **b.** 21.5 m

CHAPTER 9

E9.1 122° F   **E9.3** 293 K   **E9.5** 2500 cal
E9.7 16 000 cal   **E9.9** 2510 J
E9.11 a. 2000 J **b.** −2000 J
E9.13 −700 J (out of the system)

CP9.1 a. 45 C° **b.** 81 F° **c.** 45 K **d.** No. The Kelvin and Celsius intervals are the same.
CP9.3 a. 4250 cal **b.** 1250 cal **c.** 5500 cal
CP9.5 a. 1740 J **b.** 415 cal **c.** 1.38 C° **d.** Yes

CHAPTER 10

E10.1 40%   **E10.3 a.** 900 J **b.** 44.4%   **E10.5** 45.3%
E10.7 450 J   **E10.9** 800 watts
E10.11 No. The maximum possible efficiency is 7%.

CP10.1 a. 37.5 MJ **b.** 112.5 MJ **c.** It is used to overcome air resistance and frictional forces.
CP10.3 a. 5.0% **b.** 15 J **c.** 285 J **d.** 19 **e.** Yes

CHAPTER 11

E11.1 6.25×10^{12} electrons   **E11.3** 1.5×10^{-4} N
E11.5 a. 4.5 N   **E11.7** 12.5 N, down
E11.9 0.34 N, to the left   **E11.11** −15 000 V

CP11.1 a. 2.25×10^{10} N **b.** 5.4×10^{10} N **c.** 3.15×10^{10} N
d. 1.58×10^{10} N/C, to the left **e.** 11.1×10^{10} N, to the right
CP11.5 b. Yes, at point *A*. **c.** Yes

CHAPTER 12

E12.1 0.60 A   **E12.3** 18 Ω
E12.5 a. 0.15 A **b.** Yes **c.** 3.0 V
E12.7 a. 3.33 Ω **b.** 9.0 A **c.** 3.0 A   **E12.9** 36 watts
E12.11 a. 0.909 A **b.** 121 Ω

CP12.1 a. 1.33 Ω **b.** 1.64 A **c.** 1.09 A **d.** 16.1 watts
e. Less
CP12.3 a. 1 Ω, 0.667 Ω **b.** 3.67 Ω **c.** 1.64 A **d.** 0.55 A

CHAPTER 13

E13.1 0.667 N   **E13.3** It is half the original value.
E13.5 3.75 T   **E13.7** 2.22 T   **E13.9** 0.08 T·m²
E13.11 a. 0.60 T·m² **b.** 0 **c.** 2.4 V   **E13.13** 10 turns

CP13.1 a. 8.0×10^{-4} N/m **b.** Repulsive **c.** 2.4×10^{-4} N
d. 4.0×10^{-5} T **e.** Into the page **f.** 2.0×10^{-5} T
CP13.3 a. 60 cm² = 0.0060 m² **b.** 0.060 T·m² **c.** 0
d. 0.25 s **e.** 0.24 V

CHAPTER 14

E14.1 2.0 hz   **E14.3 a.** 10 hz **b.** 3.0 m
E14.5 a. 2 m **b.** 75 hz   **E14.7** 1.29 m
E14.9 a. 4 m **b.** 85 hz   **E14.11** 250 m
E14.13 3×10^{18} hz

CP14.1 a. 0.075 kg/m **b.** 25.8 m/s **c.** 5.16 m **d.** 3 **e.** 0.775 s
CP14.3 b. 1.20 m **c.** 283 hz **d.** 9 hz **e.** 0.60 m, 567 hz

CHAPTER 15

E15.1 1.5 m   **E15.3** 1.33 cm
E15.5 a. 30 cm **b.** Real, inverted
E15.7 a. −6 cm **b.** Virtual, upright
E15.9 a. −12 cm **b.** Virtual, upright
E15.11 a. +6 cm **b.** −5.0   **E15.13** 80 cm

CP15.1 a. 22.6 cm **b.** 23.1 cm **c.** 15.4 cm
CP15.3 a. Infinity **c.** No

CHAPTER 16

E16.1 $8 : 3 = 2.67 : 1$ **E16.3** 19   **E16.5** 6.25×10^{18}
E16.7 379.8 nm. No, it is in the near ultraviolet portion of the spectrum.   **E16.9** 1282 nm. No, it is in the infrared portion of the spectrum.
E16.11 a. 4.98×10^{14} hz **b.** 602 nm

CP16.1 a. Towards the top plate **b.** 2500 N/C
c. 4.0×10^{-16} N **d.** 4.40×10^{14} m/s^2 **e.** An upward-curved trajectory, similar to an inverted projectile-motion curve.
CP16.3 a. $n = 3 \Rightarrow n = 2$ **b.** 1.89 eV $= 3.02 \times 10^{-19}$ J
c. 4.56×10^{14} hz, 658 nm **d.** 2.46×10^{15} hz, 121.8 nm
e. The Balmer line will be visible, but not the Lyman line.

CHAPTER 17

E17.1 12   **E17.3** $_{92}U^{238} \Rightarrow {}_{90}Th^{234} + {}_{2}\alpha^{4}$, (thorium-234)
E17.5 $_{8}O^{15} \Rightarrow {}_{7}N^{15} + {}_{1}e^{0} + {}_{0}v^{0}$, (nitrogen-15)
E17.7 6 days
E17.9 $_{0}n^{1} + {}_{92}U^{235} \Rightarrow {}_{39}Y^{99} + {}_{53}I^{133} + 4\,{}_{0}n^{1}$, (yttrium-99)

CP17.1 a. C: 6, 6; N: 7, 7; O: 8, 8 **b.** 1.0
c. Ag: 61,47; Cd: 64, 48; In: 66, 49
d. 1.30, 1.33, 1.35; average = 1.33.
e. Th: 142, 90; Pa: 140,91; U:146, 92; 1.58, 1.54, 1.59;
average = 1.57.   **CP17.3 a.** 0.003 510 u **b.** 5.25×10^{-13} J
c. Yes, kinetic energy of the neutron and the He^3 nucleus.

CHAPTER 18

E18.1 11 m/s   **E18.3** 240 MPH   **E18.7** 3.46 hr
E18.9 173 km   **E18.11 a** 22.9 J **b.** 12.9 J

CP18.1 b. 3.16 m/s **c.** 10 s **d.** 10 m **e.** 31.6 m
CP18.3 b. 40.2 yr **c.** 4.02 yr **d.** 4.0 light-years **e.** 36.2 yr

CHAPTER 19

E19.1 120 Pa   **E19.3** 250 psi   **E19.5** 300 N
E19.7 20 kPa   **E19.9 a.** 300 K, 350 K **b.** 117 kPa
E19.11 0.5 g/cm^3 **E19.13 a.** Decreases **b.** $A_2 = (\tfrac{2}{3})A_1$

CP19.1 a. 0.015 m, 0.15 m
b. 7.07×10^{-4} m^2, 7.07×10^{-2} m^2 **c.** 100 **d.** 11 800 N
e. 118 N   **CP19.3 a.** 1.25×10^{-4} m^3 **b.** 0.975 kg
c. 9.56 N **d.** 1.23 N **e.** 8.33 N
CP19.5 a. 78.5 cm^2, 19.6 cm^2 **b.** 6.0 m/s
c. Yes, halving the diameter decreases the area by a factor of 4, and the velocity increases by a factor of 4.
d. Less than. By Bernoulli's principle, an increase in velocity is associated with a decrease in pressure.

CHAPTER 20

E20.1 500 s   **E20.3 a.** 110 **b.** 10 011

CP20.1 a. 3.15×10^7 s **b.** 9.45×10^{15} m **c.** 3.78×10^{16} m
d. 40 yr

APPENDIX A

EA.1 $a = F / m$ **EA.3** $g = h + f$ **EA.5** $d = (a - b) / c$
EA.7 $c = (a + bd) / b = (a / b) + d$ **EA.9** $d = c - a - b$
EA.11 $a = (dt / b) - c$
EA.13 $a = 2(x - v_0 t) / t^2 = 2x / t^2 - 2v_0 / t$

APPENDIX B

EB.1 a. 0.7 **b.** 0.37 **c.** 0.567 **d.** 0.0004
EB.3 a. 0.75 **b.** 0.625 **c.** 0.712 **d.** 0.560
EB.5 a. 75% **b.** 62.5% **c.** 71.2% **d.** 56.0%
EB.7 a. 48 **b.** 93 **c.** 97.5 **d.** 549   **EB.9 a.** 6.435×10^3
b. 5.2×10^6 **c.** 4.3×10^4 **d.** 7.9×10^{11}
EB.11 a. 2.4×10^{-3} **b.** 5.77×10^{-4} **c.** 1.2×10^{-6} **d.** 7×10^{-9}
EB.13 a. 0.0025 **b.** 0.048 **c.** 0.000 057 7 **d.** 0.000 322
EB.15 a. 4.5×10^9 **b.** 2.0×10^3 **c.** 4.2×10^2 **d.** 6.0×10^{17}

APPENDIX C

EC.1 32 m, 72° north of east
EC.3 9.4 m/s^2, 24° north of east
EC.5 32 m, 72° north of west
EC.7 22 m/s, 65° south of west
EC.9 26 m/s, 15 m/s
EC.11 –12 N, 33 N

CREDITS

Chapter 1

Figure 1.1: © Craig Tuttle/The Stock Market;
1.3: Hebrew University of Jerusalem, Courtesy of American Institute of Physics/Niels Bohr Library;
1.5: © Hank Morgan/VHSID Lab/ECE Dept. of U of MA/Science Source/Photo Researchers, Inc.;
1.6: ©NASA/Science Source/Photo Researchers, Inc.;
1.7: © Astrid and Hanns-Frieder Michler/Science Photo Library/Photo Researchers, Inc.; **1.8:** © Jean Collombet/Science Photo Library/Photo Researchers, Inc.

Chapter 2

Figure 2.16: © Patrick Watson; **BOX 2.1:** © Fred Roe/ Fundamental Photographs

Chapter 3

Figure 3.1: © Tony Freeman/Photo Edit; **3.2:** © David Young-Wolff/Photo Edit; **3.3:** © Richard Megna/ Fundamental Photographs; **3.5:** © Tony Freeman/Photo Edit; **3.15:** © Andres Aquino/Picture Perfect Photos

Chapter 4

Figure 4.1: © Tony Freeman/Photo Edit; **4.3:** © Bill Sanderson/Science Photo Library/Photo Researchers, Inc.

Chapter 5

Figure 5.1: © Richard B. Duppold/Unicorn Stock Photos; **5.2:** © Russ Kinne/Comstock Photos;
5.9: © Takeshi Takahara/Photo Researchers, Inc.;
5.12: © Science/Astronomy/Photo Researchers, Inc.;
5.15: © Tycho Brahe/The Image Works Archives;
5.23A: © Joe Sohm/The Image Works; **5.23B:** © Bob Daemmrich/The Image Works; **5.23C:** © Tony Freeman/ Photo Edit

Chapter 6

Figure 6.1: © Brian Drake/Marilyn Gartman Agency;
6.3: © Russ Kinne/Comstock Photos; **6.5:** © Russ Kinne/ Comstock Photos; **6.15:** © Richard Megna/ Fundamental Photographs; **6.16:** © Mike and Carol Werner/Comstock Photos; **6.20:** © Richard Megna/ Fundamental Photographs

Chapter 7

Figure 7.1: © David Young-Wolff/Photo Edit;
7.6, 7.7, 7.8, 7.10, 7.12: © Wm. C. Brown Publishers/ Jim Ballard Photographer; **7.14A:** © David Young-Wolff/Photo Edit; **7.16:** © Wm. C. Brown Publishers/ Jim Ballard Photographer; **7.18:** © Vandystadt/ Allsport; **7.20:** © Wm. C. Brown Publishers/Jim Ballard Photographer; **7.24:** © Bob Coyle;
7.26: © Wm. C. Brown Publishers/Jim Ballard Photographer; **BOX 7.1:** © Lawrence Migdale/Stock Boston

Chapter 8

Figure 8.1: © Tony Freeman/Photo Edit; **8.2:** © Wm. C. Brown Publishers/Jim Ballard Photographer;
8.9: © Mark Antman/The Image Works;
BOX 8.1: © Bob Daemmrich/The Image Works

Chapter 9

Figure 9.1: © Wm. C. Brown Publishers/Jim Ballard Photographer; **9.2, 9.3, 9.7:** © Tony Freeman/Photo Edit; **9.11:** The Granger Collection; **9.14:** © Wm. C. Brown Publishers/Jim Ballard Photographer; **9.17:** © Tony Freeman/Photo Edit; **9.19, BOX 9.1:** © Wm. C. Brown Publishers/Jim Ballard Photographer

Chapter 10

Figure 10.1: © David Conklin/Photo Edit;
10.11: © Patrick Watson; **10.16:** © J. Zoiner/Peter Arnold, Inc.; **10.17:** © Tom Myers/Photo Researchers, Inc.; **10.18:** © Tony Freeman/Photo Edit

Chapter 11

Figure 11.1: © Ogust/The Image Works; **11.2, 11.3, 11.4, 11.7, 11.8, 11.11:** © Wm. C. Brown Publishers/ Jim Ballard Photographer; **BOX 11.1:** © Deeks-Barber/ Photo Researchers, Inc.

Chapter 12

Figure 12.1: © Chris Luneski/Cascade Images;
BOX 12.3: © Tony Freeman/Photo Edit

Chapter 13

Figure 13.1: © Ogust/The Image Works; **13.2:** © Tony Freeman/Photo Edit; **13.4:** © Russ Kinne/Comstock Photos; **13.22:** © Wm. C. Brown Publishers/Jim Ballard Photographer; **13.24A,13.24B:** © Mark Antman/The Image Works; **13.24C:** © Tony Freeman/Photo Edit

Chapter 14

Figure 14.1: © Felecia Martinez/Photo Edit;
14.15: © Tony Freeman/Photo Edit; **14.19:** © Bob Coyle

Chapter 15

Figure 15.1: © Tony Freeman/Photo Edit;
15.10: © Bob Coyle; **15.13:** © David Parker/SPL/ Science Source Photo Researchers, Inc.; **15.27:** © Larry Mulvehill/Science Source/Photo Researchers, Inc.;
15.30: © Wm. C. Brown Publishers/Jim Ballard Photographer

Chapter 16

Figure 16.1: © Michael Gilbert/SPL/Photo Researchers, Inc.; **16.2:** © Russ Kinne/Comstock Photos; **16.3:** © Myrleen Ferguson/Photo Edit;
16.10: © Comstock Photos; **16.11:** © Wm. C. Brown Publishers/Jim Ballard Photographer; **16.16:** © Richard Megna/Fundamental Photographs; **BOX 16.1:** © Spencer Grant/Photo Edit, Inc.

Chapter 17

Figure 17.1: © Will McIntyre/Photo Researchers, Inc.;
17.16: Courtesy of G. E. Nuclear Energy, San Jose, California; **17.22:** Courtesy of Princeton University;
BOX 17.1, BOX 17.2: Tass/Sovfoto

Chapter 18

Figure 18.2: © Tony Freeman/Photo Edit;
18.18: © Richard Megna/Fundamental Photographs;
18.24: © Tony Freeman/Photo Edit

Chapter 19

Figure 19.1A: © Anna Zuckerman/Photo Edit;
19.1B: © Chris Luneski/Cascade Images; **19.6:** © Tony Freeman/Photo Edit; **19.9:** The Granger Collection;
19.15: © Tony Freeman/Photo Edit; **19.17A:** © Russ Kinne/Comstock Photos; **19.17B:** © R. Krubner/H. Armstrong Roberts; **19.22:** © Tony Freeman/Photo Edit; **19.23:** © NASA/Photo Edit; **19.26, 19.28:** © Tony Freeman/Photo Edit; **BOX 19.1:** © Donald Dietz/Stock Boston

Chapter 20

Figure 20.1: © Shelly Katz/Blackstar LA 1990 The Time Inc. Magazine Company/Reprinted by permission.; **20.3:** © Parker/SPL/Photo Researchers, Inc.; **20.6:** © Chris Bjornberg/Photo Researchers, Inc.;
20.7: © Ronald Royer/Science Photo Library/Photo Researchers, Inc.; **20.13:** © Wm. C. Brown Publishers/ Bob Coyle Photographer. Wafer courtesy Intel Corporation; **20.14:** © Michael Abbey/Science Source/ Photo Researchers, Inc.; **20.15:** © David Parker/ Science Photo Library/Photo Researchers, Inc.;
20.16: © David Young-Wolff/Photo Edit; **20.19:** © H.R. Bramag/Peter Arnold, Inc.; **BOX 20.1:** © Bob Coyle

INDEX

A

Absolute zero, 176-77
Acceleration, 25-28, 62
  angular. *See* Angular acceleration
  average. *See* Average acceleration
  centripetal. *See* Centripetal acceleration
  and constant acceleration applications, 43-45
  constant or uniform in time, 39
  and deceleration, 27
  and direction, 27
  of gravity, 38-41
  instaneous. *See* Instaneous acceleration
  rotational. *See* Rotational acceleration
  and uniformly accelerated motion, 41-43
Acceleration detector, 26
Action/reaction principle, 66
Adiabatic expansion, 198
Adiabatic process, 184
Airfoil, 431-35
Airplane wing, 431-35
Air resistance, 69
Alcohol, and specific heat capacity, 178
Algebra, using, 464-66
Alkali metals, 342
Alpha, 127
Alpha decay, 369
Alpha particles, 349
Alternating current (ac), 253-57
Alternating voltage, 278, 279
Aluminum, and specific heat capacity, 178
American Journal of Physics, 435
American Revolutionary War, 180
Ammeter, 249
Ampere, 241, 270, 274, 304
Ampere, Andre-Marie, 270
Analog meter, 249
Analyses
  Fourier or harmonic, 292
  of motion, 46
Angle, critical, 318, 319
Angular acceleration, 127-28
Angular magnification, 330
Angular momentum
  conservation of, 136-39
  definition of, 136
  vector nature of, 139
Angular velocity, 127
Answers to odd-numbered exercises and
  challenge problems, 475-77
Antineutrino, 370
Antinodes, 296

Antiparticle, 370
Aperture, 331
Appliances, power and current ratings, 255
Application programs, 455
Arago, Dominique, 274
Archimedes' principle, 416, 425-28
Aristotle, 40-41, 58, 60, 62, 69, 70
Artificial intelligence, and computers, 454-56
Atmospheric pressure, and gas behavior, 420-25
Atomic mass, 342
Atomic number, 369
Atomic physics, 8
Atomic spectra, 351-55
Atomic structure, 358-59
Atomic weights, and Dalton, 341-42
Atoms
  Bohr model of, 351-55
  and elements, 340, 341, 342
  existence of, 340-43
  and molecules, 341, 342
  and nucleus, 337-88
  quarks and other elementary particles, 444-47
  structure of, 338-63
  and weight, 340-41
Automobiles
  and braking, 18
  collision of, 115
  and Doppler effect, 301-2
  and heat engines, 194
  rounding a curve, 83-84
  stopping, 45
Average acceleration, 26
Average speed, 20-22
Average velocity, 24-25

B

Balance
  simple, 129-30
  and torque, 129-32
  torsion, 223
Ball
  and curveballs, 434-35
  on a string, 82-83
  throwing a, 69-70
Balloon
  and boat, 427-28
  and expanding universe, 449
  gases, and first law of thermodynamics, 183-84
  Pascal's, 422, 424

Balmer, J. J., 352-53
Balmer line, 355
Banked curve, 84
Bardeen, J., 453
Barium, 374
Barium titanate, 458
Barstow, California, 206
Baryons, 445, 446
Baseball
  and curveballs, 434-35
  throwing a, 48-49
Base two, 454
BASIC, 455
Basketball, shooting a, 50-51
Batavia, King of, 180
Baton, 134
Battery, dead, 243-44
Becquerel, Henri, 348, 349
Bell laboratories, 453
Bending of light, 409
Bernoulli, Daniel, 431
Bernoulli's principle, 431-35
Berylium, 368
Beta decay, 370
Beta particles, 349
Bias, 451
Bicycle, riding a, 139-41
Big bang, 449-50
Binary form, 454
Binoculars, 331
Bipolar transistors, 452
Bismuth, 370
Bit, 454
Blackbody, 353
Black holes, 410-11
Boats
  and balloons, 427-28
  floating of, 416
Bohr, Niels, 351, 352, 353, 354, 357, 358
Bohr model, 351-55
Boltzmann's constant, 185
Bottom quarks, 445
Bowling ball
  and momentum, 105
  and pendulum, 161
  stopping a, 64
Box, rope pulling, 153
Boyle, Robert, 423
Boyle's law, 423-25
Brahe, Tycho, 87
Braking, and automobile, 18
Brattain, W. H., 453
Bromine, 342

C

D